Solid

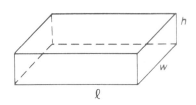

Rectangular Solid
Surface Area $S = 2\ell w + 2wh + 2\ell h$
Volume $V = \ell wh$

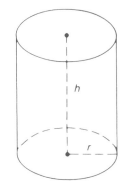

Right Circular Cylinder
Surface Area $S = 2\pi rh + 2\pi r^2$
Volume $V = \pi r^2 h$

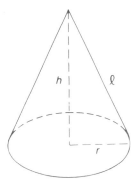

Right Circular Cone
Surface Area $S = \pi r\ell + \pi r^2$
Volume $V = \dfrac{1}{3}\pi r^2 h$

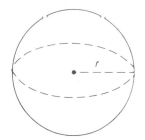

Sphere
Surface Area $S = 4\pi r^2$
Volume $V = \dfrac{4}{3}\pi r^3$

Right Pyramid
Volume $V = \dfrac{1}{3}bh$
(b is the area of the base)

ALGEBRA

For College Students

A *Student's Solutions Manual* that provides you with additional worked-out solutions to problems from this text is available for you to purchase. It also contains a self-test for each chapter with solutions so you can check your progress. Ask your bookstore manager to order a copy for you. (ISBN 0–697–01347–2.)

Math Lab, an interactive software tutorial, available for Apple and IBM personal computers, is also available for purchase. You may use this software in a lab or on your own for review, extra help, or enrichment. To order this software use ISBN 0–697–06701–7 for Apple and 0–697–06704–1 for IBM.

ALGEBRA
For College Students

Terry H. Wesner
Henry Ford Community College

Harry L. Nustad
Henry Ford Community College, Retired

Second Edition

wcb
Wm. C. Brown Publishers
Dubuque, Iowa

Book Team

Editor *Earl McPeek*
Developmental Editor *Nova A. Maack*
Designer *Julie E. Anderson*
Production Editor *Eugenia M. Collins*
Marketing Manager *Matt Shaughnessy*

Cover photo © Michael A. Decamp/The Image Bank

Copyright © 1985, 1988 by Wm. C. Brown Publishers. All rights reserved

Library of Congress Catalog Card Number: 87–71638

ISBN 0–697–10022–7

Printed in the United States of America by Wm. C. Brown Publishers
2460 Kerper Boulevard, Dubuque, IA 52001

10   9   8   7   6   5   4   3   2   1

To my sons—Tim, Tom, and Rob
Dad

To my children Gary and Cheryl
Dad

Contents

4

Rational Expressions

5

Exponents, Roots, and Radicals

6

Quadratic Equations and Inequalities

7

Graphing Linear Equations and Inequalities in Two Variables

8

Systems of Linear Equations

9

Conic Sections

10

Functions

11

Exponential and Logarithmic Functions

12

Sequences and Series

Appendixes

Preface

Algebra for College Students is designed to be used as an intermediate level text for students who have had some prior exposure to beginning algebra in either high school or college. In this second edition we have maintained our philosophy of explaining the why's of algebra, rather than simply expecting students to imitate examples. Sections are presented in such a way that, as topics progress, students realize they are actually extending properties they have already learned.

Our goal has been to make the text as easy to read as possible by presenting the material much as an instructor does in the classroom. As the title of the text indicates, we have included a cross section of applications, mainly in the exercises. These not only make the learning process more interesting for the students but also give them a reason for learning algebra.

Problem-Solving Orientation

The skill of problem solving has been integrated throughout the text. The emphasis on problem solving begins in chapter 1 with word problems that have simple arithmetic solutions. The student learns to change word phrases into algebraic expressions. In chapter 2 and throughout the rest of the text, the student is shown how to form and solve equations from word problems.

Pedagogical Features

Examples illustrate the concepts and clearly show all of the steps in solving the problem.

Concept Boxes include definitions, theorems, or properties along with our explanation in easy-to-understand language. The color shading makes these boxes easy for students to locate for review.

Notes to the students highlight important ideas and point out potential errors the students might make.

Mastery Points are listed before each exercise set. In essence, they are objectives for that section. They are specifically placed in this location to alert students to the particular skills they must know to successfully work the problems.

Exercise Sets provide abundant opportunities for students to check their understanding of the concepts being presented.

Trial Exercise Problems are located in the exercise sets. A trial exercise problem is denoted by a box around the problem number. This indicates that the solution is shown in its entirety at the end of the text. The trial exercise problems serve as additional examples to enhance the student's understanding of the concept or as step-by-step checks of the problems.

Chapter Summaries synthesize the important ideas of each chapter and are placed at the end of each chapter.

A **Chapter Review** is placed at the end of each chapter. This problem set follows the same organization as the chapter. Each problem is keyed to the section from which it was drawn. It is designed to help students determine whether they need additional study on any section of the chapter.

Cumulative Tests give students the opportunity to work problems that are randomly drawn from the chapter and from preceding chapters. Not only do students demonstrate their full understanding of the particular chapter, but they get to show their ability to retain information from previous chapters. If students need refreshing, they can use the section references to review the concept.

Answers to all problems in the chapter reviews and cumulative tests are provided in the Appendix along with answers to all the odd-numbered problems in this text.

New to the Second Edition

1. A thorough integration of problem-solving skills throughout the text.
2. An increased emphasis on word problems with real-world applications.
3. Chapter tests have been expanded to become Cumulative Tests that provide students continual review.
4. Solving first-degree equations and inequalities has been placed in chapter 2 so students can use this knowledge when solving word problems.
5. A streamlined reorganization produces a more manageable quantity of material.

Integrated Instructional Package of Supplements

We have assembled a comprehensive array of supplements that are fully coordinated with our text. They have been developed to complement each other in an effort to enhance both the teaching and learning aspects of this book.

For the Instructor

The *Instructor's Manual* has been expanded to include ten challenge problems for each chapter in a form that you can photocopy to supplement your course. Five new tests (three problem solving and two multiple choice) and five final exams have been written for this edition.

wcb *Math TestPak* was developed expressly for this second edition. It is a free, computerized testing service with two convenient options. You may use your own Apple® IIe, IIc, Macintosh®, or IBM PC to produce your test or you may use the call-in service offered by the publisher. Contact your local **wcb** sales representative for details.

wcb *Math TestPak* is menu-driven with on-line help screens to guide you through the test-making process. You may select items from the bank, edit the existing items, add new items of your own, or have the system randomly select items for you.

The printed *Test Item File* in an 8½″ × 11″ format contains all of the questions on the **wcb** *Math TestPak*. It will serve as a ready-reference if you use your own computer to generate tests. The items in the *Test Item File* are different from those in the prepared tests in the *Instructor's Manual*. Hence you will have even more items to choose for your tests.

wcb *GradePak* is a computerized grade book that holds data for classes of up to 500 students with 60 scores per student. *GradePak* allows numeric input as well as letter grades. It is available for the IBM PC and the Apple® II family of computers.

For the Student

The *Student's Solutions Manual* contains summaries of each section of every chapter of the text, solutions to selected odd-numbered exercise problems, and chapter self-tests with solutions. It is available for student purchase.

wcb *Math Tutor* is an interactive two-part computerized study guide that helps students master the text material. Part 1 includes a guided review and quiz for each section of the chapter. The objectives correspond to the mastery points in the text. Part 2 features chapter quizzes with questions randomly selected from Part 1. **wcb** *Math Tutor* is available for Apple® IIe and IIc and the IBM PC. It is free to qualified adopters and may be reproduced for computer labs or individual students.

Math Lab is a widely used software program developed by Chris Avery and Chris Barker of De Anza College that serves as a supplement to our text. It is based on a mastery learning approach to instruction. Students have the option to erase scores and to work new randomly generated problems until they get a perfect score. *Math Lab* provides students with immediate feedback. Through hints, it channels the student into the correct problem-solving procedure. *Math Lab* develops student confidence through successful practice. It is available for individual student purchase or laboratory use.

On the **Videotapes** the instructor introduces a concept, provides detailed explanations of example problems that illustrate the concept, including applications, and concludes with a summary. The tapes are available in ½″ VHS or ¾″ U-Matic format and are free to qualified adopters.

The **Audiotapes** are closely tied to the text. Here we start with a brief synopsis of the section and work out sample problems, explaining each step.

Study tips

When you work to your full capacity, you can hope to attain the knowledge and skills that will enable you to create your future and control your destiny. If you do not, you will have your future thrust upon you by others.

A Nation at Risk*

There are certain study skills that you as an algebra student need to have, or develop, to assure your success in this course. In addition to the following items listed, acquaint yourself with the text by reading the preface material that precedes these study tips. Then—

1. For every hour spent in class, plan to spend at least two hours studying outside class.
2. Before going to class, read the material to be covered. This will help you more easily understand the instructor's presentation.

---

*The National Commission on Excellence in Education. *A Nation at Risk*. Washington, D.C.: U.S. Government Printing Office, 1983.

3. Review the material related to each exercise set *before* attempting to work the problems. Be sure you understand the underlying concepts in the worked-out examples and the reason for each step.
4. Carefully read the instructions to the exercise set. Look at the examples and determine what is being asked. Remember, these same instructions will most likely appear on tests.
5. When working the exercise set, take your time, think about what you are doing in each step, and ask yourself why you are performing that step. As you become more confident, increase your speed to better prepare yourself for test situations.
6. When working the exercise sets, compare examples to see in what ways they are alike and in what ways they are different. Problems often *look* similar but are not.

If you do not know how to begin a problem, or you get partway through and are unable to proceed, (a) look back through your notes or (b) look for an exercise you can do that has the answer given and try to analyze the similarities. If doing these things does not work, put the problem aside. Often getting away from it for a time will "open the door" when you try it again. Finally, if you need to, consult your instructor and show him/her the work you have done.

The fact that you will be "using tomorrow what you are doing today" makes it imperative that you learn each concept as you go along. Most concepts, especially the ones that give you the most difficulty, need constant review.

The practice of checking your work will aid you in two ways:

1. It will develop confidence, knowing you have done the problem correctly.
2. It will help you discover your errors on an exam that might otherwise have gone undetected had you not checked your work.

When checking your work, use a different method from the one you used to solve the problem. If the same procedure is used, a tendency to make the same mistake exists. Develop methods for checking your work as you do the practice exercises. This checking then becomes automatic when taking a test.

The following hints will aid you in preparing for an exam:

1. Begin studying and reviewing a number of days prior to the exam. This will enable you to contact your instructor for help if you need it. "All-night" sessions the night before the exam seldom (if ever) yield good results.
2. Take periodic breaks—10 to 15 minutes for each hour of study. Study for no longer than four hours at a time.
3. Work to develop understanding as well as skills. Memorization is seldom useful in an algebra course, so concentrate on understanding the methods and concepts. However do not ignore skill development, since doing so can often lead to what students call "stupid mistakes."

Prior to taking an exam, use the exercise sets, chapter reviews, and/or *Student's Solutions Manual* to make out a practice test, determine where your errors lie, and retake the test to be sure that you have corrected the mistakes. Allot the same amount of time you will be allowed on test day.

When taking the algebra exam you should:

1. Look over the exam to locate the easiest problems.
2. Work these problems first.
3. Work the more difficult and time-consuming problems next. Remember, when stuck on a problem, go on to other problems and return to those giving you difficulty *only after* completing all that you can.
4. Use what time remains to check your answers or to rework those problems that you found most difficult.

Don't panic should you "draw a blank." Avoid thoughts of failure. Should you feel this happening, relax and try to clear your mind. Search out the problems you feel most confident about and begin again. Should you be unable to complete the exam, be sure to check the problems that you have completed. Always be aware of the time remaining. *Do not hurry* and do not be intimidated by other students completing the exam early.

One final bit of advice. Show your work neatly. Develop this habit when working on your practice problems. There is a close correlation between neatly laid-out work and the correct answer. Your instructor will appreciate this and be more inclined to give you more credit if the answer is wrong.

Acknowledgments

We wish to express our sincere thanks for the many comments and suggestions given to us during the preparation of the first edition. In particular, we wish to thank Lynne Hensel, William Lakey, and Douglas Nance for their excellent effort in reviewing each stage of the first edition and supplying us with numerous valuable comments, suggestions, and constructive criticisms.

For their help in typing the manuscript we thank Amy Miyazaki and Debbie Miyazaki, and a special thank you goes to Lisa Miyazaki for her superb aid in preparing the manuscript and for working all of the problems.

Because of his invaluable help and advice in the area of marketing, our sincere thanks go to Harold Elliott.

The authors would like to acknowledge the contribution of Philip Mahler, who introduced to them the idea of using the tabular format to list all possible combinations of factors in factoring trinomials. Mr. Mahler was also responsible for the idea of using the sign of the product "mn" as an operation in the second column of the table. The chief virtue of this method is that it is algorithmic. The authors have modified the method slightly by listing the greater factor first.

Throughout the development, writing, and production of this text, three people have been of such great value that we are truly indebted to them for their excellent work on our behalf. We wish to express our utmost thanks to Suresh Ailawadi, Eugenia M. Collins, and Nova A. Maack.

Error Check

Because of the careful proofreading and checking of all the examples and answers by a large number of very competent people working independently, the authors and the publisher believe this book to be virtually error-free. We wish to express our sincere thanks to Suresh Ailawadi, Terry Baker, Donald Bellairs, Harry Datsun, Bruce Harold, Lisa Miyazaki, and John Snyder. Their hard work is greatly appreciated.

Reviewers

Daniel Anderson
Department of Mathematics
University of Iowa
Dorothy Batta
Department of Mathematics
Wilber Wright College
Philip Beckman
Black Hawk College
James Blackburn
Science & Health Division
Tulsa Junior College
Nancy Bray
San Diego Mesa College
Dan Burns
Sierra College
Helen Burrier
Kirkwood Community College
Vern Byer
Department of Mathematics
University of Maine—Farmington
William Chatfield
Department of Mathematics
University of Wisconsin—Platteville
Duane Deal
Department of Mathematics
Ball State University
Mark Dugopolski
Department of Mathematics
Southeastern LA University
Ray Fartch
Department of Mathematics
Eastern Oklahoma State College
Richard L. Francis
Southeast Missouri State University
Margaret J. Greene
Department of Mathematics
Florida Junior College

Pamela Hager
Department of Mathematics
College of the Sequoias
Shelby Hawthorne
Thomas Nelson Community College
Joyce Huntington
Walla Walla Community College
500 Tausick Way
James Johnson
Department of Mathematics
Modesto Junior College
Glen Just
Department of Mathematics
Mount Saint Clare College
Chris Kolaczewski
University of Akron
Carolyn Likins
Department of Mathematics
Millikin University
Karla Martin
Middle Tennessee State University
Alfred W. Milligan
Department of Mathematics
Western New Mexico University
Ronald Milne
Department of Mathematics
Goshen College
Jesse Moore
Department of Mathematics
John A. Logan College
Ardash Ozsogomonyan
Mathematics and Science Division
College of San Mateo
Thomas Radin
San Joaquin Delta College

ALGEBRA
For College Students

Chapter 1

Basic Concepts and Properties

1-1 Sets and real numbers

Set symbolism

We begin our study with a very simple, but important, mathematical concept—the idea of a **set.*** *A set is any collection of objects or things.* We want the sets that we deal with to be *well defined;* that is, given any object, we can determine whether the object is in a given set. For example, the set of old people is not well defined because the meaning of old is not clear. Whereas the set of people whose ages are greater than seventy years is a well-defined set. In mathematics the idea of a set is used primarily to denote a group of numbers or the set of answers to a problem.

Any one of the things that make up a set is called a **member** or an **element** of that set. One way of writing a set is by listing the elements, separating them by commas, and including this listing within a pair of braces, { }. This way of representing a set is called the **listing** or **roster method.**

*George Cantor (1845–1918) is credited with the development of the ideas of set theory. He described a set as a grouping together of single objects into a whole.

Example 1–1 A	1. Using set notation, write the days of the week that begin with the letter T.

1. Using set notation, write the days of the week that begin with the letter T.

{Tuesday, Thursday}

2. Using set notation, write the set of digits that make up the telephone number for information, 555–1212.

{5,1,2}

> **Note**
> When we form a set, the elements within the set are never repeated and they can appear in any order.

We use capital letters $A,B,C,D,$ and so on, to represent a set. When we wish to show that an element belongs to a particular set, we use the symbol ϵ, which is read "is an element of" or "is a member of." Consider the set $A = \{1,4,5\}$, which is read "the set A whose elements are 1, 4, and 5." If we want to say that 4 is an element of the set A, we write $4 \in A$.

A slash mark, $/$, is used in mathematics to negate a given symbol. Therefore if ϵ means "is an element of," then $\notin$ means "is *not* an element of." To express the fact that 2 is not an element of the set A, we write $2 \notin A$.

Example 1–1 B

Using mathematical symbols, write the following statements.

1. 5 is an element of the set C. $5 \in C$

2. 4 is not an element of the set B. $4 \notin B$

Subsets

Suppose that P is the set of people in a class and W is the set of women in the same class. It is obvious that the members of W are also members of P. We say that W is a **subset** of P and use the symbol $\subseteq$ to indicate "is a subset of."

> ■ **Definition of $A \subseteq B$**
> The set A is a subset of the set B, written $A \subseteq B$, if every element of A is also an element of B.

Example 1–1 C

Given the sets $A = \{1,2,3\}$, $B = \{1,2,3,4,5\}$, $C = \{2,3,5\}$, and $D = \{2,3,1\}$, determine if the following statements are true or false.

1. $A \subseteq B$. True, since all the elements in A are also in B.

2. $A \nsubseteq C$. True, since not all of the elements in A are contained in C.

> **Note**
> $A \nsubseteq C$ is read "A is not a subset of C" and implies that there is at least one element in A that is not in C.

3. $C \subseteq D$. False, since not all of the elements in C are contained in D.

4. $A \subseteq D$. True, since all of the elements in A are contained in D.

In example 4, since the order of the elements within the set does not change the set, we can conclude that the sets A and D are the same set. Therefore we observe that the definition of subset allows a set to be a subset of itself. That is, for any set A, $A \subseteq A$. We will now use the definition of a subset to define when two sets are equal.

> ■ **Definition of $A = B$**
> The set A is said to be equal to the set B, written $A = B$, if and only if $A \subseteq B$ and $B \subseteq A$.

Suppose that we wanted to form the set of the months of the year that begin with the letter X. Since there are no months that begin with the letter X, this set has no elements and is called the **empty set** or the **null set.** *A set that contains no elements is called the empty set or null set and is denoted by the symbol* $\emptyset$.

By the definition of a subset, a set A is *not* a subset of a set B if there is at least one element in A that is not in B. The empty set, by definition, has no elements; therefore it can contain no elements that are not in another given set. We must then conclude that *the empty set is a subset of every set.*

Union and intersection of sets

If we wished to form the set of all the students who have blue eyes *or* blond hair, we would be combining the set of students with blue eyes with the set of students with blond hair. This new set that we have formed is the **union** of the two original sets.

> ■ **Definition of $A \cup B$**
> The union of the sets A and B, written $A \cup B$, is the set of all elements that are in A or in B or in both A and B.

| Example 1–1   D | Given the sets $A = \{2,3,5\}$, $B = \{1,3,7\}$, and $C = \{2,4,6\}$, form the following sets. |

1. $A \cup B$

   A union B consists of those elements that appear in A, $\{2,3,5\}$, or those in B, $\{1,3,7\}$, or those in both A and B, $\{3\}$. Therefore

   $A \cup B = \{1,2,3,5,7\}$.

2. $B \cup C = \{1,3,7\} \cup \{2,4,6\} = \{1,2,3,4,6,7\}$

If we wished to form the set of all the students who have blue eyes *and* blond hair, we would be forming a new set of students that possess the characteristic of having *both* blue eyes and blond hair. This new set that we have formed is the **intersection** of the two original sets.

> ■ **Definition of $A \cap B$**
> The intersection of the sets A and B, written $A \cap B$, is the set of only those elements that are in both A and B.

| Example 1–1   E | Given the sets $A = \{1,2,3\}$, $B = \{2,3,4,5\}$, and $C = \{4,5\}$, form the following sets. |

1. $A \cap B$

   A intersection B consists of only those elements that appear in both A and B. Hence

   $A \cap B = \{1,2,3\} \cap \{2,3,4,5\} = \{2,3\}$.

2. $A \cap C = \{1,2,3\} \cap \{4,5\}$

   There are no elements common to both A and C; therefore the intersection is the empty set.

   $A \cap C = \emptyset$

3. $A \cup (B \cap C) = \{1,2,3\} \cup (\{2,3,4,5\} \cap \{4,5\})$

   Performing the intersection of B and C, we have

   $= \{1,2,3\} \cup \{4,5\}$

   and then performing the union of $B \cap C$ with A

   $= \{1,2,3,4,5\}$.

When the intersection of two sets is the empty set, as in example 2 above, we say that the two sets are **disjoint**. That is, *the sets* A *and* B *are disjoint if and only if* A $\cap$ B $= \emptyset$.

The set of real numbers

Our study of algebra will be primarily concerned with the set of **real numbers** and its properties. We shall now review the set of real numbers and its major subsets.

In each of the previous examples, we could determine the exact number of elements in a set. This type of set is called a **finite** set. Each of the major sets of numbers that make up the set of real numbers has an unlimited number of elements. This is called an **infinite** set.

The set of real numbers is made up of the following sets of numbers.

I. The set of **natural numbers,** or counting numbers, is defined by

$$\{1,2,3,4, \cdot \ \cdot \ \cdot\}$$

and denoted by N.

Note

The three dots tell us to continue this pattern indefinitely. That is, there is no last natural number.

II. The set of **whole numbers** is defined by

$$\{0,1,2,3,4, \cdot \ \cdot \ \cdot\}$$

and denoted by W.

III. The set of **integers** is defined by

$$\{\cdot \ \cdot \ \cdot, \ -2,-1,0,1,2, \cdot \ \cdot \ \cdot\}$$

and denoted by J.

To be able to express certain sets of numbers, we need to define the mathematical concept of a **variable.** *A variable is a symbol (generally a lowercase letter) that represents an unspecified element of a set that contains two or more elements.* When we use a variable to define a set of numbers, the set of numbers that the variable represents is called its **replacement set.** For example,

$$\{x \,|\, x \text{ is a natural number less than } 6\}$$

is read "the set of all elements x such that x is a natural number less than 6." This notation defines the set $\{1,2,3,4,5\}$. The bar is read "such that" and the notation is called **set-builder notation.** In general, set-builder notation has the pattern shown in figure 1.1.

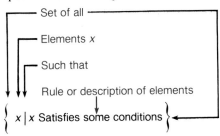

Figure 1.1

Example 1-1 F   List the elements of the following sets.

1. $\{x \mid x$ is an integer between -4 and $2\} = \{-3, -2, -1, 0, 1\}$
2. $\{x \mid x$ is a whole number greater than $10\} = \{11, 12, 13, 14, \cdots\}$

> **Note**
> Since this is an infinite set, we set up a pattern for the numbers and place three dots after the last number to indicate that this pattern continues indefinitely.

3. $\{x \mid x$ is a natural number less than $1\}$
   Since there are no natural numbers less than 1, the set is empty, $\emptyset$.

IV. The set of **rational numbers** is defined by

$$\left\{ \frac{p}{q} \;\middle|\; p \text{ and } q \,\epsilon\, J,\, q \neq 0 \right\}$$

and denoted by Q. The set-builder notation symbolizes that the set of rational numbers, Q, will consist of all the numbers that can be represented by the quotient of two integers where the denominator is not zero. A second way that the set of rational numbers can be defined is $\{x \mid$ the decimal representation of x is either terminating or repeating$\}$. Examples of terminating and repeating decimals are

$$\frac{1}{2} = 0.5, \quad \frac{1}{3} = 0.\overline{3}, \quad -\frac{1}{6} = -0.1\overline{6}, \quad -\frac{5}{4} = -1.25,$$

where a bar placed over a number or group of numbers indicates that the number(s) repeat indefinitely.

V. The set of **irrational numbers** is defined by $\{x \mid$ the decimal representation of x is nonterminating and nonrepeating$\}$ and denoted by H. Examples of irrational numbers are

$$\sqrt{3}, \quad -\sqrt{5}, \quad \pi, \quad \frac{\sqrt{2}}{2}.$$

The decimal representation of an irrational number will never terminate. We cannot find a repeating pattern of digits, no matter how many digits we write past the decimal point.

> **Note**
> By using a calculator, the above numbers can be represented by the following approximations to three decimal places:
>
> $$\sqrt{3} \approx 1.732, \quad -\sqrt{5} \approx -2.236, \quad \pi \approx 3.142, \quad \frac{\sqrt{2}}{2} \approx 0.707.$$
>
> The symbol $\approx$ is read "is approximately equal to." π is the distance around a circle (circumference) divided by the distance across the circle through the center (diameter).
> Common approximations for π are 3.14 and $\dfrac{22}{7}$.

VI. The set of **real numbers** is defined by

$$\{x \mid x \in Q \text{ or } x \in H\}$$

and denoted by R.

Whenever we encounter a problem and a specific replacement set for the variable is not indicated, it will be understood that we are dealing with the set of real numbers. All of the sets that we have considered thus far are subsets of the set of real numbers. Figure 1.2 shows this relationship.

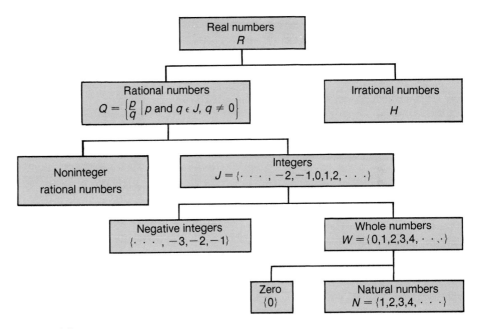

Figure 1.2

The real number line

To visualize the set of real numbers, we use a diagram called a **number line.** Any real number can be located on the number line (see figure 1.3).

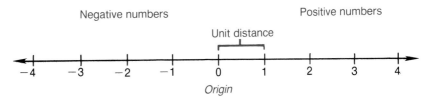

Figure 1.3

The number associated with each point on the number line is called the **coordinate** of the point. The point associated with each number is called the **graph** of that number. In figure 1.4 the numbers -2, $-\frac{1}{3}$, $\frac{3}{4}$, $\sqrt{3}$, and 3 are the coordinates of the points indicated on the line by solid circles. These solid circles are the graphs of the numbers -2, $-\frac{1}{3}$, $\frac{3}{4}$, $\sqrt{3}$, and 3.

Figure 1.4

Note
The coordinate of $\sqrt{3}$ represents an irrational number. To graph this point, we can use a decimal approximation from a calculator, $\sqrt{3} \approx 1.732$. The word "number" used in this book means "a real number."

Order and absolute value

The direction we move on the number line is also important. If we move to the right, we move in a positive direction and our numbers **increase.** If we move to the left, we move in a negative direction and our numbers **decrease**.

If we choose arbitrary points on the number line and represent them by a and b, where a and b represent some *unspecified* numbers, we observe that there is an **order** relationship between a and b.

Since the point associated with a is to the left of the point associated with b, we say that a is **less than** b, which in symbols is $a < b$. We might also say that b is **greater than** a, which in symbols is $b > a$. The symbols $<$ (less than) and $>$ (greater than) are inequality symbols called **strict inequalities** and denote an **order relationship** between numbers.

Example 1–1　G

Graph the following pairs of numbers and insert the correct inequality symbol between the numbers.

1. 3 and 5
The graph would be

Since 3 lies to the left of 5, we say that 3 is less than 5, $3 < 5$.

Note
$3 < 5$, read "3 is less than 5," can also be stated as $5 > 3$, read "5 is greater than 3." No matter which inequality symbol we use, the inequality symbol always points to the *lesser* number.

2. -90 and -100

The graph would be

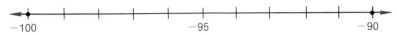

Since -90 lies to the right of -100, we say that -90 is greater than -100, $-90 > -100$.

There are two other inequality symbols, called **weak inequalities,** which are **less than or equal to,** $\leq$, and **greater than or equal to,** $\geq$. The weak inequality symbol $\leq$, less than or equal to, combines the relationship of less than ($<$) with the relationship of equality ($=$). The weak inequality symbol $\geq$, greater than or equal to, combines the relationship of greater than ($>$) with the relationship of equality ($=$).

As we study the number line, we see a very useful property called **symmetry.** The numbers are symmetrical with respect to the origin. That is, if we go two units to the right of 0, we come to the number 2, and if we go two units to the left of 0, we come to the *opposite* of 2, which is -2. This idea of how far a given number is from the origin is called the **absolute value** of that number. *The absolute value of a number is the undirected distance that the number is from the origin.* The symbol for absolute value is $|\quad|$.

Example 1–1 H

Find the indicated absolute values.

1. $|2| = 2$

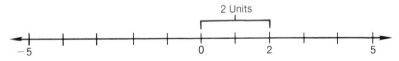

2. $|-3| = 3$

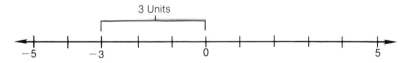

Note

The absolute value of a number is *never* negative. That is, for every $x \in R, |x| \geq 0$. This means that the value of $|x|$ is never negative.

When we wish to determine the absolute value of a positive number or zero, $x \geq 0$, the absolute value of the number is simply the original number, x. When we wish to determine the absolute value of a negative number, $x < 0$, we find that the absolute value of a negative number is the opposite of the original number. We symbolically write this as $-x$, read "the opposite of x." We can now state the definition of absolute value symbolically.

1-1 Sets and Real Numbers

■ **Definition of $|x|$**

$$|x| = \begin{cases} x \text{ if } x \geq 0 \\ -x \text{ if } x < 0 \end{cases}$$

Note

The symbol $-x$ is not necessarily a negative number. It is symbolically stating that we want the opposite of the value that x is representing.

Example 1–1 I

Find the indicated absolute values.

1. $|4| = 4$ **2.** $|-9| = -(-9) = 9$ **3.** $-|-7| = -(7) = -7$

Note

The minus sign in front of the absolute value in example 3 indicates that we want the opposite of the absolute value and, therefore, the answer is -7.

Mastery points
Can you
- List the elements of a set?
- Use set symbolism?
- Determine when a set is a subset of another set?
- Determine when sets are equal?
- Perform the operations of union and intersection on sets?
- Use set-builder notation?
- Graph a number?
- Determine the coordinate of a point?
- Determine which of two real numbers is greater?
- Find the absolute value of a number?
- Use mathematical symbols?

Exercise 1–1

Directions Write each set by listing the elements. See example 1–1 A.

1. The days of the week that begin with the letter S
2. The months of the year that begin with the letter A

3. The even integers between 9 and 15
4. The months of the year with less than 28 days

5. The months of the year that begin with C
6. The days of the week that begin with A

Directions Use mathematical symbols to write the following statements. See example 1–1 B.

7. A is a subset of D.
8. B is not a subset of C.

9. The null set is a subset of B.

10. The null set is not an element of C.

11. The set whose elements are 6, 7, 8

12. The set whose elements are 1, 4, 6

Directions Given the sets $A = \{2,4,6\}$, $B = \{1,3,5\}$, $C = \{1,2,3,4,5,6\}$, $D = \{3,1,5\}$, and $E = \{1,2,3,4,5,6,7,8\}$, determine if the following statements are true or false. See example 1–1 C.

13. $2 \in A$ **14.** $1 \in B$ **15.** $A \subseteq C$ **16.** $B \subseteq E$ **17.** $A \not\subseteq E$

18. $C \subseteq E$ **19.** $B = D$ **20.** $A = D$ **21.** $\emptyset \subseteq A$ **22.** $A \subseteq D$

Directions Given the sets $A = \{1,2,3,4\}$, $B = \{2,3,4,5,6\}$, $C = \{2,3,4\}$, and $D = \{7,8,9\}$, form the following sets. See examples 1–1 D and E.

23. $A \cup C$ **24.** $B \cup C$ **25.** $A \cup D$ **26.** $B \cup D$ **27.** $A \cap D$

28. $C \cap D$ **29.** $(C \cup B) \cap A$ **30.** $(D \cap C) \cup B$ **31.** $A \cup (C \cap D)$ **32.** $(B \cap C) \cup A$

Directions Graph the following sets of numbers on a number line. See example 1–1 G.

33. $\{-3,-1,2,4,5\}$ **34.** $\{-4,-3,-2,1,3\}$ **35.** $\left\{-\dfrac{7}{4},0,\dfrac{5}{4},3,5\right\}$

36. $\{-\sqrt{3},0,\sqrt{2},3,6\}$ **37.** $\{-5,-\sqrt{5},\sqrt{3},3,5\}$ **38.** $\{-4,-2,0,\sqrt{5},\sqrt{7}\}$

Directions Write the value of the following numbers. See examples 1–1 H and I.

39. $|-3|$ **40.** $|-5|$ **41.** $|0|$ **42.** $\left|\dfrac{1}{4}\right|$ **43.** $-|-2|$

44. $-|-7|$ **45.** $-|4|$ **46.** $-|5|$

Directions Replace the comma with the proper strict inequality symbol, $<$ or $>$, between the following numbers. See example 1–1 G.

47. 4, 9   **48.** $-4, -8$ **49.** $-3, -11$ **50.** $-12, -5$ **51.** $-15, -10$

52. $0, -5$ **53.** $-4, 0$

Directions Use mathematical symbols to write each of the following statements.

Example

0 is not an element of the set of natural numbers.

Solution

$0 \notin N$

54. The set of whole numbers is not a subset of the set of natural numbers.

55. The set of irrational numbers intersected with the set of rational numbers is the null set.

56. The union of the set of rational numbers with the set of irrational numbers is the set of real numbers.

57. $\{8\}$ is a subset of the set of natural numbers.

58. 8 is not a subset of the set of natural numbers.

59. $\{8\}$ is not an element of the set of natural numbers.

60. 8 is an element of the set of natural numbers.

Directions Determine from the given information whether the answer to each of the following statements is yes or no. The sets *A*, *B*, and *C* are general sets and are not the same as the sets in the previous exercises.

Example
If $A \subseteq B$ and $2 \in A$, must 2 be an element of *B?*

Solution
Yes. From the definition of subset, every element in *A* must be contained in *B*. Therefore if $2 \in A$ and $A \subseteq B$, then $2 \in B$.

61. If $A \subseteq B$ and $3 \in A$, must 3 be an element of *B?*

62. If $A \subseteq B$ and $7 \in B$, must 7 be an element of *A?*

63. If $A \nsubseteq B$ and $6 \in A$, does this mean that $6 \notin B$?

64. If $A \nsubseteq B$ and $1 \notin A$, does this mean that 1 must he an element of *B?*

65. If $A \subseteq B$ and $B \subseteq C$ and $4 \in A$, must 4 be an element of *C?*

66. If $A \subseteq B$ and $B \subseteq A$, does this mean that $A = B$?

67. If $A \subseteq B$ and $B \subseteq C$, must *A* be a subset of *C?*

68. If $A \subseteq B$ and $B \subseteq C$ and $5 \in A$, must 5 be an element of *C?*

69. If $A \subseteq B$ and $B \subseteq A$ and $8 \notin A$, can 8 be an element of *B?*

70. If $A \nsubseteq B$, can $B \subseteq A$?

71. Is $\emptyset \subseteq \emptyset$?

72. Is $\emptyset \subseteq B$?

73. Does $\emptyset \cap A = \emptyset$?

74. Does $\emptyset \cup A = A$?

75. Does $A \cap A = A$?

76. Does $A \cup A = A$?

1–2 Operations with real numbers

Addition

In this section we will review the rules for addition and subtraction of real numbers. We use the minus sign ($-$) to indicate a negative number and the plus sign ($+$) to indicate a positive number. A **signed number** consists of two parts: Its **absolute value** and its **sign** (see figure 1.5).

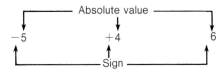

Figure 1.5

Basic Concepts and Properties

We can summarize the rules for adding real numbers.

1. If the signs are the same, we add their absolute values and prefix the sum by their common sign.
2. If the signs are different, we subtract the lesser absolute value from the greater absolute value. The answer has the sign of the number with the greater absolute value.

Example 1–2 A

Add the following.

1. $(+3) + (+6) = (+9)$

Absolute value
Common sign

2. $(-4) + (-9) = (-13)$

Absolute value
Common sign

3. $(-8) + (+5) = (-3)$

Absolute value
Sign of the number with the greater absolute value

4. $(-6) + (+10) = (+4)$

Absolute value
Sign of the number with the greater absolute value

Subtraction

We are already familiar with the operation of subtraction in problems such as $12 - 8 = 4$. If we were to add (-8) instead of subtracting $(+8)$, we observe that the results are the same, that is

Opposite of $+8$

$(+12) - (+8) = (+12) + (-8) = (+4).$

Change to addition

This leads us to the following definition of subtraction.

■ **Definition of subtraction**
For any two real numbers a and b, the difference of a and b is
$$a - b = a + (-b).$$

Concept
a minus b is equal to a plus the opposite of b.

Our steps to carry out the subtraction are as follows:

Step 1 We change the operation from subtraction to addition.

Step 2 We change the sign of the number that follows the subtraction symbol.

Step 3 We perform the addition using our rules for adding signed numbers.

Example 1–2 B

Subtract the following.

1. $(+7) - (+4) = (+7) + (-4) = (+3)$

2. $(-11) - (+3) = (-11) + (-3) = (-14)$

3. $(-14) - (-10) = (-14) + (+10) = (-4)$

4. $(-2) - (-12) = (-2) + (+12) = (+10)$

When a series of numbers involving addition and subtraction are written horizontally, we can mentally change the operation of subtraction to addition and perform the operations in order from left to right as they occur. For example, consider

$$10 - (-2) + 7 + (-4) - 8.$$

Changing the indicated subtractions to addition, we obtain the indicated sum.

$$10 + (+2) + 7 + (-4) + (-8) = 7$$

Many times, part of a problem will have a group of numbers enclosed within grouping symbols such as parentheses ( ), brackets [ ], or braces { }. If a quantity is enclosed within a grouping symbol, we treat the quantity within as a single number. Thus given

$$14 - (5 - 7) + (4 + 8) - (6 - 2),$$

we perform the operations within the parentheses first to get

$$14 - (-2) + (12) - (4),$$

and changing the subtraction to addition

$$14 + (+2) + (12) + (-4),$$

and adding from left to right

$$= 16 + (12) + (-4)$$
$$= 28 + (-4)$$
$$= 24.$$

Example 1-2 C

Perform the indicated operations.

1. $(-14) - (-8) + (-7) - (4)$

Change the subtraction to addition.

$$= (-14) + (+8) + (-7) + (-4)$$

Perform the addition from left to right.

$$= (-6) + (-7) + (-4)$$
$$= (-13) + (-4)$$
$$= -17$$

2. $12 - (4 - 7) + (8 - 3) + (5 - 9)$ Simplify within parentheses.

$$= 12 - (-3) + (5) + (-4)$$ Change to addition.
$$= 12 + (+3) + (5) + (-4)$$ Add from left to right.
$$= 15 + (5) + (-4)$$
$$= 20 + (-4)$$
$$= 16$$

Multiplication

We are already familiar with the fact that the product of two positive numbers is positive. We can see this by considering multiplication to be repeated addition. For example, $3 \cdot 4$ means the sum of the three 4s, that is,

$$3 \cdot 4 = 4 + 4 + 4 = 12.$$

The number 12 is called the **product** of 3 and 4 and 3 and 4 are called **factors** of 12. *The numbers or variables in an indicated multiplication are referred to as the factors of the product.*

We can summarize the rules of multiplication of signed numbers.

1. When we multiply two numbers having like signs, the product will be positive.
2. When we multiply two numbers having different signs, the product will be negative.

Example 1-2 D

Multiply the following.

1. $(+4)(+8) = +32$ **2.** $(-6)(-3) = +18$

3. $(-4)(+5) = -20$ **4.** $(+2)(-8) = -16$

Exponential notation

Consider the indicated products

$$4 \cdot 4 \cdot 4 = 64$$

and

$$3 \cdot 3 \cdot 3 \cdot 3 = 81.$$

A more convenient way of writing $4 \cdot 4 \cdot 4$ is 4^3, which is read "4 to the third power" or "4 cubed." We call the number 4 the **base** of the expression and the number 3 to the upper right the **exponent.**

Thus

$$4 \cdot 4 \cdot 4 = 4^3 = 64.$$

In like fashion,

$$3 \cdot 3 \cdot 3 \cdot 3$$

may be written 3^4, where 3 is the base and 4 is the exponent, and the expression is read "3 to the fourth power." Then

$$3 \cdot 3 \cdot 3 \cdot 3 = 3^4 = 81.$$

Notice that *the exponent tells how many times the base is used as a factor in an indicated product.* We call this form of a product the **exponential form.** That is, the exponential form of the product $3 \cdot 3 \cdot 3 \cdot 3$ is 3^4.

We will now relate the idea of exponents to signed numbers. Consider the following examples.

Example 1–2 E

1. $2^4 = 2 \cdot 2 \cdot 2 \cdot 2 = 16$ (The exponent, 4, indicates how many times the base, 2, is used as a factor.)

2. $(+2)^4 = (+2)(+2)(+2)(+2) = +16$

3. $(-2)^4 = (-2)(-2)(-2)(-2) = +16$ (Even number of negative numbers)

4. $(-2)^3 = (-2)(-2)(-2) = -8$ (Odd number of negative numbers)

Remember that when we have a negative number, we place it inside parentheses. With this idea in mind, we can see that there is a definite difference between $(-2)^4$ and -2^4. In the first case, the parentheses denote that this is a negative number to a power, $(-2)^4 = (-2)(-2)(-2)(-2) = +16$. In the second case, since there are no parentheses around the number, we understand that this is *not* (-2) to a power, but rather the negative of what we get for an answer when we find 2^4 as follows:

$$-2^4 = -(2^4) = -(2 \cdot 2 \cdot 2 \cdot 2) = -(16) = -16.$$

Example 1–2 F

Perform the indicated multiplication.

1. $(-3)^3 = (-3)(-3)(-3) = -27$

2. $-3^3 = -(3 \cdot 3 \cdot 3) = -27$

3. $(-3)^4 = (-3)(-3)(-3)(-3) = +81$

4. $-3^4 = -(3 \cdot 3 \cdot 3 \cdot 3) = -81$

Division

We studied subtraction of signed numbers by defining subtraction in terms of addition. We shall use a similar approach for division of signed numbers by defining division in terms of multiplication.

> ■ **Definition of division**
> The quotient of any two real numbers a and b $(b \neq 0)$ is the unique real number q such that $a = bq$. In symbols,
> $$\frac{a}{b} = q \text{ if and only if } a = bq.$$

Note
In the definition, a is called the dividend, b is the divisor, and q is the quotient.

We can summarize our rules for multiplication and division of signed numbers.

1. When we multiply or divide two numbers having like signs, our answer will be positive.
2. When we multiply or divide two numbers having different signs, our answer will be negative.

Example 1–2 G

Perform the indicated division.

1. $\dfrac{(-14)}{(-7)} = +2$

2. $\dfrac{(-24)}{(+8)} = -3$

3. $\dfrac{(+36)}{(-9)} = -4$

4. $\dfrac{(-2)(-8)}{(-4)} = \dfrac{(+16)}{(-4)} = -4$

5. $\dfrac{(+6)(-3)}{(-9)} = \dfrac{(-18)}{(-9)} = +2$

Division involving zero

In section 1–1 we defined a rational number to be any real number that can be expressed as a quotient of two integers where the divisor is not zero. The number zero, 0, is the only number that we cannot use as a divisor. To see why we exclude zero as a divisor, recall that we check a division problem by multiplying the divisor times the quotient to get the dividend. If we apply this idea in connection with zero as a divisor, we observe the following. Suppose there were a number q, such that $3 \div 0 = q$. Then $q \cdot 0$ would have to be equal to 3 for our answer to check, but this product is zero regardless of the value of q. Therefore we cannot find an answer for this problem and we say that the answer is *undefined*. If we try to divide zero by zero and again call our answer q, we have

$0 \div 0 = q$, and when we check our work, $0 \cdot q = 0$, we see that any value for q will work. In this situation we say our answer is *indeterminate*. We therefore decide that **division by zero is not allowed.**

It is important to note that, although division by zero is not allowed, this does not extend to the division of zero by some other number. We can see that $0 \div (-4) = 0$ since $(-4) \cdot 0 = 0$. Thus *the quotient of zero divided by any number other than zero is always 0*.

Example 1–2 H

Perform the division, if possible.

1. $\dfrac{0}{5} = 0$ **2.** $\dfrac{2}{0}$ is undefined   **3.** $\dfrac{-7}{0}$ is undefined

4. $\dfrac{0}{-7} = 0$ **5.** $\dfrac{0}{0}$ is indeterminate

---

Problem solving

To solve the following word problems, we will represent gains by positive real numbers and losses by negative real numbers.

Example 1–2 I

Choose a variable to represent the unknown quantity and find its value.

1. A board that is 8 feet long is joined with a board that is 5 feet long. What is the total length of the board?
Let x equal the total length of the boards.
To find the total length we must *add* the individual lengths. Thus

$x = 8 + 5 = 13$.

The total length of the board is 13 feet.

2. On a given winter's day in Chicago, the temperature was 14° in the afternoon. By 9 P.M. the temperature was −4°. How many degrees did the temperature fall from afternoon to 9 P.M.?
Let $x =$ the number of degrees fall in temperature.
We must find the difference between 14° and −4°. Thus

$x = 14 - (-4) = 14 + 4 = 18$.

There was an 18° drop in temperature.

3. On Monday the high temperature was 57°. If the high temperature dropped 6° on each of the next 3 days, what was the high temperature on Thursday?
Let $t =$ the high temperature on Thursday.
A drop of 6 degrees would be represented by −6 and we must multiply by 3 to find the total change. Thus

$t = 57 + 3(-6) = 57 + (-18) = 39$.

The high temperature on Thursday was 39°.

4. If $13.58 is spent on 14 audio tapes, how much did each tape cost?

Let $c =$ the cost of each tape.

We must divide $13.58 by 14. Thus

$$c = \frac{13.58}{14} = 0.97.$$

Each audio tape costs 97¢.

Mastery points

Can you
- Add signed numbers?
- Subtract signed numbers?
- Perform addition and subtraction, in order, from left to right?
- Treat a quantity within a grouping symbol as a single number?
- Multiply signed numbers?
- Divide signed numbers?
- Use exponents?
- Apply the rules of division involving zero?

Exercise 1–2

Directions Find each sum or difference. See examples 1–2 A, B, and C.

1. $(+6) - (+4)$

2. $(+7) - (-5)$

3. $(-8) - (+6)$

4. $(+7) - (+12)$

5. $(+10) + (-6) + (-3)$

6. $(-5) - (-1)$

7. $(+10) - (-8)$

8. $(-12) + 0$

9. $(-7) - 0$

10. $(+16) + (-7)$

11. $(+9) + (-13)$

12. $0 + (-4)$

13. $(-20) - 0$

14. $6 - 9 + 11 - 8$

15. $-9 - 8 - 7 + 12$

16. $-10 - 12 + 18 + 4$

17. $9 - 4 + 6 - 5 + 7$

18. $(8 - 3) - 6 + (7 + 5)$

19. $-6 - 5 + 8 - (12 - 6)$

20. $14 - [(-4) - (-6)] + (7 - 9)$

21. $[(-6) - 5] + 4 - (16 - 6)$

22. $[4 - (-8)] - 6 - (2 - 11)$

Directions Perform the indicated operations if possible. See examples 1–2 D, E, F, G, and H.

23. $(-2)(-5)$

24. $(-6)(-8)$

25. $(+4)(-7)$

26. $(+5)(-6)$

27. $(-9)(+3)$

28. $(-11)(+4)$

29. $(-3)(-1)(-8)$

30. $(-7)(-2)(-6)$

31. $(+4)(-2)(-6)$

32. $(+5)(-3)(+2)$

33. $(-6)(+10)(+4)$

34. $(+7)(-3)(-2)$

35. $(-4)(-5)(-1)(-3)$

36. $(-9)(+2)(0)(-4)$

37. $(-8)(-10)(+6)(0)$

38. -5^2 **39.** -2^6 **40.** $(-4)^2$ **41.** $(-6)^2$ **42.** $(-3)^3$

43. $(-4)^3$ **44.** -5^3 **45.** -2^3 **46.** -6^3 **47.** $\dfrac{(-20)}{(-10)}$

48. $\dfrac{(+32)}{(-4)}$ **49.** $\dfrac{(+32)}{(-8)}$ **50.** $\dfrac{(-22)}{(-11)}$ **51.** $\dfrac{(-21)}{(-7)}$ **52.** $\dfrac{(-34)}{(-17)}$

53. $\dfrac{(-27)}{(+3)}$ **54.** $\dfrac{0}{(-8)}$ **55.** $\dfrac{0}{0}$ **56.** $\dfrac{(-4)(-3)}{(-6)}$ **57.** $\dfrac{(-20)(+2)}{(-4)}$

58. $\dfrac{(+18)(+2)}{(-6)}$ **59.** $\dfrac{(-6)(0)}{(-2)}$ **60.** $\dfrac{(-8)(0)}{(-2)(+4)}$ **61.** $\dfrac{(-18)(+12)}{(-9)(+6)}$ **62.** $\dfrac{(-8)(-6)}{(-2)(0)}$

63. $\dfrac{(-4)(-3)}{(0)(-2)}$ **64.** $\dfrac{(-6)^2}{(-2)(-2)}$ **65.** $\dfrac{-10^2}{(-5)(-5)}$ **66.** $\dfrac{(-12)(0)}{(0)(-3)}$

67. Harry owes Kay and Bill $153 and $56, respectively, and Donna owes Harry $121. In terms of positive and negative symbols, how does Harry stand monetarily?

68. If Helen has $46 just after paying off a debt of $37, how much money did she have before paying off the debt?

69. The temperatures on January 15 in Chicago for the last six years are 14, -6, -4, 19, 7, and -12. What is the average temperature for the last six years? (*Hint:* The average is found by adding all of the values and then dividing that sum by the number of values.)

70. The temperatures on February 1 in Ogema, Wisconsin, for the last eight years are -20, -11, 0, -6, -9, -11, 4, and -3. What is the average temperature for the last eight years? (Refer to exercise 69.)

71. Eric Dickerson carried the ball six times as follows: 14-yard gain, 4-yard gain, 6-yard gain, 8-yard loss, 10-yard gain, and 4-yard gain. Represent the gains as positive integers and the loss as a negative integer and find his average for the six carries. (Refer to exercise 69.)

72. Over a six-day period, the price of a particular stock suffered losses of $5 the first two days and $4 the last four days. If the stock originally sold for $88, what was its price after the six-day period?

73. Jeff acquires a debt of $10 each day for seven days. If we represent a $10 debt by (-10), write a statement of the change in his assets after seven days. What is the change?

74. Eight days ago Joyce had $48 more in assets than she has today. Let eight days ago be represented by (-8). Write a statement for her daily change in assets if the change is the same each day. What is this daily change?

75. The temperature on a given day in Anchorage, Alaska, was $-10°$ F. The temperature went down $14°$. What was the final temperature that day?

76. A TWA flight is flying at an altitude of 27,000 feet. It suddenly hits an air pocket and drops 1,800 feet. What is its new altitude?

77. Mary Ann has $125 in her checking account. She deposits $28, $14, and $17. She then writes checks for $52 and $64. What is her final balance in the checking account?

78. An auditorium contains 68 rows of seats. If each row contains 40 seats, how many people can be seated in the auditorium?

79. In a classroom there are 4 rows of desks. If each row contains 8 desks, how many students will the classroom hold?

80. Tom received money on his birthday from four different people. He received $10, $6, $8, and $5. How much money did Tom receive for his birthday?

81. Nancy sold 12 glasses of lemonade at her corner stand. If she charged 20 cents a glass, how much did she make in sales?

82. Lisa was born in 1963. How old will she be in the year 2000?

83. A chemist now has 125 ml (milliliters) of acid and she needs 415 ml of the acid. How much more is needed?

84. Beth has $175 and wants to buy a television for $485. How much will she owe?

85. The top of Mt. Everest is 29,028 feet above sea level and the top of Mt. McKinley is 20,320 feet above sea level. How much higher is the top of Mt. Everest than the top of Mt. McKinley?

86. Death Valley is 282 feet below sea level (-282). What is the difference in the altitude between Death Valley and Mt. McKinley? (See exercise 85.)

87. Mrs. Spencer paid $25.20 for 14 watermelons for her fruit market. How much did each watermelon cost her?

88. A carpenter wishes to cut a 12-foot board into 4 pieces that are all the same length. Find the length of each piece.

89. A man drove 374 miles and used 11 gallons of gasoline. How many miles did he drive on each gallon of gasoline?

90. A grocer averages selling 68 gallons of milk each day. How many gallons of milk does he sell in two weeks? (Assume the grocery is open seven days per week.)

91. Rob's blood pressure was 127 over 82. If the first number rose by 6 and the second number dropped by 4, what is the new reading?

92. If Martha drove 252 miles in six hours, how many miles did she travel each hour (in miles per hour) if she drove at a constant speed?

93. Alice can work 62 math problems per hour. How many hours would it take her to work 1,674 problems?

94. Al can type 2,436 words in 29 minutes. How many words did he type per minute?

95. The barometric pressure rose 8 mb (millibars) and then dropped 12 mb. Later that day the pressure dropped another 4 mb and then rose 10 mb. What was the gain or loss in barometric pressure that day?

1-3 Properties of real numbers

In mathematics we begin the development of a set of properties of numbers by making certain assumptions. These assumptions are called **axioms** and are formal statements about numbers that we assume to be true. Such assumptions may arise from the observation of a number of instances of a specific situation or may just be statements that we assume are always valid. This appears to give us a freedom to introduce as many such assumptions as we wish, but it is desirable that all such axioms lead to useful consequences. The fewer axioms that we use, the more powerful will be our system.

The first of our assumptions has to do with equality. An **equality** is a mathematical statement that two symbols, or groups of symbols, are names for the same number. For example, $2 + 3$ and $4 + 1$ are different symbols for the number 5, and we can state this by the equality

$$2 + 3 = 4 + 1.$$

We shall assume that real numbers have the following properties of equality.

> **Equality properties of real numbers**
> For all real numbers a, b, and c
>
> ■ **Reflexive property of equality**
> $a = a$.
>
> ■ **Symmetric property of equality**
> If $a = b$, then $b = a$.
>
> ■ **Transitive property of equality**
> If $a = b$ and $b = c$, then $a = c$.
>
> ■ **Substitution property of equality**
> If $a = b$, then a may be replaced by b or b may be replaced by a in any statement without changing the truth or falsity of the statement.

Example 1–3 A

The following statements are examples of the properties of equality.

1. $10 = 10$ Reflexive property
2. If $3 = x$, then $x = 3$.   Symmetric property
3. If $m = n$ and $n = 5$, then $m = 5$.   Transitive property
4. If $x = 2$ and $x + 1 = 3$, then $2 + 1 = 3$.   Substitution property

In section 1–1 we discussed inequalities and the idea of order on the real number line. We will now state two properties of inequalities of real numbers.

> **Inequality properties of real numbers**
> For all real numbers a, b, and c
>
> ■ **Trichotomy property**
> Exactly one of the following is true.
> $a < b$, $a = b$,   or $a > b$.
>
> ■ **Transitive property of inequality**
> If $a < b$ and $b < c$, then $a < c$.

Example 1–3 B   The following statements are examples of the properties of inequality.

1. $a \not< b$. The statement is read "a is not less than b." From the trichotomy property, we know that if a is not less than b, then a could be equal to b or a could be greater than b. We write this mathematically as $a \geq b$.

2. If $a < b$ and $b < 3$, then $a < 3$.   Transitive property

The following properties of real numbers govern the operations of addition and multiplication.

Properties of real numbers
For all real numbers a, b, and c

■ **Closure property of addition**
$a + b \in R$.

■ **Closure property of multiplication**
$ab \in R$.

■ **Commutative property of addition**
$a + b = b + a$.

■ **Commutative property of multiplication**
$ab = ba$.

■ **Associative property of addition**
$(a + b) + c = a + (b + c)$.

■ **Associative property of multiplication**
$(ab)c = a(bc)$.

■ **Identity property of addition**
There is a unique real number 0 such that
$a + 0 = 0 + a = a$.

■ **Identity property of multiplication**
There is a unique real number 1 such that
$a \cdot 1 = 1 \cdot a = a$.

■ **Additive inverse property**
For every real number a, there is a unique real number $-a$ (read "the opposite of a" or "the negative of a") such that
$a + (-a) = 0$ and $(-a) + a = 0$.

■ **Multiplicative inverse property (reciprocal)**
For every real number a, $a \neq 0$, there is a unique real number $\dfrac{1}{a}$ such that
$a \cdot \dfrac{1}{a} = 1$ and $\dfrac{1}{a} \cdot a = 1$.

■ **Distributive property of multiplication over addition**
$a(b + c) = ab + ac$.

Example 1–3 C

The following statements are applications of the properties of real numbers.

1. $5 + 4 = 4 + 5$

$9 = 9$

Commutative property of addition

2. $3 \cdot 6 = 6 \cdot 3$

$18 = 18$

Commutative property of multiplication

3. $(2 + 3) + 4 = 2 + (3 + 4)$

$5 + 4 = 2 + 7$

$9 = 9$

Associative property of addition

4. $(2 \cdot 3)4 = 2(3 \cdot 4)$

$6 \cdot 4 = 2 \cdot 12$

$24 = 24$

Associative property of multiplication

5. $5 + 0 = 5$

Identity property of addition

6. $6 \cdot 1 = 6$

Identity property of multiplication

7. $(-7) + 7 = 0$

Additive inverse property

8. $6 \cdot \dfrac{1}{6} = 1$

Multiplicative inverse property

9. $2(3 + 4) = (2 \cdot 3) + (2 \cdot 4)$

$2(7) = 6 + 8$

$14 = 14$

Distributive property

10. $3(8 - 10) = 3\,[8 + (-10)]$

$3(-2) = 3 \cdot 8 + 3(-10)$

$-6 = 24 + (-30)$

$-6 = -6$

Distributive property

11. $(-8) + (-4) \, \epsilon \, R$

$-12 \, \epsilon \, R$

Closure property of addition

12. $(-10) \cdot 3 \, \epsilon \, R$

$-30 \, \epsilon \, R$

Closure property of multiplication

An application of the distributive property

Using the symmetric property of equality and the commutative property, we can write the distributive property of multiplication over addition as

$$ax + bx = (a + b)x.$$

Consider the example

$$4x + 5x.$$

Using the distributive property, the expression can be written

$$4x + 5x = (4 + 5)x = 9x.$$

In this expression $4x$ and $5x$ are terms that we wish to add. **A term** is any constant, variable, or indicated product, quotient, or root* of constants and variables. Terms are separated by the operations of addition and subtraction.

Simplifying expressions, as in this example, is called *combining like terms*. **Like terms** are terms that may differ only in their numerical factor. For two or more terms to be called like terms, the variable factors of the terms along with their respective exponents must be identical. However the numerical factors may be different.

Example 1–3 D	**1.** $5x^2y^3$, $-3x^2y^3$, and x^2y^3 are like terms because they differ only in their numerical factors.
	2. $5a^3b^2$ and $5a^2b^3$ both contain the same variables but are not like terms because the exponents of the respective variables are not the same.

Example 1–3 E	Perform the indicated addition or subtraction.

1. $6a + 8a = (6 + 8)a = 14a$ Distributive property

Note
The process of addition or subtraction is performed only with the numerical factors, the variable factors remain unchanged.

2. $4b + b + 3b = (4 + 1 + 3)b = 8b$

Note
If no numerical factor appears in a term, the factor is understood to be 1. That is,
$$b = 1 \cdot b.$$

3. $3z - 7z + z - 6 + 15 = (3 - 7 + 1)z + (-6 + 15) = -3z + 9$

There are many other properties of real numbers that we have not listed. These other properties, called **theorems,** are implied or can be shown to be true from the previous properties and definitions. A theorem is a property that we can prove is true. We will now state several theorems that follow logically from the properties already listed.

*Roots will be discussed in chapter 5.

> **■ Addition property of equality**
> For all real numbers a, b, and c, if $a = b$, then
> $$a + c = b + c.$$
>
> **Concept**
> We can add the same quantity to both members (also called sides) of an equality without altering the equality.

To prove that this theorem is true for all real numbers, we must start with the given information $a = b$, and, by applying the properties of real numbers, show that the desired result, $a + c = b + c$, is true.

Steps	Reasons
1. a, b, and c are real numbers	Given
2. $a + c$ is a real number	Closure property of addition
3. $a + c = a + c$	Reflexive property of equality
4. $a = b$	Given
5. $a + c = b + c$	Substitution property of equality

To discuss the steps in the previous proof, we start with the given information that a, b, and c are real numbers and because of the closure property of addition, $a + c$ must be a real number. We use the reflexive property of equality to write the equality $a + c = a + c$. Finally, since it was given that $a = b$, the substitution property of equality allows us to replace a with b in the right member and finish the proof.

The following theorems are provided in the exercise set, where we will be asked to supply the reasons.

> **■ Multiplication property of equality**
> For all real numbers a, b, and c, if $a = b$, then
> $$a \cdot c = b \cdot c.$$
>
> **Concept**
> We can multiply both members of an equality by the same number without altering the equality.

> **■ Zero factor property**
> For any real number a,
> $$a \cdot 0 = 0 \text{ and } 0 \cdot a = 0.$$
>
> **Concept**
> The product of zero and any number is zero.

Example 1–3 F

The following statements are applications of the previous theorems.

1. If $a = 3$, then $a + 4 = 3 + 4$ Addition property of equality
$$a + 4 = 7.$$

2. If $a = 5$, then $4 \cdot a = 4 \cdot 5$ Multiplication property of equality
$$4a = 20.$$

3. $4 \cdot 0 = 0$ Zero factor property

4. $-(-4) = 4$ Double-negative property

Mastery points
Can you
- Apply the equality properties for real numbers?
- Apply the inequality properties for real numbers?
- Apply the properties of real numbers?
- Combine like terms?

Exercise 1–3

Directions Apply the indicated property to write a new expression that is equal to the given expression. Assume that all variables represent real numbers. See example 1–3 C.

Examples
(a) Commutative, $5 + y$

(b) Distributive, $3(x + y)$

Solutions
(a) $y + 5$

(b) $3x + 3y$

1. Commutative, $4x + 3y$

2. Distributive, $4(a - b)$

3. Associative, $(4 + 2) + 6$

4. Identity, $4 + 0$

5. Commutative, $5a$

6. Distributive, $12(3 - y)$

7. Associative, $(2x)y$

8. Identity, $5 \cdot 1$

9. Inverse, $(-4) + 4$

10. Inverse, $4 \cdot \left(\dfrac{1}{4}\right)$

Directions Replace each question mark with the correct symbol to make the given statement an application of the given property.

Example
If $b = 4$ and $b + 3 = c$, then $? + 3 = c$; substitution property of equality.

Solution
$4 + 3 = c$

11. $x + 5 = ?$; reflexive property of equality.

12. Either $x < 0$, $x = 0$, or $x > ?$; trichotomy property.

13. If $2a - 3 = x$, then $x = ?$; symmetric property of equality.

14. If $a = b$ and $b = 5$, then $a = ?$; transitive property of equality.

15. If $a = 7$ and $b + 5 = a$, then $b + 5 = ?$; substitution property of equality.

16. If $5 < x$ and $x < y$, then $? < y$; transitive property of inequality.

Directions Rewrite each relation without the slash mark. See example 1–3 B.

17. $a \not< 5$ **18.** $3 \not\leq b$ **19.** $a \not\geq b$ **20.** $b \not> c$

21. $x \neq 5$ **22.** $4 \not< y$ **23.** $a \not\leq 6$ **24.** $b \not\geq 4$

Directions Identify which property is being used. See example 1–3 C.

Examples
(a) $5 + a = a + 5$

(b) $(-7) + 7 = 0$

Solutions
(a) Commutative property of addition

(b) Additive inverse

25. $6 + 7$ is a real number.

26. $8(a + b) = 8a + 8b$

27. $8 \cdot 1 = 8$

28. $5(x - y) = (x - y)5$

29. $\dfrac{1}{4} \cdot 4 = 1$

30. $(-4) + 4 = 0$

31. $0 + 9 = 9$

32. $(ab)(cd) = (cd)(ab)$

33. $(6 + 7) + 8 = 6 + (7 + 8)$

34. $(6 \cdot 7)8 = 6(7 \cdot 8)$

35. $(2x)y = 2(xy)$

36. $3 + (2 + b) = (2 + b) + 3$

37. $14 + 0 = 14$

38. $\left(\dfrac{2}{3}\right)\left(\dfrac{3}{2}\right) = 1$

39. $x + 6 = 6 + x$

40. $4 \cdot 5$ is a real number.

41. $(5 + 4) + y = 5 + (4 + y)$

42. $x + (-x) = 0$

43. $(a + b) + 0 = (a + b)$

44. $5(2 + b) = 10 + 5b$

Directions Fill in the missing values. Assume that all variables represent nonzero real numbers.

	Number	Additive inverse	Multiplicative inverse
Example		-4	
Solution	4		$\frac{1}{4}$
45.	6		
46.	10		
47.		5	
48.		2	
49.	x		
50.	y		
51.			$\frac{1}{7}$
52.			$\frac{1}{-5}$
53.		-3	
54.		-8	

Directions Perform the indicated addition or subtraction. See example 1–3 E.

55. $7x + 9x$ **56.** $8y + 4y$ **57.** $3a + a + 4a$

58. $b + 9b + 2b$ **59.** $3z - 2z$ **60.** $8c - 11c$

61. $5x + 3x + 4 + 9$ **62.** $8a + 6a + 12 - 7$ **63.** $4x - 3x + 8x - 7 + 3$

64. $9y + 3y - 8y + 6 - 14$

Directions Give a reason for each step in the following proofs.

65. Prove: For all real numbers a, b, and c, if $a = b$, then $a \cdot c = b \cdot c$.

a, b, and c are real numbers   a. _____

$a \cdot c$ is a real number   b. _____

$a \cdot c = a \cdot c$ c. _____

$a = b$ d. _____

$a \cdot c = b \cdot c$ e. _____

66. Prove: For any real number a, $a \cdot 0 = 0 \cdot a = 0$.

a is a real number   a. _____

$a = a \cdot 1$ b. _____

$a = a(1 + 0)$ c. _____

$a = a \cdot 1 + a \cdot 0$ d. _____

$a = a + a \cdot 0$ e. _____

$0 = a \cdot 0$ f. _____

$a \cdot 0 = 0$ g. _____

$0 \cdot a = 0$ h. _____

67. Prove: For any real number a, $-(-a) = a$.

a is a real number   a. _____

$-a + a = 0$ b. _____

$-a + [-(-a)] = 0$ c. _____

$a = -(-a)$ d. _____

1–4 Order of operations

When we perform several different types of arithmetic operations within an expression or a formula, we need a standard order in which the operations will be performed. Consider the following example.

If a woman deposits $1,000 in a savings account at 6% simple interest, the amount (A) in her account after 1 year can be found by evaluating

$$A = 1,000 + 1,000(0.06).$$

More than one answer is possible depending on the order in which we perform the operations. For example, if we first add the 1,000 and 1,000 and then multiply by (0.06) we have

$$1,000 + 1,000(0.06) = 2,000(0.06) = 120.$$

However if we multiply the 1,000 and (0.06) and to this product add the first 1,000 we have

$$1,000 + 1,000(0.06) = 1,000 + 60 = 1,060.*$$

To standardize the answer, we agree to the following order of operations, or priorities:

1. Perform any operations within a grouping symbol such as parentheses ( ), brackets [ ], braces { }, absolute value | |, radicals $\sqrt{}$, and so on, or in the numerator or the denominator of a fraction.

Note
Within a grouping symbol, the order of operations will still apply. If there are several grouping symbols intermixed, remove them by starting with the innermost one and working outward.

*This is the correct answer.

2. Perform indicated powers and roots.
3. Perform multiplication and division from left to right.
4. Perform addition and subtraction from left to right.

Note
Within any priority, every operation is of equal importance (that is, multiplication is no more important than division in priority 3). Therefore when performing a particular priority, we start at the left and proceed to the right, doing only those operations within the priority that we are performing, as we come to them, without skipping around.

Example 1–4   A

Perform the indicated operations in the proper order and simplify.

1. $9 + 12 \cdot 3 \div 2 = 9 + 36 \div 2$ Priority 3
$$= 9 + 18 \qquad \text{Priority 3}$$
$$= 27 \qquad \text{Priority 4}$$

2. $(19 - 1) \div 2 + 3 \cdot 4 = 18 \div 2 + 3 \cdot 4$ Priority 1
$$= 9 + 12 \qquad \text{Priority 3}$$
$$= 21 \qquad \text{Priority 4}$$

3. $\dfrac{1}{4} + \dfrac{3}{4} \div \dfrac{5}{8} = \dfrac{1}{4} + \dfrac{3}{\overset{}{\underset{1}{\cancel{4}}}} \cdot \dfrac{\overset{2}{\cancel{8}}}{5}$

$$= \dfrac{1}{4} + \dfrac{6}{5} \qquad \text{Priority 3}$$
$$= \dfrac{5}{20} + \dfrac{24}{20}$$
$$= \dfrac{5 + 24}{20} \qquad \text{Priority 4}$$
$$= \dfrac{29}{20} \text{ or } 1\dfrac{9}{20}$$

4. $2^2 \cdot 5 - 6 \cdot 4 = 4 \cdot 5 - 6 \cdot 4$ Priority 2
$$= 20 - 24 \qquad \text{Priority 3}$$
$$= -4 \qquad \text{Priority 4}$$

5. $\dfrac{3}{4} - \dfrac{3}{4} \cdot \dfrac{4}{9} = \dfrac{3}{4} - \dfrac{1}{3}$ Priority 3
$$= \dfrac{9}{12} - \dfrac{4}{12}$$
$$= \dfrac{9 - 4}{12} \qquad \text{Priority 4}$$
$$= \dfrac{5}{12}$$

6. $(8.21 + 12.43) \div 2.4 - (6.1)(4.7)$

$= (20.64) \div 2.4 - (6.1)(4.7)$ Priority 1

$= 8.6 - 28.67$ Priority 3

$= -20.07$ Priority 4

7. $\left(\dfrac{1}{3} + \dfrac{5}{8}\right) \div \dfrac{5}{6} = \left(\dfrac{8}{24} + \dfrac{15}{24}\right) \div \dfrac{5}{6}$

$= \dfrac{8 + 15}{24} \div \dfrac{5}{6}$ Priority 1

$= \dfrac{23}{24} \div \dfrac{5}{6}$

$= \dfrac{23}{\cancel{24}} \cdot \dfrac{\cancel{6}}{5}$
   (with $\cancel{24}$ → 4, $\cancel{6}$ → 1)

$= \dfrac{23}{20}$ or $1\dfrac{3}{20}$ Priority 3

8. $(9.3)^2 - 3(2.1)(4.6)$

$= 86.49 - 3(2.1)(4.6)$ Priority 2

$= 86.49 - 28.98$ Priority 3

$= 57.51$ Priority 4

9. $\dfrac{3(2 + 4)}{11 - 8} - \dfrac{4 + 6}{2} = \dfrac{3(6)}{11 - 8} - \dfrac{4 + 6}{2}$ Priority 1

$= \dfrac{18}{3} - \dfrac{10}{2}$ Priority 3

$= 6 - 5 = 1$ Priority 4

10. $5[4 + 3(10 - 12)]$

We first evaluate within the grouping symbol, applying the order of operations.

$5[4 + 3(-2)] = 5[4 + (-6)]$

$= 5[-2]$

$= -10$

Problem solving

Solve the following word problems using the order of operations.

Example 1–4   B

Choose a letter to represent the unknown quantity and find its value by performing the indicated operations.

1. Mrs. Miyazaki purchased 8 cans of oil at $1.10 per can and 2 air filters at $3.75 each. What was her total bill?

Let $x =$ Mrs. Miyazaki's total bill.

8 cans at $1.10 per can cost $8 \cdot \$1.10$.

2 air filters at $3.75 each cost $2 \cdot \$3.75$.

The total bill is given by

$$x = 8 \cdot 1.10 + 2 \cdot 3.75$$
$$= 8.80 + 7.50 = 16.30.$$

Mrs. Miyazaki's total bill was $16.30.

2. A woman worked a 40-hour week at $8 per hour. If she worked 6 hours of overtime at time and a half, how much money did she receive for the 46 hours of work?

Let W = the woman's total wages for the week.
40 hours at $8 per hour are $40 \cdot \$8$.

6 hours at time and a half are $6 \cdot 1\frac{1}{2} \cdot \$8$.

Thus

$$W = 40 \cdot 8 + 6 \cdot 1\frac{1}{2} \cdot 8$$
$$= 320 + 72$$
$$= 392.$$

The woman's total wages for the week was $392.

Mastery points
Can you
- Simplify expressions and formulas according to the order of operations?

Exercise 1–4

Directions Perform the indicated operations and simplify. See example 1–4 A.

1. $(-3) \cdot 4 + 14$

2. $3 \cdot 4 + (-8)$

3. $5 \cdot 6 - (-7)$

4. $6 - (-4)(3)$

5. $12 - (-5)(-4)$

6. $\frac{8 + 4}{6} + 5$

7. $\frac{18 - 6}{2} + 7 \cdot 5$

8. $3 - 12 \cdot \left(\frac{1}{4}\right)$

9. $5 - 15 \cdot \left(\frac{1}{3}\right)$

10. $8 + 0(5 - 7)$

11. $10 - 0(8 - 4)$

12. $8 + 6 - 4 \cdot 3 + 12$

13. $10 + 10 \div 10 - 10$

14. $14 \div 7 - 2 + 8 \cdot 4$

15. $24 \div 6 + 6 - 4 \cdot 3$

16. $9 - 18 \div 2 + 7 - 3^2$

17. $14 + 27 \div 3 \cdot 4 - 5^2$

18. $4 - (8 - 6)^2 + 3 \cdot 5$

19. $(4 - 9)^2 \cdot 2 + 7 - 4^2$

20. $12 + 3 \cdot 16 \div 4^2 - 5$

21. $18 - 3^2 \cdot 4 \div 6 + 7$

22. $8 - (12 - 9)^2 \cdot 2 + 5$

23. $26.4 - (3.7)(4.6)$

24. $\frac{2}{3} \div \left(\frac{5}{6} - \frac{4}{9}\right)$

25. $\dfrac{3}{8} - \dfrac{1}{2} \div \dfrac{3}{4}$

26. $\dfrac{1}{8} + \dfrac{7}{12} \cdot \dfrac{9}{14}$

27. $(14.13 + 11.4) \div 3.7 - (3.6) \cdot (4.9)$

28. $(4.6 + 3.1) \cdot (2.7) - (5.4) \cdot (7.3)$

29. $(4.7)^2 \cdot 5 - (14.64) \div (6.1)$

30. $(2.4)^2 + 5(1.9)^2 - 12.6$

31. $\left(\dfrac{11}{12} - \dfrac{5}{6}\right) \div \left(\dfrac{1}{3} + \dfrac{3}{8}\right)$

32. $\left(\dfrac{1}{4} - \dfrac{1}{6}\right) \cdot \left(\dfrac{1}{8} + \dfrac{3}{8}\right)$

33. $5[20 - 3(4 - 6) + 5]$

34. $9 + 2[5 - 3(8 - 3) + 5]$

35. $12 + [14 - 5(7 - 10) + 4]$

36. $(9 - 7)[15 - 4(3 - 6) + 9]$

37. $\dfrac{8(6 - 4)}{2} - \dfrac{27}{-3}$

38. $\dfrac{3(9 - 4)}{5} - \dfrac{10}{-2}$

39. $\dfrac{5 \cdot 6 - 4}{13} - \dfrac{(-18)}{6}$

40. $\dfrac{8 - 3 \cdot 4 + 10}{6 - 3} + \dfrac{(-18) + 6}{-4}$

41. $\left[\dfrac{10 + (-2)}{2(-1)}\right]\left[\dfrac{(-10)(-4)}{(-2)}\right]$

42. $\left[\dfrac{(-11) + (-7)}{(-9)}\right]\left[\dfrac{(-5)(-12)}{10}\right]$

43. $\dfrac{4(3 + 2) - 4^2 + 11}{(-5)(-3)}$

44. $\dfrac{(8 - 4)^2 + (-2)(-3)}{10 - (-4)(3)}$

Directions Perform the indicated operations and simplify. See example 1–4 A.

45. The area of a trapezoid whose bases are 12 meters and 8 meters and whose height is 10 meters is
$$\dfrac{1}{2} \cdot 10(12 + 8).$$
Find the area in square meters.

46. The surface area of a flat ring whose inside radius is 3 centimeters and whose outside radius is 5 centimeters is approximately
$$\dfrac{22}{7} \cdot 5^2 - \dfrac{22}{7} \cdot 3^2.$$
Find the area in square centimeters.

47. The water pressure exerted on 1 square foot (144 square inches) of surface area of a diving suit 50 feet below the surface of the water is given by
$$144\,[14.7 + 50(0.444)].$$
Perform the indicated operations to find this pressure in pounds per square inch.

48. The surface area of a ring section whose inside diameter is 19 inches and whose outside diameter is 26 inches is approximately
$$\dfrac{22}{7} \cdot \dfrac{26 + 19}{2} \cdot \dfrac{26 - 19}{2}.$$
Find the area of this surface in square inches.

49. The surface area of a rectangular solid whose length is 8 feet, width is 5 feet, and height is 7 feet is given by
$$2 \cdot 8 \cdot 7 + 2 \cdot 8 \cdot 5 + 2 \cdot 5 \cdot 7.$$
Find the surface area in square feet.

50. To convert 59 degrees Fahrenheit to Celsius temperature, we use
$$\dfrac{5}{9}\,(59 - 32).$$
Find the temperature in degrees Celsius.

51. To convert 25 degrees Celsius to Fahrenheit temperature, we use
$$\left(\dfrac{9}{5}\right)25 + 32.$$
Find the temperature in degrees Fahrenheit.

52. Norbert makes $6 per hour for a 40-hour week and receives time and a half for every hour he works over 40 hours a week. How much did he earn if he worked 47 hours in one week?

53. Royetta has $120 in her checking account. She deposits three twenty-dollar checks and then writes a check for $87. What is her balance?

54. A man purchased a case of beans (12 cans) at 53¢ per can, 4 pounds of apples at 59¢ per pound, and 3 quarts of milk at 69¢ per quart. What was his total bill (a) in cents, (b) in dollars and cents?

55. A car dealer sold 18 cars at $8,200 each and 4 trucks at $7,100 each. What is the total sale of all the cars and trucks?

56. Forest Road School has 3 first-grade classes, each with 27 students; 2 second-grade classes, each with 31 students; and 2 third-grade classes, each with 29 students. How many students are there all together in first, second, and third grade?

57. While away at school for nine months, a student spends $140 per month on housing, $95 per month on food, and $48 per month on miscellaneous expenses. How much money did the student spend on these expenses during the nine months?

58. A woman wants to carpet two rooms. One room is 4 yards by 5 yards and the second room is 4 yards by 7 yards. How many square yards of carpet are needed? If the carpet costs $8 per square yard and the pad under the carpet costs $2 per square yard, how much will the carpet and pad for these rooms cost?

59. Nancy, Alice, and Jane are typists in an office. If Nancy can type 90 words per minute, Alice can type 85 words per minute, and Jane can type 70 words per minute, how many words can they type together in 20 minutes?

60. Susan charges $5 to wash a car, $11 to mow a lawn, and $27 to wax a car. If in one week, she washed six cars, mowed three lawns, and waxed two cars, how much did she earn that week?

1–5 Terminology and evaluation

Terminology

Before we study the different types of algebraic operations, we must first define a few concepts. *An* **algebraic expression** *is any meaningful collection of variables, constants, grouping symbols, and symbols of operations.* Examples of algebraic expressions are

$$3x^2y, \quad \frac{ab}{c}, \quad \pi r^2, \quad \frac{x^2 + 2x - 1}{x^2 + 1}, \quad 2x^2 - x + 3, \quad \sqrt{x^2 + y^2}, \quad x, \quad -4.$$

In an algebraic expression *a* **term** *is any constant, variable, or indicated product, quotient, or root of constants and variables.* Terms in an algebraic expression are separated by the operations of addition and subtraction.

Note

Recall that *a constant is a symbol that does not change its value.* In the expression πr^2, π is a symbol that represents only one value (the circumference of a circle divided by its diameter equals 3.14159 · · ·) and therefore is a constant.

| Example 1–5 A | Identify the number of terms in the algebraic expressions. |
| | |

1. $3x^2 - 2x + 1$ There are three terms.

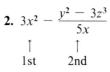

2. $3x^2 - \dfrac{y^2 - 3z^3}{5x}$ There are two terms since the fraction bar acts as a grouping symbol.

In the expression $3xy$, *each factor or group of factors is called the* **coefficient** *of the remaining factors.* That is, 3 is the coefficient of xy, x is the coefficient of $3y$, $3x$ is the coefficient of y, and so on. The 3 is called the **numerical coefficient,** and it tells us how many xy's we have in the expression.

Since we often talk about the numerical coefficient of a term, we will eliminate the word "numerical" and just say "coefficient." It will be understood that we are referring to the numerical coefficient. If no numerical coefficient appears in a term, the coefficient is understood to be 1.

Example 1–5 B

In the algebraic expression $3x^2 - 2y + z$, 3 is the coefficient of x^2, -2 is the coefficient of y, and 1 is understood to be the coefficient of z.

Note

The coefficient of a term in an algebraic expression includes the sign that precedes it. In our example the coefficient of y is -2 and not 2.

A special kind of algebraic expression is a *polynomial*. The following are characteristics of a polynomial.

1. It has real number coefficients.
2. All variables in a polynomial are raised only to natural number exponents.
3. The operations performed by the variables are limited to addition, subtraction, and multiplication.

A polynomial that contains just one term is called a *monomial;* one that contains two terms is called a *binomial;* and a polynomial that contains three terms is called a *trinomial.* Any polynomial that contains more than one term is called a *multinomial,* but no special names will be given to polynomials that contain more than three terms.

Note

We should simplify any expression before identifying it. Also, in an expression the combining of all the constant terms is understood to be a single term. For example, $x + 3 + \pi$ is thought of as $x + (3 + \pi)$ and is a binomial.

Example 1–5 C	1. x^2, $5x$, 7, and $-4x^2y^3$ are monomials.
	2. $2x - 4$, $x^2 + y^2$, and $27x^3 - y^3$ are binomials.
	3. $4x^2 + 3x - 2$ and $6x^2y^2 - 4xy + 3y^2$ are trinomials.
	4. $8x^3 - 4x^2 + 3x - 2$ has no special name and is referred to as a polynomial of four terms.
	5. $\dfrac{x}{y + z}$ and $5\sqrt{x} + y$ are not polynomials since they contain variables in the denominator or under a radical symbol.

We also identify different types of polynomials by the degree of the polynomial. *The **degree** of a polynomial in one variable is the greatest exponent of that variable in any one term.*

Example 1–5 D	Determine the degree of the polynomial.
	1. $5x^3$ Third degree because the exponent of x is 3
	2. $x^4 - 2x^3 + 3x - 5$ Fourth degree because the greatest exponent of x in any one term is 4

Note
In example 2 the polynomial has been arranged in *descending powers* of the variable. This is the form that we will use when we write polynomials in one variable.

Evaluating an algebraic expression

An extremely important process in algebra is that of calculating the numerical value of an expression when we are given specific replacement values for the variables. This process is called **evaluation.** To carry out this evaluation, we need the following **property of substitution.**

> ■ **Property of substitution**
> If $a = b$, then a may be replaced by b or b may be replaced by a in any expression without altering the value of the expression.
>
> **Concept**
> When two quantities are equal, we can replace one quantity with the other quantity without altering the value of the statement.

We frequently need to evaluate algebraic expressions. For example, the area A of a rectangle is found by multiplying the length ℓ times the width w, $A = \ell w$. If the length is 8 feet and the width is 4 feet, substituting 8 for the length, ℓ, and 4 for the width, w, the expression becomes

$$A = (8)(4) = 32,$$

and the area is 32 square feet. We substituted the respective values for length and width into the expression and carried out the indicated arithmetic.

Note

When replacing variables with the numbers they represent, it is a good procedure to put each of the numbers inside parentheses.

Example 1-5 E

Evaluate the following expressions for the given real number replacement for the variable or variables.

1. $x^2 + 3x - 4$, when $x = 3$. The expression would be $(\;)^2 + 3(\;) - 4$ without the x.

Substituting 3 in place of each x, we have

$(3)^2 + 3(3) - 4.$

Using our order of operations, we have

$$
\begin{aligned}
(3)^2 + 3(3) - 4 &= 9 + 3(3) - 4 \\
&= 9 + 9 - 4 \\
&= 14.
\end{aligned}
$$

Therefore the expression $x^2 + 3x - 4$ evaluated for $x = 3$ is 14.

2. If we know the temperature in degrees Fahrenheit (F), the temperature in degrees Celsius (C) can be found by the formula $C = \dfrac{5}{9}(F - 32)$. Find the temperature in degrees Celsius if the temperature is 77 degrees Fahrenheit.

$C = \dfrac{5}{9}(F - 32)$	Original formula
$C = \dfrac{5}{9}[(\;) - 32]$	Formula ready for substitution
$C = \dfrac{5}{9}[(77) - 32]$	Substituting
$C = \dfrac{5}{9}[45]$	Order of operations
$C = 25$	Solution

Hence 77 degrees Fahrenheit is equivalent to 25 degrees Celsius.

3. If a company sells n_1 shirts at d_1 dollars each and n_2 ties at d_2 dollars each, the total revenues R from the shirts and ties would be expressed as $R = n_1 d_1 + n_2 d_2$. Find R if $n_1 = 20$, $d_1 = 26$, $n_2 = 15$, and $d_2 = 8$.

Note

In this formula, **subscripts** are used to denote two different values for the number of articles sold (n_1 and n_2) and the price per article (d_1 and d_2). These are read "n sub-one," "n sub-two," "d sub-one," and "d sub-two."

$R = n_1 d_1 + n_2 d_2$	Original formula
$R = (\ )(\ ) + (\ )(\ )$	Formula ready for substitution
$R = (20)(26) + (15)(8)$	Substituting
$R = 520 + 120$	Order of operations
$R = 640$	Solution

Therefore the revenues from the sale of shirts and ties is $640.

Polynomial notation

Polynomials can be denoted by a single symbol such as P, Q, R, or by P_1, P_2, P_3, and so on. If we are dealing with a polynomial in one variable, the symbol for the polynomial can be used in conjunction with the variable in naming the polynomial. For example, $P(x)$ (read "P of x" or "P at x") could be used to denote the polynomial $x^2 + 3x - 4$ from number 1 of example 1–5 E. Therefore we can write $P(x) = x^2 + 3x - 4$.

The symbol inside the parentheses denotes the variable in the polynomial, and we can use this notation to indicate a specific value at which we want to evaluate the polynomial. In example 1–5 E we evaluated $x^2 + 3x - 4$ at 3. We could have represented this as $P(3)$, which means the polynomial P evaluated at 3, not P times 3.

Example 1–5 F

If $P(x) = x^2 - x + 6$ and $Q(y) = y + 2$, find the indicated values.

1. $Q(-5) = (-5) + 2$ Substitute -5 for y.
$\qquad\quad = -3$

2. $P(-3) = (-3)^2 - (-3) + 6$ Substitute -3 for x.
$\qquad\quad = 9 - (-3) + 6$
$\qquad\quad = 18$

3. $P(-4) \cdot Q(-6) = [(-4)^2 - (-4) + 6] \cdot [(-6) + 2]$
$\qquad$ Substitute -4 for x in $P(x)$ and -6 for y in $Q(y)$.
$\qquad\qquad\qquad\quad = [16 - (-4) + 6] \cdot [-4]$
$\qquad\qquad\qquad\quad = [26] \cdot [-4]$
$\qquad\qquad\qquad\quad = -104$

Note

$P(-4)$ and $Q(-6)$ represent specific real numbers and we must evaluate each of them at the indicated value before we can perform the indicated multiplication.

Algebraic notation

Many problems that we encounter are stated verbally. These need to be translated into algebraic expressions. While there is no standard procedure for changing a verbal statement into an algebraic expression, these guidelines should be of use.

1. Read the problem carefully, determining useful prior knowledge. Note what information is given and what information we are asked to find.
2. Let some letter represent one of the unknowns. Then express any other unknowns in terms of it.
3. Use the given conditions in the problem and the unknowns from step 2 to write an algebraic expression.

When translating verbal statements into equations, we look for phrases that involve the basic operations of addition, subtraction, multiplication, and division. Table 1.1 shows some examples of phrases that we commonly encounter. We let x represent the unknown number.

Phrase	Algebraic expression
Addition	
6 more than a number	
the sum of a number and 6	
6 plus a number	$x + 6$
a number increased by 6	
6 added to a number	
Subtraction	
6 less than a number	
a number diminished by 6	
the difference of a number and 6	
a number minus 6	$x - 6$
a number less 6	
a number decreased by 6	
6 subtracted from a number	
a number reduced by 6	
Multiplication	
a number multiplied by 6	
6 times a number	$6x$
the product of a number and 6	
Division	
a number divided by 6	
the quotient of a number and 6	$\dfrac{x}{6}$
$\dfrac{1}{6}$ of a number	

Table 1.1

Example 1-5 G

Write an algebraic expression for each of the following.

1. The product of x and y $\qquad$ $x \cdot y$

2. The sum of n and 7 $\qquad$ $n + 7$

3. y decreased by 3 $\qquad$ $y - 3$

4. a divided by 4 $\qquad$ $a \div 4$ or $\dfrac{a}{4}$

5. A number diminished by 6, let n represent the number. $\qquad$ $n - 6$

6. Three times a number and that product increased by 9, let x represent the number. $\qquad$ $3x + 9$

7. A number divided by 4 and that quotient decreased by 2, let a represent the number. $\qquad$ $\dfrac{a}{4} - 2$

8. Five times the sum of a number and 6, let n represent the number. $\qquad$ $5(n + 6)$

Mastery points
Can you
- Identify terms in an expression?
- Identify a polynomial by name?
- Determine the degree of a polynomial?
- Evaluate an algebraic expression?
- Evaluate a polynomial using polynomial, $P(x)$, notation?
- Write an algebraic expression?

Exercise 1-5

Directions Specify the number of terms in each expression and determine if the expression is a polynomial. See examples 1–5 A and C.

1. $5x^2 - 2x + 3$ **2.** $\dfrac{7x}{2}$ **3.** $3xy - \dfrac{3y}{5} + 7x$ **4.** $\dfrac{4a^2 - b^2}{10}$

5. $\sqrt{3x^2 - 2x + 1}$ **6.** $\dfrac{x^2 + y^2}{z^2} - a^2 + 2ab$

Directions Determine the numerical coefficients of each term in the following expressions. See example 1–5 B.

7. $4x^2 - 7x + y$ **8.** $3x^3 - 2x^2 + x$ **9.** $5y^3 + y^2 - 7y$ **10.** $4z^4 - z^3 + z^2 + 6z$

Directions Identify each polynomial as a monomial, binomial, or trinomial and determine the degree of the polynomial. See examples 1–5 C and D.

11. $5a^3 - 4a^2 + 3$ **12.** $4a^2 - a$ **13.** $2x^4 - 7x^2$ **14.** $6x^2 - 2x + 1$ **15.** $4y^5$

Directions Evaluate the following expressions if $a = -3$, $b = 2$, $c = -2$, and $d = 4$. See example 1–5 E.

16. $2(a + 3b)$

17. $2a - 3b - (2c + d)$

18. $b - 3(a - 2d)$

19. $(4a + 3b)(c - 2d)$

20. $5ab^2 + 3cd$

21. $2a^2 - 3a + 4$

22. $c^2 - d^2$

23. $a^2 - 4c^2$

24. $(3a - 2b) - (2a + b)(c + d)$

25. $a^3b - c^3d$

Directions Evaluate the following polynomials for the specific values of the variable. See example 1–5 F.

26. $P(x) = x^2 + x$, $P(3)$, $P(-2)$, $P(0)$

27. $Q(x) = x^2 - 3x + 4$, $Q(2)$, $Q(-3)$, $Q(0)$

28. $R(x) = 2x^2 - x - 1$, $R(-2)$, $R(1)$, $R(0)$

29. $P(y) = y^3 - 1$, $P(1)$, $P(-2)$, $P(0)$

30. $Q(z) = z^2 - 5z + 6$, $Q(3)$, $Q(2)$, $Q(0)$

Directions If $P(x) = x^2 + 2x + 1$, $Q(x) = 2x - 1$, and $R(x) = x^2 + x - 6$, find the indicated values. See example 1–5 F.

Example
$R[Q(2)]$

Solution
$= R[2(2) - 1]$
$= R[4 - 1]$
$= R(3)$
$= (3)^2 + (3) - 6$
$= 9 + 3 - 6$
$= 6$

> **Note**
> In this example we had to evaluate the inner polynomial, $Q(x)$, first so that we could determine the value for which we would evaluate the outer polynomial, $R(x)$.

31. $P(2) + Q(-1) - R(2)$

32. $P(0) \cdot Q(5)$

33. $P(4) - R(0)$

34. $P(-1) \cdot Q(-1) + R(-1)$

35. $Q(4) \cdot R(-1)$

36. $P[Q(2)]$

37. $P[Q(-1)]$

38. $R[P(1)]$

39. $Q[R(-2)]$

Directions Evaluate the following formulas. See example 1–5 E.

40. $V = e^3$, $e = 5$

41. $F = ma$, $m = 12$ and $a = 5$

42. $I = prt$; $p = 2{,}000$; $r = 0.09$; and $t = 3$

43. $V = k + gt$, $k = 22$, $g = 11$, and $t = 3$

44. $H = \dfrac{D^2N}{2}$, $D = 4$ and $N = 6$

45. $\ell = a + (n - 1)d$, $a = 7$, $n = 20$, and $d = 4$

46. $V_2 = V_1 + at$, $V_1 = 80$, $a = 4$, and $t = 6$

47. $S = \dfrac{1}{2}gt^2$, $g = 32$ and $t = 3$

48. $S = V_0t + \dfrac{1}{2}at^2$, $V_0 = 22$, $t = 4$, and $a = 32$

49. $A = \dfrac{n_1P_1 + n_2P_2}{n_1 + n_2}$, $n_1 = 80$, $P_1 = 3$, $n_2 = 110$, and $P_2 = 5$

50. $R_x = \dfrac{R_1 R_3}{R_2}$, $R_1 = 8$, $R_2 = 5$, and $R_3 = 15$

51. $V_2 = \dfrac{V_1 P_1}{P_2}$, $V_1 = 14$, $P_1 = 30$, and $P_2 = 35$

52. $V_1 = \dfrac{V_2 P_2}{P_1}$, $V_2 = 12$, $P_1 = 22$, and $P_2 = 33$

53. $V_2 = \dfrac{V_1 T_2}{T_1}$, $V_1 = 14$, $T_1 = 18$, and $T_2 = 27$

Directions Evaluate the following formulas. See example 1–5 E.

54. The volume V of a rectangular solid is found by multiplying length ℓ times width w times height h, $V = \ell w h$. Find the volume in cubic feet if $\ell = 12$ feet, $w = 4$ feet, and $h = 5$ feet.

55. The perimeter P of a rectangle is found by the formula $P = 2\ell + 2w$, where ℓ is the length of the rectangle and w is the width. Find the perimeter of the rectangle in meters if $\ell = 12$ meters and $w = 7$ meters.

56. It is necessary to drag a box 600 feet across a level lot in 3 minutes. The force required to pull the box is 2,000 pounds. What is the horsepower (h) needed to do this if h is defined by $h = \dfrac{L \cdot W}{33,000 \cdot t}$, where L = distance to be moved, W = force exerted, and t = time in minutes required to move the box through L?

57. A pulley 12 inches in diameter that is running at 320 revolutions per minute (rpm) is connected by a belt to a pulley 9 inches in diameter. How many revolutions per minute will the smaller pulley make if $s = \dfrac{SD}{d}$, where s = speed of smaller pulley and d = diameter of smaller pulley, given S = speed of larger pulley and D = diameter of larger pulley?

58. In a gear system the velocity V of the driving gear is defined by $V = \dfrac{vn}{N}$, where v = velocity of the follower gear, n = number of teeth of the follower gear, and N = number of teeth of the driving gear. Find V when $v = 90$ rpm, $n = 30$ teeth, $N = 65$ teeth.

59. A formula in electricity is $I = \dfrac{E}{R}$, where I represents the current measured in amperes in a certain part of a circuit, E is the potential difference in voltage across that part of the circuit, and R is the resistance in ohms of that part of the circuit. Find I in amperes if $E = 110$ volts and $R = 44$ ohms.

Directions Write an algebraic expression for each of the following. See example 1–5 G.

60. The sum of x and 8

61. 3 times x

62. 5 less than n

63. 8 more than x

64. x increased by 5 and that sum divided by 6

65. 4 times the sum of x and 7

66. n decreased by 8

67. x decreased by 4

68. a divided by 3 and that quotient increased by 4

69. A number decreased by 15

70. A number added to 6

71. 8 times a number and that product increased by 14

72. A number divided by 5 and that quotient decreased by 2

73. 4 times the sum of a number and 3

74. A number decreased by 6 and that difference divided by 9

Chapter summary

1. A **set** is any collection of objects or things.

2. The set A is a **subset** of the set B, written $A \subseteq B$, if every element of A is also an element of B.

3. The set A is said to be **equal** to the set B, written $A = B$, if and only if $A \subseteq B$ and $B \subseteq A$.

4. A set that contains no elements is called the **empty set** or the **null set** and is denoted by the symbol $\emptyset$.

5. The **union** of the sets A and B, written $A \cup B$, is the set of all elements that are in A or in B or in both A and B.

6. The **intersection** of the sets A and B, written $A \cap B$, is the set of only those elements that are in both A and B.

7. The sets A and B are **disjoint** if and only if $A \cap B = \emptyset$.

8. A **variable** is a symbol (generally a lowercase letter) that represents an unspecified element of a set that contains two or more elements.

9. The **absolute value** of a number is the undirected distance that the number is from the origin. In symbols
$$|x| = \begin{cases} x \text{ if } x \geq 0 \\ -x \text{ if } x < 0 \end{cases}$$

10. The numbers or variables in an indicated multiplication are referred to as the **factors** of the product.

11. The **exponent** tells how many times the base is used as a factor in an indicated product.

12. **Division by zero is not allowed.**

13. The quotient of *zero* divided by any number other than zero is always zero.

14. A **term** is any constant, variable, or indicated product, quotient, or root of constants and variables.

15. **Like terms** are terms that may differ only in their numerical factors.

16. *Order of Operations*

    a. Perform operations within grouping symbols or in the numerator or the denominator of a fraction.

    b. Perform indicated powers and roots.

    c. Perform multiplication and division from left to right.

    d. Perform addition and subtraction from left to right.

17. An **algebraic expression** is any meaningful collection of variables, constants, grouping symbols, and symbols of operations.

18. Each factor or group of factors is called the **coefficient** of the remaining factors. In the expression $5x$, 5 is called the **numerical coefficient.**

19. A **polynomial** is an algebraic expression that consists of one or more terms, has real number coefficients, all variables are raised only to natural number exponents, and the operations involving variables are limited to addition, subtraction, and multiplication.

20. The **degree** of a polynomial in one variable is the greatest exponent of that variable in any one term.

Chapter review

[1–1]

Directions Write each set by listing the elements.

1. The letters in the word *college*

2. The first 3 months of the year

3. The months that begin with the letter L

Directions Use mathematical symbols to write the following.

4. B is a subset of D.

5. The null set is a subset of A.

6. 5 is an element of C.

7. B is not a subset of C.

Directions Use the following set to determine if exercises 8–10 are true or false and to form the indicated sets in exercises 11–13. $A = \{1,2,3,4\}$, $B = \{2,3,4,5,6\}$, $C = \{5,6,7,8\}$, $D = \{3,4\}$.

8. $4 \in A$
9. $D \subseteq B$
10. $A \subseteq B$
11. $A \cup B$
12. $A \cap B$
13. $A \cap C$

Directions Replace the comma with the proper strict inequality symbol, $<$ or $>$, between the following numbers.

14. $0, 4$
15. $-3, -5$
16. $-3, 0$

[1–2]

Directions Find each sum or difference.

17. $(-5) + (-8)$
18. $0 - (-8)$
19. $8 - 11$

20. $0 + (-2) - 4 + (-3)$
21. $(-4) - (-10)$
22. $6 - 8 + 7 - 11 - 4$

Directions Perform the indicated operations if possible.

23. $\dfrac{(-35)}{(-7)}$
24. $\dfrac{(-18)(-2)}{(-6)}$
25. $\dfrac{(-24)}{(+8)}$
26. $\dfrac{(-8)(0)}{(-5)(-4)}$

27. -2^4
28. $\dfrac{(-14)(+12)}{(-4)(+7)}$
29. $\dfrac{(+30)}{(-6)}$

[1–3]

Directions Identify which property is being used in the problem.

30. $(-6) + 6 = 0$
31. $5(a + b) = 5a + 5b$

32. $\left(\dfrac{3}{4}\right)\left(\dfrac{4}{3}\right) = 1$
33. $(3a)b = 3(ab)$

Directions Perform the indicated addition and subtraction.

34. $5a + 6a$
35. $11x + 3x$
36. $2b + b + 6b$

37. $y + 5y + 2y$
38. $3x + 5x + 8 + 9$
39. $7a + a - 4a - 7 + 3$

40. $3x - 5x + 7x - 6 + 4$
41. $12c + 8c - 9c + 8 - 14$

[1–4]

Directions Perform the indicated operations and simplify.

42. $12 - 3 \cdot 4 - 6 + 8$

43. $10 - 16 \div 4 + 7 - 3^2$

44. $26 \div 13 + 8 - 2 \cdot 7$

45. $8 - (4 - 7)^2 + 5 \cdot 6$

46. $15 - (18 - 11)^2 \cdot 2 + 12$

47. $10 - 2[18 - 2(4 - 6) + 11] - 16$

48. $\dfrac{5(4 - 9)}{3} - \dfrac{(-24)}{8}$

49. $\left[\dfrac{(-12) + (-9)}{(-7)}\right]\left[\dfrac{(-8)(-6)}{12}\right]$

50. $\dfrac{5}{8} + \dfrac{1}{2} \div \dfrac{3}{4}$

51. $(2.6) \cdot (5.4) \div (1.8) + 5.7$

52. $(18.47 + 26.89) \div 5.6 + (1.3)^2$

[1–5]

Directions Specify the number of terms in each expression and determine if the expression is a polynomial.

53. $5x^3 - 2x^2 + 7x - 4$

54. $\dfrac{5x^2 - 2x}{4}$

Directions Evaluate the following expressions if $a = -2$, $b = 5$, $c = 3$, and $d = -4$.

55. $2a^2b - cd^2$

56. $(b - 2c)^2$

57. $(a - 2b)(3c + d)$

58. $(ac)^2 - (bd)^2$

59. If $P(x) = x^2 - x - 1$,
find (a) $P(-2)$, (b) $P(0)$, (c) $P(3)$, (d) $P(2) \cdot P(-3)$.

Directions Evaluate the following formulas.

60. $A = \dfrac{1}{2}bh$, $b = 20$ and $h = 7$

61. $A = \dfrac{1}{2}h(b_1 + b_2)$, $h = 10$, $b_1 = 6$, and $b_2 = 9$

62. $R_t = \dfrac{R_1 R_2}{R_1 + R_2}$, $R_1 = 4$ and $R_2 = 6$

Directions Write an algebraic expression for each of the following.

63. The sum of y and 4

64. 12 less than a

65. 7 times the sum of a and 6

66. x times 5 and that product decreased by 3

67. 5 times the sum of a number and 12

68. A number increased by 4 and that sum divided by 8

Chapter 2

First-Degree Equations and Inequalities

2–1 Solving equations

Terminology

A **mathematical statement** is a sentence that can be labeled true or false. Consider the statements

$$7 + 3 = 10 \quad \text{and} \quad 2 + 2 = 5.$$

By inspection we see that the first mathematical statement is true whereas the second one is false. Other mathematical sentences such as

$$2x + 2 = 11 \quad \text{and} \quad 5 - x = 2x + 3$$

cannot be labeled as true or false. Such sentences are called **open sentences.** The truth or falsity of the sentence is "open" since we do not know the value that the variable represents.

An **equation** is a statement of equality. If two expressions represent the same number, we place an equality sign, $=$, between them to form an equation. We use the following example to show the parts of an equation.

$$\underbrace{2x + 9} \qquad\qquad = \qquad\qquad \underbrace{7 - 4x}$$
$$\qquad\qquad\qquad\quad \uparrow$$

| Left member | Equality | Right member |
| of the equation | sign | of the equation |

A replacement value for the variable that forms a true statement is called a **root,** or a **solution,** of the equation. We say that a root of the given equation *satisfies* that equation. The set of all those values for the variable that causes the equation to be a true statement is called the **solution set** of the equation. We use S to denote the solution set.

Consider the equation

$$x - 4 = 10.$$

We can determine by inspection that the root of the equation is 14 because if we *replace* x with 14, we form a true statement.

$$14 - 4 = 10$$
$$10 = 10 \text{ (true)}$$

The solution set for this equation is $S = \{14\}$. The solution set in this example contains just one element. If there is at least one element in the replacement set of the variable that is not in the solution set of the equation, the equation is called a **conditional equation.** If the equation is true for every permissible value of the variable, it is called an **identical equation,** or **identity.** For example,

$$4(a - 2) = 4a - 8$$

is true for any real number replacement for a and is thus called an identity. Properties such as

$$a \cdot (b \cdot c) = (a \cdot b) \cdot c \text{ and } a + b = b + a$$

are further examples of identities. We will be concerned only with conditional equations in this chapter.

In this chapter we are concerned with **first-degree conditional equations,** also called **linear equations.** In a first-degree conditional equation in one variable, the exponent of the unknown is 1 and the solution set will contain at most one root.

If we wish to solve an equation such as

$$5(2x - 1) = 7x + 10,$$

we go through a series of steps whereby we form equations that are *equivalent, having identical solution sets,* to the original equation. *Our goal is to form equivalent equations until we isolate the unknown in one member of the equation and our equation is in the form* $x = n$, n *being some real number.* The following are equivalent equations whose solution set is $S = \{5\}$.*

$$5(2x - 1) = 7x + 10$$
$$10x - 5 = 7x + 10$$
$$3x - 5 = 10$$
$$3x = 15$$
$$x = 5$$

*The formation of these equivalent equations is achieved by using the addition and multiplication properties of equality.

Since an equation is a statement of equality between the two members of the equation, identical quantities added to or subtracted from each expression will produce an equivalent equation. This is called the **addition property of equality**.

> ■ **Addition property of equality**
> For any algebraic expressions A, B, and C, if $A = B$, then
> $$A + C = B + C.$$
>
> **Concept**
> The same expression can be added to each member of an equation and the result will be an equivalent equation.

In chapter 1 we defined subtraction in terms of addition, therefore *we can use the addition property of equality to* **subtract** *the same expression from both members of an equation.*

Example 2–1 A

Find the solution set.

1.
$$4x - 2 = 3x + 7$$
$$4x - 3x - 2 = 3x - 3x + 7 \qquad \text{Subtract } 3x \text{ from both members.}$$
$$x - 2 = 7$$
$$x - 2 + 2 = 7 + 2 \qquad \text{Add 2 to both members.}$$
$$x = 9$$
$$S = \{9\}$$

2.
$$2(3x - 1) = 5x + 2x - 3$$
$$6x - 2 = 7x - 3 \qquad \text{Simplify.}$$
$$6x - 6x - 2 = 7x - 6x - 3 \qquad \text{Subtract } 6x \text{ from both members.}$$
$$-2 = x - 3$$
$$-2 + 3 = x - 3 + 3 \qquad \text{Add 3 to both members.}$$
$$1 = x$$
$$S = \{1\}$$

Multiplication property of equality

The addition property of equality along with the properties of real numbers is sufficient to solve many of the equations that we encounter. However they are not sufficient to solve equations such as

$$4x = 12 \quad \text{or} \quad \frac{3}{4}x = 15.$$

Recall that we want our equation to be of the form $x = n$. This means that the coefficient of x must be 1. To achieve this, we need the **multiplication property of equality**.

> **■ Multiplication property of equality**
> For any algebraic expressions A, B, and C, where $C \neq 0$, if $A = B$, then
> $$A \cdot C = B \cdot C.$$
>
> **Concept**
> An equivalent equation is obtained when we multiply both members of the equation by the same nonzero expression.

In chapter 1 we defined division in terms of multiplication, therefore *we can use the multiplication property of equality to* **divide** *both members of an equation by the same nonzero expression.*

Example 2–1　B

Find the solution set.

1. $4x = 12$

$\dfrac{4x}{4} = \dfrac{12}{4}$　　Divide both members by 4.

$x = 3$

$S = \{3\}$

Note
We could have multiplied by the reciprocal of 4 to solve the equation. That is,

$4x = 12$

$\dfrac{1}{4} \cdot 4x = \dfrac{1}{4} \cdot 12$　　Multiply both members by $\dfrac{1}{4}$.

$x = 3.$

We should be familiar with the idea that to divide by a number is the same as to multiply by the reciprocal of that number.

2.　　$\dfrac{3}{4}x = 15$

$\dfrac{4}{3} \cdot \dfrac{3}{4}x = \dfrac{4}{3} \cdot 15$　　Multiply both members by the reciprocal of the coefficient.

$x = 20$

$S = \{20\}$

Recall that when we divide by a fraction, we invert and multiply. Therefore if the coefficient is a fraction, we will multiply both members of the equation by the reciprocal of the coefficient.

3. $2.6x = 10.4$

$\dfrac{2.6x}{2.6} = \dfrac{10.4}{2.6}$　　Divide both members by 2.6.

$x = 4$

$S = \{4\}$

Using the given theorems and the properties of real numbers, there are four basic steps to solving a linear equation. We shall now apply them to the equation

$$5(2x - 1) = 7x + 10.$$

Step 1 *Simplify the equation.* Perform all indicated addition, subtraction, multiplication, and division. Remove all grouping symbols. In our example, step 1 would be to carry out the indicated multiplication in the left member.

$$5(2x - 1) = 7x + 10$$
$$10x - 5 = 7x + 10$$

Step 2 *Use the addition property of equality to form an equivalent equation where all the terms involving the unknown are in one member of the equation.* By subtracting $7x$ from *both* members of the equation, we have

$$10x - 5 = 7x + 10$$
$$10x - 7x - 5 = 7x - 7x + 10$$
$$3x - 5 = 10.$$

Step 3 *Use the addition property of equality to form an equivalent equation where all the terms not involving the unknown are in the other member of the equation.* Adding 5 to *both* members of the equation, we have

$$3x - 5 = 10$$
$$3x - 5 + 5 = 10 + 5$$
$$3x = 15.$$

Step 4 *Use the multiplication property of equality to form an equivalent equation where the coefficient of the unknown is 1.* That is, $x = n$. By dividing *both* members of the equation by 3, we have

$$3x = 15$$
$$\frac{3x}{3} = \frac{15}{3}$$
$$x = 5.$$

Our solution set is $S = \{5\}$.

To check our answer, we substitute the solution in place of the unknown in the original equation. If we get a true statement, we say the solution "satisfies" the equation.

In our example, $5(2x - 1) = 7x + 10$, we found that $x = 5$. Substituting 5 in place of x in the original equation, we have

$$5[2(5) - 1] = 7(5) + 10$$
$$5[10 - 1] = 35 + 10$$
$$5[9] = 45$$
$$45 = 45 \text{ (true)}$$

and our solution checks.

Example 2–1 C   Find the solution set.

1. $-2(y + 3) + 3(2y - 1) = 10$

$\qquad -2y - 6 + 6y - 3 = 10$ Carry out the multiplication.

$\qquad\qquad\quad 4y - 9 = 10$ Combine like terms.

$\qquad\quad 4y - 9 + 9 = 10 + 9$ Add 9 to both members.

$\qquad\qquad\qquad 4y = 19$

$$\frac{4y}{4} = \frac{19}{4}$$ Divide both members by 4.

$$y = \frac{19}{4}$$

$$\text{Check: } -2\left[\left(\frac{19}{4}\right) + 3\right] + 3\left[2\left(\frac{19}{4}\right) - 1\right] = 10$$

$$-2\left[\frac{19}{4} + \frac{12}{4}\right] + 3\left[\frac{19}{2} - \frac{2}{2}\right] = 10$$

$$-2\left[\frac{31}{4}\right] + 3\left[\frac{17}{2}\right] = 10$$

$$-\frac{31}{2} + \frac{51}{2} = 10$$

$$\frac{20}{2} = 10$$

$$10 = 10 \text{ (true)}$$

and

$$S = \left\{\frac{19}{4}\right\}$$

At this point we will no longer show the check of our solution, but we should realize that a check of our work is an important final step.

    The following equations contain several fractions. When this occurs it is usually easier to **clear the equation of all fractions.** We do this by multiplying both members by the least common denominator (LCD) of all the fractions. Clearing all the fractions is considered a means of simplifying the equation and will be done as a first step when necessary. Equations containing fractions will be studied more completely in chapter 4.

2. $\dfrac{1}{2}x + 3 = \dfrac{2}{3}$

$\qquad 6\left(\dfrac{1}{2}x + 3\right) = 6 \cdot \dfrac{2}{3}$ Multiply both members by the least common denominator, 6.

$\qquad\qquad 3x + 18 = 4$

$\qquad 3x + 18 - 18 = 4 - 18$ Subtract 18 from both members.

$\qquad\qquad\qquad 3x = -14$

$$\frac{3x}{3} = -\frac{14}{3}$$ Divide both members by 3.

$$x = -\frac{14}{3}$$

$$S = \left\{-\frac{14}{3}\right\}$$

3.
$$\frac{3}{4}x - \frac{1}{2} = \frac{1}{3}x + 2$$
$$12\left(\frac{3}{4}x - \frac{1}{2}\right) = 12\left(\frac{1}{3}x + 2\right)$$ Multiply both members by the least common denominator, 12.
$$9x - 6 = 4x + 24$$
$$9x - 4x - 6 = 4x - 4x + 24$$ Subtract $4x$ from both members.
$$5x - 6 = 24$$
$$5x - 6 + 6 = 24 + 6$$ Add 6 to both members.
$$5x = 30$$
$$\frac{5x}{5} = \frac{30}{5}$$ Divide both members by 5.
$$x = 6$$
$$S = \{6\}$$

4.
$$3.18z + 3.526 = 2(0.73z - 2.709)$$
$$3.18z + 3.526 = 1.46z - 5.418$$ Carry out the multiplication.
$$3.18z - 1.46z + 3.526 = 1.46z - 1.46z - 5.418$$ Subtract $1.46z$ from both members.
$$1.72z + 3.526 = -5.418$$
$$1.72z + 3.526 - 3.526 = -5.418 - 3.526$$ Subtract 3.526 from both members.
$$1.72z = -8.944$$
$$\frac{1.72z}{1.72} = \frac{-8.944}{1.72}$$ Divide both members by 1.72.
$$z = -5.2$$
$$S = \{-5.2\}$$

5. $5(x - 4) - 2x = 3x + 7$
$$5x - 20 - 2x = 3x + 7$$ Carry out the multiplication.
$$3x - 20 = 3x + 7$$ Combine like terms.
$$3x - 3x - 20 = 3x - 3x + 7$$ Subtract $3x$ from both members.
$$-20 = 7 \text{ (false)}$$

The statement $-20 = 7$ is false and this means that there is *no solution* to the equation.

$$S = \emptyset$$

Problem solving

The following sets of word problems are designed to help us interpret verbal statements and write expressions for them in algebraic symbols. For each problem we will write an algebraic expression that changes the words into mathematical symbols. These will not be equations. We will use our experience from these problems to help us translate word problems into equations in section 2–3.

Example 2–1 D

Write an algebraic phrase for each of the following verbal statements.

1. Phil can type 75 words per minute. How many words can he type in m minutes?

If Phil can type 75 words in one minute, then we multiply

$$75 \cdot m$$

to obtain the number of words he can type in m minutes.

2. If Debbie has d dollars in her savings account and on successive days she deposits \$55 and then withdraws \$25 to make a purchase, write an expression for the balance in her savings account.

We *add* the deposits and *subtract* the withdrawals. Thus

$$d + 55 - 25 = d + 30$$

represents the balance in dollars in Debbie's savings account after the two transactions.

3. A woman paid d dollars for 20 pounds of ground beef. How much did the beef cost her per pound?

The price per pound is found by dividing the total cost by the number of pounds. Thus the price in dollars per pound of the beef is represented by

$$\frac{d}{20}.$$

Mastery points
Can you
• Apply the addition property of equality?
• Apply the multiplication property of equality?
• Solve linear equations?
• Determine when a linear equation has no solution?
• Check your answers?

Exercise 2–1

Directions Find the solution set of the following equations. See examples 2–1 A, B, and C.

1. $5x = 15$ **2.** $7x = 28$ **3.** $4y = 10$ **4.** $6x = 27$

5. $a + 5 = 11$ **6.** $x + 4 = -3$ **7.** $z - 6 = -3$ **8.** $x - 9 = 8$

9. $\dfrac{x}{3} = 4$ **10.** $\dfrac{y}{2} = 11$ **11.** $\dfrac{3a}{4} = 8$ **12.** $4.1x = 15.17$

13. $1.8y = 21.6$ **14.** $0.7a = 11.2$ **15.** $\dfrac{2y}{3} = 9$ **16.** $3b - 1 = 8$

17. $5y - 2 = 13$ **18.** $4x + 5 = 5$ **19.** $6x + 4 = -2$ **20.** $9a - 3 = -3$

21. $7x - 4x + 3 = 8$

22. $2a + 3a - 7 = 5$

23. $3(y + 1) = 4$

24. $2(2z - 3) = 7$

25. $4(3b - 1) + 2b = 11$

26. $3(2x + 1) = 7x - 3x + 4$

27. $5(2 - x) + 1 = 3x - 4 + x$

28. $7x + 3 - 2x = 4(3 - 2x) + 5$

29. $2(3a - 1) = 3(2a + 5)$

30. $5(x - 4) + 12 = 3x + 2x - 6$

31. $\frac{1}{3}x + 2 = \frac{5}{6}$

32. $\frac{1}{2}x - 1 = \frac{1}{3}x + 3$

33. $\frac{3}{4}x + 3 = \frac{5}{8}x + 4$

34. $\frac{3}{8}x + \frac{1}{4} = \frac{1}{4}x + 3$

35. $\frac{5}{12}x + 2 = \frac{2}{3}x - 4$

36. $\frac{7}{12}x - 4 = \frac{5}{6}x - 5$

37. $-4(y - 3) + 2(3y - 5) = 11$

38. $-2(a + 3) + 5(a - 1) = -4$

39. $3(2y + 4) - 3y = 6(y + 1)$

40. $3(2x + 5) - x = 4(x - 3) + 7$

41. $3[2a - (a + 7)] = 10$

42. $4[a - (3a - 4)] = a + 5$

43. $-2[3x - (x - 5)] = 3x - 4$

44. $5.6z - 22.15 = 24.33$

45. $9.3y - 27.9 + 4.6y = 55.5$

46. $7.6a + 18.4 - 3.2a = 66.8$

47. $6.8x + 5.7 = 4.3x - 15.3$

48. $2.6(x - 6.3) = 8.9x - 81.9$

49. $6.7 - 4.1(x + 1) = 1.5x - 42.2$

50. $3(2x - 1) = 6x + 7$

51. $4a + 3a - 7 = 2(3a + 1) + a$

52. $6(z + 3) - 2z = 2(2z + 1)$

Directions Find the solution set of the following equations. See examples 2–1 A, B, and C.

53. The surface area S of a rectangular solid of length ℓ, width w, and altitude h is given by $S = 2(\ell w + \ell h + hw)$. Find ℓ when $S = 236$, $w = 5$, and $h = 6$.

54. To convert Celsius temperature to Fahrenheit, we use $F = \frac{9}{5}C + 32$. Find C when $F = 77$.

55. The amount A of a principal P invested for t years at simple interest with rate r per year is given by $A = P(1 + rt)$. Solve for t if $A = 4,080$; $P = 3,000$; and $r = 9\%$.

Directions Write an algebraic expression for the following verbal statements. See example 2–1 D.

56. Rita can enter 80 words per minute on a word processor. How many words can she enter in m minutes?

57. Express the cost in cents of c cans of oil if each can costs $1.15.

58. A 10-pound bag of dog food costs d dollars. How much does the dog food cost per pound?

59. It costs Pat $18 to rent a posthole digger for h hours. What did it cost him per hour to rent the posthole digger?

60. Megan has n nickels, d dimes, and q quarters in her purse. Express in cents the amount of money she has in her purse. (*Hint: n* nickels are represented in cents by $5n$.)

61. Roger has h half dollars, q quarters, d dimes, and n nickels. Express in cents the amount of money Roger has.

62. Colleen is y years old now. Express her age (a) 21 years from now, (b) 7 years from now.

63. Richard is 3 years old. If Dick is n times as old as Richard, express Dick's age. Express Dick's age 8 years ago.

64. Edna's savings account has a current balance of $457. If she makes a withdrawal of a dollars and then makes a deposit of b dollars, express her new balance.

65. Dale has a balance of d dollars in his checking account. He makes a deposit of $464 and then writes 5 checks for m dollars each. Express his new balance in dollars in terms of d and m.

66. Marty has c cents, all in quarters. Write an expression for the number of quarters Marty has.

67. If w represents a whole number, write an expression for the next greater whole number.

68. If j represents an even integer, write an expression for the next greater even integer.

69. If j represents an odd integer, write an expression for the next greater odd integer.

70. If Joan is f feet and t inches tall, how tall is Joan in inches?

71. Sue earns $500 more than twice the amount Jon earns in a year. If Jon earns d dollars in a year, write an expression for Sue's annual salary.

72. Lisa's annual salary is $1,000 more than n times Bonnie's salary. If Bonnie earns $9,000 per year, express Lisa's annual salary.

73. Express the total cost in cents of purchasing c cans of soda pop at 59¢ per can and b loaves of bread at $1.15 per loaf.

74. A gallon of primer coat costs $9.95 per gallon and a gallon of latex-base paint costs $12.99. Express the cost of p gallons of primer and q gallons of latex-base paint.

75. Norm can type w words per minute and Rich can type 11 words per minute more than Norm. Write an expression for how many words Rich can type in 20 minutes.

2–2 Formulas and literal equations

In mathematics a **literal equation** is an equation that contains more than one variable. A **formula** is a mathematical equation that states the relationship between two or more physical conditions.

When we repeatedly use a formula to determine the value of the same variable, it is convenient to solve the equation for that variable in terms of the remaining variables and constants. In the case of the relationship between Celsius and Fahrenheit, it is useful to have one formula for Celsius in terms of Fahrenheit and another formula for Fahrenheit in terms of Celsius. Using our procedure for solving linear equations, we will now solve the Celsius formula for Fahrenheit F.

$$C = \frac{5}{9}(F - 32)$$

$$9C = 9 \cdot \frac{5}{9}(F - 32) \qquad \text{Multiply both members by the least common denominator, 9.}$$

$$9C = 5(F - 32)$$

$$9C = 5F - 160 \qquad \text{Distributive property}$$

$$9C + 160 = 5F \qquad \text{Add 160 to both members.}$$

$$\frac{1}{5}(9C + 160) = \frac{1}{5} \cdot 5F \qquad \text{Multiply both members by } \frac{1}{5}.$$

$$\frac{9}{5}C + 32 = F \qquad \text{Distributive property}$$

The formula for finding the temperature in degrees Fahrenheit, given the temperature in degrees Celsius, is

$$F = \frac{9}{5}C + 32.$$

The two formulas

$$C = \frac{5}{9}(F - 32) \quad \text{and} \quad F = \frac{9}{5}C + 32$$

may not look the same, but they express the same relationship between C and F. The first formula is solved for C in terms of F, and the second formula is solved for F in terms of C.

The following steps are a restatement of the procedure for solving linear equations. We will now apply these steps to solve formulas and literal equations for a specific variable.

Step 1 Simplify the equation.

Step 2 Obtain all terms with the variable for which we are solving in one member of the equation.

Step 3 Obtain all terms that do not have the variable for which we are solving in the other member of the equation.

Step 4 Determine the coefficient of the variable for which we are solving, and then divide both members of the equation by that coefficient.

Example 2–2 A

Solve for the specified variable.

1. The formula for the perimeter of a rectangle is $P = 2\ell + 2w$. Solve for ℓ.

$$P = 2\ell + 2w$$

$$P - 2w = 2\ell + 2w - 2w \qquad \text{Subtract } 2w \text{ from both members.}$$

$$P - 2w = 2\ell$$

$$\frac{P - 2w}{2} = \frac{2\ell}{2} \qquad \text{Divide both members by 2.}$$

$$\ell = \frac{P - 2w}{2} \qquad \text{Symmetric property}$$

2. The area of a trapezoid is given by $A = \frac{1}{2}h\,(b_1 + b_2)$. Solve for b_1.

$$A = \frac{1}{2}h(b_1 + b_2)$$

$$2A = 2 \cdot \frac{1}{2}h(b_1 + b_2) \qquad \text{Multiply both members by 2.}$$

$$2A = h(b_1 + b_2)$$

$$2A = b_1 h + b_2 h \qquad \text{Distributive property}$$

$$2A - b_2 h = b_1 h \qquad \text{Subtract } b_2 h \text{ from both members.}$$

$$\frac{2A - b_2 h}{h} = b_1 \qquad \text{Divide both members by } h.$$

$$b_1 = \frac{2A \quad b_2 h}{h} \qquad \text{Symmetric property}$$

3. Solve the literal equation $8x + 4 = 5x + y$ for x.

$$8x + 4 = 5x + y$$

$$8x - 5x + 4 = 5x - 5x + y \qquad \text{Subtract } 5x \text{ from both members.}$$

$$3x + 4 = y$$

$$3x + 4 - 4 = y - 4 \qquad \text{Subtract 4 from both members.}$$

$$3x = y - 4$$

$$\frac{3x}{3} = \frac{y - 4}{3} \qquad \text{Divide both members by 3.}$$

$$x = \frac{y - 4}{3}$$

4. Solve the literal equation $2(3x - y) = 3(x + 3y) + 2$ for x.

$$2(3x - y) = 3(x + 3y) + 2$$

$$6x - 2y = 3x + 9y + 2 \qquad \text{Simplify each member.}$$

$$6x - 3x - 2y = 3x - 3x + 9y + 2 \qquad \text{Subtract } 3x \text{ from both members.}$$

$$3x - 2y = 9y + 2$$

$$3x - 2y + 2y = 9y + 2y + 2 \qquad \text{Add } 2y \text{ to both members.}$$

$$3x = 11y + 2$$

$$\frac{3x}{3} = \frac{11y + 2}{3} \qquad \text{Divide both members by 3.}$$

$$x = \frac{11y + 2}{3}$$

Note

Although we have not stated any restrictions on the variables, it is understood that the values that the variables can take on must be such that *no denominator is ever zero.* That is: example 1 has no restrictions; example 2, $h \neq 0$; example 3 has no restrictions; and example 4 has no restrictions.

Exercise 2-2

Directions Find the value of the variable whose replacement value is not given.

1. $W = I^2R; I = 6, W = 324$

2. $L = \dfrac{V}{WH}; W = 12, H = 6, L = 20$

3. $A = P + Pr; A = 3{,}240; r = (0.08)$

4. $V = k + gt; k = 22, t = 3, V = 55$

5. $\ell = a + (n - 1)d; a = 7, d = 4, \ell = 83$

6. $A = \dfrac{1}{2}h(b_1 + b_2); A = 54, b_2 = 10, h = 6$

7. $a = \dfrac{V_2 - V_1}{t}; a = 4, V_2 = 83, t = 8$

8. $T = \dfrac{R - R_0}{aR_0}; T = 2, R_0 = 4, a = 3$

Directions Solve the following formulas or literal equations for the specified variable. Assume that no denominator is equal to zero. See example 2–2 A.

9. $I = prt; t$

10. $E = IR; R$

11. $E = mc^2; m$

12. $V = \ell wh; h$

13. $F = ma; m$

14. $K = PV; V$

15. $A = bh; h$

16. $A = bh; h$

17. $W = I^2R; R$

18. $V = \ell wh; w$

19. $P - 2\ell + 2w; w$

20. $V = k + gt; k$

21. $V = k + gt; g$

22. $A = P + Pr; r$

23. $D = dq + R; q$

24. $m = -p(\ell - x); x$

25. $m = -p(\ell - x); \ell$

26. $R = W - b(2c + b); c$

27. $R = W - b(2c + b); W$

28. $2S = 2Vt - gt^2; V$

29. $V - r^3(u - b); u$

30. $S = \dfrac{n}{2}[2a + (n - 1)d]; a$

31. $S = \dfrac{n}{2}[2a + (n - 1)d]; d$

32. $V = \dfrac{1}{3}\pi h^2(3R - h); R$

33. $2S = 2Vt - gt^2; g$

34. $P = n(P_2 - P_1) - c; P_1$

35. $\ell = a + (n - 1)d; d$

36. $2x + 3y = 12; y$

37. $2x + 3y = 12; x$

38. $2x - y = 5x + 6y; x$

39. $2x - y = 5x + 6y; y$

40. $3(4x - y) = 2x + y + 6; y$

41. $3(4x - y) = 2(x + y + 3); x$

42. $ax + 4 = by - 3; x$

43. $ay + 7 - x = 3x + 4; y$

44. $a(x + 2) = by; x$

45. $b(y - 4) = a(x + 3); y$

46. The distance s that a body projected downward with an initial velocity of v falls in t seconds because of the force of gravity is given by

$s = \dfrac{1}{2}gt^2 + vt$. Solve for g.

47. Solve the formula in exercise 46 for v.

48. The net profit P on sales of n identical cars is given by $P = n(P_2 - P_1) - c$, where P_2 is the selling price, P_1 is the cost to the dealer, and c is the operating expense. Solve for P_2.

49. Solve the formula in exercise 48 for P_1.

50. The perimeter of an isosceles triangle with base b and sides s is given by $P = 2s + b$. Solve for s.

2–3 Verbal problems

Many problems that we encounter are written or stated verbally. We need to translate these verbal statements into equations that we can solve algebraically. When translating verbal statements into equations, we should look for phrases involving the basic operations of addition, subtraction, multiplication, and division. Table 1.1 in chapter 1 showed some examples of phrases that are commonly encountered.

We shall now combine our ability to write an expression and our ability to solve an equation and apply them to solving a verbal problem. While there is no standard procedure for solving a verbal problem, the following guidelines should be useful.

1. Read the problem carefully. Determine useful prior knowledge and note what information is given and what information we are asked to find.
2. Whenever possible, draw a picture or use a diagram to represent the information in the problem.
3. Let some letter represent one of the unknowns, then express other unknowns in terms of it.
4. Use the given conditions in the problem and the unknowns from step 3 to write an algebraic equation.
5. Solve the equation for the unknown. Relate this answer to any other unknowns in the problem.
6. Check the results in the original statement of the problem.

Example 2–3 A

Number problems

Write an equation for the problem and solve for the unknown quantities.

1. One number is 18 more than a second number. If their sum is 62, find the two numbers.

If we knew the value of the second number, then the first number would be 18 more. Therefore let x be the second number (second number $= x$). Then the first number is $x + 18$. Since these two numbers add up to 62, we write the equation as

first number $(x + 18)$	sum $+$	second number x	is $=$	62 62,

where $x =$ the second number and $x + 18 =$ the first number. Solving for x, we have

$$(x + 18) + x = 62$$
$$x + 18 + x = 62$$
$$2x + 18 = 62$$
$$2x = 44$$
$$x = 22$$

and
$$x + 18 = (22) + 18 = 40.$$

Hence the second number is 22 and the first number is 40.

To check our answers, we must determine whether they satisfy the conditions stated in the original problem. Since 40 is 18 more than 22 and since the sum of 22 and 40 is 62, we know that our answers are correct. We will not show the checks of the following problems, but we should realize that a check of our work is an important final step.

2. If a number is divided by 4 and this quotient is increased by 6, the result is 13. Find the number.

Let x be the number. Then the number divided by 4 is $\dfrac{x}{4}$ and that quotient increased by 6 is $\dfrac{x}{4} + 6$. Since this equals 13, we have

number divided by 4 $\dfrac{x}{4}$	increased $+$	by 6 6	is $=$	13 13,

where $x =$ the number.
Solving for x,

$$\frac{x}{4} + 6 = 13$$
$$\frac{x}{4} = 7$$
$$x = 28.$$

Hence the number is 28.

3. The sum of three numbers is 63. The first number is twice the second number, and the third number is three times the first number. Find the three numbers. We must know the value of the second number to find the first number, and we must know the value of the first number to find the third number.

Therefore everything depends on the second number. If x equals the second number, we have

second number $= x$
first number $= 2x$ (twice the second)
third number $= 3(2x) = 6x$ (three times the first).

Since their sum is 63, we have

first number	sum	second number	sum	third number	is	63
$2x$	$+$	x	$+$	$6x$	$=$	$63.$

Solving for x,

$$2x + x + 6x = 63$$
$$9x = 63$$
$$x = 7.$$

Hence $2x = 2(7) = 14$ and $6x = 6(7) = 42$. Therefore the second number (x) is 7, the first number ($2x$) is 14, and the third number ($6x$) is 42.

Interest problem

4. Phil had \$20,000, part of which he invested at 8% interest and the rest at 6%. If his total income from the two investments for one year was \$1,460, how much did he invest at each rate?

To solve this problem we must understand how interest is computed. If we invested \$5,000 at 8% interest, after one year we would have 8% of \$5,000 or $(0.08)(\$5,000) = \$400$ interest. Thus the amount of interest earned in one year is the product of the rate times the principal (the amount invested).

In the given problem we have two principals and two rates, and our formula will involve the sum of the two earned interests, which is equal to \$1,460.

Let x represent the amount of money invested at 8%. Since this amount and the amount invested at 6% total \$20,000, we can describe the 6% principal as the remainder after the x amount has been invested at 8%, that is, $20,000 - x$ is invested at 6% interest. We now can write the equation

amount of interest at 8%	total	amount of interest at 6%	was	total interest
$(0.08)x$	$+$	$(0.06)(20,000 - x)$	$=$	$1,460.$

Solving for x,

$$(0.08)x + (0.06)(20,000 - x) = 1,460$$
$$(0.08)x + 1,200 - (0.06)x = 1,460$$
$$(0.02)x + 1,200 = 1,460$$
$$(0.02)x = 260$$
$$x = 13,000$$

and $20,000 - x = 20,000 - (13,000) = 7,000$.

Hence the amount invested at 8% (x) is \$13,000 and the amount invested at 6% ($20,000 - x$) is \$7,000.

When we know the sum of two numbers, we let x equal one number, then the sum minus x (sum $- x$) will be the other number.

Geometry problem

5. The length of a rectangle is 9 feet more than its width. If the perimeter of the rectangle is 58 feet, find the dimensions.

We need the prior knowledge that the perimeter of a rectangle (distance around) is found by the formula $P = 2\ell + 2w$. If we let x represent the width of the rectangle, then the length would be 9 feet longer, or $x + 9$. We can now substitute into the formula.

$$P = 2\ell + 2w$$
$$58 = 2(x + 9) + 2(x)$$

Solving for x,

$$58 = 2(x + 9) + 2(x)$$
$$58 = 2x + 18 + 2x$$
$$58 = 4x + 18$$
$$40 = 4x$$
$$10 = x$$

and $x + 9 = (10) + 9 = 19$.

Hence the width (x) is 10 feet and the length $(x + 9)$ is 19 feet.

Mastery points
Can you
- Translate verbal statements into equations?
- Solve for the unknown quantities?

Exercise 2–3

Directions Write an equation for the problem and solve for the unknown quantities. See example 2–3 A–1, 2, and 3.

Number problems

Example
One number is 9 times a second number and their sum is 120. Find the numbers.

Solution
Let x represent the second number, then the first number is 9 times the second number, or $9x$. Since their sum is 120, the equation is

first number	sum	second number	is	120
$9x$	$+$	x	$=$	120.

Solving for x,

$$9x + x = 120$$
$$10x = 120$$
$$x = 12$$

and $9x = 9(12) = 108$.

Therefore the second number (x) is 12 and the first number ($9x$) is 108.

1. What number added to its double gives 51?

2. Find a number such that twice the sum of that number and 7 are 38.

3. Find two numbers whose sum is 81 and whose difference is 35.

4. One number is 11 more than twice a second number. If their sum is 53, what are the numbers?

5. Six times a number, increased by 10, gives 88. Find the number.

6. One number is seven times another. If their difference is 28, what are the numbers?

7. The sum of the number of teeth on two gears is 64 and their difference is 12. How many teeth are on each gear?

8. Two gears have a total of 83 teeth. One gear has 15 less teeth than the other. How many teeth are on each gear?

9. Two electrical voltages have a total of 126 volts (V). If one voltage is 32 V more than the other, find the voltages.

10. The sum of two voltages is 85 and their difference is 32. Find the voltages.

11. The sum of two resistors in a series is 24 ohms and their difference is 14 ohms. How many ohms are in each resistor?

12. One resistor exceeds another resistor by 25 ohms and their sum is 67 ohms. How many ohms are in each resistor?

Example

The sum of three consecutive odd integers is 51. Find the integers.

Solution

First, we shall examine consecutive odd integers in general. Consider the list 1, 3, 5, 7 or 215, 217, 219, 221. We observe that the next odd integer on either list is found by adding 2 to the previous integer. Therefore if we let x be the first (least) of the three consecutive odd integers, then $x + 2$ would be the second and $(x + 2) + 2 = x + 4$ must be the third. Since their sum is 51, the equation would be

first odd integer	sum	second odd integer	sum	third odd integer	is	51
x	$+$	$x + 2$	$+$	$x + 4$	$=$	51.

Solving for x,

$$x + x + 2 + x + 4 = 51$$
$$3x + 6 = 51$$
$$3x = 45$$
$$x = 15.$$

Therefore $x + 2 = (15) + 2 = 17$ and $x + 4 = (15) + 4 = 19$.

Hence the first consecutive odd integer (x) is 15, the second ($x + 2$) is 17, and the third ($x + 4$) is 19.

13. The sum of three consecutive integers is 69. Find the integers.

14. The sum of three consecutive even integers is 48. Find the integers.

15. The sum of three consecutive odd integers is 87. Find the integers.

16. Four times the first of three consecutive integers is 1 less than three times the sum of the second and third. Find the integers.

17. Five times the first of three consecutive even integers is 4 less than twice the sum of the second and third. Find the integers.

18. One-fourth of the middle integer of three consecutive even integers is 24 less than one-half of the sum of the other two integers. Find the three integers.

19. One number is 27 more than another. The smaller number is one-fourth of the larger number. Find the numbers.

20. A number plus one-half of the number plus one-third of the number equal 33. Find the number.

21. A number is decreased by 7 and twice this result is 34. What is the number?

22. The sum of three numbers is 38; the second number is twice the first number and the third number is 2 more than the first number. Find the three numbers.

23. The sum of three numbers is 49. The second number is three times the first number and the third number is 6 less than the first number. Find the three numbers.

24. One number is 7 more than another number. Find the two numbers if three times the larger number exceeds four times the smaller number by 13.

Interest problems

See example 2–3 A–4.

Example

Lynne made two investments totaling $25,000. She made an 18% profit on one investment, but she took an 11% loss on the other investment. If her net gain was $2,180, how much was invested at each rate?

Solution

Let x be the amount invested at 18%, then the amount invested at 11% was what was left over from the $25,000 or $25,000 - x$. Her profit of 18% on the one investment is denoted by $(0.18)x$ and her loss of 11% on the other investment is $-(0.11)(25,000 - x)$; the loss is denoted as a negative amount. Since her net gain was $2,180, the equation would be

18% profit	net	11% loss	was	2,180
$(0.18)x$	$-$	$(0.11)(25,000 - x)$	$=$	$2,180.$

Solving for x,

$$(0.18)x \quad (0.11)(25,000 \quad x) - 2,180$$
$$(0.18)x - 2,750 + (0.11)x = 2,180$$
$$(0.29)x - 2,750 = 2,180$$
$$(0.29)x = 4,930$$
$$x = 17,000$$

and $25,000 - x = 25,000 - (17,000) = 8,000.$

So the amount invested at an 18% profit (x) was $17,000 and the amount invested at an 11% loss $(25,000 - x)$ was $8,000.

25. Terry has $15,000. He invests part of this money at 8% and the rest at 6%. His income for one year from these investments totals $1,100. How much is invested at each rate?

26. Harry invested $26,000, part at 10% and the rest at 12%. If his income for one year from these investments totals $2,860, how much was invested at each rate?

27. Margot invests a total of $12,000, part at 10% and part at 12%. Her total income for one year from the investments totals $1,340. How much is invested at each rate?

28. Jill invests a total of $12,000, part at 6% and part at 10%. If her total income for one year is $956, how much is invested at each rate?

29. Andrew invested a total of $18,000, part at 5% and part at 9%. If his income for one year from the 9% investment was $100 less than his income from the 5% investment, how much was invested at each rate?

30. Peter has invested $5,000 at an 8% rate. How much more must he invest at 10% to make the total income for one year from both sources a 9% rate?

31. Jeremy has $6,000 invested at 6%; how much more must he invest at 10% to realize a net return of 9%?

32. Dick has $26,000, part of which he invests at 10% interest and the rest at 14%. If his income for one year from the 14% investment is $760 more than that from the 10% investment, how much is invested at each rate?

33. Nina made two investments totaling $18,000. She made a 14% profit on one investment, but she took a 9% loss on the other investment. If her net gain was $680, how much was each investment?

34. Jim made two investments totaling $34,000. One investment made him a 12% profit, but on the other investment he took a 21% loss. If his net loss was $870, how much was each investment?

Geometry problems

See example 2–3 A–5.

Example
The length of a rectangle is 1 inch less than three times the width. Find the dimensions if the perimeter is 70 inches.

Solution
If x represents the width of the rectangle, then by multiplying the width by 3 ($3x$) and subtracting 1, ($3x - 1$), we have the representation of the length. Using the formula for the perimeter of a rectangle, $P = 2\ell + 2w$, and the fact that the perimeter P is 70, we can write the equation

$P = 2\ell + 2w$
$70 = 2(3x - 1) + 2x.$

Solving for x,

$70 = 6x - 2 + 2x$
$70 = 8x - 2$
$72 = 8x$
$9 = x$

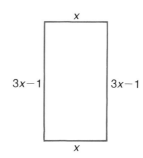

and $3x - 1 = 3(9) - 1 = 27 - 1 = 26.$
Therefore the width of the rectangle (x) is 9 inches and the length ($3x - 1$) is 26 inches.

35. The length of a rectangle is 5 feet more than its width. If the perimeter is 102 feet, find the length and width.

36. The width of a rectangle is 3 meters less than the length. If the perimeter of the rectangle is 126 meters, find the dimensions of the rectangle.

37. When the length of a side of a square is increased by 4 inches, the area is increased by 64 square inches. Find the original length of a side.

38. When the length of a side of a square is decreased by 3 meters, the area is diminished by 33 square meters. Find the original length of a side.

39. One side of a triangle is twice as long as the second side and the third side is 4 less than three times the second side. If the perimeter is 38 centimeters, find the lengths of the sides.

40. One side of a triangle is three times as long as the second side and the third side is 1 more than two times the second side. If the perimeter is 37 meters, find the lengths of the sides.

Mixture problems

Example
What quantities of 65% pure silver and 45% pure silver must be mixed together to give 100 grams of 50% pure silver?

Solution
To solve this mixture problem, we need to understand that the amount of silver in any given mixture is found by multiplying the percent of silver in the mixture times the amount of mixture. If x represents the amount of 65% pure silver in the final mixture, then $100 - x$ will represent the amount of 45% pure silver. Since the amount of silver present before mixing must be the same as the amount after mixing, the equation would be

amount of silver in 65% pure	mixed together	amount of silver in 45% pure	to give	100 grams of 50% pure silver
$(0.65)x$	$+$	$(0.45)(100 - x)$	$=$	$(0.50)(100)$.

Solving for x,

$$(0.65)x + (0.45)(100 - x) = (0.50)(100)$$
$$(0.65)x + 45 - (0.45)x = 50$$
$$(0.20)x + 45 = 50$$
$$(0.20)x = 5$$
$$x = 25$$

and $100 - x = 100 - (25) = 75$.

So the amount of 65% pure silver (x) is 25 grams and the amount of 45% pure silver $(100 - x)$ is 75 grams.

41. An auto mechanic has two bottles of battery acid. One contains a 10% solution and the other a 4% solution. How many cubic centimeters of each solution must be used to make 120 cm³ of a solution that is 6% acid?

42. A metallurgist wishes to form 2,000 kg of an alloy that is 80% copper. This alloy is to be obtained by fusing some alloy that is 68% copper and some alloy that is 83% copper. How many kilograms of each alloy must be used?

43. If a jeweler wishes to form 12 ounces of 75% pure gold from sources that are 60% and 80% pure gold, how much of each substance must be mixed together to produce this?

44. A chemist wishes to make 1,000 liters of a 3.5% acid solution by mixing a 2.5% solution with a 25% solution. How many liters of each solution is necessary?

45. A pharmacist wishes to fill a total of 200 3-grain and 2-grain capsules using 500 grains of a certain drug. How many capsules of each kind does he fill?

46. A solution that is 38% silver nitrate is to be mixed with a solution that is 3% silver nitrate to obtain 100 centiliters of solution that is 5% silver nitrate. How many centiliters of each solution should be used in the mixture?

47. A druggist has two solutions, one 60% hydrogen peroxide and the other 30% hydrogen peroxide. How many liters of each should she mix to obtain 30 liters of a solution that is 40% hydrogen peroxide?

2–4 Linear inequalities

Representing the solution set of a linear inequality

When we replace the equals sign in a linear equation with one of the inequality symbols ($<$ less than, $>$ greater than, $\leq$ less than or equal to, $\geq$ greater than or equal to), we form a **linear inequality.** The following are examples of linear inequalities.

$$5x + 3 \leq 4, \quad 2(3x - 1) > 5x - 7, \quad 5x - 2x + 1 < x - 3$$

A major difference between a linear equation and a linear inequality is the solution set. The solution set of a linear equation has at most one value, whereas the solution set of a linear inequality usually consists of an unlimited number of values. Consider the inequality $3x \geq 9$. We see by inspection that if we substitute $3, 3\frac{1}{3}, 4$, or 5 for x, the inequality would be true. In fact, we see that if we substitute any number greater than or equal to 3, that is ($x \geq 3$), the inequality would be true. This demonstrates that our solution set has an unlimited number of solutions. We would state it in set-builder notation as $S = \{x \mid x \geq 3\}$, which is read "the set of all elements x such that x is greater than or equal to 3."

There are other ways to indicate the solution set of an inequality. One way is to graph the solution set. To graph the solution set $S = \{x \mid x \geq 3\}$, we simply draw a number line (as we did in chapter 1), place a solid circle (dot) at 3 on the number line, and draw an arrow extending from the circle to the right, as in figure 2.1.

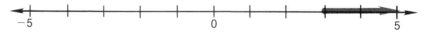

Figure 2.1

The solid circle at 3 represents the fact that 3 is included in the solution set.

Example 2–4 A

Represent the following solution sets graphically.

1. $\{x \mid x < -1\}$
Here x is representing all real numbers less than -1, but not -1 itself. To denote the fact that x cannot equal -1, we put a hollow circle at -1.

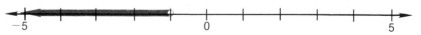

2. $\{x \mid x \geq 2\}$

The greater than or equal to symbol, $\geq$, indicates that we contain the point 2, so we place a solid circle at 2.

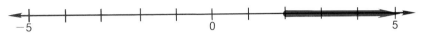

3. $\{x \mid -2 \leq x < 1\}$

The statement $-2 \leq x < 1$ is called a **compound inequality** and is read "-2 is less than or equal to x and x is less than 1." We place a solid circle at -2 to denote that -2 is included, and we place a hollow circle at 1 to show that 1 is not included. We then draw a line segment between the two circles.

Note

When we graph inequalities, a strict inequality, $<$ or $>$, is represented by a hollow circle at the number. A weak inequality, $\leq$ or $\geq$, is represented by a solid circle at the number.

Interval notation

Another way to represent the solution set of an inequality is to use **interval notation.** In interval notation we use a parenthesis, (or), to represent that the endpoint is *not* part of the interval and a bracket, [or], to represent that the endpoint is part of the interval.

Consider the solution set $S = \{x \mid -3 < x \leq 4\}$. To write this in interval notation, we first write down the endpoints, writing the lesser one first, and then separate them by a comma.

$$-3, 4$$

If the endpoint is indicated by a strict inequality symbol, $<$ or $>$, place a parenthesis next to that endpoint. If the endpoint is indicated by a weak inequality symbol, $\leq$ or $\geq$, place a bracket next to that endpoint. So,

Strict inequality—Parenthesis

$\{x \mid -3 < x \leq 4\}$ is equivalent to $(-3, 4]$.

Weak inequality—Bracket

Example 2-4 B

Represent the following solution sets with interval notation.

1. $\{x \mid -2 \le x \le 3\}$

Since the endpoints are included, we use brackets in interval notation.

$[-2,3]$

2. $\{x \mid 0 < x < 6\}$

Since the endpoints are not included, we use parentheses in interval notation.

$(0,6)$

3. $\{x \mid x \ge 1\}$

To indicate that our solution set continues indefinitely, we use $+\infty$, read "positive infinity." We place a parenthesis next to the $+\infty$ symbol because there is no endpoint to be contained.

$[1,+\infty)$

4. $\{x \mid x < -3\}$

Interval $(-\infty,-3)$

The symbol $-\infty$ is read "negative infinity."

Note

In interval notation the lesser number must come first. In example 4, $(-3,-\infty)$ would be incorrect.

Solving a linear inequality

The properties that we will use to solve a linear inequality are similar to those that we used to solve linear equations. They are as follows:

■ **Addition property of inequality**

For any algebraic expressions A, B, and C, if $A < B$, then
$$A + C < B + C.$$

■ **Multiplication property of inequality**

For any algebraic expressions A, B, and C, if $A < B$, then
1. if $C > 0$ (positive), then
$$A \cdot C < B \cdot C,$$
2. if $C < 0$ (negative), then
$$A \cdot C > B \cdot C.$$

Note
The two properties are stated in terms of the less than ($<$) symbol. These properties also apply for any of the other inequality symbols ($>$, $\leq$, or $\geq$).

Just as with equations, we can use the addition property to subtract the same expression from both members of an inequality. The multiplication property allows us to divide both members of an inequality by the same nonzero expression.

To demonstrate these operations, consider the inequality $8 < 12$.

1. If we add or subtract 4 in each member, we still have a true statement.

$$
\begin{array}{ccc}
8 < 12 & \text{or} & 8 < 12 \\
8 + 4 < 12 + 4 & & 8 - 4 < 12 - 4 \\
12 < 16 & & 4 < 8
\end{array}
$$

2. If we multiply or divide by 4 in each member, we still have a true statement.

$$
\begin{array}{ccc}
8 < 12 & \text{or} & 8 < 12 \\
8 \cdot 4 < 12 \cdot 4 & & \dfrac{8}{4} < \dfrac{12}{4} \\
32 < 48 & & 2 < 3
\end{array}
$$

3. But if we multiply or divide by -4 in each member, we have to reverse the direction of the inequality before we have a true statement.

$$
\begin{array}{ccc}
8 < 12 & \text{or} & 8 < 12 \\
8(-4) > 12(-4) & & \dfrac{8}{-4} > \dfrac{12}{-4} \\
-32 > -48 & & -2 > -3
\end{array}
$$

Note
When we reverse the direction of the inequality symbol, we say that we **reversed the sense or order** of the inequality.

To summarize our operations, we see that, with one exception, they are the same as the operations for linear equations. **Whenever we multiply or divide both members of an inequality by a negative number, we must reverse the direction of the inequality symbol.**

We shall now solve a linear inequality. The procedure for solving a linear inequality is the same four steps that we used to solve a linear equation.

Consider the inequality

$$3(4x - 1) \geq 5x + 3x + 5.$$

Step 1 *We simplify* the problem by carrying out the indicated multiplication in the left member and the addition in the right member.

$$3(4x - 1) \geq 5x + 3x + 5$$
$$12x - 3 \geq 8x + 5$$

Step 2 *We want all terms containing the unknown,* x, *in one member of the inequality.* Therefore we subtract $8x$ from both members of the inequality.

$$12x - 3 \geq 8x + 5$$
$$12x - 8x - 3 \geq 8x - 8x + 5$$
$$4x - 3 \geq 5$$

Note

A negative coefficient of the unknown can be avoided if we form equivalent inequalities where the unknown appears only in the member of the inequality that has the greater coefficient of the unknown.

Step 3 *We want all terms not involving the unknown in the other member of the inequality.* Therefore we add 3 to both members of the inequality.

$$4x - 3 \geq 5$$
$$4x - 3 + 3 \geq 5 + 3$$
$$4x \geq 8$$

Step 4 *We form an equivalent inequality where the coefficient of the unknown is 1.* Hence we divide both members of the inequality by 4.

$$4x \geq 8$$
$$\frac{4x}{4} \geq \frac{8}{4}$$
$$x \geq 2$$

Our solution set is $S = \{x \mid x \geq 2\}$.

Note

In step 4 we must be careful to observe whether we are multiplying or dividing by a positive or negative number so that we will form the correct inequality.

Example 2–4 C

Find the solution set of the following inequalities. Leave the answer in (1) set-builder notation, (2) graphical notation, and (3) interval notation.

1. $2(1 - 2x) \geq 4(2 - 3x) + 2x$ Simplify.

$$2 - 4x \geq 8 - 12x + 2x$$
$$2 - 4x \geq 8 - 10x \qquad \text{Add } 10x \text{ to both members.}$$
$$6x + 2 \geq 8 \qquad \text{Subtract 2 from both members.}$$
$$6x \geq 6 \qquad \text{Divide both members by 6.}$$
$$x \geq 1$$

Hence the solution set is

$$S = \{x \mid x \geq 1\},$$

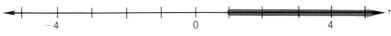

or $[1, +\infty)$.

2. $4(3 - 2x) + 3x > 2$ Simplify.

$$12 - 8x + 3x > 2$$
$$12 - 5x > 2 \qquad \text{Subtract 12 from both members.}$$
$$-5x > -10$$

We now divide both members by -5 and *reverse* the direction of the inequality symbol.

$$\frac{-5x}{-5} < \frac{-10}{-5}$$
$$x < 2$$

The solution set is

$$S = \{x \mid x < 2\},$$

or $(-\infty, 2)$.

3. $-5 \leq 2x - 1 < 3$

When solving a compound inequality, the solution must be such that the unknown appears only in the middle member of the inequality. We can still use all of our properties, if we apply them to all *three* members. We must reverse the direction of *all* the inequality symbols when multiplying or dividing by a negative number.

$$-5 \leq 2x - 1 < 3 \qquad \text{Add 1 to all three members.}$$
$$-5 + 1 \leq 2x - 1 + 1 < 3 + 1$$
$$-4 \leq 2x < 4 \qquad \text{Divide all three members by 2.}$$
$$\frac{-4}{2} \leq \frac{2x}{2} < \frac{4}{2}$$
$$-2 \leq x < 2$$

The solution set is

$$S = \{x \mid -2 \le x < 2\},$$

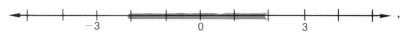

or $[-2,2)$.

4. $-12 < 8 - 4x \le 16$ Subtract 8 from all three members.
    $-12 - 8 < 8 - 8 - 4x \le 16 - 8$
        $-20 < -4x \le 8$

We now divide all three members by -4 and *reverse* the direction of *all* the inequality symbols.

$$\frac{-20}{-4} > \frac{-4x}{-4} \ge \frac{8}{-4}$$
$$5 > x \ge -2$$

The solution set is

$$S = \{x \mid -2 \le x < 5\},$$

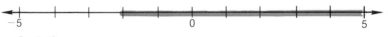

or $[-2,5)$.

Note
When we state the answer in set-builder notation, it is customary to have compound inequalities stated with less than inequality symbols. This makes changing from one form of notation to another easier.

Mastery points
Can you
- Solve inequalities and compound inequalities?
- Represent the solution set of an inequality in set-builder notation, graphical notation, or interval notation?

Exercise 2–4

Directions   Represent the following solution sets both graphically and with interval notations. See examples 2–4 A and B.

1. $\{x \mid -3 \le x \le 1\}$ **2.** $\{x \mid 0 < x < 4\}$ **3.** $\{x \mid -5 \le x < -1\}$ **4.** $\{x \mid x > -3\}$

5. $\{x \mid x \ge 4\}$ **6.** $\{x \mid x < 2\}$ **7.** $\{x \mid x \le -1\}$ **8.** $\{x \mid x \le 0\}$

Directions Find the solution set of the following inequalities. Leave the answer in both set-builder notation and interval notation. See example 2–4 C.

9. $2x > 18$

10. $3x < 12$

11. $\frac{2}{3}x \geq 8$

12. $\frac{3}{4}x \geq 9$

13. $-4x \leq 20$

14. $-3x > 27$

15. $3x + 2x < x + 6$

16. $6x - 2x > 5x - 3$

17. $2x + (4x - 1) > 6 - x$

18. $3(2x + 1) < 9$

19. $4(2x - 5) \leq 10x + 7$

20. $4 - 2(3x + 1) > 8x - 12$

21. $7 - 3(5x - 4) \leq 12 - 9x$

22. $2(x - 4) - 14 \leq 3(5 - 3x)$

23. $2(4x + 3) \geq 5 - 4(x - 1)$

24. $6(2x - 3) \leq 9(x - 1) - 4$

25. $4(2 - x) + 7 > 3x - 3(x - 1)$

26. $4x - 4(x + 2) < 3(2 - x)$

27. $14 \geq 5(1 - x) + 2(x + 3)$

28. $7 < 4(1 - 2x) + 3(x - 4)$

29. $8.2x - 3.6x + 7.1 \geq 1.8x + 23.9$

30. $2.1(x - 6) < 0.4x + 7.8$

31. $4.3(x - 2) \geq 3.1x - 25.4$

32. $7.3(3 - x) < 4.9(2 - x) - 4.7$

33. $12.6(3x - 2) + 8.9 \leq 8.9(2x - 4) - 0.7$

34. $-2 < 3x + 1 < 4$

35. $-3 < 3x - 5 < 4$

36. $-2 \leq 4x + 2 \leq 8$

37. $0 \leq 7x - 2 \leq 6$

38. $-5 < 4x + 5 \leq 5$

39. $-4 \leq 3x + 6 \leq 6$

40. $2 \leq 1 - x \leq 6$

41. $3 < 5 - 2x < 7$

42. $-2 < 4 - 3x \leq 0$

43. $-5 \leq 8 - 2x \leq 0$

44. $0 \leq 1 - 4x < 7$

45. $-4 \leq 3 - 2x \leq -1$

46. $-7 \leq 4 - 2x \leq -4$

47. $0 \leq 2 - 3x \leq 4$

48. $4 < 4 - 3x \leq 6$

Directions Write an inequality to represent the following statements.

Example
A student's score must be below 60 to fail the examination.

Solution
Let x = the student's score. Then the inequality would be
$x < 60$.

49. A student's score must be at least 90 to receive an A on the exam.

50. A student must score at least 75 on the final exam to pass the course.

51. The temperature today will not get above 42.

52. The temperature today will be at least 80.

53. A salesperson needs to sell at least 10 new cars to make a bonus.

54. The temperature today will range from a low of 18 to a high of 41.

55. On a partly sunny day, there will be at least 96 minutes of sunlight but at most 384 minutes of sunlight.

56. The selling price P must be more than the cost c but less than twice the cost.

57. The selling price P must be at least one and one-half times the cost c but at most three times the cost.

Directions Write an inequality using the given information and solve.

Example

Two times a number added to 7 is greater than 19. Find all numbers that satisfy this condition.

Solution

Let x = the number. Then the inequality would be

$2x + 7 > 19$ Subtract 7 from both members.

    $2x > 12$ Divide both members by 2.

      $x > 6$.

Hence $S = \{x \mid x > 6\}$ or $(6, +\infty)$.

58. When 4 is added to three times a number, the result is at least 12. Find all numbers that satisfy this condition.

59. When 6 is subtracted from five times a number, the result is less than 17. Find all numbers that satisfy this condition.

60. Four times a number plus 6 is at least 21. Find all numbers that satisfy this condition.

61. If one-half of a number is added to 16, the result is greater than 24. Find all numbers that satisfy this condition.

62. Three times a number is subtracted from 11 and this result is less than 6. Find all numbers that satisfy this condition.

63. Two times a number is subtracted from 19 and this result is at most 8. Find all numbers that satisfy this condition.

64. A student has scores of 7, 10, and 8 on three quizzes. What must she score on the fourth quiz to have an average of 8 or higher?

65. A student has scores of 66, 71, and 84 on three exams. If an average of 75 is required to pass the course, what is the minimum score he must have on the fourth test to pass?

66. Two times a number minus 6 is greater than 4 but less than 19. Find all numbers that satisfy these conditions.

67. Three times a number plus 2 is greater than 12 but less than 23. Find all numbers that satisfy these conditions.

68. The perimeter of a rectangle must be less than 100 feet. If the length is known to be 30 feet, find all numbers that the width could be. (*Note:* The width of a real rectangle must be a positive number.)

69. The perimeter of a square must be greater than 16 inches but less than 84 inches. Find all values of a side that satisfy these conditions. (*Hint:* The perimeter of a square is given by $P = 4s$, where s represents the length of a side.)

2–5 Equations and inequalities involving absolute value

Absolute value equations

In chapter 1 we defined the absolute value of a number x to be

$$|x| = \begin{cases} x, \text{ if } x \geq 0, \\ -x, \text{ if } x < 0. \end{cases}$$

The absolute value of a number represents the undirected distance from that number to the origin on the number line, that is, the distance from x to 0.

Consider the equation

$$|x| = 2.$$

The right member, 2, denotes the distance that the graph of x is located from the origin. The equation directs us to find all numbers that are 2 units from the origin.

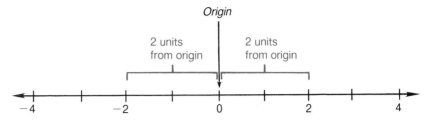

Figure 2.2

We see from figure 2.2 that 2 and -2 are both 2 units from the origin and, therefore, satisfy the equation $|x| = 2$. The solution set is then given as $S = \{-2,2\}$.

■ $|x| = a$
For any real number x and $a \geq 0$,
$$|x| = a \text{ is equivalent to } x = a \text{ or } x = -a.$$

Concept
An equation of the form $|x| = a$, called an **absolute value equation,** is equivalent to the equations $x = a$ or $x = -a$.

Note
Recall that $|x| \geq 0$. We realize from this that a in our generalizations must be nonnegative, $a \geq 0$. This means that in an equation of this type, there is no solution if a is negative. For example, if $|x| = -2$, then the solution set is $\emptyset$.

Example 2 5 A

Find the solution set of the following absolute value equations.

1. $|x| = 5$
We form the two equations equivalent to the absolute value equation

$x = 5$ or $x = -5$,

and the solution set is

$S = \{5, -5\}$.

To check an absolute value equation, simply substitute the solutions into the original equation. In example 1 this would be

for 5	for -5
$|5| = 5$ (true)	$|-5| = 5$. (true)

2. $|x| + 4 = 10$

We must first *isolate* the absolute value in one member of the equation so that we can form our equivalent equations.

$$|x| + 4 = 10$$

Subtract 4 from both members.

$$|x| = 6$$

Form the equivalent equations.

$$x = 6 \text{ or } x = -6$$

Hence the solution set is $S = \{6, -6\}$.

3. $|3a - 2| = 7$

$$\begin{array}{lll} 3a - 2 = 7 \text{ or} & 3a - 2 = -7. & \text{Form the equivalent equations.} \\ 3a = 9 \text{ or} & 3a = -5 & \text{Add 2 to both members of both equations.} \\ a = 3 \text{ or} & a = -\dfrac{5}{3} & \text{Divide both members of both equations by 3.} \end{array}$$

The solution set is $S = \left\{ 3, -\dfrac{5}{3} \right\}$.

4. $|2a - 3| = |a + 4|$

The solution set to this equation will be those values that satisfy the condition that $2a - 3$ and $a + 4$ are either equal to each other or are opposites of each other. Hence we have the equations

Equal to each other

$$2a - 3 = a + 4 \qquad \text{or}$$

Opposites of each other

$$2a - 3 = -(a + 4).$$

Solving each of the equations, we have

$$\begin{array}{ll} 2a - 3 = a + 4 & \text{Subtract } a. \\ a - 3 = 4 & \text{Add 3.} \\ a = 7 \end{array} \qquad \left| \begin{array}{ll} & \text{Remove the} \\ 2a - 3 = -(a + 4) & \text{parentheses.} \\ 2a - 3 = -a - 4 & \text{Add } a. \\ 3a - 3 = -4 & \text{Add 3.} \\ 3a = -1 & \text{Divide by 3.} \\ a = -\dfrac{1}{3}. \end{array} \right.$$

Therefore the solution set is $S = \left\{ 7, -\dfrac{1}{3} \right\}$.

Note

In example 4 the solution set of the absolute value equation is the same whether we take the opposite of the first expression or the opposite of the second expression. That is, our solution set would be the same if we had solved $-(2a - 3) = a + 4$ instead of $2a - 3 = -(a + 4)$.

Absolute value inequalities

If the equals sign, $=$, is replaced with an inequality symbol, $<$, $>$, $\leq$, or $\geq$, an absolute value equation becomes an **absolute value inequality.** Consider the absolute value inequality

$$|x| < 2.$$

This inequality states that the distance between x and the origin is less than 2 units.

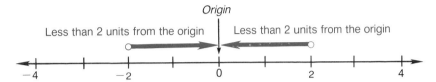

Figure 2.3

We see from figure 2.3 that all numbers between -2 and 2 satisfy the inequality $|x| < 2$. The solution set can be given in any one of the three forms that we studied in section 2–4. That is, in set-builder notation the solution set would be $S = \{x \mid -2 < x < 2\}$, in interval notation $(-2,2)$, and in graphical notation as shown in figure 2.4.

Figure 2.4

We can generalize this observation in the following theorem.

■ $|x| < a$
For any real number x and $a > 0$,
$$|x| < a \text{ is equivalent to } -a < x < a.$$

Concept
Given $|x| < a$, then x will be any real number between the opposite of a and a.

Note
The theorem is stated in terms of the strict inequality $<$, but it is still true if we replace the strict inequality symbol with the weak inequality symbol $\leq$.

Example 2–5 B

Find the solution set of the following absolute value inequalities and leave the answer in (1) set-builder notation, (2) interval notation, and (3) graphical notation.

1. $|x + 3| \leq 6$

To solve this inequality, we form the equivalent compound inequality where the expression inside the absolute value symbol is set greater than or equal to -6 and less than or equal to 6.

$$-6 \leq x + 3 \leq 6$$

Subtract 3 from all three members.

$$-9 \leq x \leq 3$$

Hence the solution set is

$S = \{x \mid -9 \leq x \leq 3\}$ or $[-9,3]$ or

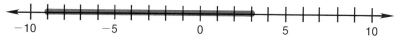

2. $|5 - 2x| \leq 3$

$-3 \leq 5 - 2x \leq 3$ Form the equivalent compound inequality.

$-8 \leq -2x \leq -2$ Subtract 5 from all three members.

Divide all three members by -2 and reverse the direction of all inequality symbols.

$$\frac{-8}{-2} \geq \frac{-2x}{-2} \geq \frac{-2}{-2}$$

$$4 \geq x \geq 1$$

Therefore the solution set is

$S = \{x \mid 1 \leq x \leq 4\}$ or $[1,4]$ or

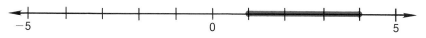

We have only considered those absolute value inequalities that involve less than, $<$, or less than or equal to, $\leq$. Consider the absolute value inequality

$$|x| > 2.$$

This inequality states that the distance between x and the origin is more than 2 units.

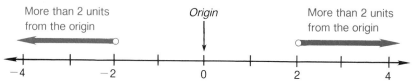

Figure 2.5

We see from figure 2.5 that all numbers greater than 2 or less than -2 satisfy the inequality $|x| > 2$. The solution set is the union of the two intervals, $x > 2$ or $x < -2$, and is given by

$$S = \{x \mid x < -2 \text{ or } x > 2\} \text{ or } (-\infty, -2) \cup (2, +\infty)$$

or as shown in figure 2.6.

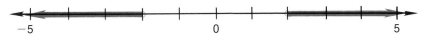

Figure 2.6

We can generalize this observation in the following theorem.

■ $|x| > a$
For any real number x and $a > 0$, $|x| > a$ is equivalent to $x < -a$ or $x > a$.

Concept
Given $|x| > a$, then x will be any real number less than the opposite of a or greater than a.

Note
The theorem is stated in terms of the strict inequality $>$, but it is still true if we replace the strict inequality symbol with a weak inequality symbol. That is, $|x| \geq a$ is equivalent to $x \leq -a$ or $x \geq a$.

Example 2–5 C

Find the solution set of the following absolute value inequalities and leave the answer in (1) set-builder notation, (2) interval notation, and (3) graphical notation.

1. $|3x - 4| \geq 2$
   Form the two equivalent inequalities.

   $3x - 4 \leq -2$ or $3x - 4 \geq 2$

   Add 4 to both members of both inequalities.

   $3x \leq 2$ or $3x \geq 6$

   Divide both members of both inequalities by 3.

   $x \leq \dfrac{2}{3}$ or $x \geq 2$

   Hence the solution set is
   $$S = \left\{x \mid x \leq \frac{2}{3} \text{ or } x \geq 2\right\} \text{ or } \left(-\infty, \frac{2}{3}\right] \cup [2, +\infty) \text{ or}$$

2. $|4x + 3| - 4 > 7$

$|4x + 3| > 11$ Isolate the absolute value by adding 4 to both members.

$4x + 3 < -11$ or $4x + 3 > 11$ Form the two equivalent inequalities.

$4x < -14$ or $4x > 8$ Subtract 3 from both members of both inequalities.

$x < -\dfrac{7}{2}$ or $x > 2$ Divide both members of both inequalities by 4.

Therefore the solution set is

$$S = \left\{ x \mid x < -\frac{7}{2} \text{ or } x > 2 \right\} \text{ or } \left(-\infty, -\frac{7}{2}\right) \cup (2, +\infty) \text{ or}$$

Note

A common error in solving absolute value inequalities is to apply the wrong procedure. Once we have the absolute value isolated in one member of the inequality, we identify the type of inequality and apply the appropriate theorem.

Mastery points
Can you
- Solve an absolute value equation?
- Solve an absolute value inequality?

Exercise 2–5

Directions Find the solution set of the following absolute value equations. See example 2–5 A.

1. $|x| = 9$

2. $|a| = 4$

3. $|b| + 2 = 6$

4. $|x| - 5 = 7$

5. $|x + 4| = 6$

6. $|a - 3| = 2$

7. $|3x - 4| = 8$

8. $|2a + 7| = 9$

9. $|5x - 3| = -4$

10. $|4a + 8| = -7$

11. $|4b - 3| + 2 = 8$

12. $|5a + 2| - 7 = 4$

13. $|2.1x - 6.3| = 8.4$

14. $|1.8x - 10.8| = 5.4$

15. $\left|\dfrac{1}{2}x + 5\right| = 7$

16. $\left|\dfrac{3}{4}a - 2\right| = 6$

17. $\left|\dfrac{2}{3}b - 6\right| = 4$

18. $|5 - 3x| - 4 = 8$

19. $|3 - 4a| + 2 = 5$

20. $|3x - 7| = |5x + 3|$

21. $|2a + 5| = |6a + 7|$

22. $|3 - 2a| = |4a + 6|$

23. $|3y - 4| = |3y + 8|$

24. $|4b - 3| = |4b + 7|$

     First-Degree Equations and Inequalities

Directions Solve the following absolute value inequalities. For exercises 25–34, leave the answer in (1) set-builder notation, (2) interval notation, and (3) graphical notation. For exercises 35–63, leave the answer in both set-builder and interval notation. See examples 2–5 B and C.

25. $|x| < 4$

26. $|x| \leq 3$

27. $|x| \geq 2$

28. $|x| > 4$

29. $|x| - 2 \leq 1$

30. $|x| - 1 < 2$

31. $|x| + 3 > 8$

32. $|x| + 5 \geq 7$

33. $|3x - 4| < 6$

34. $|2x + 5| \leq 4$

35. $|5x - 3| \geq 7$

36. $|4x - 5| > 9$

37. $|3x - 4| \leq 13$

38. $|2x + 7| > 11$

39. $|5x + 7| < 12$

40. $|1 - 2x| \leq 5$

41. $|4 - 3x| \leq 13$

42. $|6 - 3x| < 12$

43. $|5 - 2x| < 15$

44. $|4x - 9| < 0$

45. $|3x + 6| \leq -1$

46. $|5 - 8x| < -3$

47. $|4x - 6| \geq -2$

48. $|3x + 7| > -9$

49. $|7x - 4| + 5 < 7$

50. $|3x - 11| + 6 < 9$

51. $|1 - 3x| - 4 \geq 3$

52. $|2 - 7x| - 3 \geq 4$

53. $|3x - 4| + 5 < 4$

54. $|7x - 8| + 6 \leq 3$

55. $|4x - 9| + 6 \geq 4$

56. $4 + |3x + 1| > 6$

57. $8 + |5x - 3| > 10$

58. $4 + |3 - 2x| \geq 7$

59. $|4.8x - 18.42| > 11.34$

60. $|3.2x - 12.2| > 13.4$

61. $|8.2x - 6.15| \leq 10.25$

62. $|10.8x - 10.8| \leq 5.4$

63. $|2.1x - 6.3| < 8.4$

Directions Write an absolute value inequality for the following statements and solve for the unknown. Leave the answer in both set-builder and interval notation.

Example
The absolute value of a number is more than 3.

Solution
If we let x represent the unknown number, then the absolute value inequality would be
$|x| > 3$.
Changing this to the equivalent inequalities, we have
$x < -3$ or $x > 3$,
and the solution set would be
$S = \{x \mid x < -3 \text{ or } x > 3\}$ or $(-\infty, -3) \cup (3, +\infty)$.

64. The absolute value of a number is less than 4.

65. The absolute value of a number is at most 6.

66. The absolute value of a number is at least 8.

67. The absolute value of twice a number is less than 14.

68. The absolute value of three times a number decreased by 4 is at least 11.

69. The absolute value of twice a number increased by 5 is more than 15.

Chapter summary

1. A **mathematical statement** is a sentence that can be labeled true or false.

2. An **equation** is a statement of equality.

3. A replacement value for the variable that forms a true statement (satisfies that equation) is called a **root,** or **solution,** of that equation.

4. The set of all those values for the variable that causes the equation to be a true statement is called the **solution set** of the equation and is denoted by S.

5. If there is at least one element in the replacement set of the variable that is not in the solution set of the equation, the equation is called a **conditional equation.**

6. An equation that is true for every permissible value of the variable is called an **identical equation,** or **identity.**

7. In a **first-degree conditional equation** in one variable, also called a **linear equation,** the exponent of the unknown is 1 and the solution set will contain at most one root.

8. The **addition property of equality** enables us to add to or subtract the same amount from each member of an equation and the result will be an equivalent equation.

9. The **multiplication property of equality** enables us to multiply or divide both members of an equation by the same nonzero number.

10. Whenever we multiply or divide both members of an inequality by a negative number, we must **reverse the direction of the inequality symbol.**

11. In **interval notation** we use a parenthesis, (or), to represent that the endpoint is *not* part of the interval and a bracket, [or], to represent that the endpoint is part of the interval.

12. If $|x| = a$ and $a \geq 0$, then $x = a$ or $x = -a$.

13. If $|x| < a$ and $a > 0$, then $-a < x < a$.

14. If $|x| > a$ and $a > 0$, then $x < -a$ or $x > a$.

Chapter review

[2–1]
Directions   Find the solution set of the following equations.

1. $4x = 32$

2. $x + 11 = 17$

3. $\dfrac{a}{7} = 4$

4. $\dfrac{5b}{2} = 10$

5. $7a - 2 = 13$

6. $2(3z - 4) = 12$

7. $4(3 - 2x) + 3 = 4x - 5$

8. $4(3a + 2) - 2a = 5(a - 1)$

9. $7.8a - 16.9 = 4.3a + 14.6$

[2–2]
Directions   Solve the following formulas or literal equations for the specified variable. Assume that no denominator is equal to zero.

10. $v = \ell wh;\ w$

11. $v = k + gt;\ t$

12. $D = dq + R;\ d$

13. $v = r^2(a - b);\ b$

14. $2s = 2vt - gt^2;\ v$

15. $\ell = a + (n - 1)d;\ n$

16. $3x - y = 5x - 4y;\ x$

[2–3]
Directions Write an equation for the problem and solve for the unknown quantities.

17. If three times a number is increased by 15 and the answer is 51, what is the number?

18. One-third of a number is 6 less than one-half of the number. Find the number.

19. The sum of three numbers is 27. The second number is three times the first and the third number is 6 less than the first. Find the three numbers.

20. The length of a rectangle is 8 feet more than twice the width. The perimeter is 82 feet. Find the dimensions.

21. Mary Ann has $24,000, part of which she invests at 10% interest and the rest at 8%. If her income for one year from the two investments was $2,220, how much did she invest at each rate?

22. A solution that is 42% hydrochloric acid is to be mixed with a solution that is 12% hydrochloric acid to obtain 100 centiliters of solution that is 24% hydrochloric acid. How many centiliters of each solution should be used in the mixture?

[2–4]
Directions Find the solution set of the following inequalities. Leave the answer in (1) set-builder notation and (2) interval notation.

23. $5x \leq 30$

24. $\frac{3}{4}x > 12$

25. $-2x < 9$

26. $2(3x - 4) \leq 1 - 2x$

27. $10 - 2(3x - 4) > 9 - 12x$

28. $5(2x - 3) \leq 7(x + 1) + 3$

29. $-4 < 2x + 3 < 5$

30. $0 \leq 5x + 4 \leq 4$

31. $5 < 3 - x < 8$

32. $-6 \leq 4 - 3x < 2$

33. $6 < 6 - 4x \leq 10$

Directions Write an inequality using the given information and solve.

34. When 5 is subtracted from four times a number, the result is at least 19. Find all numbers that satisfy this condition.

35. Three times a number plus 7 is greater than 22 but less than 34. Find all numbers that satisfy these conditions.

[2–5]
Directions Find the solution set of the following absolute value equations.

36. $|x| - 4 = 11$

37. $|3a + 5| = 12$

38. $|7 - 2x| = 10$

39. $|4c - 6| + 12 = 18$

40. $|3a + 4| = |2a - 3|$

41. $|4y + 6| = |2y - 5|$

Directions Solve the following absolute value inequalities. Leave the answer in both set-builder and interval notation.

42. $|x| - 4 \geq 6$

43. $|2x + 5| < 6$

44. $|5x - 1| \leq 7$

45. $|4x + 7| > 9$

46. $|1 - 3x| \geq 5$

47. $|3 - 4x| < 12$

48. $|5x + 1| > -3$

49. $6 + |1 - 2x| < 10$

50. $|4x + 5| - 5 \geq 8$

51. $|6x + 3| < -4$

Chapter 2 cumulative test

Directions Determine if the following statements are true or false.

[1-1] **1.** $\frac{1}{2} \in J$ [1-1] **2.** $5 \subseteq R$ [1-1] **3.** $0 \in W$

[1-1] **4.** $Q \cup H = R$ [1-1] **5.** $|-4| < |-6|$ [1-1] **6.** $\{4\} \subseteq \{1,2,3,4\}$

[1-1] **7.** $Q \cap H = \emptyset$

Directions Perform the indicated operations if possible and simplify.

[1-2] **8.** $(-4)(+3)(-2)$ [1-2] **9.** $(-8) - (-7)$ [1-2] **10.** $(-6)(-2)$

[1-2] **11.** $\frac{(-5) + 5}{(-2)}$ [1-2] **12.** $25 - (5 - 11) + 6$ [1-2] **13.** $\frac{(-4)(-9)}{(2)(-3)}$

[1-2] **14.** -7^2

Directions Identify which property of real numbers is being used.

[1-3] **15.** $a(b + c) = (b + c)a$ [1-3] **16.** $(xy)z = z(xy)$

Directions Form the following sets.

[1-1] **17.** $\{1,2,3\} \cup \{2,3,4,5\}$ [1-1] **18.** $\{8,10,11\} \cup \{4,6,9\}$ [1-1] **19.** $\{10,11,12,13\} \cap \{10,12\}$

Directions Solve the following literal equations and find the solution set for the equations and inequalities. Assume that no denominator is equal to zero.

[2-1] **20.** $6x + 5 = 2x + 12$ [2-2] **21.** $M = -P(\ell - x); P$

[2-2] **22.** $P = n(P_2 - P_1) - c; P_2$ [2-4] **23.** $-3x \le 15$

[2-5] **24.** $|4x + 6| \ge 5$ [2-1] **25.** $5(2 - 3x) + 4 = 2x - 7$

[2-4] **26.** $7(3 - 4x) + 6 \le 6(2 - 4x)$ [2-5] **27.** $|2a + 3| = |3a + 2|$

[2-1] **28.** $6(3a - 5) - 4a = 7(a - 2)$ [2-1] **29.** $3(2x - 4) + 6 = 6x + 8$

[2-5] **30.** $|4 - 3x| \le 7$

[2-3] **31.** The sum of three numbers is 72. The first number is twice the second number and the third number is three times the first number. Find the three numbers.

[2-3] **32.** Harold has $40,000, part of which he invests at 11% and the rest at 8%. If his income for one year from the 11% investment is $1,740 more than that from the 8% investment, how much did he invest at each rate?

Chapter 3

Exponents and Polynomials

3-1 Sums and differences of polynomials

Like terms

In section 1–5 we learned by the process of evaluation that a polynomial is a symbol representing a real number. Therefore the ideas and properties that apply to operations with real numbers also apply to polynomials. Let us now examine the operations of addition and subtraction of polynomials.

Recall that we can add or subtract only like terms and that like terms can differ only in their numerical coefficients (numerical factors).

Example 3–1 A   Perform the indicated addition or subtraction.

1. $5xy^2 + 3xy^2 = (5 + 3)xy^2 = 8xy^2$

2. $3a^2b + 5a^2b + 2a^2b = (3 + 5 + 2)a^2b = 10a^2b$

Note
The numerical coefficient of a term includes the sign that precedes it. Therefore we consider any addition and subtraction of terms as a sum of terms in which the sign that precedes the term is taken as the sign of the numerical coefficient.

like terms

3. $x^3y + 4xy^2 - 3x^3y + 2xy^2 = [1 + (-3)]x^3y + (4 + 2)xy^2$

like terms

$$= -2x^3y + 6xy^2$$

Removing grouping symbols

We learned in chapter 1 that any quantity enclosed within a grouping symbol is treated as a single number. Now we are going to use the distributive property to remove grouping symbols such as ( ), [ ], and { }. Consider the following examples:

1. The quantity $(2x + y)$ can be written as $1 \cdot (2x + y)$ because if there is no numerical coefficient, then 1 is understood to be the coefficient. Applying the distributive property:
$$1(2x + y) = 1 \cdot 2x + 1 \cdot y = 2x + y.$$

2. The quantity $+(2x + y)$ can be written as $+1 \cdot (2x + y)$, giving
$$+1 \cdot (2x + y) = (+1) \cdot 2x + (+1) \cdot y = 2x + y.$$

3. The quantity $-(2x + y)$ can be written as $-1 \cdot (2x + y)$, giving
$$-1 \cdot (2x + y) = (-1) \cdot 2x + (-1) \cdot y = -2x - y.$$

We observe from examples 1 and 2 that *if an expression inside a grouping symbol is preceded by no symbol or by a + sign, the grouping symbol can be dropped and the enclosed terms remain unchanged.* From example 3 we observe that *if an expression inside a grouping symbol is preceded by a − sign, then when the grouping symbol is dropped, we change the sign of each enclosed term.*

Example 3–1 B   Remove all grouping symbols and perform the indicated addition or subtraction.

1. $(3x^2 + 2xy - 3y^2) + (x^2 - 5xy + 2y^2)$
$= 3x^2 + 2xy - 3y^2 + x^2 - 5xy + 2y^2$
$= (3x^2 + x^2) + (2xy - 5xy) + (-3y^2 + 2y^2)$
$= 4x^2 - 3xy - y^2$

Note
We can carry out the addition of these same two polynomials by lining them up in columns such that the like terms appear in the same columns.

$$
\begin{array}{l}
(3x^2 + 2xy - 3y^2) \\
+\ (\ x^2 - 5xy + 2y^2) \\
\end{array}
\qquad
\begin{array}{l}
3x^2 + 2xy - 3y^2 \\
\underline{x^2 - 5xy + 2y^2} \\
4x^2 - 3xy - y^2
\end{array}
$$

2. $(2x^2 - 7x + 6) - (x^2 - 5x + 9) = 2x^2 - 7x + 6 - x^2 + 5x - 9$
$= (2x^2 - x^2) + (-7x + 5x) + (6 - 9)$
$= x^2 - 2x - 3$

Note
This example may also be worked in the column form, but remember that when we remove the second pair of parentheses, we must change the sign of each enclosed term and then combine the like terms.

$$
\begin{array}{l}
(2x^2 - 7x + 6) \\
-\ (\ x^2 - 5x + 9) \\
\end{array}
\qquad
\begin{array}{l}
2x^2 - 7x + 6 \\
\underline{-x^2 + 5x - 9} \\
x^2 - 2x - 3
\end{array}
$$

3. $(5a^2 + 3a^2b - 3b^2) - (2a^2 - 3ab^2 - 4b^2)$
$= 5a^2 + 3a^2b - 3b^2 - 2a^2 + 3ab^2 + 4b^2$
$= (5a^2 - 2a^2) + 3a^2b + 3ab^2 + (-3b^2 + 4b^2)$
$= 3a^2 + 3a^2b + 3ab^2 + b^2$

By the column method we have

$$
\begin{array}{l}
(5a^2 + 3a^2b - 3b^2) \\
-\ (2a^2 - 3ab^2 - 4b^2) \\
\end{array}
\qquad
\begin{array}{l}
5a^2 + 3a^2b - 3b^2 \\
-\ 2a^2 + 3ab^2 + 4b^2
\end{array}
$$

$$
\begin{array}{l}
5a^2 + 3a^2b - 3b^2 \\
-\ 2a^2 + 3ab^2 + 4b^2 \\
\hline
3a^2 + 3a^2b + 3ab^2 + b^2
\end{array}
$$
rearranging so that like terms appear in the same columns.

Note
In future examples we will mentally regroup the like terms and use the horizontal method to add and subtract polynomials.

There are many situations in which there will be grouping symbols within grouping symbols. In these cases *it is usually easier to remove the innermost grouping symbol first.*

Example 3–1 C

Remove all grouping symbols and perform the indicated addition or subtraction.

1. $2a^2 - [a + (a^2 - 3a)] = 2a^2 - [a + a^2 - 3a]$
$$= 2a^2 - [a^2 - 2a]$$
$$= 2a^2 - a^2 + 2a$$
$$= a^2 + 2a$$

After removing the parentheses, we combine the like terms before removing the brackets. *Simplify within a group whenever possible.*

2. $3x - \{6x - [y - (3x + 2y)] + 2x\}$
$$= 3x - \{6x - [y - 3x - 2y] + 2x\}$$
$$= 3x - \{6x - [-y - 3x] + 2x\}$$
$$= 3x - \{6x + y + 3x + 2x\}$$
$$= 3x - \{11x + y\}$$
$$= 3x - 11x - y$$
$$= -8x - y$$

Mastery points
Can you
• Identify like terms?
• Perform addition and subtraction of polynomials?
• Remove grouping symbols?

Exercise 3–1

Directions Perform the indicated addition and subtraction. See example 3–1 A.

1. $4x + 6x^2 - 2x^2 + 3x - 5x^2 + x$

2. $5y^2 - 12y + 6y^5 - 4y^2 + y$

3. $-a^2 + a - 5a^3 + 2a^2 + 4a$

4. $2x^2y - 4xy^2 + 3x^2y$

5. $5x^2y - 3xy + 6xy - x^2y$

6. $6a^2b - 2ab^2 + 3a^2b - 4a^2b^2$

Directions Remove all grouping symbols and combine like terms. See example 3–1 B.

7. $(3x^2y - 2xy^2 + 9xy) + (2xy^2 - 4x^2y + 3xy)$

8. $(5ab^2 - 2a^2b^2 + 3a) - (4a^2b^2 + 3ab^2)$

9. $(5a^2 - b^2) - (4a^2 + 3b^2) - (a^2 - 7b^2)$

10. $(7x^3 - 2x^2y + 4xy^2 - 6y^3) - (4x^3 - 3x^2y - 2xy^2 - y^3)$

11. $(12x - 24yz) + (46yz - 16x - 26z)$

12. $(19x + 3y) - (-22x - 6y)$

13. $-(5a^2b - 6ab + 16c) + (7ab - 4a^2b)$

14. $(6ab + 11b^2c) - (11ab - 6bc)$

15. $-(14x - 31y) - (14x - 6y + 3z)$

16. $(x^2 - 3x + z) - (x^2 + 5x - 8) + (3x^2 - 4x)$

17. $(2x^2 - 7x + 3) + (5x^2 - 6) - (4 - 3x^2)$

18. $(7x^2 - 2xy + y^2) + (3x^2 - 6y^2 + 3xy) - (4y^2 - 6x^2 - xy)$

19. $(5xy - y^2) - (3yz + 2xy) + (3y^2 - 4xy)$

20. $(5ab + 2b^2) - (4a^2 - 3b^2) + (2ab - 7b^2)$

Directions Perform the following addition and subtraction in column form. See example 3–1 B.

21.
$(3x^2 + 4xy - 5y^2)$
$+ \underline{(\ x^2 - 7xy + 3y^2)}$

22.
$(4a^2 + 2ab + 3b^2)$
$+ \underline{(\ a^2 - 5ab + \ b^2)}$

23.
$(3x^2y - 2x^2y^2 + 3xy^2)$
$- \underline{(2x^2y + xy - 5xy^2)}$

24.
$(4t^2s + 3ts - 2ts^2)$
$- \underline{(2t^2s + 3ts - 2ts^2)}$

25.
$(-6a^2b^2 + 3ab - 4)$
$- \underline{(-4a^2b^2 + 2ab - 7)}$

26.
$(-9x^2y^2 - 4xy + 13)$
$- \underline{(-11x^2y^2 + 5xy - 6)}$

Directions Set up the following problems and perform the indicated addition and subtraction.

Example
Subtract $5a^2 - 2a + 4$ from $6a^2 - 7a + 3$.

Solution
$(6a^2 - 7a + 3) - (5a^2 - 2a + 4)$
$= 6a^2 - 7a + 3 - 5a^2 + 2a - 4$
$= a^2 - 5a - 1$

27. Subtract $3x^2 - 2x + 1$ from $5x^2 + 3x - 7$.

28. Subtract $2a^2 - 7a + 3$ from $a^2 - 2a + 4$.

29. Subtract $5a^2 - 6a + 7$ from $5a^2 + 2a - 3$.

30. Subtract $-3t^2 + 4t + 5$ from $8t^2 - 11t + 12$.

31. Subtract $-2x^2 + 3x - 7$ from $7x^2 + 9x - 4$.

32. From $6t^2 - 7t + 14$, subtract $8t^2 - 11t + 6$.

33. From $-7y^2 + 3y + 11$, subtract $2y^2 - 13y + 3$.

34. From $6z^2 + 5z - 4$, subtract $-4z^2 + 3z - 9$.

35. From $4a^2 + 7a - 12$, subtract $-8a^2 + 7a - 5$.

36. Subtract $2x^2 - 9x + 4$ from the sum of $6x^2 + 3x$ and $5x^2 - 4x + 2$.

37. Subtract $a^2 - 7a + 11$ from the sum of $3a^2 - 4$ and $-5a^2 + 6a$.

38. Subtract $-4t^2 - 3t + 11$ from the sum of $8t^2 - 6$ and $-5t^2 + 11t + 5$.

39. Subtract $6xy - y^2$ from the sum of $5x^2 + 3xy$ and $2x^2 - 7xy - y^2$.

40. From the sum of $8a^2 + 11$ and $2a^2 - 7a + 6$, subtract $4a^2 - a - 1$.

41. From the sum of $5x^2 - 11x - 7$ and $-2x^2 + 3x - 4$, subtract $3x^2 - 7x + 5$.

42. From the sum of $-3t^2 + 2t - 6$ and $-5t^2 - 4t - 8$, subtract $-8t^2 + 2t + 4$.

43. From the sum of $-t^2 - 7t + 1$ and $-3t^2 + 4t - 8$, subtract $-4t^2 - 6t - 5$.

Directions Remove all grouping symbols and combine like terms. See example 3–1 C.

44. $5a - [3a - (2a - 3)]$

45. $a - 4 + [3a - (2a + 1)]$

46. $4x + [3x - (x + y)]$

47. $7x - [4x + 3y + (x - 2y)]$

48. $3x - [x - y - (7x - 3y)]$

49. $-(3x - 2y) - (x + y) - [2x - y + (3x - 4y)]$

50. $a - [b + (2a - 4b)] + [5a - 3b]$

51. $(2x - 7y) - \{3x - [4y - (2x + 5y)]\}$

52. $6x^2 - [5x + 3x^2 - (4x^2 - 7x)]$

53. $a - \{a - [a - (2a - b) + 3a]\}$

54. $-\{5x - 3y + [2x - (5x - 7y)] + 4y\}$

55. $(3x + x^2) - \{4x^2 - 3x + [2x^2 - x - (5x^2 + 4x)]\}$

56. $2x + [x - (x^2 - y)] - [2x - (x^2 + y)]$

57. $5a - [6a - (2b - 3c)] - [4b - 3a]$

58. $5a - \{(ab + 4c) + 8b - [7a - (3b - 5c)] + 2a\}$

59. $5x^2 - [3x - (2x^2 + x)] - \{x - [3x^2 - (2x^2 + 3x) - x]\}$

60. $-[-3a^2 + (4a - 3) + 5a] - \{7a^2 + [4a - (a^2 - 3) + 5a^2]\}$

61. $-\{4x^2 - (5x + 3) - 2x\} + \{3x^2 - [7x^2 + (2x - 4) - 5x]\}$

62. $4x^2y - \{3xy^2 - [7x^2y^2 + 3xy^2 - (11x^2y + 2x^2y^2)] - 5x^2y\}$

63. $(9x^2y^2 - 4xy^2) - \{8x^2y^2 - [2xy^2 - (-3x^2y + 7x^2y^2)]\}$

Directions If $P(x) = 2x^2 - x + 3$, $Q(x) = 5x - 4$, and $R(x) = 3x^2 + 4x - 7$, express each of the following in terms of x and perform the indicated operations. See example 1–5 F.

Example

$[P(x) - Q(x)] + R(x)$

Solution

$= [(2x^2 - x + 3) - (5x - 4)] + (3x^2 + 4x - 7)$

$= [2x^2 - x + 3 - 5x + 4] + 3x^2 + 4x - 7$

$= [2x^2 - 6x + 7] + 3x^2 + 4x - 7$

$= 2x^2 - 6x + 7 + 3x^2 + 4x - 7$

$= 5x^2 - 2x$

64. $P(x) + Q(x) + R(x)$

65. $P(x) - Q(x) + R(x)$

66. $P(x) - [Q(x) + R(x)]$

67. $Q(x) - [P(x) + R(x)]$

68. $R(x) - [Q(x) - P(x)]$

69. $[Q(x) - R(x)] - P(x)$

70. $-R(x) + [Q(x) - P(x)]$

71. $-P(x) + [Q(x) - R(x)]$

3–2 Properties of exponents

Exponential form

In chapter 1 we discussed exponents as related to real numbers. Since variables are symbols for real numbers, we shall now apply the properties of exponents to them. The expression a^3 (read "a to the third power") is called the **exponential form** of the product

$$a \cdot a \cdot a.$$

We call a the **base** and 3 the **exponent**.

Exponents and Polynomials

> **■ Definition of an exponent**
>
> $$a^n = \underbrace{a \cdot a \cdot a \cdots a,}_{n \text{ factors of } a}$$
>
> where n is a positive integer.
>
> **Concept**
>
> The exponent tells us how many times the base is used as a factor in an indicated product.

Note

An exponent acts only on the symbol immediately to its left. That is, in xy^3 the exponent 3 applies only to the base y, whereas $(xy)^3$ would mean the exponent applies to both the x and the y.

Example 3–2 A

Write each expression in exponential form and identify the base and the exponent.

1. $x \cdot x \cdot x \cdot x = x^4$ Base x, exponent 4

2. $4 \cdot 4 \cdot 4 = 4^3$ Base 4, exponent 3

3. $(x^2 + y)(x^2 + y)(x^2 + y) = (x^2 + y)^3$ Base $(x^2 + y)$, exponent 3

4. $(-3) \cdot (-3) \cdot (-3) \cdot (-3) = (-3)^4$ Base -3, exponent 4

5. $-(3 \cdot 3 \cdot 3 \cdot 3) = -3^4$ Base 3, exponent 4

Note

In examples 4 and 5 we review the ideas of exponents related to signed numbers. Recall that $(-3)^4 = 81$, whereas $-3^4 = -81$.

Multiplication property of exponents

Consider the indicated product of $a^2 \cdot a^3$. If we rewrite a^2 and a^3 by using the definition of exponents, we have

$$a^2 \cdot a^3 = \overbrace{a \cdot a}^{a^2} \cdot \overbrace{a \cdot a \cdot a}^{a^3},$$

and again using the definition of exponents, this becomes

$$a^2 \cdot a^3 = \overbrace{a \cdot a \cdot a \cdot a \cdot a}^{5 \text{ factors of } a} = a^5.$$

This leads us to the observation that $a^2 \cdot a^3 = a^{2+3} = a^5$.

> **■ Multiplication property of exponents**
> For all real numbers a and positive integers m and n,
> $$a^m \cdot a^n = a^{m+n}.$$
>
> **Concept**
> When multiplying factors having **like bases,** add the exponents to get the exponent of the common base.

Example 3–2 B

Find each of the following products.

1. $y^4 \cdot y^5 = y^{4+5} = y^9$

2. $3^2 \cdot 3^4 = 3^{2+4} = 3^6 = 729$

> **Note**
> When we multiply expressions of the same base, we add the exponents; we do not multiply the bases.
> $$3^2 \cdot 3^4 \neq 9^6$$

3. $x \cdot x^7 \cdot x^3 = x^{1+7+3} = x^{11}$

> **Note**
> If there is no visible exponent associated with a numeral or a variable, the exponent is understood to be 1.

4. $x^{3n} \cdot x^{2n}$
We multiply like bases by adding the exponents.
$= x^{3n+2n}$
$= x^{5n}$

Power of a power property of exponents

A second property of exponents can be derived by applying the definition of exponents and the multiplication property of exponents. Consider the expression $(a^4)^3$.

$$(a^4)^3 = \underbrace{a^4\, a^4\, a^4}_{\substack{\text{3 factors} \\ \text{of } a^4}} = \underbrace{a^{4+4+4}}_{\substack{\text{adding the} \\ \text{exponent 4} \\ \text{three times}}} = a^{12}$$

From arithmetic we know that multiplication is repeated addition of the same number. Therefore adding the exponent 4 three times is the same as $3 \cdot 4$. Thus

$$(a^4)^3 = a^{3 \cdot 4} = a^{12}.$$

Exponents and Polynomials

> ■ **Power of a power property of exponents**
> For all real numbers a and positive integers m and n,
> $$(a^m)^n = a^{nm}.$$
>
> **Concept**
> A power of a power is found by multiplying the exponents.

Example 3–2 C

Perform the indicated operations.

1. $(x^3)^5 = x^{5 \cdot 3} = x^{15}$

2. $(2^3)^4 = 2^{4 \cdot 3} = 2^{12} = 4{,}096$

3. $(a^{2n})^{3n}$
Multiply the exponents.
$= a^{3n \cdot 2n}$
$= a^{6n^2}$

Group of factors to a power property of exponents

A third property of exponents can be derived using the definition of exponents and the commutative and associative properties of multiplication. Observe that

$$(ab)^3 = ab \cdot ab \cdot ab$$
$$\underbrace{}_{\text{3 factors of } ab}$$

$$= \underbrace{a \cdot a \cdot a}_{\substack{\text{3 factors} \\ \text{of } a}} \cdot \underbrace{b \cdot b \cdot b}_{\substack{\text{3 factors} \\ \text{of } b}}$$

$$= a^3 b^3.$$

> ■ **Group of factors to a power property of exponents**
> For all real numbers a and b and positive integers n,
> $$(ab)^n = a^n b^n.$$
>
> **Concept**
> When a group of factors is raised to a power, we will raise each of the factors in the group to this power.

Example 3–2 D

Perform the indicated operations.

1. $(xy)^5 = x^5 y^5$

2. $(2ab)^4 = 2^4 a^4 b^4 = 16a^4 b^4$

Note
A common error is to forget to raise the numerical coefficient to the appropriate power, in example 2, 2 is raised to the fourth power.

3–2 Properties of Exponents

3. $(2 + 3)^3 = (5)^3 = 125$

Note

The quantity $(2 + 3)^3 \neq 2^3 + 3^3$ because 2 and 3 are *terms,* not *factors,* as the property requires. If we consider $(2 + 3)$ to be a single number, then by the definition of exponents we have

$$(2 + 3)^3 = (2 + 3)(2 + 3)(2 + 3)$$

or, in general,

$$(a + b)^3 = (a + b)(a + b)(a + b).$$

We will discuss the method of multiplying this product in section 3–3.

The following examples illustrate some problems in which more than one property is used within the problem.

Example 3–2 E

Perform the indicated operations.

1. $(2x^3y^4)^3 = 2^3(x^3)^3(y^4)^3 = 8x^9y^{12}$

2. $(3a^4b^6)^2 = 3^2(a^4)^2(b^6)^2 = 9a^8b^{12}$

3. $(-3a^2)(2ab^3)(-4a^3b^5) = [(-3)(2)(-4)]\,(a^2aa^3)(b^3b^5) = 24a^6b^8$

4. $(3x^2y^5)^2(-2x^4y)^3 = 3^2(x^2)^2(y^5)^2\,(-2)^3(x^4)^3y^3$

$\qquad\qquad = 9x^4y^{10} \cdot (-8)x^{12}\,y^3$

$\qquad\qquad = [9 \cdot (-8)]\,(x^4x^{12})(y^{10}y^3)$

$\qquad\qquad = -72x^{16}y^{13}$

5. $(3x^2)^2x^3 + (2x^3)^2x = 3^2(x^2)^2x^3 + 2^2(x^3)^2x$

$\qquad\qquad = 9x^4x^3 + 4x^6x$

$\qquad\qquad = 9x^7 + 4x^7$

$\qquad\qquad = 13x^7$

6. $(5a^3)^2a^2 + (3a^4)^2a = 5^2(a^3)^2a^2 + 3^2(a^4)^2a$

$\qquad\qquad = 25a^6a^2 + 9a^8a$

$\qquad\qquad = 25a^8 + 9a^9$

Note

In example 5 we were able to carry out the addition because we had like terms. In example 6 the addition was not performed because a^8 and a^9 are not like terms.

Exponents and Polynomials

Exercise 3–2

Directions Write each expression in exponential form and then identify the base and the exponent. See example 3–2 A.

1. $(-2)(-2)(-2)(-2)$

2. $5 \cdot 5 \cdot 5 \cdot 5 \cdot 5 \cdot 5$

3. $x \cdot x \cdot x \cdot x \cdot x$

4. $y \cdot y \cdot y$

5. $(2x)(2x)(2x)(2x)$

6. $(ab^2)(ab^2)(ab^2)$

7. $(x^2 + 3y)(x^2 + 3y)(x^2 + 3y)$

8. $(2a^2 - b)(2a^2 - b)$

Directions Perform the indicated operations. Assume all variables used as exponents represent positive integers. See examples 3–2 B, C, D, and E.

9. $a^5 \cdot a^4$

10. $x^3 \cdot x^9$

11. $y \cdot y^2$

12. $b \cdot b^4$

13. $(-2)^3(-2)^3$

14. $(-3)(-3)^3$

15. $(-2)(-2)^3$

16. $(-2)(-2^2)$

17. $(-2^2)(3^2)$

18. $(-3^2)(2^2)$

19. $x^2 \cdot x \cdot x^5$

20. $y^4 \cdot y^2 \cdot y$

21. $(x^2y^2)(x^5y^3)$

22. $(a^5b)(a^2b^3)$

23. $(2ab^2)(3a^2)$

24. $(-2a^2)(4a^5)$

25. $(-3x^3)(-2x^2)$

26. $(5ab^2)(2a^5b^4)$

27. $(8x^2y^5)(3xy^4)$

28. $(-6a^3b^7)(4ab^4)$

29. $(2^2)^3$

30. $(-2^2)^3$

31. $(-3^2)^3$

32. $(-3^3)^2$

33. $(-2^3)^2$

34. $(x^4)^7$

35. $(a^2)^6$

36. $(a^3b^6)^4$

37. $(x^2yz^3)^5$

38. $(5R^2S^5)^2$

39. $(-7s^4t^2)^2$

40. $(-3a^9b^7)^3$

41. $(-x^9y^{12}z^8)^4$

42. $(3x^2y)^2(2xy^3)$

43. $(a^2b^3)^4(ab^5)$

44. $(2a^3b^2)(a^5b^3)^4$

45. $(x^2y^6)^2(x^3y^4)^3$

46. $(-a^2b)(a^5b^2)^3$

47. $(x^4y^5)^2(-x^2y^8)$

48. $(-2a^2b^4)^3(-3ab^5)^2$

49. $(-5x^4y^2)^2(-3x^2y^7)$

50. $(2a^2)^2a^3 + (3a)^3a^4$

51. $(3x^3)^2x^3 + 2x^5(2x^2)^2$

52. $(5b^2)^22b^7 - (3b^4)^25b^3$

53. $(4a^5)^22a^3 - (3a^4)^34a$

54. $(-2x^2)^33x^4 + (5x^5)^2$

55. $(3x^5)^2 - (2x^2)^3x^2$

56. $(x^4)^2x^3 + (x^3)^3x$

57. $(2a^3)^33a^4 + (3a^5)^22a$

58. $(b^5)^34b^3 - (2b^3)^4b^2$

59. $(x^4)^33x^5 + (2x^2)^53x^4$

60. $x^{5n} \cdot x^{4n}$

61. $a^{2b} \cdot a^{7b}$

62. $x^{6n} \cdot x^n$

63. $a^{5b} \cdot a^{4b}$

64. $x^{2n + 1} \cdot x^{n + 4}$

65. $a^{2b + 5} \cdot a^{3b - 2}$

66. $R^{3S} \cdot R^{2S + 3}$

67. $x^{y - 4} \cdot x^{2y + 9}$

68. $(a^{3n})^{4n}$

69. $(x^{3y})^{5y}$

70. $(R^{2S})^S$

71. $(x^{3n})^{n + 1}$

72. $(a^{2b})^{b + 3}$

73. $(x^{n + 1})^{n + 2}$

74. $(a^{2b + 1})^{b + 2}$

75. The amount A accumulated in a savings account earning 12% interest per year compounded monthly is given by $A = P(1.01)^n$, where P represents the amount of deposit and n is the number of months the money is left on deposit. Find A, if $P = 5,000$ and $n = 12$.

76. Find A in exercise 75, if $P = 2,000$ and $n = 6$.

77. The amount A of a radioactive substance remaining after time t can be found using the formula $A = A_0(0.5)^{\frac{t}{n}}$, where A_0 represents the

original amount of radioactive material and n is the half-life given in the same units of time as t. If the half-life of nitrogen 13 is 10 minutes, how much radioactive nitrogen will remain after 20 minutes if we start with 10 grams?

78. If the half-life of uranium 229 is 58 minutes, how much radioactive uranium will remain after 174 minutes if we start with 80 ounces? (Refer to exercise 77.)

3–3 Products of polynomials

Extended distributive property

In chapter 1 we stated the distributive property as

$$a(b + c) = ab + ac.$$

Many problems have more than two terms inside the grouping symbol. We will now extend the distributive property to more than two terms and will use subscripts to state the *extended distributive property of multiplication over addition.*

> ■ **Extended distributive property**
> $$a(b_1 + b_2 + \cdots + b_n) = ab_1 + ab_2 + \cdots + ab_n.$$
>
> **Concept**
> When we multiply a multinomial by a monomial, we multiply each term of the multinomial by the monomial.

Example 3–3 A

Perform the indicated multiplication.

1. $2x^3(3x^2 - 5x + 7)$
We multiply the monomial $2x^3$ times each term in the trinomial to get

$$(2x^3)(3x^2) + (2x^3)(-5x) + (2x^3)(7).$$

In each indicated product, we multiply the coefficient and add the exponents of the like bases to get

$$2x^3(3x^2 - 5x + 7) = 6x^5 - 10x^4 + 14x^3.$$

2. $4a^2b^3(2a^2 - 3ab + 4b^2) = (4a^2b^3)(2a^2) + (4a^2b^3)(-3ab) + (4a^2b^3)(4b^2)$
$$= 8a^4b^3 - 12a^3b^4 + 16a^2b^5$$

Multiplication of multinomials

The product of two binomials will require the use of the distributive property several times. That is, in the product

$$(x + y)(2x + y),$$

we consider $(x + y)$ as a single factor and apply the distributive property.

$$(x + y)(2x + y) = (x + y) \cdot 2x + (x + y) \cdot y$$

We now apply the distributive property again.

$$(x + y) \cdot 2x + (x + y) \cdot y = x \cdot 2x + y \cdot 2x + x \cdot y + y \cdot y$$
$$= 2x^2 + 2xy + xy + y^2$$

The last step in the problem is to combine like terms, if there are any.

$$2x^2 + (2xy + xy) + y^2 = 2x^2 + 3xy + y^2$$

Notice that in this product each term of the first factor is multiplied by each term of the second factor. We can generalize our procedure as follows:

When multiplying two multinomials, we multiply each term of the first multinomial by each term of the second multinomial and then combine like terms.

Example 3–3 B

Perform the indicated multiplication and simplify.

1. $(a + 4)(a + 1)$

When we perform the multiplication of two binomials, the four products that are obtained can be seen more clearly by using arrows to indicate the multiplication that is being carried out.

$$(a + 4)(a + 1) = a \cdot a + a \cdot 1 + 4 \cdot a + 4 \cdot 1$$

$$= a^2 + a + 4a + 4$$
$$= a^2 + 5a + 4 \qquad \text{Combine like terms.}$$

Note
We have drawn arrows to indicate the multiplication that is being carried out. This is a convenient way for us to indicate the multiplication to be performed.

2. $(2x + 3)(5x - 2) = 10x^2 - 4x + 15x - 6$

$$= 10x^2 + 11x - 6$$

Note
A word that is useful for remembering the multiplication to be performed when multiplying two binomials is **FOIL.** Foil is an abbreviation signifying **F**irst times first, **O**uter times outer, **I**nner times inner, and **L**ast times last.

3. $(3a - 2b)(2a - 5b) = 6a^2 - 15ab - 4ab + 10b^2$

$$= 6a^2 - 19ab + 10b^2$$

In arithmetic we multiply numbers stated vertically. We can use this same procedure to multiply two multinomials. To perform the example 3 multiplication vertically, we would proceed as follows:

$$
\begin{array}{r}
2a - 5b \\
\underline{3a - 2b.}
\end{array}
$$

Multiply $-2b$ and $2a - 5b$.

$$
\begin{array}{r}
2a - 5b \\
\underline{3a - 2b} \\
-4ab + 10b^2
\end{array}
$$

Multiply $3a$ and $2a - 5b$. Line up any like terms in the same columns.

$$
\begin{array}{r}
2a - 5b \\
\underline{3a - 2b} \\
-4ab + 10b^2 \\
\underline{6a^2 - 15ab \qquad}
\end{array}
$$

Add like terms.

$$
\begin{array}{r}
2a - 5b \\
\underline{3a - 2b} \\
-4ab + 10b^2 \\
\underline{6a^2 - 15ab \qquad} \\
6a^2 - 19ab + 10b^2
\end{array}
$$

Special products

Three special products appear so often that we should be able to write the answer without computation. Consider the product

$$(a + b)^2 = (a + b)(a + b),$$

which becomes

$$a^2 + ab + ab + b^2,$$

and combining like terms, we get

$$a^2 + 2ab + b^2.$$

This is called the **square of a binomial** and has certain characteristics. Inspection shows us that in

$$(a + b)^2 = a^2 + 2ab + b^2,$$

the three terms of the product can be obtained in the following manner:

1. The first term of the product is the *square of the first term* of the binomial $[(a)^2 = a^2]$.
2. The second term of the product is *two times the product of the two terms of the binomial* $[2(a \cdot b) = 2ab]$.
3. The third term of the product is the *square of the second term* of the binomial $[(b)^2 = b^2]$.

■ **Square of a binomial**
For real numbers a and b,
$$(a + b)^2 = a^2 + 2ab + b^2.$$

Concept
$$(1\text{st term} + 2\text{nd term})^2 =$$
$$(1\text{st term})^2 + 2(1\text{st term} \cdot 2\text{nd term}) + (2\text{nd term})^2$$

Example 3–3 C

Perform the indicated multiplication and simplify.

1. $(x + 3)^2$
The first term of the binomial is x and the second term of the binomial is 3. Substituting into the formula, we have

$$(x + 3)^2 = (x)^2 + 2(x \cdot 3) + (3)^2$$

1st   1st 2nd   2nd
term   term term term
$$= x^2 + 6x + 9.$$

Note
$(x + 3)^2 = x^2 + 6x + 9$, *not* $x^2 + 9$. This is a common error.

2. $(2a + b)^2 = (2a)^2 + 2(2a \cdot b) + (b)^2$
$$= 4a^2 + 4ab + b^2$$

3. $(5x + 4y)^2 = (5x)^2 + 2(5x)(4y) + (4y)^2$
$$= 25x^2 + 40xy + 16y^2$$

Our second special product is

$$(a - b)^2 = (a - b)(a - b),$$

which becomes

$$a^2 - ab - ab + b^2$$

and simplifies to

$$a^2 - 2ab + b^2.$$

This also is called the square of a binomial. We can apply the previous procedure to this special product if we take the sign of the operation ($+$ or $-$) as the sign of the coefficient.

$$(a - b)^2 = [a + (-b)]^2 = (a)^2 + 2[a \cdot (-b)] + (-b)^2$$
$$= a^2 - 2ab + b^2$$

Example 3–3 D

Perform the indicated multiplication and simplify.

1. $(2x - y)^2$

The first term of the binomial is $2x$ and the second term of the binomial is $-y$. Substituting into the formula, we have

$$(2x - y)^2 = (2x)^2 + 2[(2x)(-y)] + (-y)^2$$

$$\begin{array}{cccc} \uparrow & \uparrow & \uparrow & \uparrow \\ \text{1st} & \text{1st} & \text{2nd} & \text{2nd} \\ \text{term} & \text{term} & \text{term} & \text{term} \end{array}$$

$$= 4x^2 - 4xy + y^2.$$

2. $(4a - 3b)^2 = (4a)^2 + 2[(4a)(-3b)] + (-3b)^2$
$$= 16a^2 - 24ab + 9b^2$$

The third special product is obtained by multiplying the sum and difference of the same two terms. Consider the product

$$(a + b)(a - b) = a^2 - ab + ab - b^2$$
$$= a^2 - b^2.$$

Special characteristics are evident in this product also; namely, the product may be obtained by

1. squaring the first term of the factors, and
2. subtracting the square of the second term of the factors.

We can consider b or $-b$ the second term since the square of each is b^2. That is,

$$(b)^2 = (-b)^2 = b^2.$$

This special product is called the **difference of two squares.**

> ■ **Difference of two squares**
> In general, for real numbers a and b,
> $$(a + b)(a - b) = a^2 - b^2.$$
>
> **Concept**
> (1st term $+$ 2nd term) (1st term $-$ 2nd term) $=$
> (1st term)^2 $-$ (2nd term)^2

Example 3–3 E

Perform the indicated multiplication and simplify.

1. $(x + 5)(x - 5)$

The first term is x and the second term is 5. Substituting into the formula, we have

$$(x + 5)(x - 5) = (x)^2 - (5)^2$$
$$\uparrow \qquad \uparrow$$
$$\text{1st} \qquad \text{2nd}$$
$$\text{term} \qquad \text{term}$$
$$= x^2 - 25.$$

2. $(2a + b)(2a - b) = (2a)^2 - (b)^2$
$$= 4a^2 - b^2$$

In all of the examples that we have considered, whether they were special products or not, a single procedure is sufficient. When multiplying two multinomials, *we multiply each term of the first multinomial by each term of the second multinomial and then combine like terms.*

Example 3–3 F

Perform the indicated operations and simplify.

1. $(a + 3)(2a^2 - a + 4)$

Although there are three terms in the second parentheses, we still follow the procedure of every term in the first factor times every term in the second factor.

$(a + 3)(2a^2 - a + 4)$
$= (a)(2a^2) + (a)(-a) + (a)(4) + (3)(2a^2) + (3)(-a) + (3)(4)$
$= 2a^3 - a^2 + 4a + 6a^2 - 3a + 12$ Combine like terms.
$= 2a^3 + 5a^2 + a + 12$

2. $(x + y)(x + 2y)(x - y)$

When there are three quantities to be multiplied, we apply the associative property to multiply two of them together first and take that product times the third.

$$[(x + y)(x + 2y)](x - y) = [x^2 + 2xy + xy + 2y^2](x - y)$$
$$= [x^2 + 3xy + 2y^2](x - y)$$
$$= x^3 - x^2y + 3x^2y - 3xy^2 + 2xy^2 - 2y^3$$
$$= x^3 + 2x^2y - xy^2 - 2y^3$$

3. $(2a - 5)^2 - (a + 3)(a - 4)$

We must take care of all multiplication before we can perform the indicated subtraction.

$$(2a - 5)^2 - (a + 3)(a - 4)$$
$$= [(2a)^2 + 2(2a)(-5) + (-5)^2] - [a^2 - 4a + 3a - 12]$$
$$= [4a^2 - 20a + 25] - [a^2 - a - 12]$$
$$= 4a^2 - 20a + 25 - a^2 + a + 12 \qquad \text{Remove the brackets.}$$
$$= 3a^2 - 19a + 37 \qquad \text{Combine like terms.}$$

> **Note**
> In example 3 we placed grouping symbols around the quantities being multiplied. This is an important step. Without the grouping symbols, our second line would have been $4a^2 - 20a + 25 - a^2 - a - 12$, and we would have subtracted only the first term of the product and not the entire product as was indicated.

4. $-2\{3x + 2[x - (3x + 1)]\}$
$$= -2\{3x + 2[x - 3x - 1]\} \qquad \text{Remove the parentheses.}$$
$$= -2\{3x + 2[-2x - 1]\} \qquad \text{Combine like terms.}$$
$$= -2\{3x - 4x - 2\} \qquad \text{Distributive property}$$
$$= -2\{-x - 2\} \qquad \text{Combine like terms.}$$
$$= 2x + 4 \qquad \text{Distributive property}$$

Mastery points
Can you
- Multiply a monomial by a multinomial?
- Multiply two multinomials?
- Multiply two multinomials vertically?
- Find the square of a binomial and the difference of two squares using the special products?

Exercise 3-3

Directions Perform the indicated multiplication and simplify. See example 3-3 A.

1. $a(2a^2 - 3a + 4)$ **2.** $2x(5x^2 - 3x + 7)$ **3.** $-2y(3y^2 - 5y + 4)$

4. $-4x(2x^2 - 3x - 9)$ **5.** $3a^2(4a^2 - 2ab + 3b^2)$ **6.** $5x^3(2x^2 - 3x + 4)$

7. $6xy(5x^2y - 4xy^3 + 2xy)$ **8.** $-a^3b(3a^2b^5 - ab^4 - 7a^2b)$

9. $-5x^3y^4(2x^2y + 7xy^6 - 3x^2y^5)$ **10.** $6a^3b^5(-4a^2b + 5a^3b^3 - 8a^5b + 2)$

Directions Perform the indicated multiplication and simplify (a) by multiplying horizontally and (b) by multiplying vertically. See example 3-3 B.

11. $(a + 5)(a + 3)$ **12.** $(x + 4)(x + 6)$ **13.** $(b + 4)(b - 5)$ **14.** $(x + 7)(x - 8)$

15. $(2x - y)(x + y)$ **16.** $(3a + b)(a + b)$ **17.** $(5x - y)(2x + y)$ **18.** $(3x + y)(4x - y)$

19. $(7x - 5y)(6x + 4y)$ **20.** $(4a - 7b)(3a - 5b)$

Directions Perform the indicated multiplication and simplify. Use any of the special products where possible. See examples 3-3 C, D, E, and F.

21. $(x + 3)^2$ **22.** $(2a + b)^2$ **23.** $(3x + y)^2$

24. $(5x + 2y)^2$ **25.** $(4x + 3y)^2$ **26.** $(2x - 3)^2$

27. $(2a - 5)^2$ **28.** $(3a - b)^2$ **29.** $(4x - 3y)^2$

30. $(a - 3b)(a + 2b)$ **31.** $(x - 3y)(x + 3y)$ **32.** $(2x - 3yz)(2x + 3yz)$

33. $(5a + 2bc)(5a - 2bc)$ **34.** $(a + 2b)(a^2 - ab + b^2)$ **35.** $(x + y)(x^2 - 3xy + 2y^2)$

36. $(x - 2y)(3x^2 + 4xy - 2y^2)$ **37.** $(3a - 2b)(5a^2 - 7ab + 3b^2)$ **38.** $(x^2 - 2x + 1)(x^2 + 3x + 2)$

39. $(x^2 + 3x + 4)(x^2 - 2x - 3)$ **40.** $(2a^2 - 3a + 5)(a^2 - 4a + 2)$ **41.** $(5b^2 - 3b + 7)(b^2 + b + 3)$

42. $(a - 3b)^3$ **43.** $(x + 2y)^3$ **44.** $(2x - 3y)^3$

45. $(4a - b)^3$ **46.** $(x - 2y)(2x + y)(x - y)$ **47.** $(a - 2b)(2a - b)(a + 2b)$

Directions Perform the indicated operations, remove all grouping symbols, and simplify. See example 3-3 F.

48. $(x + 2)^2 + (x - 4)^2$ **49.** $(a + 5)^2 + (a - 2)^2$

50. $(2x - 1)(x + 3) - (x + 2)^2$ **51.** $(3x - 2)(x + 5) - (x - 4)^2$

52. $(2x + 3)^2 - (3x + 1)(x - 2)$ **53.** $(y + 3)(2y - 4) - (y + 5)(y - 4)$

54. $(a + 3)(a - 4) - (3a - 2)^2$ **55.** $3[b - (3b + 2) + 5]$

56. $-2[5a - (2a + 3) - 3(3a + 7)]$ **57.** $2x[3x - 2(5x + 1)]$

58. $3y[-2y + 3(5y - 4)]$ **59.** $-2\{3x - 2[5x - (7 - 3x)]\}$

60. $-5x\{2x - 4[x + (3x - 1)]\}$

Directions Perform the indicated multiplication and simplify. Assume that all variables used as exponents represent positive integers.

Examples

(a) $x^2(x^n + 1)$

(b) $(x^n - 1)(x^n + 3)$

Solutions

(a) $= x^2 \cdot x^n + x^2 \cdot 1$
$= x^{n+2} + x^2$

(b) $= x^n \cdot x^n + x^n \cdot 3 + (-1) \cdot x^n + (-1) \cdot 3$
$= x^{n+n} + 3x^n - x^n - 3$
$= x^{2n} + 2x^n - 3$

61. $a^n(a^2 + 3)$ **62.** $b^{2n}(b^3 - 1)$ **63.** $x^n(3x^n + 1)$ **64.** $x^{n+2}(x^{n+1} + x)$

65. $x^{n-2}(x^{n+5} - x^2)$ **66.** $(a^n + 1)(a^n - 2)$ **67.** $(b^n - 3)(b^n - 2)$ **68.** $(2a^n - b^n)^2$

69. $(3x^n + y^n)^2$ **70.** $(x^{2n} + y^n)(x^{2n} - y^n)$ **71.** $(a^n + b^{3n})(a^n - b^{3n})$ **72.** $(3x^{2n} - 2y^{3n})^2$

Directions Solve the following verbal problems.

73. The area of the shaded region between two circles is $\pi(R + r)(R - r)$. Perform the indicated multiplication.

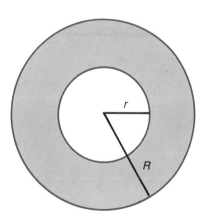

74. When squares of c units on a side are cut from the corners of a square sheet of metal x units on a side and folded up into a tray, its volume is $c(x - 2c)(x - 2c)$ cubic units. Perform the indicated multiplication.

75. The total area of the surface of a cylinder is determined by $A = 2\pi r(h + r)$. Simplify the right member. (π is the lowercase Greek letter pi.)

76. The equation for the distance traveled by a rocket fired vertically upward into the air is given by $S = 16t(35 - t)$, where the rocket is S feet from the ground after t seconds. Simplify the right member.

77. In engineering the equation for the deflection of a beam is given by
$$Y = \frac{Wx}{48EI}(2x^3 - 3\ell x^2 - \ell^3).$$ Simplify the right member.

78. In engineering the equation of transverse shearing stress in a rectangular beam is given in two forms:

(a) $T = \dfrac{V}{8I}(h + 2V_1)(h - 2V_1)$ and

(b) $T = \dfrac{3V}{2A}\left(1 + \dfrac{2V_1}{H}\right)\left(1 - \dfrac{2V_1}{H}\right)$.

Simplify the right member of each equation.

3–4 Further properties of exponents

Quotient property of exponents

Another useful property of exponents can be seen from the following example. Consider the expression

$$\frac{a^5}{a^3}.$$

We use the definition of exponents to write the fraction as

$$\frac{a^5}{a^3} = \frac{a \cdot a \cdot a \cdot a \cdot a}{a \cdot a \cdot a}.$$

We reduce the fraction and get

$$\frac{a \cdot a \cdot a \cdot a \cdot a}{a \cdot a \cdot a} = \frac{a \cdot a}{1} = a \cdot a = a^2.$$

We reduced by three factors of a, leaving $5 - 3 = 2$ factors of a in the numerator. Therefore

$$\frac{a^5}{a^3} = a^{5-3} = a^2.$$

■ **Quotient property of exponents**
For all real numbers a, $a \neq 0$, and integers m and n,

$$a^n \div a^m = \frac{a^n}{a^m} = a^{n-m}.$$

Concept
To divide quantities having **like bases,** subtract the exponent of the denominator from the exponent of the numerator to get the exponent of the common base in the quotient.

Note
If $a = 0$, we have an expression that has no meaning. Therefore $a \neq 0$ indicates that we want our variable to assume no value that would cause the denominator to be zero.

Example 3–4 A

Perform the indicated operations and simplify.

1. $\dfrac{a^9}{a^6} = a^{9-6} = a^3$

2. $b^{12} \div b^4 = b^{12-4} = b^8$

3. $\dfrac{2^8}{2^3} = 2^{8-3} = 2^5 = 32$

When dividing numbers raised to a power, the division is carried out by means of subtracting exponents. *The base is unaltered.*

4. $\dfrac{3^3 x^{12} y^9}{3 x^4 y^8} = 3^{3-1} \cdot x^{12-4} \cdot y^{9-8}$
$= 3^2 x^8 y^1$
$= 9x^8 y$

Negative exponents

Until now, we have considered only those problems in which the exponent of the numerator is greater than the exponent of the denominator. Consider the example

$$\frac{a^3}{a^5}.$$

By the definition of exponents, this becomes

$$\frac{a^3}{a^5} = \frac{a \cdot a \cdot a}{a \cdot a \cdot a \cdot a \cdot a},$$

and reducing the fraction,

$$\frac{a \cdot a \cdot a}{a \cdot a \cdot a \cdot a \cdot a} = \frac{1}{a \cdot a} = \frac{1}{a^2}.$$

Again we reduced by three factors of a, leaving $5 - 3 = 2$ factors of a in the denominator. Hence

$$\frac{a^3}{a^5} = \frac{1}{a^2}.$$

However if we use the quotient property of exponents to carry out the division,

$$\frac{a^3}{a^5} = a^{3-5} = a^{-2}.$$

Since we should arrive at the same answer regardless of which procedure we use, then a^{-2} must be equivalent to $\frac{1}{a^2}$, that is, $a^{-2} = \frac{1}{a^2}$. This leads us to the definition of a^{-n}.

■ **Definition of a negative exponent**
For all real numbers a, $a \neq 0$, and positive integers n,

$$a^{-n} = \frac{1}{a^n}.$$

Concept
When a symbol is raised to a negative exponent, we can rewrite it as the reciprocal of that symbol to the positive exponent.

Example 3–4 B

Write the following without negative exponents.

1. $a^{-4} = \dfrac{1}{a^4}$

2. $3^{-2} = \dfrac{1}{3^2} = \dfrac{1}{9}$

Note
In example 2, the value of 3^{-2} cannot be determined until the exponent is made positive.

3. $-b^{-3} = -(b^{-3}) = -\left(\dfrac{1}{b^3}\right) = -\dfrac{1}{b^3}$

Note
Only the sign of the exponent changes, not the sign of the base.

Zero as an exponent

We now examine the situation involving the division of like bases raised to the same power. For example, consider

$$\frac{a^3}{a^3}.$$

By the definition of exponents, we have

$$\frac{a^3}{a^3} = \frac{a \cdot a \cdot a}{a \cdot a \cdot a} = \frac{1}{1} = 1,$$

and by the quotient property of exponents,

$$\frac{a^3}{a^3} = a^{3-3} = a^0.$$

Since

$$\frac{a^3}{a^3} = 1 \text{ and } \frac{a^3}{a^3} = a^0,$$

then for consistency, we make the following definition for a^0.

> ■ **Definition of zero as an exponent**
> For all real numbers a, $a \neq 0$,
> $$a^0 = 1.$$
>
> **Concept**
> Any number other than zero raised to the zero power is equal to 1.

Note
All the properties of exponents stated so far now apply for any integer used as an exponent.

Example 3–4 C

Write the following expressions without using zero as an exponent.

1. $x^0 = 1$

2. $8^0 = 1$

3. $(-5)^0 = 1$

4. $-5^0 = -(5^0) = -(1) = -1$

5. $(2x^2 + y)^0 = 1$

6. $4x^0 = 4 \cdot x^0 = 4 \cdot 1 = 4$

Example 3–4 D   Perform the indicated operations. Leave the answer with only positive exponents.

1. $\dfrac{a^3b^2c^4}{ab^5c^4} = a^{3-1}b^{2-5}c^{4-4} = a^2b^{-3}c^0 = a^2 \cdot \dfrac{1}{b^3} \cdot 1 = \dfrac{a^2}{b^3}$

2. $\dfrac{x^3y^4}{x^5y^9} = x^{3-5} \cdot y^{4-9} = x^{-2}y^{-5} = \dfrac{1}{x^2y^5}$

3. $b^5 \cdot b^{-3} = b^{5+(-3)} = b^2$

4. $\dfrac{x^{-3}y^4}{x^{-7}y^9} = x^{(-3)-(-7)}y^{4-9} = x^4y^{-5} = \dfrac{x^4}{y^5}$

5. $(3a^5)^{-2} = \dfrac{1}{(3a^5)^2} = \dfrac{1}{3^2(a^5)^2} = \dfrac{1}{9a^{10}}$

Alternate solution:

$(3a^5)^{-2} = 3^{-2}(a^5)^{-2} = 3^{-2} \cdot a^{-10} = \dfrac{1}{3^2a^{10}} = \dfrac{1}{9a^{10}}$

In example 5 we see that there can be alternate methods of working problems involving exponents, depending on the order in which the properties are applied.

Fraction to a power property of exponents

Our last property of exponents can be derived from the definition of exponents. Consider the expression

$$\left(\dfrac{a}{b}\right)^3.$$

$$\left(\dfrac{a}{b}\right)^3 = \underbrace{\dfrac{a}{b} \cdot \dfrac{a}{b} \cdot \dfrac{a}{b}}_{\substack{3 \text{ factors} \\ \text{of } \frac{a}{b}}} = \dfrac{\overbrace{a \cdot a \cdot a}^{\substack{3 \text{ factors} \\ \text{of } a}}}{\underbrace{b \cdot b \cdot b}_{\substack{3 \text{ factors} \\ \text{of } b}}} = \dfrac{a^3}{b^3}.$$

Thus

$$\left(\dfrac{a}{b}\right)^3 = \dfrac{a^3}{b^3}.$$

Example 3–4 E

Perform the indicated operations and simplify.

1. $\left(\dfrac{2a}{b}\right)^3 = \dfrac{(2a)^3}{b^3} = \dfrac{2^3 a^3}{b^3} = \dfrac{8a^3}{b^3}$

2. $\left(\dfrac{3x}{y^3}\right)^2 \left(\dfrac{y^4}{x^3}\right)^3 = \dfrac{(3x)^2}{(y^3)^2} \cdot \dfrac{(y^4)^3}{(x^3)^3} = \dfrac{3^2 x^2}{y^6} \cdot \dfrac{y^{12}}{x^9} = \dfrac{9x^2 y^{12}}{x^9 y^6}$

$\qquad = 9 \cdot x^{2-9} \cdot y^{12-6} = 9x^{-7}y^6 = \dfrac{9y^6}{x^7}$

3. $\left(\dfrac{6ab^{-2}}{3a^{-3}b^4}\right)^2$

$\qquad = (2 \cdot a^{1-(-3)} \cdot b^{-2-4})^2 \qquad$ Simplify inside the parentheses.

$\qquad = (2a^4 b^{-6})^2 = \left(\dfrac{2a^4}{b^6}\right)^2$

$\qquad = \dfrac{(2a^4)^2}{(b^6)^2} = \dfrac{2^2(a^4)^2}{b^{12}}$

$\qquad = \dfrac{4a^8}{b^{12}}$

Scientific notation

An important use of integer exponents is in scientific areas where we deal with very large or very small numbers. For example, the mass of a hydrogen atom is 0.000 000 000 000 000 000 000 001 67 gram; the mass of an electron is 0.000 000 000 000 000 000 000 000 000 91 gram; the half-life of lead 204 is 14,000,000,000,000,000,000 years.

Working with such numbers becomes quite difficult; therefore we convert the number into a more manageable form called **scientific notation.** We define the scientific notation form of a number Y to be the product

$$Y = a \times 10^n,$$

where a is a number greater than or equal to 1 and less than 10, and n is an integer. Use the following steps to achieve this form of the number Y.

Step 1 Move the decimal point to a position immediately following the first nonzero digit in Y.

Step 2 Count the number of places the decimal point has been moved. This is the exponent, n, to which 10 is raised.

Step 3 If (a) the decimal point is moved to the *left*, n is *positive*;
(b) the decimal point is moved to the *right*, n is *negative*;
(c) the decimal point already follows the first nonzero digit, n is *zero*.

Example 3–4 F

Express the following in scientific notation.

1. $9{,}000{,}000 = 9.000000. \times 10^6 = 9 \times 10^6$

2. $0.00000467 = 0.000004.67 \times 10^{-6} = 4.67 \times 10^{-6}$

3. $4.37 = 4.37 \times 10^0$

4. $-0.00341 = -0.003.41 \times 10^{-3} = -3.41 \times 10^{-3}$

Sometimes it is necessary to convert a number in scientific notation to its standard form. To do this we apply the properties in reverse. Therefore when n, the exponent of 10, is

1. *positive,* the decimal point is moved to the *right* n places;
2. *negative,* the decimal point is moved to the *left* n places; and
3. *zero,* the decimal point is not moved.

Example 3–4 G

Express the following in standard form.

1. $6.37 \times 10^4 = 6.3700. = 63{,}700$

2. $4.81 \times 10^{-5} = 0.00004.81 = 0.0000481$

3. $8.59 \times 10^0 = 8.59$

4. $-3.48 \times 10^{-1} = -0.3.48 = -0.348$

Scientific notation can be used to simplify numerical calculations when the numbers are very large or very small. We first change the numbers to scientific notation and then use the properties of exponents to help perform the indicated operations.

Example 3–4 H   Perform the indicated operations using scientific notation.

1. $(198,000,000)(0.00347)$
   $= (1.98 \times 10^8)(3.47 \times 10^{-3})$ Scientific notation
   $= (1.98 \cdot 3.47) \times (10^8 \cdot 10^{-3})$ Commutative and associative
                                                          properties
   $= 6.8706 \times 10^5$ Multiply.
   $= 687,060$ Standard form

2. $\dfrac{(92,000,000)(0.0036)}{(0.018)(4,000)}$

   $= \dfrac{(9.2 \times 10^7)(3.6 \times 10^{-3})}{(1.8 \times 10^{-2})(4.0 \times 10^3)}$ Scientific notation

   $= \dfrac{(9.2)(3.6)10^7 \cdot 10^{-3}}{(1.8)(4.0)10^{-2} \cdot 10^3}$ Commutative and associative
                                                                                properties

   $= \dfrac{(9.2)(3.6)}{(1.8)(4.0)} \cdot 10^3$ Properties of exponents

   $= 4.6 \times 10^3$ Division
   $= 4,600$ Standard form

Mastery points
Can you
- Perform division on factors involving exponents?
- Perform operations involving negative exponents?
- Perform operations involving zero as an exponent?
- Raise a fraction to a power?
- Use scientific notation?

Exercise 3–4

Directions Assume that all variables or groupings in this exercise represent nonzero real numbers and that all variables used as exponents represent positive integers.

Directions Write each expression with only positive exponents. See examples 3–4 B and C.

1. 5^0
2. $(-3)^0$
3. $(2a^3 - b)^0$
4. $(5x^2 + 4y)^0$
5. $7x^0$

6. $9y^0$
7. a^{-4}
8. b^{-5}
9. $3a^{-2}$
10. x^2y^{-5}

11. $ab^{-4}c^{-3}$
12. $\dfrac{1}{4a^{-3}}$
13. $\dfrac{1}{2x^{-5}}$
14. $\dfrac{1}{x^2y^{-4}}$

Directions Perform all indicated operations and leave the answer with only positive exponents. See examples 3–4 A and E.

15. $\dfrac{a^3}{a^4 a^7}$

16. $\dfrac{x^5 y^3}{xy^5}$

17. $\dfrac{2^3 x^4 y^9}{2xy^3}$

18. $\dfrac{3^3 a^4 b^5}{3^2 a^2 b^3}$

19. $\dfrac{3x^2 y^5}{3^4 x^5 y^5}$

20. $\dfrac{5^2 a^4 b}{5^3 a^9 b^4}$

21. 2^{-2}

22. -2^{-2}

23. -3^{-4}

24. $(-2)^{-3}$

25. $\dfrac{2^{-4}}{2^{-2}}$

26. $\dfrac{3^{-6}}{3^{-4}}$

27. $x^{-5} x^2$

28. $(2a^{-3})(3a^{-5})$

29. $(5x^2)(4x^{-7})$

30. $(3x^{-4} y^3)(2x^2 y^{-1})$

31. $(5a^{-3} b^{-4})(2a^2 b^5)$

32. $(2a^3)^{-3}$

33. $(3x^4)^{-2}$

34. $(5a^2 b^{-3})^{-2}$

35. $(4a^{-3} b^4)^{-3}$

36. $(x^{-3} y^0)^2$

37. $(2a^{-2} b^0)^3$

38. $(5x^{-4} y^0 z^2)^{-2}$

39. $(3a^2 b^{-3} c^0)^{-3}$

40. $\left(\dfrac{2x}{y^3}\right)^2$

41. $\left(\dfrac{3a^2}{b}\right)^3$

42. $\left(\dfrac{5a^2 b}{c^4}\right)^2$

43. $\left(\dfrac{3x^4 y^5}{z^9}\right)^3$

44. $\left(\dfrac{2x^5 y^2}{4x^3 y^7}\right)^4$

45. $\left(\dfrac{9a^5 b^4}{18ab^{11}}\right)^3$

46. $\dfrac{x^{-4} y^2}{x^5 y^{-3}}$

47. $\dfrac{a^2 b^{-2}}{a^{-4} b^5}$

48. $\left(\dfrac{6y^{-3}}{12y^{-5}}\right)^3$

49. $\dfrac{x^{-4} y^2}{x^{-6} y^5}$

50. $\left(\dfrac{y^{-4} z^{-3}}{y^{-5} z^4}\right)^2$

51. $\left(\dfrac{3x^{-2}}{12x^{-4}}\right)\left(\dfrac{16x^{-5}}{x}\right)$

52. $\left(\dfrac{a^{-2} b^3 c^0}{a^3 b^{-4} c^2}\right)\left(\dfrac{a^{-4} b^{-1} c^2}{a^5 b^2}\right)$

53. $\left(\dfrac{3^2 a^{-3} b^3}{6a^{-5} b^{-2}}\right)\left(\dfrac{12a^{-7} b^{-3}}{a^2 b^5}\right)$

54. $\left(\dfrac{4a^{-3}}{12a^{-5}}\right)^{-3}$

55. $\left(\dfrac{xy^0}{z^{-2}}\right)^{-4}$

56. $\left(\dfrac{2a^3 b^{-4}}{6a^5 b^2}\right)^{-2}$

57. $\left(\dfrac{4x^2 y^3 z^{-2}}{12x^5 y^{-2} z^0}\right)^{-3}$

58. $(3x^2 y^{-2})^2 (x^{-4} y^3)^3$

59. $(2a^3 b^{-4})^3 (a^{-2} b^5)^3$

60. $(x^{-2} y^4)^{-3}(x^{-1} y^{-2})^4$

61. $(a^3 b^{-2} c^{-4})^{-2}(a^{-2} b^3 c^0)^3$

62. $(ab^{-4} c^2)^{-5}(a^{-2} b^0 c^3)^{-3}$

63. $(x^2 y^{-4})^{-3}(xy^3 z^4)^{-2}$

Directions Perform the indicated operations and leave the answer as a product with no fractions and each variable occurring only once.

Examples

(a) $(a^{1-n})^{-3}$

(b) $\dfrac{x^{2n-7}}{x^{n-6}}$

Solutions

(a) $= a^{(1-n)(-3)}$
$= a^{-3+3n}$
$= a^{3n-3}$

(b) $= x^{(2n-7)-(n-6)}$
$= x^{2n-7-n+6}$
$= x^{n-1}$

64. $x^{2n+3} x^{2-n}$

65. $a^{b-4} a^{2b-1}$

66. $(a^{1-2n})^{-2}$

67. $(x^{4-3n})^{-2}$

68. $\dfrac{x^{n-6}}{x^{n-3}}$

69. $\dfrac{a^{2n-1}}{a^{n-5}}$

70. $\dfrac{x^n y^{2n+1}}{x^{n-3} y^{n-4}}$

71. $\dfrac{a^{2n} b^{3n-2}}{a^{n-1} b^{n-5}}$

72. $\left(\dfrac{x^n}{x^{2n-1}}\right)^{-2}$

73. $\left(\dfrac{x^{2n}}{x^{n+1}}\right)^{-3}$

Directions Express the following numbers in scientific notation. See example 3–4 F.

74. 65,000,000   **75.** 155,000   **76.** 0.00012   **77.** 0.0863   **78.** −0.0567

79. The mass of the moon is 8,060,000,000,000,000,000,000 tons.

80. General Motors reported a gross earnings of $7,230,000,000.

81. A cu in. equals 0.0005787 cu ft.

82. 43,560 sq ft equals an acre.

83. A gram equals 0.0022046 pound.

84. 46,656 cu in. equals a cu yd.

85. A kilogram equals 0.001102 ton.

86. A particular gauge of copper wire has a resistance of 0.00006203 ohms/lb.

87. A light-year, the distance light travels in a year, is 5,872,000,000 mi.

88. 3,785.3 milliliters equals a gal.

Directions Convert the following numbers in scientific notation to their standard form. See example 3–4 G.

89. -4.37×10^{-2} **90.** 7.61×10^{-7} **91.** 4.99×10^{6} **92.** 7.23×10^{0} **93.** 4.83×10^{4}

Directions Multiply or divide the following using scientific notation. Leave the answer in scientific form, $a \times 10^{n}$, where a is rounded to two decimal places. See example 3–4 H.

94. $5.23 \times 10^{9} \cdot 1.073 \times 10^{6}$ **95.** $5.12 \times 10^{6} \cdot 6.2 \times 10^{4}$ **96.** $8.473 \times 10^{3} \cdot 8.4 \times 10^{-4}$

97. $1.673 \times 10^{-4} \cdot 7.5 \times 10^{-6}$ **98.** $(6.23 \times 10^{2}) \div (5.73 \times 10^{4})$ **99.** $(1.47 \times 10^{3}) \div (8.03 \times 10^{1})$

100. $(7.82 \times 10^{-2}) \div (5.6 \times 10^{5})$ **101.** $(7.23 \times 10^{5}) \div (6.075 \times 10^{-9})$ **102.** $0.876 \cdot 21.46$

103. $0.000476 \cdot 0.0053$ **104.** $0.000000089 \cdot 0.145$ **105.** $473 \cdot 0.0000579$

106. $0.00053 \cdot 0.634 \cdot 0.0000000146$ **107.** $243 \cdot 0.00073 \div 65,000$

108. $6,184,000 \div 0.0079 \cdot 52.1$

Directions Solve the following verbal problems, leave the answer in scientific notation rounded to two decimal places.

109. The amount of stress on a piece of metal is given by stress $= \dfrac{\text{force producing the stress}}{\text{area of the surface}}$ Find the stress on a piece of metal where the force is 452.6 kg and the area is 0.000763 sq cm.

111. A unit of light intensity is a lumen. If 1 lumen equals 0.001496 watts, how many lumens are there in 4,760,000 watts?

110. If the volume of a right circular cylinder is given by $V = \pi r^{2} h$, find the volume of a right circular cylinder that has radius $r = 0.0073$ m and height $h = 12$ m.

3–5 Common factors and factoring by grouping

Prime numbers

In section 3–3 we studied the procedure for multiplying polynomials. Now we will study the reversal of that process, which is called **factoring.** Factoring polynomials is necessary to deal with algebraic fractions, since, as we have seen with arithmetic fractions, the denominators of the fractions must be in factored form to reduce fractions or to find the **least common denominator** (LCD). Factoring is also useful as a means of finding the solution of certain types of equations.

We saw in chapter 1 that whenever we multiply any two real numbers, a and b, a and b are called **factors** of the **product** ab. For example, 2 and 3 are factors of 6 because $2 \cdot 3 = 6$. The first type of factoring that we will consider is factoring a natural number into its prime factors.

The set of natural numbers greater than 1 can be divided into two types of numbers, prime numbers and composite numbers. First, a **prime number** *is any natural number greater than 1 that is divisible only by itself and by 1.* For example,

$$2, 3, 5, 7, 11, 13$$

are all prime numbers since each is a natural number greater than 1 and divisible only by 1 and the number itself. Second, a **composite number** *is any natural number greater than 1 that is not a prime number.* For example,

$$4, 9, 28, 42$$

are composite numbers since 4 is divisible by 2; 9 is divisible by 3; 28 is divisible by 2, 4, 7, and 14; 42 is divisible by 2, 3, 6, 7, 14, and 21.

When a composite number is stated as a product of only prime numbers, we call this the **prime factor form** or **completely factored form** of the composite number. We can also state negative integers in a completely factored form by first factoring out -1 and then factoring the resulting positive integer into its prime factors.

Example 3–5 A

Express the following numbers in prime factor form.

1. $28 = 4 \cdot 7 = 2 \cdot 2 \cdot 7 = 2^2 \cdot 7$

> **Note**
> Whenever a prime number appears as a factor more than once, we write it in exponential form.

2. $42 = 6 \cdot 7 = 2 \cdot 3 \cdot 7$

3. $23 = 23$

> **Note**
> The number 23 is a prime number, therefore it cannot be expressed as a product of only prime factors. It would be incorrect to write $23 \cdot 1$ since 1 is not a prime number.

4. $120 = 10 \cdot 12 = 2 \cdot 5 \cdot 2 \cdot 2 \cdot 3 = 2^3 \cdot 3 \cdot 5$

The greatest common factor

Remember that the distributive property states

$$a(b + c) = ab + ac.$$

The left member of the equation is the indicated product of the two factors, a and $(b + c)$, and is referred to as the factored form of $ab + ac$. Therefore if we start with the indicated sum $ab + ac$ and then write it as $a(b + c)$, we are factoring the expression (changing the expression to an indicated multiplication form).

$$\xrightarrow{\text{Multiplying}}$$
$$a(b + c) = ab + ac$$
$$\xleftarrow{\text{Factoring}}$$

We now use the distributive property to write $2x + 6$ in a factored form. First, we see that $2x + 6$ can be written

$$2 \cdot x + 2 \cdot 3.$$

By applying the distributive property we have

$$2x + 6 = 2 \cdot x + 2 \cdot 3 = 2(x + 3).$$

$2(x + 3)$ is called the factored form of $2x + 6$.

This type of factoring, as its name implies, involves looking for numbers or variables that are **common factors** in *all* of the original terms of the expression. In our example, 2 was the greatest common factor.

We were able to determine the greatest common factor of our example by inspection. In some problems this may not be possible, and the following procedure will be necessary. Factor the polynomial $15x^5y^2 + 30x^3y^4$.

Step 1 Factor each term such that it is the product of primes and variables to powers.

$$15x^5y^2 \qquad\qquad\qquad 30x^3y^4$$
$$3 \cdot 5 \cdot x^5 \cdot y^2 \qquad\qquad 2 \cdot 3 \cdot 5 \cdot x^3 \cdot y^4$$

Step 2 Write down all factors that are common to *every* term.

$$3 \cdot 5 \cdot x \cdot y$$

Note

We do not have 2 as part of our greatest common factor since it does not appear in *all* of the terms.

Step 3 Take the factors in step 2 and raise them to the *least* power to which they were raised in any of the terms.

$$3^1 \cdot 5^1 \cdot x^3 \cdot y^2 = 15x^3y^2$$

This is the greatest common factor (GCF).

Step 4 We can now write the polynomial in its factored form.

$$15x^5y^2 + 30x^3y^4$$
$$= 15x^3y^2 \cdot x^2 + 15x^3y^2 \cdot 2y^2$$
$$= 15x^3y^2(x^2 + 2y^2)$$

Note

If we do not see by inspection what the greatest common factor is multiplied by in each term, we simply divide the GCF into each term. The answer is what the GCF is multiplied by. In the previous example we could have divided each term by $15x^3y^2$,

$$\frac{15x^5y^2}{15x^3y^2} = x^2 \text{ and } \frac{30x^3y^4}{15x^3y^2} = 2y^2,$$

and our factorization would have been the same as in step 4.
$$15x^5y^2 + 30x^3y^4 = 15x^3y^2(x^2 + 2y^2)$$

Notice that $15x^5y^2 + 30x^3y^4$ could also be factored into

$$15(x^5y^2 + 2x^3y^4) \text{ or } 30\left(\frac{1}{2}x^5y^2 + x^3y^4\right) \text{ or } x^3(15x^2y^2 + 30y^4).$$

This allows room for a given polynomial to be factored in many ways, unless some restrictions are placed on the procedure. We wish to factor each polynomial in a unique manner that will not permit such variations in the results. Thus it is customary to adopt the following criteria for a completely factored polynomial.

A polynomial with integer coefficients is considered to be in *completely factored form* when it is expressed as the product of polynomials with integer coefficients and when none of the factors except a monomial can still be written as the product of two polynomials with integer coefficients.

We see that $15x^3y^2(x^2 + 2y^2)$ is the *completely factored form* of the expression $15x^5y^2 + 30x^3y^4$. All of the coefficients are integers and none of the factors other than the monomial, $15x^3y^2$, can still be written as the product of two polynomials with integer coefficients.

In general, whenever we factor a monomial out of a polynomial, we factor the monomial in such a way that it has a positive coefficient. We should realize that we can also factor out the opposite, or negative, of this common factor. In our previous example we could have factored out $-15x^3y^2$ and the completely factored form would have been

$$-15x^3y^2(-x^2 - 2y^2).$$

Observe that the only change in our answer when we factor out the opposite of the common factor is that the signs of all the terms inside the parentheses change.

| Example 3–5 B | Write the following polynomials in completely factored form.

1. $9r^3s^2 + 15r^2s^4 + 3rs^2 = 3^2 \cdot r^3 \cdot s^2 + 3 \cdot 5 \cdot r^2 \cdot s^4 + 3 \cdot r \cdot s^2$.
We have 3, r, and s in common and the GCF is $3rs^2$. The completely factored form is

$$3rs^2 \cdot 3r^2 + 3rs^2 \cdot 5rs^2 + 3rs^2 \cdot 1 = 3rs^2(3r^2 + 5rs^2 + 1).$$

Note
In example 1, the last term in the factored form is 1. This occurs when a term and the GCF are the same, that is, whenever we are able to factor all the numbers and variables out of a given term.

2. $a(x + 2y) + b(x + 2y)$
The quantity $(x + 2y)$ is common to both terms. We then factor the common quantity out of each term and place the remaining factors from each term in a second parentheses.

$$a(x + 2y) + b(x + 2y) = \underset{\substack{\text{Common} \\ \text{factor}}}{(x + 2y)}\underset{\substack{\text{Remaining} \\ \text{factors}}}{(a + b)}$$

Factoring by grouping

Consider the example $ax + ay + bx + by$. We observe that this is a four-term polynomial and we will *try* to factor it by grouping.

$$ax + ay + bx + by = (ax + ay) + (bx + by)$$

There is a common factor of a in the first two terms and a common factor of b in the second two terms.

$$(ax + ay) + (bx + by) = a(x + y) + b(x + y)$$

The quantity $(x + y)$ is common to both terms, and factoring it out, we have

$$a(x + y) + b(x + y) = (x + y)(a + b).$$

Therefore we have factored the polynomial by grouping.

| Example 3–5 C | Factor completely.

1. $2ac - ad + 4bc - 2bd = (2ac - ad) + (4bc - 2bd)$
Grouping the first two terms and the last two terms, the first two terms have a as a common factor and the last two terms have $2b$ as a common factor.

$$a(2c - d) + 2b(2c - d)$$

Factoring out the common quantity and placing the remaining factors in a second set of parentheses, we have

$$(2c - d)(a + 2b).$$

2. $6ax + by + 2ay + 3bx = 6ax + 2ay + 3bx + by$
$$= (6ax + 2ay) + (3bx + by)$$
$$= 2a(3x + y) + b(3x + y)$$
$$= (3x + y)(2a + b)$$

Note
Sometimes the terms must be rearranged such that the pairs will have a common factor, as we did in example 2.

Mastery points
Can you
- Determine the greatest common factor of a polynomial?
- Factor the common factors from a polynomial?
- Factor a four-term polynomial by grouping?

Exercise 3–5

Directions Express the following numbers in prime factor form. If a number is prime, so state. See example 3–5 A.

1. 12   **2.** 15   **3.** 24   **4.** 28   **5.** 56   **6.** 60   **7.** 41   **8.** 43   **9.** 39

Directions Supply the missing factors or terms.

Examples
(a) $-3a + a^3b = -a(? - ?)$ (b) $x^2y - z^3 = -(? + ?)$ (c) $3a^2 - 9b = ?(-a^2 + 3b)$

Solutions
(a) $= -a(3 - a^2b)$ (b) $= -(-x^2y + z^3)$ (c) $= -3(-a^2 + 3b)$

10. $5x - 3 = -(? + ?)$ **11.** $2a - 3b = -(? + ?)$

12. $2x^2 - 6y = -2(? + ?)$ **13.** $3a^3 - 9b^2 = -3(? + ?)$

14. $6a - 8b - 12c = 2(? - ? - ?)$ **15.** $-4x^3 - 36xy + 16xy^2 = -4(? + ? - ?)$

16. $5x^2y + 15x^3y^2 + 25x^2y^2 = -5x^2y(? - ? - ?)$ **17.** $3a^3b^2 + 12a^2b^3 + 15a^4b^2 = -3a^2b^2(? - ? - ?)$

18. $-ab^2 - ac^2 = ?(b^2 + c^2)$ **19.** $3x^2 - 9xy = ?(-x + 3y)$

20. $-10a^2b^2 + 15ab - 20a^3b^3 = ?(-2ab + 3 - 4a^2b^2)$

21. $-24RS - 16R + 32R^2 = ?(3S + 2 - 4R)$ **22.** $3x^n - 2x^ny = ?(3 - 2y)$

Directions Write in completely factored form. Assume that all variables used as exponents represent positive integers. See examples 3–5 B and C.

23. $5x^2 + 10xy - 20y$ **24.** $8a - 12b + 16c$ **25.** $18ab - 27a + 3ac$

26. $15x^2 - 27y^2 + 12$ **27.** $15R^2 - 21S^2 + 36T$ **28.** $8x^3 + 4x^2$

29. $3R^2S - 6RS^2 + 12RS$ **30.** $16x^3y - 3x^2y^2 + 24x^2y^3$ **31.** $12x^4y^3 - 8x^4y^2 + 16x^3y$

32. $24a^3b^2 - 3a^2b^2 + 12a^2b^5$ **33.** $3x^2y - 6xy^4 + 15x^3y^2$ **34.** $a(x - 3y) + b(x - 3y)$

35. $3a(x - y) + b(x - y)$ **36.** $5a(3x - 1) + 10(3x - 1)$ **37.** $14b(a + c) - 7(a + c)$

38. $21x(1 + 2z) - 35y(1 + 2z)$ **39.** $4ab(2x + y) - 8ac(2x + y)$ **40.** $2x(y - 16) - (y - 16)$

41. $5a(b + 3) - (b + 3)$ **42.** $15a^3b^2(x + y) + 30a^2b^5(x + y)$ **43.** $x^ny^2 + x^nz$

44. $x^{3n} + x^{2n} + x^n$ **45.** $x^{2n}y^{2n} - x^ny^n$ **46.** $x^{n+4} + x^4$

47. $y^{n+3} - y^3$ **48.** $ac + ad - 2bc - 2bd$ **49.** $2ax + 6bx - ay - 3by$

50. $2ax + bx - 4ay - 2by$ **51.** $2ax + 3bx + 8ay + 12by$ **52.** $5ax - 3by + 15bx - ay$

53. $2ax^2 - 3b - bx^2 + 6a$ **54.** $20x^2 - 3yz + 5xz - 12xy$ **55.** $3ac - 2bd - 6ad + bc$

56. $6ac + 3bd - 2ad - 9bc$ **57.** $2x^3 + 15 + 10x^2 + 3x$ **58.** $3x^3 - 6x^2 + 5x - 10$

59. $8x^3 - 4x^2 + 6x - 3$

Directions Solve the following verbal problems.

60. The area of the surface of a cylinder is determined by $A = 2\pi rh + 2\pi r^2$, where r is the radius and h is the height. Factor the right member.

61. The total surface area of a right circular cone is given by $A = \pi rs + \pi r^2$, where r is the base radius and s is the slant height. Factor the right member.

62. The equation for the height of a rocket fired vertically upward into the air is given by $S = 560t - 16t^2$, where the rocket is S feet from the ground after t seconds. Factor the right member.

63. If you have P dollars in a savings account where the yearly compounded rate is r, then the amount of money in the account at the end of one year can be written as $A = P + Pr$. Factor the right member.

64. In engineering the equation for deflection of a beam is given by
$$Y = \frac{2wx^4}{48EI} - \frac{3\ell wx^3}{48EI} - \frac{\ell^3wx}{48EI}.$$
Factor the right member.

3–6 Trinomials

Determining when a trinomial will factor

We are going to factor trinomials of the form $ax^2 + bx + c$. This is called the **standard form** of a trinomial, where we have a single variable and the terms of the polynomial are arranged in descending powers of that variable. The a, b, and c in our standard form represent integer constants. For example,

$$4x^2 + 20x + 21$$

is a trinomial in standard form, where $a = 4$, $b = 20$, and $c = 21$.

*The **trinomial** will factor into factors with integer coefficients only if we can find a pair of integers, which we will call* m *and* n, *whose product is* a · c *and whose sum is* b. In our specific trinomial we want $m \cdot n$ to be equal to $a \cdot c = 4 \cdot 21 = 84$ and $m + n$ to be equal to b, which is 20. Therefore this trinomial will factor only if we can find m and n values satisfying the conditions $m \cdot n = 84$ and $m + n = 20$. The values satisfying these conditions are 6 and 14 because $6 \cdot 14 = 84$ and $6 + 14 = 20$. A method used for finding these integers is shown. Use this method only when you cannot determine the integers by inspection.

Our first step in factoring a trinomial will be to determine if there are two integers, m and n, satisfying the conditions $m \cdot n = a \cdot c$ and $m + n = b$. Often we are able to determine the two integers by inspection, but for those problems where the m and n values are not obvious, the following procedure will be useful.

Determine if the trinomial $4x^2 + 20x + 21$ will factor.

Step 1 Make a table of two columns. The heading of the first column will be the absolute value of the product of m and n and the heading of the second column will be the sign of this product.

84	+

Step 2 Take the natural numbers, $\{1,2,3,4, \cdot \cdot \cdot\}$, and divide them into this product. Those numbers that divide in evenly we write in factored form with the quotient times the divisor.

84	+
84 · 1	
42 · 2	
28 · 3	
21 · 4	
14 · 6	
12 · 7	
7 · 12	
6 · 14	
4 · 21	
3 · 28	
2 · 42	
1 · 84	

Note that the top six factorizations are the same as the bottom six factorizations, therefore we need only perform this procedure until the factors repeat.

$$12 \cdot 7$$
$$7 \cdot 12 \qquad \text{The factors repeat.}$$

Hence we have

84	+
84 · 1	
42 · 2	
28 · 3	
21 · 4	
14 · 6	
12 · 7	

Step 3 Using the sign in the second column as an operation symbol, perform this operation on the factors in the first column.

Product	Sum
84	+
84 · 1	84 + 1 = 85
42 · 2	42 + 2 = 44
28 · 3	28 + 3 = 31
21 · 4	21 + 4 = 25
14 · 6	14 + 6 = 20
12 · 7	12 + 7 = 19

Step 4 From the second column, find the value whose absolute value is equivalent to the $m + n$ value in the beginning of the problem.
$m + n = 20$ and $14 + 6 = 20$ satisfies that condition in our problem.

Step 5 Give the two integers found in step 4 the proper signs that will yield the correct product and sum. Our product was positive, therefore the two factors have the same sign, and since their sum is positive, both integers must be positive. Our answer is 14 and 6, since $14 \cdot 6 = 84$ and $14 + 6 = 20$.

Since we were able to find these m and n values, we are *assured that this trinomial will factor.* If we had not been able to find m and n values, we would say that the *trinomial will not factor.*

Factoring trinomials

We have just learned how to determine if a trinomial is factorable. We will now factor trinomials by combining that procedure with the ideas of factoring out what is common.

Consider the product

$$(2x + 3)(2x + 7).$$

By multiplying these two quantities together and combining like terms, we get a trinomial.

$$(2x + 3)(2x + 7) = 4x^2 + 14x + 6x + 21$$
$$= 4x^2 + 20x + 21$$

To completely factor the trinomial $4x^2 + 20x + 21$ entails reversing the procedure to get

$$(2x + 3)(2x + 7).$$

We are able to determine that the trinomial is factorable if we can find two integer values such that $m \cdot n = 4 \cdot 21 = 84$ and $m + n = 20$. We have determined that the values are 14 and 6. Since we could find these two integers, we are sure that this trinomial will factor. If we observe the multiplication process in our example, we see that m and n appear as the coefficients of the middle terms that are to be combined for our final answer.

$$(2x + 3)(2x + 7) = 4x^2 + 14x + 6x + 21$$
$$= 4x^2 + 20x + 21$$

This is precisely what we do with the m and n values. We replace the coefficient of the middle term in the trinomial with these values. In our example, m and n are 14 and 6 and we replace the 20 with them.

$$\overbrace{20x}$$
$$4x^2 + 20x + 21 = 4x^2 + 14x + 6x + 21$$

Our next step is to group the first two terms and the last two terms.

$$(4x^2 + 14x) + (6x + 21)$$

Now we factor out what is common in each pair. We see that the first two terms contain the common factor $2x$ and the last two terms contain the common factor 3.

$$2x(2x + 7) + 3(2x + 7)$$

When we reach this point, what is inside the parentheses in each term will be the same. Since the quantity, $(2x + 7)$, represents just one number and this number is common to both terms, we can factor it out.

$$2x\underbrace{(2x + 7)} + 3\underbrace{(2x + 7)}$$
Common to both terms

Having factored out what is common, what is left in each term is placed in a second parentheses.

$$2x(2x + 7) + 3(2x + 7)$$
$$(2x + 7)(2x + 3)$$
Common   Remaining
factor   factors

The trinomial is factored.

We can now summarize the steps for factoring trinomials:

Step 1 Determine if the trinomial $ax^2 + bx + c$ is factorable by finding m and n such that $m \cdot n = a \cdot c$ and $m + n = b$. If m and n do not exist, we conclude that the trinomial will not factor.

Step 2 Replace the middle term, bx, by the sum of mx and nx.

Step 3 Place parentheses around the first and second terms and around the third and fourth terms. Factor out what is common to each pair.

Step 4 Factor out the common quantity of each term and place the remaining factors from each term in a second parentheses.

Example 3–6 A

Express the following trinomials in completely factored form. If the trinomial will not factor, so state.

1. $6x^2 + 13x + 6$

Step 1 $m \cdot n = 6 \cdot 6 = 36$ and $m + n = 13$
We determine by inspection that m and n are 9 and 4.

Note
If we cannot determine the m and n values by inspection, then we should use the procedure for finding the values.

$$\overset{13x}{\overbrace{}}$$

Step 2 $= 6x^2 + 9x + 4x + 6$

Step 3 $= (6x^2 + 9x) + (4x + 6)$

$= 3x(2x + 3) + 2(2x + 3)$

Step 4 $= (2x + 3)(3x + 2)$

Note
The order in which we place m and n into the problem will not change the answer.

(Alternate)

$$\overset{13x}{\overbrace{}}$$

Step 2 $= 6x^2 + 4x + 9x + 6$

Step 3 $= (6x^2 + 4x) + (9x + 6)$
$= 2x(3x + 2) + 3(3x + 2)$

Step 4 $= (3x + 2)(2x + 3)$

We see that the outcome in step 4 is the same regardless of the order of m and n in the problem.

2. $3x^2 + 5x + 2$

Step 1 $m \cdot n = 3 \cdot 2 = 6$ and $m + n = 5$
m and n are 2 and 3.

$$\overbrace{}^{5x}$$

Step 2 $= 3x^2 + 2x + 3x + 2$

Step 3 $= (3x^2 + 2x) + (3x + 2)$
$= x(3x + 2) + 1(3x + 2)$

We observe in the last two terms that the greatest common factor is only 1 or -1.

Step 4 $= (3x + 2)(x + 1)$

3. $4x^2 - 11x + 6$

Step 1 $m \cdot n = 4 \cdot 6 = 24$ and $m + n = -11$
m and n are -3 and -8.

$$\overbrace{}^{-11x}$$

Step 2 $= 4x^2 - 3x - 8x + 6$

Step 3 $= (4x^2 - 3x) + (-8x + 6)$
$= x(4x - 3) - 2(4x - 3)$

We have 2 or -2 as the greatest common factor in the last two terms. We factor out -2 so that we will have the same quantity inside the parentheses.

Step 4 $= (4x - 3)(x - 2)$

Note
If the third term in step 2 is preceded by a minus sign, we will usually factor out the negative factor.

4. $6x^2 + 9x + 4$

Step 1 $m \cdot n = 6 \cdot 4 = 24$ and $m + n = 9$
Our m and n values are not obvious by inspection, therefore we use our procedure to determine them.

Product	Sum
24	**+**
24 · 1	24 + 1 = 25
12 · 2	12 + 2 = 14
8 · 3	8 + 3 = 11
6 · 4	6 + 4 = 10
4 · 6 ◄	Repeats

We observe that in the second column, we never obtained a sum of 9, therefore $6x^2 + 9x + 4$ *will not factor.*

Factoring by inspection—an alternate approach

In the beginning of this section we studied a systematic procedure for determining if a trinomial will factor and how to factor it. In many instances we can determine how the trinomial will factor by inspecting the problem rather than by applying this procedure.

Factoring by inspection is accomplished as follows. Factor $7x + 2x^2 + 3$.

Step 1 Write the trinomial in standard form.

$$2x^2 + 7x + 3$$

Step 2 Determine the possible combinations of first-degree factors of the first term.

$$(2x\quad)(x\quad)$$

Step 3 Combine with the factors of step 2 all the possible factors of the last term.

$$(2x\quad3)(x\quad1)$$
$$(2x\quad1)(x\quad3)$$

Step 4 Determine the possible symbol, $(+$ or $-)$, between the terms in each binomial.

$$(2x + 3)(x + 1)$$
$$(2x + 1)(x + 3)$$

The rules of signed numbers given in chapter 1 provide the answer to step 4.

1. If the third term is preceded by a $+$ sign and the middle term is preceded by a $+$ sign, then the symbols will be

$$(\quad+\quad)(\quad+\quad).$$

2. If the third term is preceded by a $+$ sign and the middle term is preceded by a $-$ sign, then the symbols will be

$$(\quad-\quad)(\quad-\quad).$$

3. If the third term is preceded by a $-$ sign, then the symbols will be

$$(\quad+\quad)(\quad-\quad)$$

or

$$(\quad-\quad)(\quad+\quad).$$

Note
It is assumed that the first term is preceded by a $+$ sign or no sign. If it is preceded by a $-$ sign, these rules could still be used if (-1) is first factored out of all the terms.

Step 5 Determine which factors, if any, yield the correct middle term.

$(2x + 3)(x + 1)$

$+3x$

$+2x$

$(+3x) + (+2x) = +5x$

$(2x + 1)(x + 3)$

$+x$

$+ 6x$

$(+x) + (+6x) = {}^+7x$

The second set of factors gives us the correct middle term. Therefore $(2x + 1)(x + 3)$ is the factorization of $2x^2 + 7x + 3$.

Example 3–6 B

Factor completely the following trinomials by inspection.

1. Step 1 $x^2 + 7x + 12$

Step 2 $(x \quad)(x \quad)$

Step 3 $(x \quad 12)(x \quad 1)$
$(x \quad 6)(x \quad 2)$
$(x \quad 3)(x \quad 4)$

Step 4 $(x + 12)(x + 1)$
$(x + 6)(x + 2)$
$(x + 3)(x + 4)$

Step 5 $(x + 12)(x + 1)$

$+12x$

$+x$

$(+12x) + (+x) = +13x$

$(x + 6)(x + 2)$

$+6x$

$+2x$

$(+6x) + (+2x) = +8x$

$(x + 3)(x + 4)$

$+3x$

$+4x$

$(+3x) + (+4x) = {}^+7x$

The factorization of $x^2 + 7x + 12$ is $(x + 3)(x + 4)$.

2. Step 1 $6x^2 + 17x + 5$

Step 2 $(6x\ \ )(x\ \ )$
$(3x\ \ )(2x\ \ )$

Step 3 $(6x\ \ 5)(x\ \ 1)$
$(6x\ \ 1)(x\ \ 5)$
$(3x\ \ 5)(2x\ \ 1)$
$(3x\ \ 1)(2x\ \ 5)$

Step 4 $(6x + 5)(x + 1)$
$(6x + 1)(x + 5)$
$(3x + 5)(2x + 1)$
$(3x + 1)(2x + 5)$

Step 5 $(6x + 5)(x + 1)$

$+5x$ $(+5x) + (+6x) = +11x$

$+6x$

$(6x + 1)(x + 5)$

$+x$ $(+x) + (+30x) = +31x$

$+30x$

$(3x + 5)(2x + 1)$

$+10x$ $(+10x) + (+3x) = +13x$

$+3x$

$(3x + 1)(2x + 5)$

$+2x$ $(+2x) + (+15x) = +17x$

$+15x$

The last set of factors gives us the correct middle term. Hence $(3x + 1)(2x + 5)$ is the factorization of $6x^2 + 17x + 5$.

3. $13x - 5 + 6x^2$

Step 1 $6x^2 + 13x - 5$

Step 2 $(6x\ \ )(x\ \ )$
$(2x\ \ )(3x\ \ )$

Step 3 $(6x \quad 5)(x \quad 1)$
$(6x \quad 1)(x \quad 5)$
$(2x \quad 5)(3x \quad 1)$
$(2x \quad 1)(3x \quad 5)$

Step 4 $(6x + 5)(x - 1)$ or $(6x - 5)(x + 1)$
$(6x + 1)(x - 5)$ or $(6x - 1)(x + 5)$
$(2x + 5)(3x - 1)$ or $(2x - 5)(3x + 1)$
$(2x + 1)(3x - 5)$ or $(2x - 1)(3x + 5)$

Step 5 $(6x + 5)(x - 1)$ or $(6x - 5)(x + 1)$

$+5x \qquad\qquad -5x$

$-6x \qquad\qquad +6x$

$(+5x) + (-6x) = -x$ or $(-5x) + (+6x) = +x$

$(6x + 1)(x - 5)$ or $(6x - 1)(x + 5)$

$+x \qquad\qquad -x$

$-30x \qquad\qquad +30x$

$(+x) + (-30x) = -29x$ or $(-x) + (+30x) = +29x$

$(2x + 5)(3x - 1)$ or $(2x - 5)(3x + 1)$

$+15x \qquad\qquad -15x$

$-2x \qquad\qquad +2x$

$(+15x) + (-2x) - +13x$ or $(-15x) + (+2x) = -13x$

$(2x + 1)(3x - 5)$ or $(2x - 1)(3x + 5)$

$+3x \qquad\qquad -3x$

$-10x \qquad\qquad +10x$

$(+3x) + (-10x) = -7x$ or $(-3x) + (+10x) = +7x$

The factorization of $6x^2 + 13x - 5$ is $(2x + 5)(3x - 1)$.

Can you
- Determine two integers whose product is one number and whose sum is another number?
- Recognize when a trinomial will factor and when it will not?
- Factor trinomials of the form $ax^2 + bx + c$?
- Always remember to look for common factors before applying any of the factoring rules?

Exercise 3–6

Directions Determine if the following expressions will factor by finding m and n, if they exist.

1. $4x^2 - 5x + 1$
2. $x^2 + 10x + 25$
3. $2x^2 - 9x + 10$
4. $2a^2 + 13a + 18$
5. $a^2 + 16a + 64$
6. $10x^2 + 7x - 6$
7. $a^2 - 81$
8. $2x^2 - x - 4$
9. $2x^2 - 5x - 12$
10. $9y^2 - 30y + 25$
11. $12z^2 + 16z + 5$
12. $12a^2 - 37a + 28$

Directions Factor completely each trinomial. If a trinomial will not factor, state that as the answer. See examples 3–6 A and B.

13. $3a^2 + 7a - 6$
14. $2y^2 + y - 6$
15. $4x^2 - 5x + 1$
16. $2z^2 + 3z + 1$
17. $x^2 + 18x + 32$
18. $2a^2 - 7a + 6$
19. $2y^2 - y - 1$
20. $5x^2 - 7x - 6$
21. $8x^2 - 17x + 2$
22. $9x^2 - 6x + 1$
23. $2a^2 - 11a + 12$
24. $5y^2 + 4y + 6$
25. $2x^2 + 13x + 18$
26. $6x^2 + 13x + 6$
27. $7x^2 + 20x - 3$
28. $4a^2 + 20a + 21$
29. $4x^2 - 4x - 3$
30. $4z^2 - 2z + 5$
31. $6x^2 - 23x - 4$
32. $9x^2 - 21x - 8$
33. $10z^2 + 9z + 2$
34. $10y^2 + 7y - 6$
35. $7x^2 - 3x + 6$
36. $2x^2 - 9x + 10$
37. $5x^2 - 9x - 2$
38. $6x^2 + 21x + 18$
39. $6x^2 - 17x + 12$
40. $4y^2 + 10y + 6$
41. $3x^2 + 12x + 12$
42. $6x^2 + 7x - 3$
43. $3x^2 + 8x - 4$
44. $2x^2 + 6x - 20$
45. $6x^2 - 38x + 40$
46. $18x^2 + 15x - 18$
47. $-4x^2 - 12x - 9$
48. $-9z^2 + 30z - 25$
49. $-7x^2 + 36x - 5$
50. $-3x^2 - 2x - 4$
51. $12x^2 + 13x - 4$
52. $15x^2 + 2x - 1$
53. $4x^3 + 10x^2 + 4x$
54. $2x^3 - 6x^2 - 20x$
55. $18x^2 - 42x - 16$
56. $2x^3 + 15x^2 + 7x$
57. $8x^2 - 18x + 9$
58. $8y^2 - 14y - 15$

Directions Solve the following verbal problems.

59. In a rectangle, if the width is 2 inches shorter than the length and the area is given as 80 square inches, it is necessary to factor the trinomial $x^2 - 2x - 80$ to find the dimensions. Factor this trinomial.

60. When a stone is thrown vertically into the air, the height, S, of the stone at any instant in time, t, is given by $S = -16t^2 + 32t - 16$. Factor the right member.

61. If an open box is formed from a sheet of cardboard, 6 inches on a side, by cutting equal squares of size x from each corner and then bending up the sides, the volume of the box would be given by the formula $4x^3 - 24x^2 + 36x$. Factor this trinomial.

3–7 Other types of trinomials

In exercise set 3–6, exercise 17, we factored the trinomial $x^2 + 18x + 32$ into $(x + 16)(x + 2)$. This problem is a special case of the general trinomial because the leading coefficient a is equal to 1. When we factor a trinomial where $a = 1$, we can eliminate several of the steps in our factoring procedure. In fact, once we find the values for m and n, the problem is essentially finished.

Consider again the factorization of $x^2 + 18x + 32$. On closer inspection we see that the final factored form has the m and n values, 16 and 2, simply placed inside the parentheses, $(x + 16)(x + 2)$.

In our example, a is equal to 1 and the conditions for m and n become $m \cdot n = 1 \cdot c = c$ and $m + n = b$. Hence trinomials of the form $x^2 + bx + c$ will factor only if we are able to find two integers, m and n, such that $m \cdot n = c$ and $m + n = b$.

Example 3–7 A

Completely factor the following trinomials.

1. $x^2 + 2x - 24$
$m \cdot n = -24$ and $m + n = b = 2$. Therefore m and n are 6 and -4 and we have

$(x + 6)(x - 4)$.

2. $x^2 - 8x + 15$
$m \cdot n = 15$ and $m + n = -8$. Hence m and n are -5 and -3 and we have

$(x - 5)(x - 3)$.

There are many other types of trinomials that will factor by our m and n procedure. Let us examine the problem

$$x^2y^2 - 8xy + 15.$$

If we consider (xy) as a quantity, we have

$$(xy)^2 - 8(xy) + 15,$$

and our problem looks like the previous example. Since m and n are -5 and -3, it factors into

$$(xy - 5)(xy - 3).$$

Example 3–7 B

Completely factor the following trinomials.

1. $x^2 + xy - 6y^2 = 1 \cdot x^2 + 1 \cdot xy - 6 \cdot y^2$

    Step 1 $a = 1, b = 1$, and $c = -6$. Hence $m \cdot n = 1 \cdot (-6) = -6$ and $m + n = 1$. Therefore m and n are 3 and -2.

                            $\overbrace{xy}$

    Step 2 $= x^2 + \overbrace{3xy - 2xy} - 6y^2$

    Step 3 $= (x^2 + 3xy) + (-2xy - 6y^2)$
               $= x(x + 3y) - 2y(x + 3y)$

    Step 4 $= (x + 3y)(x - 2y)$

> **Note**
> We observe that our alternate procedure would work if we put x and y in the parentheses first and then insert the m and n values, that is, $(x \quad y)(x \quad y)$ then
> $$(x + 3y)(x - 2y).$$

2. $-4a^2 + 8ab - 4b^2 = -4(a^2 - 2ab + b^2)$

> **Note**
> Whenever the leading coefficient is negative, the problem will be easier if we factor out the greatest common negative factor.

    $m \cdot n = 1$ and $m + n = -2$. Therefore m and n are -1 and -1 and the completely factored form would be

$$-4(a - b)(a - b) = -4(a - b)^2.$$

3. $6x^2 - 11xy + 4y^2$

    Step 1 $m \cdot n = 6 \cdot 4 = 24$ and $m + n = -11$. Hence m and n are -3 and -8.

                            $\overbrace{-11xy}$

    Step 2 $= 6x^2 - \overbrace{3xy - 8xy} + 4y^2$

Step 3 $= (6x^2 - 3xy) + (-8xy + 4y^2)$
$= 3x(2x - y) - 4y(2x - y)$

Step 4 $= (2x - y)(3x - 4y)$

4. $x^2(a - b) + 6x(a - b) + 8(a - b)$

Observe that $(a - b)$ is a common factor of each term and we have

$(a - b)(x^2 + 6x + 8)$.

The trinomial $x^2 + 6x + 8$ has m and n values of 4 and 2 and the completely factored form is

$(a - b)(x + 4)(x + 2)$.

Perfect-square trinomials

Two of the special products that we studied in section 3–3 were the squares of a binomial. We will now restate those special products.

$$a^2 + 2ab + b^2 = (a + b)^2$$

and

$$a^2 - 2ab + b^2 = (a - b)^2.$$

The right members of the equations are called the squares of binomials, and the left members are called perfect-square trinomials. Perfect-square trinomials can always be factored by our m and n procedure. However if we observe that the first and last terms of a trinomial are perfect squares, we should see if the trinomial will factor as the square of a binomial. To factor a trinomial as a perfect-square trinomial, three conditions need to be met.

1. The first term must have a positive coefficient and be a perfect square, a^2.
2. The last term must have a positive coefficient and be a perfect square, b^2.
3. The middle term must be twice the product of the bases of the first and last terms, $2ab$.

We observe that

$$9x^2 + 12x + 4$$
$$= (3x)^2 + 2(3x)(2) + (2)^2.$$

Condition 1   Condition 3   Condition 2

Therefore it is a perfect-square trinomial and factors into

$$(3x + 2)^2.$$

Example 3–7 C   The following examples show the factoring of some other perfect-square trinomials.

		Condition 1		Condition 3		Condition 2	Square of a binomial
1.	$4x^2 + 20x + 25 =$	$(2x)^2$	$+$	$2(2x)(5)$	$+$	$(5)^2$	$= (2x + 5)^2$
2.	$9x^2 - 6x + 1 =$	$(3x)^2$	$-$	$2(3x)(1)$	$+$	$(1)^2$	$= (3x - 1)^2$
3.	$16x^2 + 24x + 9 =$	$(4x)^2$	$+$	$2(4x)(3)$	$+$	$(3)^2$	$= (4x + 3)^2$
4.	$9y^2 - 30y + 25 =$	$(3y)^2$	$-$	$2(3y)(5)$	$+$	$(5)^2$	$= (3y - 5)^2$

Mastery points

Can you
- Always remember to look for common factors before applying any of the factoring rules?
- Recognize special cases of the general trinomial?
- Factor perfect-square trinomials?

Exercise 3–7

Directions Write in completely factored form. Assume all variables used as exponents represent positive integers. See examples 3–7 A, B, and C.

1. $x^2 + 13x - 30$

2. $a^2 - 14a + 24$

3. $y^2 + 5y - 24$

4. $a^2 + 8a + 12$

5. $x^2 - 2x - 24$

6. $y^2 - y - 12$

7. $3x^2 + 33x - 36$

8. $3y^2 - 18y - 48$

9. $4x^2 - 4x - 24$

10. $5y^2 - 15y - 55$

11. $3y^2 - 27y + 12$

12. $4s^2 - 28s + 12$

13. $x^2y^2 - 4xy - 21$

14. $x^2y^2 - 3xy - 18$

15. $x^2y^2 + 13xy + 12$

16. $a^2b^2 + 13ab - 30$

17. $x^2y^2 - 14xy + 24$

18. $y^2z^2 + 5yz - 24$

19. $4a^2b^2 + 24ab + 32$

20. $3x^2y^2 + 21xy + 30$

21. $-2x^2y^2 + 6xy + 20$

22. $-3x^2y^2 + 6xy + 45$

23. $-4a^2b^2 - 20ab - 24$

24. $2a^2b^2 + ab - 6$

25. $3x^2y^2 + 7xy - 6$

26. $5x^2y^2 - 7xy - 6$

27. $2a^2b^2 - 7ab + 6$

28. $9a^2b^2 - 6ab + 1$

29. $2x^2y^2 - xy - 1$

30. $5a^2b^2 + 4ab + 6$

31. $6x^2y^2 + 13xy + 6$

32. $(a + b)^2 - 4(a + b) - 12$

33. $(x - 2y)^2 + 10(x - 2y) + 9$

34. $(2a - b)^2 + (2a - b) - 30$

35. $(y + 3z)^2 - 5(y + 3z) - 14$

36. $(2a + 3b)^2 + 8(2a + 3b) + 12$

37. $(R + 5S)^2 - 6(R + 5S) + 8$

38. $(x - 4y)^2 - 10(x - 4y) + 21$

39. $x^2 + 10x + 25$

40. $a^2 + 16a + 64$

41. $b^2 - 12b + 36$

42. $x^2 + 14x + 49$

43. $c^2 + 18c + 81$

44. $4x^2 + 20x + 25$

45. $9y^2 + 12y + 4$

46. $a^2 - 16ab + 64b^2$

47. $x^2 - 14xy + 49y^2$

48. $9a^2 - 12ab + 4b^2$ **49.** $4x^2 - 12xy + 9y^2$ **50.** $25a^2 + 10ab + b^2$

51. $9x^2 + 30xy + 25y^2$ **52.** $a^2 + 7ab + 10b^2$ **53.** $x^2 + 11xy + 24y^2$

54. $x^2 - 12xy + 32y^2$ **55.** $R^2 + 5RS - 24S^2$ **56.** $y^2 - 4yz - 5z^2$

57. $a^2 + 2ab - 35b^2$ **58.** $y^2 - 4yz - 12z^2$ **59.** $x^2 + 10xy + 16y^2$

60. $3x^2 + xy - 2y^2$ **61.** $4y^2 + 11yz - 3z^2$ **62.** $2a^2 - 3ab - 2b^2$

63. $3R^2 - 4RS - 4S^2$ **64.** $12x^2 - 5xy - 3y^2$

65. $x^2(a - 2b) + 8x(a - 2b) + 12(a - 2b)$ **66.** $a^2(x - 3y) + 14a(x - 3y) + 24(x - 3y)$

67. $x^2(y + 2z) - 13x(y + 2z) + 30(y + 2z)$ **68.** $a^2(2b - c) + a(2b - c) - 12(2b - c)$

69. $x^2(3y - z) + 15x(3y - z) + 36(3y - z)$ **70.** $a^2(2b + 3c) - 6a(2b + 3c) - 27(2b + 3c)$

71. $a^2(3x - y) - 7a(3x - y) - 60(3x - y)$ **72.** $x^2(2y + z) + 2x(2y + z) - 63(2y + z)$

73. $a^2(x + 5y) - 9a(x + 5y) + 14(x + 5y)$ **74.** $x^{2n} + 5x^n + 6$

75. $x^{2n} + 9x^n + 14$ **76.** $x^{2n} - 4x^n - 12$ **77.** $x^{2n} - 8x^n + 15$ **78.** $3x^{2n} - x^n - 2$

79. $2x^{2n} - 7x^n - 4$ **80.** $3x^{2n} - 4x^n - 4$ **81.** $2x^{2n} - 7x^n + 6$ **82.** $5x^{2n} - 19x^n - 4$

Directions Solve the following verbal problems.

83. If a rectangle has a perimeter of 52 inches and an area of 153 square inches, the dimensions of the rectangle can be found by factoring the expression $W^2 - 26W + 153$. Factor this expression.

84. If the length of a rectangle is 3 meters more than three times its width and its area is 168 square meters, then the dimensions of the rectangle can be found by factoring the expression $3W^2 + 3W - 168$. Factor this expression.

85. When two capacitors are connected in parallel, the total capacitance of the circuit, C_t, is given by $C_t = C_1 + C_2$. If the number of microfarads in C_1 is the square of the number in C_2 and the total capacitance is 42 microfarads, then the expression to find the capacitance C_2 is given by $C_2^2 + C_2 - 42$. Factor this expression.

3-8 Other methods of factoring

The difference of two squares

We saw in section 3–3 that the product of $(a + b)(a - b)$ is $a^2 - b^2$. We refer to the indicated product $(a + b)(a - b)$ as the *product* of the *sum* and *difference* of the same two terms. Notice that in one factor we *add* the terms and in the other we find the *difference* between these same terms. The product will *always* be the *difference of the squares of the two terms*.

To factor the *difference of two squares*, we reverse the equation from section 3–3.

> ■ **The difference of two squares factor as**
> $$a^2 - b^2 = (a + b)(a - b).$$
>
> **Concept**
> Verbally we have
> (first term)^2 — (second term)^2 factors into
> (first term + second term) · (first term − second term).

In order to use this factoring technique, we must be able to recognize **perfect squares.** For example,

$$16, \quad x^2, \quad 25a^2, \quad 9y^4$$

are all perfect squares since

$$16 = 4 \cdot 4 = (4)^2$$
$$x^2 = x \cdot x = (x)^2$$
$$25a^2 = 5a \cdot 5a = (5a)^2$$
$$9y^4 = 3y^2 \cdot 3y^2 = (3y^2)^2.$$

We see from the examples that *a perfect square is any quantity that can be written as an exact square of a rational quantity.*

This is the procedure we use for factoring the difference of two squares.

Step 1 Identify that we have a perfect square minus another perfect square.

Step 2 Rewrite the problem as a first term squared minus a second term squared.

$$\text{(first term)}^2 - \text{(second term)}^2$$

Step 3 Factor the problem into the first term plus the second term times the first term minus the second term.

$$\text{(first term + second term)(first term − second term)}$$

Example 3–8 A

Write in completely factored form.

1. $x^2 - 16 = (x)^2 - (4)^2$
$\qquad\qquad\quad = (x + 4)(x - 4)$

2. $25a^2 - 4b^2 = (5a)^2 - (2b)^2$
$\qquad\qquad\qquad = (5a + 2b)(5a - 2b)$

3. $2x^2 - 18y^2 = 2(x^2 - 9y^2)$
$\qquad\qquad\qquad = 2\,[(x)^2 - (3y)^2]$
$\qquad\qquad\qquad = 2(x + 3y)(x - 3y)$

We should see from example 3 that our first step in any factoring problem is to look for any common factors. Often an expression that does not appear to be factorable becomes more apparently so by taking out the common factor.

4. $x^4 - 81 = (x^2)^2 - (9)^2$
$$= (x^2 + 9)(x^2 - 9)$$
$$= (x^2 + 9)[(x)^2 - (3)^2]$$
$$= (x^2 + 9)(x + 3)(x - 3)$$

Our first factorization of $x^4 - 81$ yielded the factors $x^2 + 9$ and $x^2 - 9$. On further inspection we observe that $x^2 - 9$ is the difference of two squares.

5. $x^{2n} - 4 = (x^n)^2 - (2)^2$
$$= (x^n + 2)(x^n - 2)$$

The difference of two cubes

To factor problems involving the difference of two squares, we identified the two terms as squares and applied the property. Now we will factor the **difference of two cubes** in a similar fashion.

Consider the indicated product of $(a - b)(a^2 + ab + b^2)$. If we carry out the multiplication, we have

$$(a - b)(a^2 + ab + b^2) = a^3 + a^2b + ab^2 - a^2b - ab^2 - b^3$$
$$= a^3 - b^3.$$

Therefore $(a - b)(a^2 + ab + b^2) = a^3 - b^3$ and $(a - b)(a^2 + ab + b^2)$ is the factored form of $a^3 - b^3$.

■ **The difference of two cubes factors as**
$$a^3 - b^3 = (a - b)(a^2 + ab + b^2).$$

Concept
If we are able to write a two-term polynomial as a first term cubed minus a second term cubed, then it will factor as the difference of two cubes.

(1st term)³ − (2nd term)³ =
(1st term − 2nd term) [(1st term)² + (1st term · 2nd term) + (2nd term)²]

Example 3–8 B

Factor completely.

1. $a^3 - 8$

We rewrite the first term as a cube and the second term as a cube.

$a^3 - 8 = (a)^3 - (2)^3$

The first term is a and the second term is 2. Then we write down the procedure for factoring the difference of two cubes.

$(\quad - \quad) [(\quad)^2 + (\quad)(\quad) + (\quad)^2]$

  ↑   ↑   ↑   ↑   ↑   ↑
 1st 2nd   1st   1st 2nd   2nd

Now substitute a where the first term is and 2 where the second term is.

Exponents and Polynomials

$$(a - 2) [(a)^2 + (a)(2) + (2)^2]$$

$$\uparrow \quad \uparrow \quad \uparrow \qquad \uparrow \quad \uparrow \qquad \uparrow$$
1st 2nd 1st   1st 2nd   2nd

Finally, we simplify.

$$(a - 2)(a^2 + 2a + 4)$$

Therefore $a^3 - 8 = (a - 2)(a^2 + 2a + 4)$.

2. $40x^3 - 135y^3$

Factor out the common factor of 5.

$$= 5(8x^3 - 27y^3) = 5 [(2x)^3 - (3y)^3]$$

Then $5(\quad - \quad) [(\quad)^2 + (\quad)(\quad) + (\quad)^2]$

and $5(2x - 3y) [(2x)^2 + (2x)(3y) + (3y)^2]$
$$= 5(2x - 3y)(4x^2 + 6xy + 9y^2).$$

3. $a^{21} - 64b^{15} = (a^7)^3 - (4b^5)^3$

Then $(\quad - \quad) [(\quad)^2 + (\quad)(\quad) + (\quad)^2]$

and $(a^7 - 4b^5) [(a^7)^2 + (a^7)(4b^5) + (4b^5)^2]$
$$= (a^7 - 4b^5)(a^{14} + 4a^7b^5 + 16b^{10}).$$

Note

In example 3 we observe that a number raised to a power that is a multiple of 3 can be written as a cube by dividing the exponent by 3. The quotient is the exponent of the number inside the parentheses and the 3 is the exponent outside the parentheses. For example,
$$y^{12} = (y^4)^3 \text{ and } z^{24} = (z^8)^3.$$

The sum of two cubes

If we carry out the indicated multiplication in $(a + b)(a^2 - ab + b^2)$, we have

$$(a + b)(a^2 - ab + b^2) = a^3 - a^2b + ab^2 + a^2b - ab^2 + b^3$$
$$= a^3 + b^3.$$

Therefore $(a + b)(a^2 - ab + b^2) = a^3 + b^3$ and $(a + b)(a^2 - ab + b^2)$ is the factored form of $a^3 + b^3$.

> ■ **The sum of two cubes factors as**
> $$a^3 + b^3 = (a + b)(a^2 - ab + b^2).$$
>
> **Concept**
> If we are able to write a two-term polynomial as a first term cubed plus a second term cubed, then it will factor as the sum of two cubes.
> $$\text{(1st term)}^3 + \text{(2nd term)}^3 =$$
> $$\text{(1st term + 2nd term)} [\text{(1st term)}^2 - \text{(1st term)(2nd term)} + \text{(2nd term)}^2]$$

Example 3–8 C

Factor completely.

1. $x^3 + 27 = (x)^3 + (3)^3$

We now write the procedure for the sum of two cubes.

$$(\quad + \quad)[(\quad)^2 - (\quad)(\quad) + (\quad)^2]$$

$\uparrow \quad \uparrow \qquad \uparrow \qquad \uparrow \quad \uparrow \qquad \uparrow$

1st 2nd   1st   1st 2nd   2nd

Substitute.

$$(x + 3)[(x)^2 - (x)(3) + (3)^2]$$

$\uparrow \quad \uparrow \quad \uparrow \qquad \uparrow \quad \uparrow \quad \uparrow$

1st 2nd 1st   1st 2nd 2nd

Simplify.

$$(x + 3)(x^2 - 3x + 9)$$

Therefore $x^3 + 27 = (x + 3)(x^2 - 3x + 9)$.

2. $8x^3 + y^9 = (2x)^3 + (y^3)^3$

The first term is $2x$ and the second term is y^3.

Then $(\quad + \quad)[(\quad)^2 - (\quad)(\quad) + (\quad)^2]$

and $(2x + y^3)[(2x)^2 - (2x)(y^3) + (y^3)^2]$

$= (2x + y^3)(4x^2 - 2xy^3 + y^6)$.

3. $a^3b^3 + 27c^3 = (ab)^3 + (3c)^3$

Then $(\quad + \quad)[(\quad)^2 - (\quad)(\quad) + (\quad)^2]$

and $(ab + 3c)[(ab)^2 - (ab)(3c) + (3c)^2]$

$= (ab + 3c)(a^2b^2 - 3abc + 9c^2)$.

Mastery points

Can you

• Factor the difference of two squares?
• Factor the sum or difference of two cubes?

Exercise 3–8

Directions Write in completely factored form. Assume all variables used as exponents represent positive integers. See examples 3–8 A, B, and C.

1. $a^2 - 49$ 　　　**2.** $36 - R^2$ 　　　**3.** $64 - S^2$ 　　　**4.** $4a^2 - 25b^2$

5. $36x^2 - y^4$ 　　**6.** $x^4 - 4$ 　　　**7.** $16a^2 - 49b^2$ 　　**8.** $144x^2 - 121y^2$

9. $8a^2 - 32b^2$ 　　**10.** $3x^2 - 27y^2$ 　　**11.** $50 - 2x^2$ 　　　**12.** $128a^2 - 2b^2$

13. $98a^2b^2 - 50x^2y^2$ 　**14.** $x^4 - y^4$ 　　**15.** $x^4 - 81$ 　　　**16.** $16R^4 - 1$

17. $x^4 - 16y^4$ **18.** $x^{2n} - 1$ **19.** $a^{2n} - 4$ **20.** $x^{4n} - 25$

21. $x^{2n} - y^{2n}$ **22.** $x^{4n} - 16$ **23.** $x^{4n} - 81$

24. $a^2(x + 2y) - b^2(x + 2y)$ **25.** $4x^2(a + 5b) - y^2(a + 5b)$ **26.** $2a^2(x + y) - 8b^2(x + y)$

27. $3a^2(3x - y) - 27(3x - y)$ **28.** $4a(x^2 - y^2) - 8b(x^2 - y^2)$ **29.** $9x(4a^2 - b^2) - 3y(4a^2 - b^2)$

30. $(a + b)^2 - (x - y)^2$ **31.** $(2a + b)^2 - (x - 2y)^2$ **32.** $(2x - y)^2 - (x + y)^2$

33. $(3a - b)^2 - (2a - b)^2$ **34.** $(4x + 3y)^2 - (4x - 2y)^2$ **35.** $(x + 2y)^2 - (x - 3y)^2$

36. $R^3 - 8$ **37.** $27a^3 - b^3$ **38.** $8y^3 - 27$ **39.** $x^3 + y^3$ **40.** $27 + x^3$

41. $a^3 + 8b^3$ **42.** $64a^3 + b^3$ **43.** $8y^3 - 27x^3$ **44.** $64a^3 - 8$ **45.** $81a^3 - 3b^3$

46. $2a^3 + 16$ **47.** $3x^3 + 24$ **48.** $2a^3 - 16$ **49.** $64z^3 + 125$ **50.** $a^5 + 27a^2b^3$

51. $2a^3 - 54b^3$ **52.** $8x^2y^3 - x^5$ **53.** $R^5 + 64R^2S^3$ **54.** $54y^3 + 2z^3$ **55.** $3x^5 - 81x^2y^3$

56. $a^3b^9 - c^{15}$ **57.** $x^3y^{12} + z^9$ **58.** $x^{15}y^6 - 8z^9$ **59.** $a^{18}b^9 - 27c^3$ **60.** $2a^3b^3 + 16$

61. $3x^3y^6 + 81z^3$ **62.** $(a + 2b)^3 - c^3$ **63.** $(x + 3y)^3 + z^3$

64. $(2a - b)^3 + 8c^3$ **65.** $(4a + b)^3 - 27c^3$ **66.** $(2a + b)^3 - (x - y)^3$

67. $(a - 2b)^3 - (2x + y)^3$ **68.** $(3a + b)^3 - (a - 2b)^3$ **69.** $(3x - y)^3 + (2x + 3y)^3$

70. $(a - 2b)^3 - (a + b)^3$ **71.** $(2x - y)^3 + (x + y)^3$ **72.** $(5x - 2y)^3 + (3x + 2y)^3$

73. In engineering the equation of transverse shearing stress in a rectangular beam is given in two forms:

(a) $T = \dfrac{V}{8I}(h^2 - 4v_1^2)$ and (b) $T = \dfrac{3V}{2A}\left(1 - \dfrac{4V_1^2}{H^2}\right)$.

Factor the right member in each part.

3-9 Factoring: A general strategy

In this section we will review the different methods of factoring that we have studied in the previous sections. The following outline gives a general strategy for factoring polynomials.

I. Factor out any common factors.
   Examples
   1. $5x^3 - 25x^2 = 5x^2(x - 5)$
   2. $x(a - 2b) + 2y(a - 2b) = (a - 2b)(x + 2y)$

II. Count the number of terms.
   A. Two terms: Check to see if the polynomial is the difference of two squares, the difference of two cubes, or the sum of two cubes.
   Examples
   1. $x^2 - 25y^2 = (x - 5y)(x + 5y)$ Difference of two squares
   2. $8a^3 - b^3 = (2a - b)(4a^2 + 2ab + b^2)$ Difference of two cubes
   3. $m^3 + 64n^3 = (m + 4n)(m^2 - 4mn + 16n^2)$ Sum of two cubes

B. Three terms: Check to see if the polynomial is a perfect-square trinomial. If it is not, use one of the general methods for factoring a trinomial.

Examples

1. $m^2 + 6m + 9 = (m + 3)^2$ Perfect-square trinomial
2. $m^2 - 5m - 14 = (m - 7)(m + 2)$ General trinomial, leading coefficient of 1
3. $6x^2 + 7x - 20 = (2x + 5)(3x - 4)$ General trinomial, leading coefficient other than 1

C. Four terms: Check to see if we can factor by grouping.

Examples

1. $ax + 3a - 2bx - 6b = (a - 2b)(x + 3)$
2. $x^3 + 2x^2 - 3x - 6 = (x^2 - 3)(x + 2)$

III. Check to see if any of the factors we have written can be factored further. Any common factors that were missed in part I can still be factored out here.

Examples

1. $x^4 - 11x^2 + 28 = (x^2 - 4)(x^2 - 7) = (x - 2)(x + 2)(x^2 - 7)$
   Difference of two squares
2. $4a^2 - 36b^2 = (2a - 6b)(2a + 6b) = 2(a - 3b)2(a + 3b)$
   $= 4(a - 3b)(a + 3b)$ Overlooked common factor

The following examples illustrate our strategy for factoring polynomials.

Examples 3–9 A

Completely factor the following polynomials.

1. $12a^3 - 3ab^2$

   I. First, we look for any common factors.

   $3a(4a^2 - b^2)$

   II. The factor $4a^2 - b^2$ has two terms and is the difference of two squares.

   $3a(4a^2 - b^2) = 3a(2a - b)(2a + b)$

   III. After checking to see if any of the factors will factor further, we conclude that $3a(2a - b)(2a + b)$ is the completely factored form. Therefore

   $12a^3 - 3ab^2 = 3a(2a - b)(2a + b)$.

2. $3x^2 + 7xy - 6y^2$

   I. There is no common factor (other than 1 or -1).

   II. Since this trinomial is not a perfect square, we use our general method for factoring trinomials.

   $m \cdot n = -18$ and $m + n = 7$, m and n are -2 and 9.

   $3x^2 + 7xy - 6y^2$
   $= 3x^2 - 2xy + 9xy - 6y^2$
   $= x(3x - 2y) + 3y(3x - 2y)$
   $= (3x - 2y) (x + 3y)$

   III. None of the factors will factor further.

   $3x^2 + 7xy - 6y^2 = (3x - 2y)(x + 3y)$

3. $2am^2 + bm^2 - 2an^2 - bn^2$

    I. There is no common factor (other than 1 or -1).

    II. The polynomial has four terms and we factor by grouping.

$$2am^2 + bm^2 - 2an^2 - bn^2$$
$$= m^2(2a + b) - n^2(2a + b)$$
$$= (2a + b)(m^2 - n^2)$$

    III. The factor $m^2 - n^2$ is the difference of two squares. The complete factorization is

$$2am^2 + bm^2 - 2an^2 - bn^2 = (2a + b)(m - n)(m + n).$$

4. $5x^3 + 5x^2 + 20x$

    I. The greatest common factor is $5x$.

$$5x(x^2 + x + 4)$$

    II. The trinomial $x^2 + x + 4$ will not factor.

    III. None of the factors will factor further.

$$5x^3 + 5x^2 + 20x = 5x(x^2 + x + 4)$$

5. $a^8 + 8a^2b^3$

    I. The greatest common factor is a^2.

$$a^2(a^6 + 8b^3)$$

    II. The factor $a^6 + 8b^3$ is the sum of two cubes.

$$a^2(a^6 + 8b^3) = a^2(a^2 + 2b)(a^4 - 2a^2b + 4b^2)$$

    III. None of the factors will factor further.

$$a^8 + 8a^2b^3 = a^2(a^2 + 2b)(a^4 - 2a^2b + 4b^2)$$

6. $a^2(4 - x^2) + 8a(4 - x^2) + 16(4 - x^2)$

    I. The greatest common factor is $4 - x^2$.

$$(4 - x^2)(a^2 + 8a + 16)$$

    II. The factor $4 - x^2$ is the difference of two squares. The factor $a^2 + 8a + 16$ is a perfect-square trinomial.

$$(2 - x)(2 + x)(a + 4)^2$$

    III. None of the factors will factor further.

$$a^2(4 - x^2) + 8a(4 - x^2) + 16(4 - x^2) = (2 - x)(2 + x)(a + 4)^2$$

Mastery points
Can you
- Factor out the greatest common factor?
- Factor the difference of two squares?
- Factor the difference of two cubes?
- Factor the sum of two cubes?
- Factor trinomials?
- Factor a four-term polynomial?
- Use the general strategy for factoring polynomials?

Exercise 3–9

Directions Completely factor the following polynomials. If a polynomial will not factor, so state. See the outline of the general strategy for factoring polynomials and example 3–9 A.

1. $m^2 - 49$
2. $81 - x^2$
3. $x^2 + 6x + 5$
4. $a^2 + 11a + 10$
5. $7a^2 + 36a + 5$
6. $3x^2 + 13x + 4$
7. $2a^2 + 15a + 18$
8. $5b^2 + 16b + 12$
9. $a^2b^2 + 2ab - 8$
10. $x^2y^2 - 5xy - 14$
11. $27a^3 + b^3$
12. $x^3 + 64y^3$
13. $25x^2(3x + y) + 5a(3x + y)$
14. $36x^2(2a - b) - 12x(2a - b)$
15. $10x^2 - 20xy + 10y^2$
16. $6a^2 - 24ab - 48b^2$
17. $4m^2 - 16n^2$
18. $9x^2 - 36y^2$
19. $(a - b)^2 - (2x + y)^2$
20. $(3a + b)^2 - (x - y)^2$
21. $3x^6 - 81y^3$
22. $32a^3 - 4ab^9$
23. $12x^3y^2 - 18x^2y^2 + 16xy^4$
24. $9x^5y - 6x^3y^3 + 3x^2y^2$
25. $4x^2 - 9y^2$
26. $36 - a^2b^2$
27. $3ax + 6ay - bx - 2by$
28. $6am + 4bm - 3an - 2bn$
29. $27a^9 - b^3c^3$
30. $x^{12} - y^3z^6$
31. $5a^2 - 32a - 21$
32. $7a^2 + 16a - 15$
33. $a^4 - 5a^2 + 4$
34. $x^4 - 37x^2 + 36$
35. $4a^2 - 4ab - 15b^2$
36. $6x^2 + 7xy - 3y^2$
37. $y^4 - 16$
38. $a^4 - 81$
39. $4a^2 + 10a + 4$
40. $6x^2 + 18x - 60$
41. $(x + y)^2 - 8(x + y) - 9$
42. $(a - 2b)^2 + 7(a - 2b) + 10$
43. $6a^2 + 7a - 5$
44. $4x^2 + 17x - 15$
45. $4ab(x + 3y) - 8a^2b^2(x + 3y)$
46. $3x^2y(m - 4n) + 15xy^2(m - 4n)$
47. $4a^2 - 20ab + 25b^2$
48. $9m^2 - 30mn + 25n^2$
49. $80x^5 - 5x$
50. $3b^5 - 48b$
51. $3a^5b - 18a^3b^3 + 27ab^5$
52. $3x^3y^3 + 6x^2y^4 + 3xy^5$
53. $9a^2 - (x + 5y)^2$
54. $7x(a^2 - 4ab^2) + 14(a^2 - 4b^2)$
55. $3x^2 + 8x - 91$
56. $3x^2 - 32x - 91$
57. $3x^5y^9 + 81x^2z^6$
58. $a^8b^4 + 27a^2b^7$
59. $a^2(9 - x^2) - 6a(9 - x^2) + 9(9 - x^2)$
60. $x^2(16 - b^2) + 10x(16 - b^2) + 25(16 - b^2)$

Chapter summary

1. **Like terms** are terms that may differ only in their numerical coefficients.

2. In the expression a^n, a is called the **base** and n the **exponent.**

3. The following are properties and definitions involving exponents.

   $$\overbrace{}^{n \text{ factors of } a}$$

   a. $a^n = \overbrace{a \cdot a \cdot a \cdots a}^{n \text{ factors of } a}$

   b. $a^m \cdot a^n = a^{m+n}$

   c. $(a^m)^n = a^{nm}$

   d. $(ab)^n = a^n b^n$

   e. $a^m \div a^n = \dfrac{a^m}{a^n} = a^{m-n}, a \neq 0$

   f. $a^{-n} = \dfrac{1}{a^n}, a \neq 0$

   g. $a^0 = 1, a \neq 0$

   h. $\left(\dfrac{a}{b}\right)^n = \dfrac{a^n}{b^n}, b \neq 0$

4. When multiplying two multinomials, we multiply each term of the first multinomial by each term of the second multinomial and then combine like terms.

5. Three **special products** are
   $(a + b)^2 = a^2 + 2ab + b^2$
   $(a - b)^2 = a^2 - 2ab + b^2$
   $(a + b)(a - b) = a^2 - b^2$

6. The **scientific notation** form of a number Y is $Y = a \times 10^n$, where a is a number greater than or equal to 1 and less than 10, and n is an integer.

7. **A prime number** is any natural number greater than 1 that is divisible only by itself and by 1.

8. A polynomial with integer coefficients will be considered to be in **completely factored form** when it is expressed as the product of polynomials with integer coefficients and none of the factors except a monomial can still be written as the product of two polynomials with integer coefficients.

9. We try to factor **four-term polynomials** by grouping.

10. The **trinomial $ax^2 + bx + c$** will factor if we can find a pair of integers, m and n, whose product is $a \cdot c$ and whose sum is b.

11. The **trinomial $x^2 + bx + c$** will factor only if we can find a pair of integers, m and n, whose product is c and whose sum is b.

12. The **difference of two squares** factors as
    $$a^2 - b^2 = (a + b)(a - b).$$

13. The **sum or difference of two cubes** factors as
    $$a^3 + b^3 = (a + b)(a^2 - ab + b^2)$$
    and
    $$a^3 - b^3 = (a - b)(a^2 + ab + b^2).$$

Chapter review

[3–1]
Directions Perform the indicated addition and subtraction.

1. $7y^2 - 6y + 4y^2 - 3y + 2y^2$

2. $7x^2y + 2xy^2 - 3xy^2 + 7x^2y + 5x^2y^2$

Directions Remove all grouping symbols and combine like terms.

3. $(3x^2 - 2x + 7) - (x^2 + 3x - 4)$

4. $7a - [2a - (3a + 4)]$

5. $2x^2 - [4x + 5x^2 - (3x^2 - 6x)]$

[3–2]

Directions Perform the indicated operations. Assume all variables used as exponents represent positive integers.

6. $(-5x^2)(3x^3)$

7. $(2a^2b)(3ab^4)$

8. $(2a^3b^4c)^3$

9. $(xy^3z)^2(x^2y)^3$

10. $(3a^2)^2a^3 + (2a)^3a^4$

11. $(2b^5)^2 - (3b^2)^3b^2$

[3–3]

Directions Perform the indicated multiplication and simplify.

12. $5a^2b(2a^2 - 3ab + 4b^2)$

13. $(2a - b)(a + b)$

14. $(y - 7)(y + 7)$

15. $(2a + b)^2$

16. $(a + b)(a^2 - 3ab + 2b^2)$

17. $(3x - 2)(x + 4) - (x - 3)^2$

[3–4]

Directions Perform all indicated operations and leave the answers with only positive exponents. Assume that all variables represent nonzero real numbers.

18. $\dfrac{2a^2b^4}{2^3a^5b}$

19. $(3x^{-4})(2x^{-5})$

20. $(3x^{-2}y^3z^0)^2$

21. $\left(\dfrac{16x^4y^3}{4xy^8}\right)^3$

22. $\dfrac{a^5b^{-2}}{a^{-4}b}$

23. $\dfrac{2^{-3}a^0b^{-3}c^2}{4^{-1}a^{-3}b^{-2}c^{-5}}$

24. $(2x^3y^{-4})^2(x^{-2}y)^3$

[3–5]

Directions Write in completely factored form.

25. $12x^3 - 18x^2$

26. $12a^4b - 4a^2b^3 + 24a^3b^2$

27. $x(a + b) - 2y(a + b)$

28. $10x^2(x - 3z) + 5x(x - 3z)$

29. $6ax - 3ay - 2bx + by$

30. $4ax + 6by + 8ay + 3bx$

31. $4ax + 2ay + 6bx + 3by$

32. $ax + 3bx - 2ay - 6by$

33. $6ax + by - 2bx - 3ay$

34. $2ax + 3by + 6bx + ay$

[3–6]

Directions Write in completely factored form. If a polynomial will not factor, so state.

35. $2a^2 - 7a + 6$

36. $8x^2 - 14x + 5$

37. $6y^2 - 5y - 4$

38. $3a^2 + 2a - 1$

39. $4x^2 + 11x - 3$

40. $4x^2 + 4x + 1$

41. $24a^2 + 22a + 3$

42. $8x^2 - 18x + 9$

43. $2x^2 + 15x + 18$

44. $6x^2 + 51x + 24$

45. $-2x^2 + 11x + 6$

[3–7]

Directions Write in completely factored form. If a polynomial will not factor, so state.

46. $x^2 + 14x + 24$

47. $x^2 - 9x + 14$

48. $2a^3 - 8a^2 - 10a$

49. $x^3 - x^2 - 6x$

50. $x^2y^2 + 10xy + 24$

51. $a^2b^2 - 8ab - 20$

52. $4a^2b^2 + 11ab + 6$

53. $4x^2y^2 - 20xy + 9$

54. $(x + 3y)^2 + (x + 3y) - 2$

55. $(x + 2y)^2 - 14(x + 2y) + 49$

56. $x^4 - 17x^2 + 16$

57. $x^4 - 9x^2 + 20$

58. $4x^4 + 7x^2 - 36$

59. $3x^2 + 7xy - 6y^2$

60. $a^2 - ab - 2b^2$

61. $x^2 + 5xy + 6y^2$

Directions Write in completely factored form.

62. $x^2 - 81$

63. $4x^2 - 36y^2$

64. $3a^2 - 27b^2$

65. $a^4 - 81$

66. $x^2(a + 2b) - y^2(a + 2b)$

67. $4a^2(x - 3z) - 9b^2(x - 3z)$

68. $a^3 + 8b^3$

69. $27x^3 - y^3$

70. $64a^3 + 8b^3$

71. $27x^5 + x^2y^3$

72. $2x^3 - 16b^3$

73. $x^{12}y^{15} - z^9$

Directions Write in completely factored form.

74. $9x^2 + 24x + 16$

75. $a^6 - b^9$

76. $3a^3 - 75a$

77. $a^2(9 - x^2) + 12a(9 - x^2) + 36(9 - x^2)$

78. $6a^2 + 17a - 3$

79. $4mx - 8my + 3nx - 6ny$

80. $40x^4 + 5xy^3$

81. $a^3b^2 - 4a^2b^3 + 4ab^4$

82. $x^2 + 2xy - 8y^2$

83. $15a^2 + 4a - 4$

Chapter 3 cumulative test

Directions Determine if the following statements are true or false.

[1–2] **1.** $|-10| < 0$

[1–2] **2.** $-3^2 = (-3)^2$

[1–1] **3.** $J \subseteq Q$

Directions Perform the indicated operations, if possible, and simplify.

[1–2] **4.** $\dfrac{(-8)}{(-4)}$

[1–2] **5.** $\dfrac{(-7)}{0}$

[1–2] **6.** $(-2) - (-6)$

[1–2] **7.** $(-2)(4)(0)(-6)$

[1–2] **8.** $(-3)^4$

[1–4] **9.** $48 - 24 \div 6 - 3 - 2^2$

[1–4] **10.** $2[-3(10 - 7) - 12 + 4]$

Directions Solve the following equations and inequalities.

[2–4] **11.** $3x - 4 \geq x + 10$

[2–1] **12.** $2(3x - 4) + 7 = 8x - 11$

[2–5] **13.** $|3x + 4| = 5$

[2–5] **14.** $|8 - 3x| < 9$

[2–5] **15.** $|6x + 5| - 4 > 2$

[2–4] **16.** $-4 < 3 - 2x < 6$

[2–4] **17.** $-3 \leq 2x + 7 \leq 4$

[2–3] **18.** Three times a number is subtracted from 43 and this result is less than 16. Find all numbers that satisfy this condition.

[2–3] **19.** Four times a number minus 6 is greater than 6 but less than 22. Find all numbers that satisfy these conditions.

Directions Perform the indicated operations and simplify. Assume that all variables represent nonzero real numbers.

[3–2] **20.** $(2a^2b)(3a^3b^4)$ [3–2] **21.** $(2x^3y^4)^4$

[3–4] **22.** $\dfrac{27a^5b^9c}{18a^2b^4c^7}$ [3–4] **23.** $\dfrac{x^{-4}y^3}{x^2y^{-1}}$

[3–4] **24.** $\left(\dfrac{3x^2y^4}{9x^5y}\right)^3$ [3–1] **25.** $(5x^2 - 3x + 4) - (x^2 - x + 7)$

[3–4] **26.** $(3a^{-3}b^2c^{-1})^{-3}$ [3–1] **27.** $4ab - \{2a + 3b - [a - (2b - 3ab)]\}$

[3–3] **28.** $(3a - b)^2$ [3–2] **29.** $(2a^2b)^3(a^3b^2)^4$

[3–3] **30.** $6x^4y^3(2x^2 - 3x^2y + 4y^2)$

Directions Write in completely factored form. If a polynomial will will not factor, so state.

[3–8] **31.** $x^2 - 36$ [3–8] **32.** $x^3 + 27y^3$ [3–6] **33.** $2a^2 - 15a + 18$

[3–7] **34.** $y^2 + 7y + 6$ [3–7] **35.** $x^2y^2 - 2xy - 8$ [3–6] **36.** $7x^2 - 34x - 5$

[3–5] **37.** $2ax - 6ay + 3bx - 9by$ [3–5] **38.** $15x^2(2a - b) + 5x(2a - b)$

[3–7] **39.** $4x^2 - 20xy + 25y^2$ [3–8] **40.** $3a^2 - 12b^2$

Chapter 4

Rational Expressions

4–1 Fundamental properties of rational expressions

The rational expression

In chapter 1 we defined a rational number to be any number that can be expressed as a quotient of two integers where the denominator is not zero. That is, the set of rational numbers are the elements in the set

$$\left\{ \frac{p}{q} \;\middle|\; p \text{ and } q \text{ are integers and } q \neq 0 \right\}.$$

If this definition is extended to polynomials, the result is called a **rational expression.**

> ■ **Definition of a rational expression**
> A **rational expression** is any algebraic expression that can be written as a quotient of two polynomials with denominator not 0.

That is, rational expressions are elements in the set

$$\left\{ \frac{P}{Q} \,\middle|\, P \text{ and } Q \text{ are polynomials, } Q \neq 0 \right\}.$$

The following are rational expressions:

$$\frac{2x}{3}, \quad \frac{2x - 3}{x^2 + 6x - 1}, \quad \frac{5x + 1}{3}, \quad \text{and} \quad 5x^3 - 3x + 1.$$

Recall that division by zero is not defined. The rational expression becomes meaningless for those values of the variable for which $Q = 0$. For example, the rational expression

$$\frac{3x - 2}{x - 6}$$

is meaningful for every value of x *except* 6 since if $x = 6$, then $x - 6 = 6 - 6 = 0$. Thus we say that

$$\frac{3x - 2}{x - 6}, \; x \neq 6,$$

is a rational expression.

Note

The restrictions on the variable are found by setting all factors of the denominator that contain the variable equal to zero and solving the equation for the variable.

Domain of a rational expression

We call the set of all those values that the variable in a rational expression can assume the **domain** of the rational expression.

> ■ **Definition of the domain**
> The **domain** of a rational expression is the set of all replacement values of the variable for which the expression is defined.

Thus given the rational expression

$$\frac{3x - 2}{x - 6},$$

the domain is defined to be the set of all real numbers except 6. In set-builder notation we symbolize the domain by

$$\text{domain} = \{x \,|\, x \in R, \, x \neq 6\}.$$

| Example 4–1 A | Determine the domain of the following rational expressions. Express this in set-builder notation. |

1. $\dfrac{x-5}{x+7}$

Set the denominator equal to 0 and solve for x.

$x + 7 = 0$
$\quad\quad x = -7$

The number -7 cannot be used as a replacement for x.

Domain $= \{x \,|\, x \,\epsilon\, R, \, x \neq -7\}$

2. $\dfrac{4x-3}{x^2-16}$

Factor the denominator.

$x^2 - 16 = (x + 4)(x - 4)$
$x + 4 = 0 \quad \text{or } x - 4 = 0 \qquad$ Set each factor equal to 0.
$\quad\quad x = -4 \text{ or} \quad\quad x = 4$

The numbers -4 and 4 cannot be used as replacements for x.

Domain $= \{x \,|\, x \,\epsilon\, R, \, x \neq -4, \, x \neq 4\}$

3. $\dfrac{5y^2}{y^2+9}$

Since $y^2 + 9$ cannot be factored and $y^2 + 9$ is positive for *every* value of y, the domain is all real numbers.

Domain $= \{y \,|\, y \,\epsilon\, R\}$

Reducing to lowest terms

For all values of the variable that are in the domain of the rational expression, the expression denotes a quotient of two real numbers. For this reason, the properties of the quotients of real numbers must then apply to rational expressions.

We now consider a very important property of real numbers, the **fundamental principle of fractions.**

> ■ **Fundamental principle of fractions**
> If a, b, and c are real numbers, $b \neq 0$ and $c \neq 0$, then
> $$\frac{a}{b} = \frac{ac}{bc}.$$

We have the same property for rational expressions, called the **fundamental principle of rational expressions.**

By this principle, given the rational expression

$$\frac{x - 3}{x^2 - 9}$$

and the polynomial $x - 3$, we divide the numerator and the denominator by $x - 3$.

$$\frac{x - 3}{x^2 - 9} = \frac{x - 3}{(x + 3)(x - 3)} = \frac{1}{x + 3} \quad (x \neq -3, x \neq 3)$$

We use the fundamental principle of rational expressions to make the numerator and the denominator of a rational expression **relatively prime.**

The following polynomials are relatively prime:

7 and 8, $x + 3$ and $x - 4$, $(2x - 1)(x + 3)$ and $(x - 1)(3x + 4)$.

When the numerator and the denominator of a rational expression are relatively prime, the expression is said to be **reduced to its lowest terms.** To express a rational expression in lowest terms, we

1. completely factor the numerator and the denominator, and
2. divide the numerator and the denominator by the common factors, thus making them relatively prime.

Example 4–1 B

Simplify each rational expression by reducing it to lowest terms. State restrictions on the variables.

1. $\dfrac{16x^2}{24x^4}$

Find the greatest common factor of $16x^2$ and $24x^4$, $8x^2$, and factor.

$$\frac{16x^2}{24x^4} = \frac{2(8x^2)}{3x^2(8x^2)}$$

Divide the numerator and the denominator by $8x^2$.

$$\frac{16x^2}{24x^4} = \frac{2}{3x^2} \quad (x \neq 0)$$

Note

This process is commonly called **cancelling**.

2. $\dfrac{6x - 18}{5x - 15}$

Factor the numerator and the denominator.

$$\frac{6x - 18}{5x - 15} = \frac{6(x - 3)}{5(x - 3)}$$

Divide the numerator and the denominator by the common factor $x - 3$.

$$\frac{6x - 18}{5x - 15} = \frac{6}{5} \quad (x \neq 3)$$

3. $\dfrac{y - 4}{4 - y}$

We note that the numerator and the denominator are exactly the same except for the signs. That is, $4 - y = -y + 4 = (-1)(y - 4)$. We can find the common factor(s) used to reduce by multiplying the numerator and the denominator by -1.

$$\frac{y - 4}{4 - y} = \frac{(-1)(y - 4)}{(-1)(4 - y)} = \frac{(-1)(y - 4)}{y - 4} \quad \text{Multiply in the denominator.}$$
$$= \frac{-1}{1} \qquad \text{Reduce by the common factor}$$
$$\qquad\qquad\qquad\qquad y - 4.$$
$$= -1 \ (y \neq 4)$$

Note

The quotients of opposites, or negatives, always equal -1.

4. $-\dfrac{x^2 - x - 12}{20 - x - x^2}$

Factor the numerator and the denominator.

$$-\frac{x^2 - x - 12}{20 - x - x^2} = -\frac{(x - 4)(x + 3)}{(4 - x)(5 + x)}$$

Since we have factors $x - 4$ and $4 - x$, multiply the numerator and the denominator by -1.

$$= -\frac{(-1)(x - 4)(x + 3)}{(-1)(4 - x)(5 + x)}$$
$$= -\frac{(-1)(x - 4)(x + 3)}{(x - 4)(5 + x)} \qquad (-1)(4 - x) = x - 4$$
$$= -\frac{(-1)(x + 3)}{1(5 + x)} \qquad \text{Reduce by } x - 4.$$
$$= \frac{x + 3}{x + 5} \quad (x \neq 4, x \neq -5) \qquad -\frac{-a}{b} = \frac{a}{b} \text{ and } 5 + x = x + 5$$

Exercise 4–1

Directions State the domain of the given rational expression (a) in words and (b) symbolically in set-builder notation. See example 4–1 A.

1. $\dfrac{3}{x - 4}$

2. $\dfrac{5}{x - 8}$

3. $\dfrac{x}{x + 1}$

4. $\dfrac{5y}{y + 9}$

5. $\dfrac{3z}{2z - 3}$

6. $\dfrac{5x}{8x - 5}$

7. $\dfrac{x - 3}{7x + 4}$

8. $\dfrac{p + 9}{6p + 5}$

9. $\dfrac{2x - 3}{x^2 - 4x}$

10. $\dfrac{m + 5}{3m^2 + 6m}$

11. $\dfrac{4x - 7}{x^2 + 8x + 16}$

12. $\dfrac{7 - 5x}{x^2 - 14x + 49}$

13. $\dfrac{2y - 5}{4y^2 - 25}$

14. $\dfrac{3x + 4}{9x^2 - 16}$

15. $\dfrac{8x^2 + 1}{2x^2 - 5x - 3}$

16. $\dfrac{p^2 - 7p + 1}{4p^2 + p - 3}$

17. $\dfrac{4x^2 - 2x + 1}{x^2 + 4}$

18. $\dfrac{17x - 3}{2x^2 + 5}$

Directions Simplify the given rational expression by reducing to lowest terms. Assume no denominator equals zero. See example 4–1 B.

19. $\dfrac{25x^3}{15x^4}$

20. $\dfrac{36y^4}{-42y^2}$

21. $\dfrac{p^3q^5}{pq^7}$

22. $\dfrac{-m^5n^3}{-m^8n}$

23. $\dfrac{a^2b^4c^3}{-abc^7}$

24. $\dfrac{-x^4y^6z}{x^3y^5z}$

25. $-\dfrac{-27mn^3p^7}{36m^4np^5}$

26. $\dfrac{72p^3qr^4}{-63pqr}$

27. $\dfrac{3x - 6}{7x - 14}$

28. $\dfrac{8a + 12}{6a + 9}$

29. $\dfrac{15a}{5a^2 - 10a}$

30. $\dfrac{-36x}{42x^3 + 24x}$

31. $\dfrac{12x^4 - 12x^3}{6x^3}$

32. $\dfrac{24x^2 - 36x}{32x^2}$

33. $\dfrac{8y^2 - 8}{6y - 6}$

34. $\dfrac{6x + 6}{3x - 3}$

35. $\dfrac{5x - 5y}{y - x}$

36. $\dfrac{9b - 9a}{a - b}$

37. $\dfrac{a^2 - 9}{4a + 12}$

38. $\dfrac{6a + 3b}{4a^2 - b^2}$

39. $\dfrac{4y^2 - 1}{1 + 2y}$

40. $\dfrac{64 - 49p^2}{7p - 8}$

41. $\dfrac{x^3 + 8}{x + 2}$

42. $\dfrac{a^3 - 64}{a - 4}$

43. $\dfrac{3x - 3y}{2y^3 - 2x^3}$

44. $\dfrac{5a^3 + 5b^3}{7b + 7a}$

45. $\dfrac{y^2 - 49}{y^2 + 14y + 49}$

46. $\dfrac{a^2 - 10a + 25}{a^2 - 25}$

47. $\dfrac{m^2 - 4m - 12}{m^2 - m - 6}$

48. $\dfrac{n^2 - 4n + 3}{n^2 - 5n + 6}$

49. $\dfrac{y^2 - 4y - 32}{y^2 - y - 20}$

50. $\dfrac{x^2 - x - 42}{x^2 + 12x + 36}$

51. $\dfrac{2a^2 - 3a + 1}{2a^2 + a - 1}$

52. $\dfrac{3b^2 - 10b + 3}{3b^2 - 7b + 2}$

53. $\dfrac{2x^2 + 3x - 9}{4x^2 + 13x + 3}$

54. $\dfrac{8y^2 - 22y + 5}{4y^2 - 15y - 4}$

55. $\dfrac{12m^2 + 17m + 6}{8m^2 + 10m + 3}$

56. $\dfrac{6x^2 + 17x + 7}{12x^2 + 13x - 35}$

57. $\dfrac{4a^2 + 2a - 12}{6a^2 - 7a - 3}$

58. $\dfrac{6x^2 - x - 1}{4x^2 + 18x - 10}$

59. $\dfrac{5y^2 - 10y - 15}{4y^2 - 36}$

60. $\dfrac{7x^2 - 7}{3x^2 - 15x + 12}$

61. $\dfrac{3a^2 + 16a - 12}{6 - 7a - 3a^2}$

62. $\dfrac{a^2 + 2ab - 24b^2}{a^2 - 8ab + 16b^2}$

63. $\dfrac{p^2 + 7pq + 12q^2}{p^2 + 5pq + 6q^2}$

4-2 Multiplication and division of rational expressions

Multiplication

Recall that polynomials represent real numbers, so what we did with fractions applies to rational expressions. It can be proved that the following property holds.

> ■ **Multiplication of fractions**
> If a, b, c, and d are real numbers, $b \neq 0$ and $d \neq 0$, then
> $$\frac{a}{b} \cdot \frac{c}{d} = \frac{a \cdot c}{b \cdot d}.$$

To illustrate,

$$\frac{3}{4} \cdot \frac{5}{7} = \frac{3 \cdot 5}{4 \cdot 7} = \frac{15}{28}.$$

We now extend this property to rational expressions.

> ■ **Multiplication of rational expressions**
> If P, Q, R, and S are polynomials, $Q \neq 0$ and $S \neq 0$, then
> $$\frac{P}{Q} \cdot \frac{R}{S} = \frac{P \cdot R}{Q \cdot S}.$$
>
> **Concept**
> When multiplying two rational expressions, multiply the numerators to determine the numerator of the product and multiply the denominators to determine the denominator of the product.

For example, consider the indicated product

$$\frac{x - 3}{x + 1} \cdot \frac{x + 2}{x - 1} = \frac{(x - 3)(x + 2)}{(x + 1)(x - 1)} = \frac{x^2 - x - 6}{x^2 - 1}.$$

Since we want the resulting product in its simplest form, any possible reduction by common factors should be performed before the multiplication of the numerators and the denominators takes place.

Example 4-2 A

Find the indicated products reduced to lowest terms. Assume all denominators represent nonzero numbers.

1. $\dfrac{6x^2}{15} \cdot \dfrac{10}{3x^3}$

$$\frac{6x^2}{15} \cdot \frac{10}{3x^3} = \frac{6x^2 \cdot 10}{15 \cdot 3x^3} = \frac{60x^2}{45x^3} = \frac{4 \cdot 15x^2}{3x \cdot 15x^2}$$

Divide the numerator and the denominator by the common factor $15x^2$.

$$= \frac{4}{3x}$$

2. $\dfrac{3x^2 - 6x}{x + 5} \cdot \dfrac{x^2 - 25}{x^2 - 10x + 25}$

Factor where possible.

$$\dfrac{3x^2 - 6x}{x + 5} \cdot \dfrac{x^2 - 25}{x^2 - 10x + 25} = \dfrac{3x(x - 2)}{x + 5} \cdot \dfrac{(x + 5)(x - 5)}{(x - 5)^2}$$

Multiply the numerators and the denominators together.

$$= \dfrac{3x(x - 2)(x + 5)(x - 5)}{(x + 5)(x - 5)(x - 5)}$$

Divide the numerator and the denominator by the common factors $x + 5$ and $x - 5$.

$$= \dfrac{3x(x - 2)}{x - 5}$$
$$= \dfrac{3x^2 - 6x}{x - 5}$$

3. $\dfrac{3x - 1}{x - 4} \cdot \dfrac{16 - x^2}{3x^2 + 14x - 5}$

Factor where possible.

$$\dfrac{3x - 1}{x - 4} \cdot \dfrac{16 - x^2}{3x^2 + 14x - 5} = \dfrac{3x - 1}{x - 4} \cdot \dfrac{(4 - x)(4 + x)}{(3x - 1)(x + 5)}$$

Multiply the numerators and the denominators together.

$$= \dfrac{(3x - 1)(4 - x)(4 + x)}{(x - 4)(3x - 1)(x + 5)}$$

Since there are factors $4 - x$ and $x - 4$, we multiply the numerator and the denominator by -1.

$$= \dfrac{(-1)(4 - x)(3x - 1)(4 + x)}{(-1)(x - 4)(3x - 1)(x + 5)}$$
$$= \dfrac{(-1)(4 - x)(3x - 1)(4 + x)}{(4 - x)(3x - 1)(x + 5)}$$

Divide the numerator and the denominator by the common factors $4 - x$ and $3x - 1$.

$$= \dfrac{(-1)(4 + x)}{x + 5}$$
$$= -\dfrac{x + 4}{x + 5} \quad \left(4 + x = x + 4 \text{ and } \dfrac{-a}{b} = -\dfrac{a}{b}\right)$$

Rational Expressions

We now summarize the steps used in multiplying rational expressions.

1. Completely factor the numerators and the denominators where possible.
2. Multiply the numerators and the denominators as a single rational expression.
3. Divide the numerator and the denominator by any common factor(s).
4. Multiply the remaining factors in the numerator and the remaining factors in the denominator.

Division of rational expressions

Recall that two rational numbers are *reciprocals* if their product is 1. That is, $\frac{3}{4}$ and $\frac{4}{3}$ are reciprocals since $\frac{3}{4} \cdot \frac{4}{3} = 1$. We note that the reciprocal of a number is found by interchanging the numerator and the denominator. The reciprocal of a rational expression is found in the same way. Thus the reciprocal of $\frac{x-3}{x+7}$ is $\frac{x+7}{x-3}$ and the reciprocal of $y+3$ is $\frac{1}{y+3}$.

The division of rational expressions is the same as the division of rational numbers.

■ **Division of rational numbers**
If a, b, c, and d are real numbers and b, c, and $d \neq 0$, then
$$\frac{a}{b} \div \frac{c}{d} = \frac{a}{b} \cdot \frac{d}{c}.$$

To illustrate,
$$\frac{5}{8} \div \frac{3}{7} = \frac{5}{8} \cdot \frac{7}{3} = \frac{5 \cdot 7}{8 \cdot 3} = \frac{35}{24}.$$

We can extend this to rational expressions.

■ **Division of rational expressions**
If P, Q, R, and S are polynomials, $Q \neq 0$, $R \neq 0$, and $S \neq 0$, then
$$\frac{P}{Q} \div \frac{R}{S} = \frac{P}{Q} \cdot \frac{S}{R} = \frac{P \cdot S}{Q \cdot R}.$$

Concept
When dividing two rational expressions, the quotient is obtained by multiplying the first expression by the reciprocal of the second expression.

Example 4–2 B

Find each of the indicated quotients reduced to lowest terms. Assume no denominator equals zero.

1. $\dfrac{36p^3}{7q} \div \dfrac{15p}{14q^2}$

$\dfrac{36p^3}{7q} \div \dfrac{15p}{14q^2} = \dfrac{36p^3}{7q} \cdot \dfrac{14q^2}{15p}$ Multiply by the reciprocal.

$= \dfrac{36 \cdot 14 \cdot p^3 q^2}{7 \cdot 15 \cdot pq}$

$= \dfrac{24p^2q \cdot (21pq)}{5 \cdot (21pq)}$

Divide the numerator and the denominator by the common factor $21pq$.

$= \dfrac{24p^2q}{5}$

2. $\dfrac{2x + 1}{x^2 - 9} \div \dfrac{2x^2 + 7x + 3}{x - 3}$

$\dfrac{2x + 1}{x^2 - 9} \div \dfrac{2x^2 + 7x + 3}{x - 3} = \dfrac{(2x + 1)}{(x^2 - 9)} \cdot \dfrac{(x - 3)}{(2x^2 + 7x + 3)}$

$= \dfrac{(2x + 1)(x - 3)}{(x + 3)(x - 3)(2x + 1)(x + 3)}$

Divide the numerator and the denominator by $(x - 3)(2x + 1)$.

$= \dfrac{1}{(x + 3)(x + 3)}$

$= \dfrac{1}{x^2 + 6x + 9}$

3. $\dfrac{x^2 + 2x}{5x} \div (2x^2 + x - 6)$

Since $2x^2 + x - 6 = \dfrac{2x^2 + x - 6}{1}$, then

$\dfrac{x^2 + 2x}{5x} \div \dfrac{2x^2 + x - 6}{1} = \dfrac{x^2 + 2x}{5x} \cdot \dfrac{1}{2x^2 + x - 6}$.

Factor and multiply the numerators and the denominators.

$= \dfrac{x(x + 2)}{5x(2x - 3)(x + 2)}$

Divide the numerator and the denominator by $x(x + 2)$.

$= \dfrac{1}{5(2x - 3)}$

$= \dfrac{1}{10x - 15}$

Exercise 4–2

Directions Find each indicated product in lowest terms. Assume all denominators are nonzero. See example 4–2 A.

1. $\dfrac{8}{9x^2} \cdot \dfrac{12x}{4}$

2. $\dfrac{12b}{7} \cdot \dfrac{28}{4b^3}$

3. $\dfrac{5x^2}{9y^3} \cdot \dfrac{18y}{20x^4}$

4. $\dfrac{15n}{28m} \cdot \dfrac{14m^3}{30n^2}$

5. $\dfrac{16c}{42ab^2} \cdot \dfrac{3a^3b}{8c^4}$

6. $\dfrac{12xy}{25z^3} \cdot \dfrac{5z}{18x^3y^2}$

7. $16a^2b^3 \cdot \dfrac{3}{24ab}$

8. $8x^3yz^2 \cdot \dfrac{15}{36xyz}$

9. $\dfrac{4a + 8}{3a - 12} \cdot \dfrac{a - 2}{5a + 10}$

10. $\dfrac{y + 3}{8y - 16} \cdot \dfrac{3y - 6}{4y + 12}$

11. $(a + 3) \cdot \dfrac{a - 6}{6a + 18}$

12. $\dfrac{x - 3}{5x - 25} \cdot (5 - x)$

13. $\dfrac{p^2 + 2p + 1}{1 - 4p} \cdot \dfrac{16p^2 - 1}{p^2 - 1}$

14. $\dfrac{m^2 - 9}{3m + 4} \cdot \dfrac{9m^2 - 16}{m^2 + 6m + 9}$

15. $\dfrac{x^2 - 9x + 20}{x^2 - 5x + 6} \cdot \dfrac{x^2 - 3x + 2}{x^2 - 5x + 4}$

16. $\dfrac{y^2 + 3y - 4}{y^2 + 2y - 3} \cdot \dfrac{y^2 - y - 6}{y^2 + y - 12}$

17. $\dfrac{2a^2 - a - 6}{3a^2 - 4a + 1} \cdot \dfrac{3a^2 + 7a + 2}{2a^2 + 7a + 6}$

18. $\dfrac{4b^2 + 8b + 3}{2b^2 - 5b - 12} \cdot \dfrac{b^2 - 16}{2b^2 + 7b + 3}$

19. $\dfrac{x^3 + 8}{x^2 - 4} \cdot \dfrac{x^2 - 4x + 4}{2x^2 + 3x - 14}$

20. $\dfrac{4x^2 - 49}{8x^3 + 27} \cdot \dfrac{4x^2 + 12x + 9}{2x^2 - 13x + 21}$

21. $\dfrac{y}{x^2 + 3xy + 2y^2} \cdot \dfrac{x}{} \cdot \dfrac{x^2 + 2xy + y^2}{x^2 - y^2}$

22. $\dfrac{b^2 - a^2}{3a^2 + ab - 2b^2} \cdot \dfrac{6a^2 - ab - 2b^2}{a^2 - 2ab + b^2}$

Directions Find each of the indicated quotients in lowest terms. Assume all denominators are nonzero. See example 4–2 B.

23. $\dfrac{9a^2b}{8ab^3} \div \dfrac{18ab^3}{16a^2b^2}$

24. $\dfrac{7ab^3}{10x^2y^3} \div \dfrac{21a^2b^2}{15x^3y^2}$

25. $\dfrac{r + 4}{r^2 - 1} \div \dfrac{r^2 - 16}{r + 1}$

26. $\dfrac{4x + 4y}{x - 3} \div \dfrac{x + y}{x^2 - 9}$

27. $\dfrac{6 - 2x}{2x + 8} \div (9 - 3x)$

28. $\dfrac{x^2 - 25}{2x + 10} \div (x^2 - 10x + 25)$

29. $(x^2 - 2x - 3) \div \dfrac{4x - 12}{x^2 - 1}$

30. $(4x^2 - 9) \div \dfrac{4x + 6}{x + 3}$

31. $\dfrac{x^3 - 125}{2x + 1} \div \dfrac{x^2 - 25}{2x^2 - 9x - 5}$

32. $\dfrac{3x^2 + 10x - 8}{5x - 15} \div \dfrac{x^3 + 64}{x^2 - 9}$

33. $\dfrac{56 - x - x^2}{x^2 - 6x - 7} \div \dfrac{x^2 + 4x - 21}{x^2 - 4x - 5}$

34. $\dfrac{a^2 + 16a + 63}{20 - a - a^2} \div \dfrac{a^2 - 5a - 84}{a^2 - a - 12}$

35. $\dfrac{2x^2 + 5x - 12}{3x^2 - 8x - 16} \div \dfrac{2x^2 + 3x - 9}{3x^2 + 13x + 12}$

36. $\dfrac{6x^2 - 5x - 4}{8x^2 + 10x + 3} \div \dfrac{3x^2 + 14x - 24}{4x^2 - 12x - 7}$

Directions Find each indicated product or quotient in lowest terms. Assume all denominators are nonzero. See examples 4–2 A and B.

37. $\dfrac{a^2 - b^2}{2a + 4b} \cdot \dfrac{a + 2b}{a + b}$

38. $\dfrac{4x + 4y}{x - 3y} \cdot \dfrac{x^2 - 9y^2}{x + y}$

39. $\dfrac{n^2 - m^2}{2m - 3n} \div \dfrac{m - n}{4m^2 - 9n^2}$

40. $\dfrac{p^2 - q^2}{q - 3p} \div \dfrac{p^2 + 2pq + q^2}{9p^2 - q^2}$

41. $\dfrac{a^2 - 8a + 15}{a^2 + 7a + 6} \cdot \dfrac{a^2 - 6a - 7}{a^2 - 9} \div \dfrac{a^2 - 12a + 35}{a^2 - 36}$

42. $\dfrac{x^2 + 3x - 28}{x^2 - 8x + 7} \div \dfrac{x^2 - 7x + 12}{x^2 - x - 42} \cdot \dfrac{2x - 8}{x^2 - 36}$

43. $\dfrac{2x^2 + 5x - 3}{3x^2 - 10x - 8} \div \dfrac{2x^2 - 7x + 3}{3x^2 - x - 2} \cdot \dfrac{x^2 - 6x + 9}{x^2 - 2x + 1}$

44. $\dfrac{6 - x - 2x^2}{5x^2 + 7x - 6} \cdot \dfrac{20x^2 - 7x - 3}{6x^2 - 25x + 4} \div \dfrac{4x^2 - 11x - 3}{6x^2 - 19x + 3}$

45. $\dfrac{27y^3 - 8x^3}{x + y} \div \dfrac{2x - 3y}{x^3 + y^3}$

46. $\dfrac{a^3 + 64b^3}{16b^2 - a^2} \cdot \dfrac{a + 4b}{a^2 + 3ab - 4b^2}$

47. $\dfrac{ab - a + 3b - 3}{ab + a - 4b - 4} \div \dfrac{ab - 2a + 3b - 6}{ab + 2a - 4b - 8}$

48. $\dfrac{mn - n + 4m - 4}{mn - 3n + 4m - 12} \cdot \dfrac{mn - 3n + 2m - 6}{mn + 2m + n + 2}$

49. $\dfrac{xz - xw + yz - yw}{xz + xw - yz - yw} \cdot \dfrac{xz - xw - yz + yw}{xz + xw + yz + yw}$

50. $\dfrac{pr + ps + qr + qs}{pr - ps - qr + qs} \div \dfrac{xr + xs - yr - ys}{xr - xs + yr - ys}$

4–3 Addition and subtraction of rational expressions

When we add or subtract a pair of fractions that have the same denominator, such as $\dfrac{3}{4} + \dfrac{5}{4}$, we use the property of adding or subtracting fractions.

> ■ **Property of adding or subtracting fractions**
> If a, b, and c are real numbers, $b \neq 0$, then
> $$\frac{a}{b} + \frac{c}{b} = \frac{a + c}{b} \quad \text{and} \quad \frac{a}{b} - \frac{c}{b} = \frac{a - c}{b}.$$

For example,

$$\frac{3}{4} + \frac{5}{4} = \frac{3 + 5}{4} = \frac{8}{4} = 2 \quad \text{and} \quad \frac{7}{8} - \frac{1}{8} = \frac{7 - 1}{8} = \frac{6}{8} = \frac{3}{4}.$$

We can extend this property to rational expressions.

> **■ Adding or subtracting rational expressions**
> If P, Q, and R are polynomials, $Q \neq 0$, then
> $$\frac{P}{Q} + \frac{R}{Q} = \frac{P + R}{Q} \quad \text{and} \quad \frac{P}{Q} - \frac{R}{Q} = \frac{P - R}{Q}.$$
>
> **Concept**
> To add, or subtract, two rational expressions having the same denominators, add, or subtract, the numerators and place this sum over the common denominator.

Example 4–3 A

Find each indicated sum in lowest terms. State restrictions on the variables.

1. $\dfrac{3x + 2}{x - 2} + \dfrac{4x - 5}{x - 2}$

$$\frac{3x + 2}{x - 2} + \frac{4x - 5}{x - 2} = \frac{(3x + 2) + (4x - 5)}{x - 2}$$
$$= \frac{3x + 2 + 4x - 5}{x - 2}$$

Combine like terms in the numerator.

$$\frac{3x + 2}{x - 2} + \frac{4x - 5}{x - 2} = \frac{7x - 3}{x - 2} \; (x \neq 2)$$

2. $\dfrac{3x - 2y}{x^2 - y^2} - \dfrac{y - 2x}{x^2 - y^2}$

$$\frac{3x - 2y}{x^2 - y^2} - \frac{y - 2x}{x^2 - y^2} = \frac{(3x - 2y) - (y - 2x)}{x^2 - y^2}$$
$$= \frac{3x - 2y - y + 2x}{x^2 - y^2}$$
$$= \frac{5x - 3y}{x^2 - y^2} \; (x \neq y, x \neq -y)$$

Note
When we write the sum or difference of the numerators, it is important that we write each numerator in parentheses to avoid making a common mistake when we subtract. Failure to do this would have caused the numerator in example 2 to become $3x - 2y - y \ominus 2x$.

When adding, or subtracting, two rational expressions $\dfrac{P}{Q}$ and $\dfrac{R}{S}$ that *do not* have the same denominator, we can add or subtract them this way:

■ **Adding or subtracting rational expressions**

If P, Q, R, and S are polynomials, $Q \neq 0$ and $S \neq 0$, then

$$\frac{P}{Q} + \frac{R}{S} = \frac{P \cdot S + Q \cdot R}{Q \cdot S} \quad \text{and} \quad \frac{P}{Q} - \frac{R}{S} = \frac{P \cdot S - Q \cdot R}{Q \cdot S}.$$

We can illustrate the use of this property for rational expressions.

1.
$$\begin{aligned}
\frac{x+1}{x-2} + \frac{x-3}{x+4} &= \frac{(x+1)(x+4) + (x-2)(x-3)}{(x-2)(x+4)} \\
&= \frac{(x^2 + 5x + 4) + (x^2 - 5x + 6)}{(x-2)(x+4)} \qquad \text{Add the numerators.} \\
&= \frac{2x^2 + 10}{(x-2)(x+4)} \ (x \neq 2, x \neq -4) \qquad \text{Combine like terms.}
\end{aligned}$$

2.
$$\begin{aligned}
\frac{x}{x-7} - \frac{2x+1}{x-9} &= \frac{x \cdot (x-9) - (x-7)(2x+1)}{(x-7)(x-9)} \\
&= \frac{(x^2 - 9x) - (2x^2 - 13x - 7)}{(x-7)(x-9)} \qquad \text{Subtract the numerators.} \\
&= \frac{x^2 - 9x - 2x^2 + 13x + 7}{(x-7)(x-9)} \qquad \text{Remove the parentheses and change the signs.} \\
&= \frac{-x^2 + 4x + 7}{(x-7)(x-9)} \ (x \neq 7, x \neq 9)
\end{aligned}$$

This method is useful when adding or subtracting only two rational expressions and when the denominators have no common factors. We now consider the alternative for adding or subtracting rational expressions when these conditions are not present.

When rational numbers or rational expressions do not have the same denominator, we apply the fundamental principle of rational expressions to obtain equivalent expressions (expressions that name the same number) with a common denominator. To do this, we must find the *least common multiple* (LCM) of the denominators.

■ **Least common multiple**

The least common multiple (LCM) of a set of expressions is the "smallest" (least) expression that is exactly divisible by the numbers (or expressions).

To find the LCM of a set of polynomials, we

1. completely factor the polynomials, then
2. the LCM consists of the product of all distinct factors of the polynomials, each raised to the greatest power to which it appears in any of the factorizations.

Example 4–3 B

Find the least common multiple (LCM) of the following sets of polynomials.

1. $24x^3$, $15x^2$, $10x$

Completely factor each polynomial.

$24x^3 = 2^3 \cdot 3 \cdot x^3$
$15x^2 = 3 \cdot 5 \cdot x^2$
$10x = 2 \cdot 5 \cdot x$

The LCM contains each different factor that appears—2, 3, 5, and x—raised to the greatest power that it occurs in any one factorization.
The LCM $= 2^3 \cdot 3 \cdot 5 \cdot x^3 = 120x^3$.

2. $2x - 4$, $x^2 - 4$, $x^2 + 4x + 4$

Completely factor each polynomial.

$2x - 4 = 2(x - 2)$
$x^2 - 4 = (x + 2)(x - 2)$
$x^2 + 4x + 4 = (x + 2)^2$

Thus the LCM contains each different factor that occurs—
2, $(x + 2)$, and $(x - 2)$—raised to its greatest power.
The LCM $= 2(x - 2)(x + 2)^2$.

We now apply the fundamental principle of rational expressions to "build" a given expression to an equivalent rational expression. To do this, we must **multiply** the numerator and the denominator by the same nonzero polynomial.

Example 4–3 C

Write each rational expression as an equivalent expression with the indicated denominator.

1. $\dfrac{13}{3a}$, denominator $12a^2$

We obtain the factor with which to build by dividing $12a^2$ by $3a = 4a$.
Multiply the numerator and the denominator by $4a$ to get

$$\frac{13}{3a} = \frac{13 \cdot 4a}{3a \cdot 4a} = \frac{52a}{12a^2}.$$

2. $\dfrac{3x - 2}{2x - 1}$, denominator $4x^2 - 1$

Since $4x^2 - 1$ factors to $(2x - 1)(2x + 1)$, we must multiply the numerator and the denominator by $2x + 1$. Thus

$$\frac{3x - 2}{2x - 1} = \frac{(3x - 2)(2x + 1)}{(2x - 1)(2x + 1)} = \frac{6x^2 - x - 2}{4x^2 - 1}.$$

3. $\dfrac{3}{1 - x}$, denominator $x - 1$

Since $x - 1 = (-1)(1 - x)$, multiply the numerator and the denominator by -1.

$$\frac{3}{1 - x} = \frac{3(-1)}{(1 - x)(-1)} = \frac{-3}{x - 1}$$

We now add and subtract rational expressions that have unlike denominators by building the expressions to equivalent rational expressions having a common denominator—the LCM of the denominators (called the least common denominator, LCD).

| Example 4–3 D | Find the indicated sum or difference in lowest terms. Assume no denominator equals zero. |

1. $\dfrac{6}{5a} + \dfrac{7}{6b}$

The LCM of $5a$ and $6b$ is $5a \cdot 6b = 30ab$.

$$\frac{6}{5a} + \frac{7}{6b} = \frac{6}{5a} \cdot \frac{6b}{6b} + \frac{7}{6b} \cdot \frac{5a}{5a} = \frac{36b}{30ab} + \frac{35a}{30ab} = \frac{36b + 35a}{30ab}$$

2. $\dfrac{5}{16a} - \dfrac{7}{24a^2}$

$16a = 2^4 \cdot a$
$24a^2 = 2^3 \cdot 3 \cdot a^2 \qquad$ LCM $= 2^4 \cdot 3 \cdot a^2 = 48a^2.$

$$\frac{5}{16a} - \frac{7}{24a^2} = \frac{5}{16a} \cdot \frac{3a}{3a} - \frac{7}{24a^2} \cdot \frac{2}{2} = \frac{15a}{48a^2} - \frac{14}{48a^2} = \frac{15a - 14}{48a^2}$$

3. $\dfrac{10}{x - 3} + \dfrac{9}{x^2 + 3x - 18}$

Factor the denominators to find the LCM.

$$x^2 + 3x - 18 = (x - 3)(x + 6)$$

Since the denominator of the first expression is a factor of $x^2 + 3x - 18$, the LCM is $(x - 3)(x + 6)$.

$$\frac{10}{x-3} + \frac{9}{x^2 + 3x - 18} = \frac{10}{x-3} \cdot \frac{x+6}{x+6} + \frac{9}{(x-3)(x+6)}$$

$$= \frac{10(x+6)}{(x-3)(x+6)} + \frac{9}{(x-3)(x+6)}$$

$$= \frac{10x + 60 + 9}{(x-3)(x+6)}$$

$$= \frac{10x + 69}{(x-3)(x+6)}$$

4. $\dfrac{y+3}{21 - 4y - y^2} - \dfrac{y+5}{y^2 + 6y - 27}$

Factor each denominator.

$$21 - 4y - y^2 = -(y^2 + 4y - 21) = -(y-3)(y+7)$$

$$y^2 + 6y - 27 = (y+9)(y-3)$$

The LCM of the denominators is $(y+9)(y-3)(y+7)$.

$$\frac{y+3}{21 - 4y - y^2} - \frac{y+5}{y^2 + 6y - 27}$$

$$= \frac{y+3}{-(y-3)(y+7)} \cdot \frac{y+9}{y+9} - \frac{y+5}{(y+9)(y-3)} \cdot \frac{y+7}{y+7}$$

$$= \frac{(y+3)(y+9)}{-(y-3)(y+7)(y+9)} - \frac{(y+5)(y+7)}{(y+9)(y-3)(y+7)}$$

We multiply the numerator and the denominator of the first rational expression by -1 and multiply the factors in each numerator.

$$= \frac{-(y^2 + 12y + 27)}{(y-3)(y+7)(y+9)} - \frac{(y^2 + 12y + 35)}{(y+9)(y-3)(y+7)}$$

$$= \frac{(-y^2 - 12y - 27) - (y^2 + 12y + 35)}{(y-3)(y+9)(y+7)}$$

$$= \frac{-y^2 - 12y - 27 - y^2 - 12y - 35}{(y-3)(y+9)(y+7)}$$

$$= \frac{-2y^2 - 24y - 62}{(y-3)(y+9)(y+7)}$$

5. $\dfrac{x^2}{x-8} + \dfrac{7x+8}{8-x}$

Since the denominators are the negatives of each other, we multiply the numerator and the denominator of $\dfrac{7x+8}{8-x}$ by -1.

$$\frac{x^2}{x-8} + \frac{7x+8}{8-x} \cdot \frac{-1}{-1} = \frac{x^2}{x-8} + \frac{-7x-8}{x-8}$$

$$= \frac{x^2 - 7x - 8}{x-8}$$

$$= \frac{(x+1)(x-8)}{x-8} \qquad \text{Factor the numerator.}$$

$$= x + 1 \qquad \text{Reduce by } x - 8.$$

Exercise 4–3

Directions Find the least common multiple of the set of polynomials. See example 4–3 B.

1. $14x, 35$

2. $42y, 36$

3. $10k, 16k^2$

4. $18x^3, 21x^2$

5. $32a^3b, 9ab^2$

6. $48x^2y^2, 30x^4y$

7. $xy^3, 6x^2y^2, 15xy^4$

8. $4mn^4, 14m^2n^3, 35mn$

9. $4a, 2a - 4$

10. $9b, 15b + 30$

11. $x - 7, 3x - 21, 6x$

12. $x + 9, 4x + 36, 8x$

13. $p^2 - p - 12, p^2 + 6p + 9, 3p - 12$

14. $n^2 - 9, n^2 - n - 6, 8n^2 + 24n$

15. $a^2 - 25, 10a + 50, 5a - 25$

16. $x^2 + 6x + 9, 2x + 6, x^2 + x - 6$

17. $49 - q^2, q^2 + 5q - 14, q - 7$

18. $2y + 10, 25 - y^2, y^2 - 10y + 25$

19. $b - a, a^2 - b^2, 5a + 5b$

20. $m - n, 4m + 4n, n^2 - 2mn + m^2$

21. $2a^2 - 13a - 7, 6a^2 + a - 1, a^2 - 49$

22. $4m^2 + 5m - 6, 3m^2 + m - 10, 16m^2 - 9$

Directions Build the given rational expression to an equivalent expression with the given denominator. All denominators are nonzero. See example 4–3 C.

23. $\dfrac{4}{7x}$, denominator $21x^3$

24. $\dfrac{8}{9y^2}$, denominator $72y^5$

25. $\dfrac{5x}{8y}$, denominator $24x^2y^2$

26. $\dfrac{-3a}{7b^2}$, denominator $35a^2b^3$

27. $4p$, denominator $p - 3$

28. $a + 3$, denominator $2a - 1$

29. $\dfrac{p - 3}{p + 2}$, denominator $p^2 - 4$

30. $\dfrac{n + 7}{n - 9}$, denominator $n^2 - 81$

31. $\dfrac{4x}{2x - 3}$, denominator $8x^3 - 27$

32. $\dfrac{9x}{3x + 1}$, denominator $27x^3 + 1$

33. $\dfrac{2n - 1}{n + 7}$, denominator $n^2 + 2n - 35$

34. $\dfrac{4x + 3}{x - 9}$, denominator $x^2 - 4x - 45$

35. $\dfrac{2y - 5}{4y - 1}$, denominator $4y^2 + 7y - 2$

36. $\dfrac{3x + 4}{2x - 9}$, denominator $8x^2 - 30x - 27$

37. $-\dfrac{3m}{4 - 5m}$, denominator $25m^2 - 16$

38. $\dfrac{9 - a}{4 - a}$, denominator $a^2 + a - 20$

39. $\dfrac{-9}{9 - a}$, denominator $a - 9$

40. $-\dfrac{-12}{10 - b}$, denominator $b - 10$

Directions Find each indicated sum or difference in lowest terms. Place restrictions on the denominators. See example 4–3 A.

41. $\dfrac{5}{q} + \dfrac{8}{q}$

42. $\dfrac{7}{x} - \dfrac{17}{x}$

43. $\dfrac{4y}{y + 4} - \dfrac{7y}{y + 4}$

44. $\dfrac{8a}{a - 5} + \dfrac{7}{a - 5}$

45. $\dfrac{7x - 1}{3x + 4} - \dfrac{2x + 7}{3x + 4}$

46. $\dfrac{8a - 3}{5a^2 - 1} + \dfrac{3a + 5}{5a^2 - 1}$

47. $\dfrac{3y - 2}{4y + 3} - \dfrac{y + 5}{4y + 3}$

48. $\dfrac{2b + 1}{b^2 - 4} + \dfrac{1 - b}{b^2 - 4}$

Directions Find the indicated sum or difference. Assume all denominators are nonzero. See example 4–3 D.

49. $\dfrac{8}{3x} + \dfrac{5}{4x}$

50. $\dfrac{7}{5a} - \dfrac{9}{6a}$

51. $\dfrac{4a}{3a + 5} + \dfrac{2a}{2a - 3}$

52. $\dfrac{5m}{m - 8} - \dfrac{3m - 8}{2m + 7}$

53. $\dfrac{x + 2}{x - 9} - \dfrac{x - 6}{9 - x}$

54. $\dfrac{2a - 1}{2a - 3} + \dfrac{4 - a}{3 - 2a}$

55. $\dfrac{17}{5y - 10} + \dfrac{19}{2y + 4}$

56. $\dfrac{10}{4a - 6} - \dfrac{13}{3a + 9}$

57. $\dfrac{7}{x^2 - 5x - 6} + \dfrac{9}{x^2 - 1}$

58. $\dfrac{14}{a^2 - 7a - 18} - \dfrac{8}{a^2 - 4}$

59. $\dfrac{5x}{x^2 - 2xy - 3y^2} - \dfrac{2y}{x^2 + 2xy + y^2}$

60. $\dfrac{8q}{4q^2 - 9p^2} + \dfrac{5p}{4q^2 - 12pq + 9p^2}$

61. $\dfrac{a - 7}{2a^2 + 9a - 5} + \dfrac{4 - a}{4a^2 + 23a + 15}$

62. $\dfrac{b + 9}{12b^2 - 5b - 2} - \dfrac{8 - 2b}{3b^2 - 17b + 10}$

63. $(4a - 3) - \dfrac{2a + 5}{5a - 2}$

64. $\dfrac{5x + 4}{3x + 1} + (8x - 5)$

65. $\dfrac{5}{8p} + \dfrac{6p - 5}{4p^2 - 8p - 60}$

66. $\dfrac{9p - 2}{2p^2 - 2p - 84} + \dfrac{7}{6p}$

67. $\dfrac{3y}{y^2 + 5y + 6} - \dfrac{5}{4 - y^2}$

68. $\dfrac{6a}{6 + a - a^2} - \dfrac{7a}{a^2 + 7a + 10}$

69. $\dfrac{b + 3}{b + 2} + \dfrac{2b}{20 - 5b^2}$

70. $\dfrac{3}{6b^2 - 4bc} - \dfrac{4}{6c^2 - 9bc}$

71. $\dfrac{2x - 3y}{x^2 - 4xy - 12y^2} - \dfrac{2y - x}{x^2 - 12xy + 36y^2}$

72. $\dfrac{a - b}{a^2 - 3ab - 4b^2} + \dfrac{2b - 5a}{a^2 - 16b^2}$

73. $\dfrac{x - 3}{8x^2 - 26x + 15} + \dfrac{3x + 2}{6x^2 - 13x - 5}$

74. $\dfrac{2p - 3}{8p^2 - 18p - 5} - \dfrac{5 - 7p}{4p^2 - 27p - 7}$

75. $\dfrac{5m - n}{8m^2 + 15mn - 2n^2} + \dfrac{3m + n}{5m^2 + 6mn - 8n^2}$

76. $\dfrac{b - 2a}{3a^2 - 2ab - 8b^2} - \dfrac{3a - 5b}{9a^2 + 6ab - 8b^2}$

Directions Solve the following verbal problems.

77. Workers A, B, C, and D can complete a given job in p, q, r, and s hours, respectively. Working together they can complete in one hour

$$\frac{1}{p} + \frac{1}{q} + \frac{1}{r} + \frac{1}{s}$$

of the job. Obtain a single expression for what they can do together in one hour.

78. In electricity the total resistance of any parallel circuit is given by

$$\frac{1}{R_t} - \frac{1}{R_1} + \frac{1}{R_2} + \frac{1}{R_3} + \frac{1}{R_4}.$$

Combine the terms in the right member.

79. Given the focal lengths f_1 and f_2 of two thin lenses that are a distance d apart, the focal length F of the system of lenses is given by

$$\frac{1}{F} = \frac{1}{f_1} + \frac{1}{f_2} - \frac{d}{f_1 f_2}.$$

Combine the fractions in the right member.

80. In electricity the true current reading I_t of a current meter is given by

$$I_t = I_r + \frac{R_m}{R_t} \cdot I_r,$$

where I_r = the meter reading, R_m = the resistance of the current meter, and R_t = the total resistance of the circuit without the meter. Perform the indicated operations in the right member and write as a single expression in I_r, R_m, and R_t.

81. A lens maker's equation for making a lens is given by

$$\frac{1}{f} = (n - 1)\left(\frac{1}{R_1} + \frac{1}{R_2}\right),$$

where f is the focal length of the lens, n is the index of refraction, and R_1 and R_2 are the radii of curvature of the surfaces. Simplify the right member by performing the indicated operations and obtaining a single rational expression.

82. The expression

$$V_1\left(1 + \frac{T_2 - T_1}{T_1}\right) - V_2\left(\frac{T_2 - T_1}{T_2} - 1\right)$$

gives the volume change of a gas under constant pressure. Simplify the expression by performing the indicated operations.

4-4 Complex rational expressions

The previous sections of chapter 4 dealt with simple rational numbers and simple rational expressions—rational numbers and rational expressions that have a single integer, or single polynomial, in the numerator and in the denominator. We now consider a **complex rational expression**—a rational expression whose numerator or denominator, or both, contain rational expressions.

To simplify such expressions as

$$\frac{\dfrac{2}{x}}{\dfrac{3}{x^2}}, \quad \frac{\dfrac{2}{x} - 3}{4 + \dfrac{1}{x}}, \quad \frac{\dfrac{5}{y} - \dfrac{3}{y}}{y - 3}, \quad \text{and} \quad \frac{\dfrac{p + 1}{p - 3}}{\dfrac{4 - p}{p}},$$

we eliminate the rational expression within the numerator and/or the denominator. Consider the complex rational expression

$$\frac{\dfrac{p + 1}{p - 3}}{\dfrac{4 - p}{p}}.$$

For the sake of discussion we call

$$\frac{p + 1}{p - 3}$$

the **primary numerator** and

$$\frac{4 - p}{p}$$

the **primary denominator.** The expressions $p - 3$ and p are called the **secondary denominators.** Thus to simplify the complex rational expression, we must eliminate the secondary denominators. This can be accomplished in either one of two ways.

Example 4–4   A

Simplify each complex rational expression. Assume all denominators are nonzero.

1. $\dfrac{\dfrac{2}{x}}{\dfrac{3}{x^2}}$

Perform the indicated division.

$$\frac{\dfrac{2}{x}}{\dfrac{3}{x^2}} = \frac{2}{x} \div \frac{3}{x^2} = \frac{2}{x} \cdot \frac{x^2}{3} = \frac{2x^2}{3x} = \frac{2x}{3}$$

2. $\dfrac{\dfrac{2}{x} - 3}{4 + \dfrac{1}{x}}$

Perform the indicated addition (in the denominator), the indicated subtraction (in the numerator), and then divide.

$$\frac{\dfrac{2}{x} - 3}{4 + \dfrac{1}{x}} = \frac{\dfrac{2 - 3x}{x}}{\dfrac{4x + 1}{x}} = \frac{2 - 3x}{x} \div \frac{4x + 1}{x} = \frac{2 - 3x}{x} \cdot \frac{x}{4x + 1}$$

$$= \frac{x(2 - 3x)}{x(4x + 1)} = \frac{2 - 3x}{4x + 1}$$

A second method to accomplish this same end is to find the least common multiple (LCM) of the *secondary denominators* and to apply the fundamental principle of rational expressions by multiplying the primary numerator and the primary denominator by the LCM, thus eliminating the secondary denominators.

Example 4–4 B   Simplify each complex rational expression. Assume all denominators are nonzero.

1. $\dfrac{\dfrac{7}{8}}{\dfrac{5}{6}}$

The LCM of the secondary denominators, 8 and 6, is 24. We multiply the numerator and the denominator by 24.

$$\frac{\dfrac{7}{8} \cdot 24}{\dfrac{5}{6} \cdot 24} = \frac{7 \cdot 3}{5 \cdot 4} = \frac{21}{20}$$

2. $\dfrac{\dfrac{p+1}{p-3}}{\dfrac{4-p}{p}}$

The LCM of $p - 3$ and p is $p(p - 3)$. Multiply the numerator and the denominator by the LCM.

$$\frac{\dfrac{p+1}{p-3}}{\dfrac{4-p}{p}} = \frac{\dfrac{p+1}{p-3} \cdot p(p-3)}{\dfrac{4-p}{p} \cdot p(p-3)} = \frac{(p+1)p}{(4-p)(p-3)} = \frac{p^2 + p}{-p^2 + 7p - 12}$$

3. $\dfrac{\dfrac{5}{y} - \dfrac{3}{y}}{y - 3}$

The LCM of the secondary denominators is y. We multiply the primary numerator, $\dfrac{5}{y} - \dfrac{3}{y}$, and the primary denominator, $y - 3$, by y.

$$\frac{\dfrac{5}{y} - \dfrac{3}{y}}{y - 3} = \frac{\left(\dfrac{5}{y} - \dfrac{3}{y}\right) \cdot y}{(y - 3) \cdot y} = \frac{\dfrac{5}{y} \cdot y - \dfrac{3}{y} \cdot y}{(y - 3)y} = \frac{5 - 3}{y(y - 3)} = \frac{2}{y(y - 3)}$$

4. $\dfrac{(y + 2) - \dfrac{3}{y - 6}}{(y - 1) + \dfrac{4}{y - 6}}$

The LCM of the secondary denominators is $y - 6$. Multiply the primary numerator, $(y + 2) - \dfrac{3}{y - 6}$, and the primary denominator,

$(y - 1) + \dfrac{4}{y - 6}$, by $y - 6$.

$$\frac{(y + 2) - \dfrac{3}{y - 6}}{(y - 1) + \dfrac{4}{y - 6}} = \frac{\left[(y + 2) - \dfrac{3}{y - 6}\right](y - 6)}{\left[(y - 1) + \dfrac{4}{y - 6}\right](y - 6)}$$

$$= \frac{(y + 2)(y - 6) - \dfrac{3}{y - 6} \cdot (y - 6)}{(y - 1)(y - 6) + \dfrac{4}{y - 6} \cdot (y - 6)}$$

$$= \frac{y^2 - 4y - 12 - 3}{y^2 - 7y + 6 + 4}$$

$$= \frac{y^2 - 4y - 15}{y^2 - 7y + 10} \quad \text{(Cannot be simplified.)}$$

Note

A reminder again—always check the factorability of the numerator and the denominator to *reduce* the simple rational expression to lowest terms.

Mastery points

Can you

- Recognize a complex rational expression?
- Simplify a complex rational expression?

Exercise 4–4

Directions Simplify each complex rational expression to lowest terms by performing the indicated operations. Assume all the denominators are nonzero. See example 4–4 A.

1. $\dfrac{\dfrac{3}{5}}{\dfrac{4}{7}}$

2. $\dfrac{\dfrac{1}{3} + \dfrac{1}{2}}{\dfrac{2}{3} - \dfrac{5}{6}}$

3. $\dfrac{4 - \dfrac{2}{5}}{3 + \dfrac{3}{10}}$

4. $\dfrac{\dfrac{4}{x}}{\dfrac{8}{x^2}}$

5. $\dfrac{\dfrac{7}{m^2}}{\dfrac{8}{m}}$

6. $\dfrac{\dfrac{5}{b}}{\dfrac{9}{b - 3}}$

7. $\dfrac{\dfrac{-4}{y + 1}}{\dfrac{3}{y}}$

8. $\dfrac{\dfrac{8}{a - 3}}{\dfrac{-6}{a + 2}}$

9. $\dfrac{\dfrac{x}{x + 7}}{\dfrac{x}{x - 3}}$

10. $\dfrac{\dfrac{m - 3}{4}}{\dfrac{2m + 7}{6}}$

11. $\dfrac{\dfrac{3x - 4}{8}}{\dfrac{2x + 1}{10}}$

12. $\dfrac{\dfrac{4h - 1}{h}}{\dfrac{2h + 3}{h + 5}}$

Directions Simplify each complex rational expression by multiplying the numerator and the denominator by the LCM of the secondary denominators. See example 4–4 B.

13. $\dfrac{\dfrac{5}{9}}{\dfrac{3}{4}}$

14. $\dfrac{\dfrac{1}{6} - \dfrac{7}{8}}{\dfrac{2}{3} + \dfrac{1}{4}}$

15. $\dfrac{\dfrac{5}{y^2}}{\dfrac{7}{y}}$

16. $\dfrac{\dfrac{3}{x - 2}}{\dfrac{9}{x}}$

17. $\dfrac{\dfrac{7}{a + 9}}{\dfrac{-4}{a - 5}}$

18. $\dfrac{\dfrac{n-9}{n+2}}{\dfrac{n}{3n-5}}$

19. $\dfrac{\dfrac{4x^2-y^2}{3x}}{\dfrac{2x+y}{4x}}$

20. $\dfrac{\dfrac{x+3y}{y}}{\dfrac{x^2-9y^2}{6y}}$

21. $\dfrac{4-\dfrac{3}{x+3}}{5+\dfrac{6}{x-1}}$

22. $\dfrac{\dfrac{7}{y-3}+8}{9-\dfrac{1}{2y+3}}$

23. $\dfrac{\dfrac{1}{x}+\dfrac{2}{y^2}}{\dfrac{3}{x^2}-\dfrac{7}{x}}$

24. $\dfrac{\dfrac{6}{a}-\dfrac{5}{b}}{\dfrac{1}{a^2}+\dfrac{1}{b^2}}$

25. $\dfrac{m+n}{\dfrac{1}{m}-\dfrac{1}{n}}$

26. $\dfrac{x-y}{\dfrac{2}{x}+\dfrac{3}{y}}$

27. $\dfrac{\dfrac{5}{p^2}+\dfrac{4}{q}}{p-q}$

28. $\dfrac{\dfrac{1}{m^2}-\dfrac{1}{n^2}}{m+n}$

29. $\dfrac{(a+5)+\dfrac{3}{a+4}}{(a+3)-\dfrac{5}{a+4}}$

30. $\dfrac{(x-3)+\dfrac{7}{2x+1}}{(x+9)-\dfrac{3}{2x+1}}$

31. $\dfrac{y-\dfrac{3}{4y-3}}{(y+2)+\dfrac{3}{y+5}}$

32. $\dfrac{(m-3)+\dfrac{6}{2m+3}}{m-\dfrac{9}{m-6}}$

33. $\dfrac{\dfrac{t^2-2t-8}{t^2+7t+6}}{\dfrac{t^2-t-6}{t^2+2t+1}}$

34. $\dfrac{\dfrac{y^2-5y-14}{y^2+3y-10}}{\dfrac{y^2-8y+7}{y^2+6y+5}}$

35. $\dfrac{\dfrac{3}{x^2-x-6}}{\dfrac{2}{x+2}-\dfrac{4}{x-3}}$

36. $\dfrac{\dfrac{9}{a-7}+\dfrac{8}{2a+3}}{\dfrac{10}{2a^2-11a-21}}$

37. $\dfrac{5+\dfrac{4}{b-1}}{\dfrac{7}{b+5}-\dfrac{3}{b-1}}$

38. $\dfrac{\dfrac{6}{x+5}-7}{\dfrac{8}{x+5}-\dfrac{9}{x+3}}$

39. $\dfrac{\dfrac{1}{1-x}-\dfrac{1}{1+x}}{\dfrac{1+x}{1-x}-\dfrac{1-x}{1+x}}$

40. $\dfrac{\dfrac{a-1}{a-b}-\dfrac{a-b}{a-1}}{\dfrac{1}{a-1}-\dfrac{1}{a-b}}$

41. $\dfrac{\dfrac{3}{ab}+\dfrac{4}{bc}-\dfrac{2}{ac}}{\dfrac{5}{abc}}$

42. $\dfrac{\dfrac{x}{yz}-\dfrac{y}{xz}+\dfrac{z}{xy}}{\dfrac{1}{x^2y^2}-\dfrac{1}{x^2z^2}+\dfrac{1}{y^2z^2}}$

Directions Solve the following verbal problems.

43. In electricity a relationship for the current, i, in a capacitor of "size" C is given by

$$i = \dfrac{V_s - \dfrac{it}{C}}{R}. \text{ Simplify the right member.}$$

44. In electricity the voltage between two adjacent nodes, denoted by $V_{AA'}$, is given by Millman's Theorem,

$$V_{AA'} = \dfrac{\dfrac{V_{S1}}{R_1} + \dfrac{V_{S2}}{R_2} + \dfrac{V_{S3}}{R_3}}{\dfrac{1}{R_1} + \dfrac{1}{R_2} + \dfrac{1}{R_3}},$$

where V_{S1}, V_{S2}, V_{S3} represent equivalent voltages of the branches between nodes A and A', whereas R_1, R_2, and R_3 are equivalent resistances of the branches between nodes A and A'. Simplify the right member.

45. A refrigeration coefficient, CP, of performance formula for the ideal refrigerator is given by

$$CP = \dfrac{1}{\dfrac{T_2}{T_1} - 1}. \text{ Simplify the right member.}$$

46. The capacitance C of a circuit connecting three capacitances C_1, C_2, and C_3 in series is given by

$$C = \dfrac{1}{\dfrac{1}{C_1} + \dfrac{1}{C_2} + \dfrac{1}{C_3}}. \text{ Simplify the right member.}$$

47. When making a round trip whose one-way distance is d, the average rate (speed) traveled, r, is given by

$$r = \dfrac{2d}{\dfrac{d}{r_1} + \dfrac{d}{r_2}},$$

where r_1 is the average rate going and r_2 is the average rate coming back. Simplify the right member.

4–5 Quotients of polynomials

Division of a multinomial by a monomial

The first type of division that we will study is that of a multinomial divided by a monomial. Consider the indicated division

$$\frac{4a^3 - 12a^2 + 8a}{2a}.$$

Using the meaning of division, we can rewrite this as

$$(4a^3 - 12a^2 + 8a) \cdot \frac{1}{2a},$$

and applying the distributive property, we have

$$\frac{4a^3 - 12a^2 + 8a}{2a} = \frac{4a^3}{2a} - \frac{12a^2}{2a} + \frac{8a}{2a}$$
$$= 2a^2 - 6a + 4.$$

> ■ **Dividing a multinomial by a monomial**
> For monomials $a_1, a_2, \cdot \cdot \cdot, a_n$, and d, where $d \neq 0$,
> $$\frac{a_1 + a_2 + \cdot \cdot \cdot + a_n}{d} = \frac{a_1}{d} + \frac{a_2}{d} + \cdot \cdot \cdot + \frac{a_n}{d}.$$
>
> **Concept**
> To divide a multinomial by a monomial, simply divide each term of the multinomial by the monomial and write the resulting quotients.

Example 4–5 A

Perform the indicated division. Assume that no denominator equals zero.

1. $\dfrac{4a^6 + 16a^4 - 6a}{4a^2} = \dfrac{4a^6}{4a^2} + \dfrac{16a^4}{4a^2} - \dfrac{6a}{4a^2}$

$\qquad\qquad = a^4 + 4a^2 - \dfrac{3}{2a}$

2. $\dfrac{x(3y - 2) + 2(3y - 2)}{3y - 2} = \dfrac{x(3y - 2)}{3y - 2} + \dfrac{2(3y - 2)}{3y - 2}$

Divide by the quantity $3y - 2$ that is common to both the numerator and the denominator.

$\dfrac{x(3y - 2)}{3y - 2} + \dfrac{2(3y - 2)}{3y - 2} = x + 2$

Note
A common error in this type of problem is demonstrated in the example

$$\frac{x^3 + x^2}{x^2} \neq \frac{x^3 + 1}{1}.$$

It is tempting to simply "cancel" the x^2 in the numerator with the x^2 in the denominator, but the correct procedure is

$$\frac{x^3 + x^2}{x^2} = \frac{x^3}{x^2} + \frac{x^2}{x^2} = x + 1.$$

Division of a multinomial by a multinomial

The second type of division that we will study is that of a multinomial divided by another multinomial. Consider the example

$$\frac{2x^2 - x - 15}{x - 3}.$$

The example is set up so that the divisor, $x - 3$, and the dividend, $2x^2 - x - 15$, are arranged in descending powers of the variable.

$$x - 3 \overline{)2x^2 - x - 15}$$

Now we divide the first term of the dividend, $2x^2$, by the first term of the divisor, x. The result is $\frac{2x^2}{x} = 2x$, and we place the $2x$ above the division line.

$$x - 3 \overline{)2x^2 - x - 15} \quad \overset{2x}{}$$

Next, we multiply $x - 3$ and $2x$, placing the product below the dividend, and subtract.

$$
\begin{array}{r}
2x \\
x - 3 \overline{)2x^2 - x - 15} \\
2x^2 - 6x
\end{array}
\qquad 2x(x - 3) = 2x^2 - 6x
$$

Recall that when we subtract, we change the signs and add. Then
$(2x^2 - x) - (2x^2 - 6x) = 2x^2 - x - 2x^2 + 6x = 5x.$

$$
\begin{array}{r}
2x \\
x - 3 \overline{)2x^2 - x - 15} \\
\ominus\ 2x^2 \oplus 6x \\
\hline
5x
\end{array}
$$

Now we bring down the -15 and repeat the same process.

$$
\begin{array}{r}
2x \\
x - 3 \overline{)2x^2 - x - 15} \\
2x^2 - 6x \\
\hline
5x - 15
\end{array}
$$

Divide the $5x$ by x, which results in 5, and multiply 5 and $x - 3$.

$$
\begin{array}{r}
2x + 5 \\
x - 3 \overline{)2x^2 - x - 15} \\
2x^2 - 6x \\
\hline
5x - 15 \\
5x - 15
\end{array}
$$

Subtract $(5x - 15) - (5x - 15) = 5x - 15 - 5x + 15 = 0.$

$$
\begin{array}{r}
2x + 5 \\
x - 3 \overline{)2x^2 - x - 15} \\
2x^2 - 6x \\
\hline
5x - 15 \\
5x - 15 \\
\hline
0
\end{array}
$$

We subtract and get a remainder of zero, so the quotient is $2x + 5$.

$$\frac{2x^2 - x - 15}{x - 3} = 2x + 5$$

To check the problem, multiply the quotient times the divisor to get the dividend.

$$(2x + 5)(x - 3) = 2x^2 - x - 15$$

Note

It is important, when dividing a multinomial by a multinomial, that both the divisor and the dividend have their terms arranged in descending powers of the variable.

Example 4-5 B

Find the indicated quotients. Assume no divisor is equal to zero.

1. $(2x^2 - 11x + 15) \div (x - 3)$

$$
\begin{array}{r}
2x - 5 \\
x - 3 \overline{\smash{\big)}\ 2x^2 - 11x + 15} \\
\underline{(-)\ 2x^2 - 6x} \downarrow \\
-5x + 15 \\
\underline{(-)-5x + 15} \\
0
\end{array}
$$

Therefore $\dfrac{2x^2 - 11x + 15}{x - 3} = 2x - 5$.

2. $(2y^3 - y^2 + 5y + 5) \div (2y + 1)$

$$
\begin{array}{r}
y^2 - y + 3 \\
2y + 1 \overline{\smash{\big)}\ 2y^3 - y^2 + 5y + 5} \\
\underline{(-)\ 2y^3 + y^2} \\
-2y^2 + 5y \\
\underline{(-)\ -2y^2 - y} \\
6y \mid 5 \\
\underline{(-)\ 6y + 3} \\
2
\end{array}
$$

When there is a remainder, as in this example, we write the remainder over the divisor.

$$\frac{2y^3 - y^2 + 5y + 5}{2y + 1} = y^2 - y + 3 + \frac{2}{2y + 1}$$

3. $(2x^3 - x + 5) \div (x - 1)$

There is no term in the dividend that contains x^2. Therefore we will insert $0x^2$ as a placeholder so that all powers of the variable x are present in descending order. The value of the dividend has not changed since we added $0x^2$, which is another name for 0.

$$
\begin{array}{r}
2x^2 + 2x + 1 \\
x - 1 \overline{)\, 2x^3 + 0x^2 - x + 5} \\
(-)\ \underline{2x^3 - 2x^2} \\
2x^2 - x \\
(-)\ \underline{2x^2 - 2x} \\
x + 5 \\
(-)\ \underline{x - 1} \\
6
\end{array}
$$

Hence $\dfrac{2x^3 - x + 5}{x - 1} = 2x^2 + 2x + 1 + \dfrac{6}{x - 1}$.

Mastery points

Can you

- Divide a multinomial by a monomial?
- Divide a multinomial by a multinomial?

Exercise 4–5

Directions Perform the indicated divisions. Assume that no divisor is equal to zero. See examples 4–5 A and B.

1. $\dfrac{25x^2 - 15x + 10}{5}$

2. $\dfrac{2a^4 - 3a^2 + a}{a}$

3. $\dfrac{4x^4 - 8x^3 + 12x}{-4x}$

4. $\dfrac{15y^5 + 25y^2 + 10y}{-5y^2}$

5. $\dfrac{ac^2 - ac}{ac}$

6. $\dfrac{bx - b^2x^2}{bx}$

7. $\dfrac{30x^3y^4 + 21x^2y^2 - 18x^2y^4}{3x^2y^2}$

8. $\dfrac{36x^4y^2z^3 - 24x^2y^5z + 18x^2y^2z^2}{6x^2y^3z}$

9. $\dfrac{21a^7b^2c^3 - 35a^5b^5c^3 + 49a^4b^2c^3}{7abc^3}$

10. $\dfrac{x(y - 2) - z(y - 2)}{y - 2}$

11. $\dfrac{2a(b - 4) + 3c(b - 4)}{b - 4}$

12. $\dfrac{x^2y(z + 3) - 3x^4y^3(z + 3)}{xy(z + 3)}$

13. $(a^2 - 2a - 8) \div (a + 2)$

14. $(y^2 + 2y - 3) \div (y + 3)$

15. $(y^2 + 7y + 11) \div (y + 5)$

16. $(2x - 5 + x^2) \div (x + 4)$

17. $(10 - 7x + x^2) \div (x - 3)$

18. $\dfrac{a^2 - 5a + 1}{a - 1}$

19. $\dfrac{a^3 + a^2 - 2a + 12}{a + 3}$

20. $\dfrac{y^3 + 3y^2 - y - 6}{y + 2}$

21. $\dfrac{2y^3 - y^2 - 2y + 3}{y + 1}$

22. $\dfrac{3x^3 - 4x^2 - 5x - 4}{x + 2}$

23. $\dfrac{5x^2 - 11x + 2x^3 + 4}{2x - 1}$

24. $\dfrac{3a^3 - 2 - 5a - a^2}{3a + 2}$

25. $\dfrac{4a^3 + 8a^2 - 5a + 1}{2a + 1}$

26. $\dfrac{6y^3 - y^2 - 11y + 10}{3y - 2}$

27. $\dfrac{9y^3 + 11y + 6}{3y + 1}$

28. $\dfrac{27x^3 - 1}{3x - 1}$

29. $\dfrac{a^3 + 27}{a + 3}$

30. $\dfrac{x^4 - 16}{x - 2}$

31. $\dfrac{3x^4 - 2x^3 + x - 1}{x + 1}$

32. $\dfrac{2x^3 + 5x^2 + 5x + 3}{x^2 + x + 1}$

33. $\dfrac{3a^3 - 4a^2 + 10a - 3}{a^2 - a + 3}$

34. $\dfrac{a^4 - 2a^2 - 3a - 1}{a^2 - 2a - 1}$

35. $\dfrac{y^4 + 4y^3 + 3y^2 - 2y - 1}{y^2 + 3y + 1}$

36. $\dfrac{y^4 - y^3 - 11y^2 + 10y + 2}{y^2 - 4y + 2}$

37. $\dfrac{x^4 + 4x^3 + x^2 - 10x - 12}{x^2 + x - 4}$

38. $\dfrac{2x^4 - x^3 + 5x^2 - x + 3}{x^2 + 1}$

39. $\dfrac{3a^4 - 2a^3 + 2a^2 - 4a - 8}{a^2 + 2}$

40. $\dfrac{a^4 + 4a^3 + a^2 - 10a - 9}{a^2 + 3a + 2}$

41. $\dfrac{2y^4 - 3y^3 + 8y^2 - 9y + 8}{2y^2 - 3y + 2}$

42. $\dfrac{3y^4 + y^3 - 8y^2 - 3y - 5}{3y^2 + y + 1}$

Directions Solve the following verbal problems.

43. Evaluate $2y^3 - y^2 - 2y + 3$ at -1 and compare this answer to the remainder found in exercise 21.

44. Evaluate $3x^3 - 4x^2 - 5x - 4$ at -2 and compare this answer to the remainder found in exercise 22.

45. The area of a rectangle is found by multiplying the length times the width. If the area of a rectangle is $6x^2 - 17x + 12$ and the length is $3x - 4$, find the width.

46. A contractor uses the expression $x^2 + 6x + 8$ to represent the square footage of a room. If she decides that the length of the room will be represented by $x + 4$, what will the width of the room be in terms of x?

47. The volume of a box is found by multiplying the length times the width times the height. If the volume of a box is $6x^3 + 11x^2 - 19x + 6$, the height is $x + 3$, and the width is $2x - 1$, find the length.

48. An electrician uses the expression $x^2 + 5x + 6$ to determine the amount of wire to order when wiring a house. If the formula comes from multiplying the number of rooms times the number of outlets and he knows the number of rooms to be $x + 2$, find the number of outlets in terms of x.

4–6 Equations containing rational expressions

Rational equations

In chapter 3 we studied how to find the solution set of a linear equation. We now consider equations that contain at least one term that is a rational expression. We call these equations **rational equations.** The same properties of real numbers that we used in chapter 3 will apply to solve rational equations once the polynomial denominators are eliminated.

Recall that an equivalent equation is obtained when each term of an equation is multiplied by the same nonzero constant. Thus, given a rational equation, to obtain an equivalent equation and eliminate the denominators, we *multiply both members of the equation by the LCM of the denominators.*

Example 4–6 A

Find the solution set S of the following rational equation.

1. $\dfrac{x}{3} + \dfrac{x}{4} - \dfrac{x}{6} = 35$

Since the LCM of 3, 4, and 6 is 12, we multiply each term of the equation by 12.

$12 \cdot \dfrac{x}{3} + 12 \cdot \dfrac{x}{4} - 12 \cdot \dfrac{x}{6} = 12 \cdot 35$

Simplify in the left member to obtain the equivalent equation.

$4x + 3x - 2x = 420$

$5x = 420$

$x = \dfrac{420}{5} = 84$

$S = \{84\}$

When the rational equation contains rational expressions with the variable in the denominator, multiplying all terms by the LCM of the denominators does *not always* produce an equation that is equivalent to the given equation.

2. $1 - \dfrac{2a}{a + 1} = \dfrac{2}{a + 1}$

The only denominator is $a + 1$, so we multiply each term of the equation by $a + 1$.

$(a + 1) \cdot 1 - (a + 1) \cdot \dfrac{2a}{a + 1} = (a + 1) \cdot \dfrac{2}{a + 1}$

$a + 1 - 2a = 2$

$1 - a = 2$

$-a = 1$ Multiply each member by -1.

$a = -1$

However in the original equation -1 *is not in the domain of the rational expressions involved*. Therefore -1 cannot be a solution and we conclude that the equation has no solution. Hence $S = \emptyset$.

Extraneous solutions

The possible solution -1 in example 2 is called an **extraneous solution**—a solution of the equation obtained by multiplying each term by the LCM of the denominators but *not a solution of the original equation*.

Note
It is important to observe the domain of the rational expressions in the rational equation to know where extraneous solutions may occur.

Equations and formulas containing rational expressions and more than one variable are common in scientific fields. It is often desirable to solve such equations for one variable in terms of the other variables in the equation. We use the same properties that we used in solving preceding equations of this section.

Example 4–6 B

Solve each of the following equations for the indicated variable. Assume all denominators are nonzero.

1. $\dfrac{3x}{5} - \dfrac{4y}{3} = 6 \qquad$ for x

Multiply each term of the equation by the LCM of 5 and 3, that is, 15.

$$15 \cdot \frac{3x}{5} - 15 \cdot \frac{4y}{3} = 15 \cdot 6$$

$\qquad 3 \cdot 3x - 5 \cdot 4y = 90 \qquad\qquad$ Simplify each term.

$$9x - 20y = 90$$

$\qquad\qquad 9x = 90 + 20y \qquad$ Add $20y$ to each member.

$\qquad\qquad x = \dfrac{90 + 20y}{9} \qquad$ Divide each member by 9.

2. $\dfrac{3}{x} - 7 = 9y + \dfrac{2y}{3x} \qquad$ for y

Multiply each term of the equation by the LCM of x and $3x$, that is, $3x$.

$$3x \cdot \frac{3}{x} - 3x \cdot 7 = 3x \cdot 9y + 3x \cdot \frac{2y}{3x}$$

$$3 \cdot 3 - 21x = 27xy + 2y$$

$$9 - 21x = 27xy + 2y$$

Since we cannot combine the terms containing y in the right member, we must *factor* the common factor, y, from each term.

$$9 - 21x - (27x + 2)y$$

Divide each term by the coefficient of y, which is $27x + 2$.

$$\frac{9 - 21x}{27x + 2} = y$$

3. In physics the rule governing the speeds of two gears, one the driver gear A and the other the driven gear B, is given by

$$\frac{T_A}{T_B} = \frac{R_B}{R_A},$$

where $T =$ the number of teeth in the gear and $R =$ the revolutions per minute of the gear. Solve for the revolutions per minute of the driven gear, R_B.

Multiply each term by the LCM of T_B and R_A which is $T_B R_A$.

$$T_B R_A \cdot \frac{T_A}{T_B} = T_B R_A \cdot \frac{R_B}{R_A}$$

$$R_A \cdot T_A = T_B \cdot R_B$$

Divide each member by T_B.

$$R_B = \frac{R_A \cdot T_A}{T_B}$$

Exercise 4–6

Directions Find the solution set S of each of the following equations. See example 4–6 A.

1. $\dfrac{x + 7}{4} = \dfrac{2x}{12}$

2. $\dfrac{4b - 3}{10} - \dfrac{2b - 1}{6}$

3. $\dfrac{3x}{5} - \dfrac{4x}{3} = 1$

4. $\dfrac{m}{3} + 4 = \dfrac{7m}{4}$

5. $\dfrac{4}{a} - \dfrac{6}{3a} = \dfrac{3}{5}$

6. $4 - \dfrac{5}{9b} = \dfrac{7}{6b}$

7. $\dfrac{b - 3}{10} + \dfrac{2b + 1}{15} = 2$

8. $\dfrac{3y - 4}{16} - \dfrac{2 - 3y}{12} = 1$

9. $\dfrac{x - 3}{3x} = \dfrac{2x + 3}{9x}$

10. $\dfrac{5 - b}{8b} = \dfrac{3b + 7}{6b}$

11. $\dfrac{5}{x - 2} = \dfrac{4}{2x + 1}$

12. $\dfrac{-3}{x + 7} = \dfrac{2}{3x - 1}$

13. $\dfrac{5}{4a + 2} = \dfrac{7}{2a + 1}$

14. $\dfrac{11}{3y - 2} = \dfrac{8}{12y - 8}$

15. $\dfrac{6}{6x + 3} = \dfrac{3}{2x + 1} + 5$

16. $1 + \dfrac{5}{3m - 9} = \dfrac{10}{m - 3}$

17. $\dfrac{x - 1}{x^2 - 4} = \dfrac{6}{x - 2}$

18. $\dfrac{5}{y + 3} = \dfrac{2y + 1}{y^2 - 9}$

19. $\dfrac{x}{x - 2} + \dfrac{2}{3} = \dfrac{2}{x - 2}$

20. $\dfrac{3}{2} - \dfrac{1}{x - 4} = \dfrac{-2}{2x - 8}$

21. $4 - \dfrac{2x}{5 - x} = \dfrac{6}{x - 5}$

22. $\dfrac{5y}{3 - y} + \dfrac{8}{y - 3} = 4$

23. $\dfrac{8b}{b^2 - 16} = \dfrac{3}{b + 4} + \dfrac{5}{4 - b}$

24. $\dfrac{10}{5 - a} = \dfrac{7}{a + 5} - \dfrac{6}{a^2 - 25}$

25. $\dfrac{5}{a - 3} - \dfrac{1}{a + 2} = \dfrac{6a}{a^2 - a - 6}$

26. $\dfrac{7}{x + 6} = \dfrac{9}{x - 9} - \dfrac{2x - 1}{x^2 - 3x - 54}$

27. $\dfrac{4}{2x - 6} - \dfrac{12}{4x + 12} = \dfrac{12}{x^2 - 9}$

28. $\dfrac{11x - 3}{10x^2 - 3x - 4} - \dfrac{4}{5x - 4} = \dfrac{9}{2x + 1}$

29. $\dfrac{6}{q^2 + q - 6} = \dfrac{5}{q^2 + 3q - 10}$

30. $\dfrac{9}{y^2 - 6y + 8} = \dfrac{14}{y^2 - 16}$

31. $\dfrac{13}{n^2 + 2n - 15} - \dfrac{1}{n^2 + 10n + 25} = 0$

32. $\dfrac{2}{2n^2 - 7n - 4} + \dfrac{6}{6n^2 - 5n - 4} = 0$

33. $\dfrac{4}{x^2 - 9} = \dfrac{7}{x^2 - 7x + 12} - \dfrac{5}{x^2 - x - 12}$

34. $\dfrac{-2}{a^2 + 3a - 4} - \dfrac{6}{a^2 - 1} = \dfrac{1}{a^2 + 5a + 4}$

35. $\dfrac{6}{2n^2 + n - 3} - \dfrac{5}{4n^2 - 9} = \dfrac{6}{2n^2 - 5n + 3}$

36. $\dfrac{6}{8x^2 - 14x - 15} + \dfrac{1}{8x^2 - 26x + 15} = \dfrac{9}{16x^2 - 9}$

Directions Solve the given equations and formulas for the indicated variable. See example 4–6 B.

37. $\dfrac{4}{a} - \dfrac{3}{b} = 7$ for b

38. $\dfrac{6}{x} - 4 = \dfrac{7}{y}$ for y

39. $\dfrac{3}{p} + 4 = \dfrac{6q}{2p} - 3a$ for p

40. $\dfrac{a+6}{3} - \dfrac{b-2}{4} = \dfrac{c}{12}$ for a

41. $\dfrac{y-3}{x+2} = \dfrac{5}{3}$ for y

42. $\dfrac{-7}{3} = \dfrac{y+1}{x-5}$ for x

43. $S = \dfrac{a}{1-r}$ for r (Progression formula)

44. $S = \dfrac{n(a+b)}{2}$ for a (Progression formula)

45. $A = \dfrac{1}{2} h(b_1 + b_2)$ for b_1 (Area of a trapezoid)

46. $C = \dfrac{5}{9}(F - 32)$ for F (Temperature conversion)

See example 4–6 B–3.

47. The coefficient of linear expansion, k, of a solid when heated is given by
$$k = \dfrac{L_t - L_0}{L_0 t},$$
where L_t is the length at $t\,°C$, L_0 is the length at $0\,°C$, and t is any given temperature in Celsius. Solve for t. Solve for L_0.

48. A formula for unknown resistance R_x in a battery is given by
$$R_x = R_m \left(\dfrac{E_1}{E_2} - 1 \right).$$
Solve for E_1.

49. The interest rate, r, on a given amount of money, P, over a given period of time, t, which pays interest, I, is given by
$$r = \dfrac{I}{Pt}.$$
Solve the equation for t.

50. The kinetic energy of a body, KE, is computed by
$$KE = \dfrac{Wv^2}{2g},$$
where W is the weight in pounds, v is the velocity expressed in feet per second, and g is the acceleration due to gravity. Solve for g.

51. The total resistance R of a parallel circuit is given by
$$\dfrac{1}{R} = \dfrac{1}{R_1} + \dfrac{1}{R_2} + \dfrac{1}{R_3},$$
where R_1, R_2, and R_3 are the resistances of the respective circuits. Solve for R_3.

52. The capacitance of capacitors C_1, C_2, and C_3 in a series circuit is given by
$$\dfrac{1}{C} = \dfrac{1}{C_1} + \dfrac{1}{C_2} + \dfrac{1}{C_3}.$$
Solve for C.

53. Given a large piston and a small piston on which forces F and f, respectively, are applied, then
$$\dfrac{F}{f} = \dfrac{A}{a},$$
where A is the area of the large piston and a is the area of the small piston. Solve for A.

54. As gas expands when heated, the relationship between pressure P, volume V, and absolute temperature T is given by
$$\dfrac{P_1 V_1}{T_1} = \dfrac{P_2 V_2}{T_2},$$
called Charles' Law. Solve for V_2.

55. The formula
$$\dfrac{R_2}{R_1} = \dfrac{M + T_2}{M + T_1}$$
gives a relationship for the increase in the resistance of a circuit caused by a rise in temperature. Solve for T_2. Solve for R_1.

56. The lens maker's formula for the focal length, f, of a lens is given by
$$\dfrac{1}{f} = (\mu_m - 1)\left(\dfrac{1}{R_1} + \dfrac{1}{R_2} \right),$$
where μ_m is the index of refraction and R_1 and R_2 are radii of curvature. Solve for f. Solve for μ_m.

4–7 Problem solving with rational equations

Now that we can solve rational equations, let us use that skill in solving some kinds of verbal problems that result in rational equations when we use problem-solving steps.

Example 4–7 A
Work problem

1. Worker P can produce a part in 3 hours and worker Q can produce the same part in 4 hours. How long would it take them to produce the part working together?

Assuming they work at a constant rate, we must think in terms of how much each worker can do in 1 hour. Since worker P can produce the part in 3 hours and worker Q can produce the part in 4 hours, P can produce $\frac{1}{3}$ of the part in 1 hour and Q can produce $\frac{1}{4}$ of the part in 1 hour. Let $x = $ the time necessary to produce the part working together. Then they can produce $\frac{1}{x}$ of the part in 1 hour and the equation is given by

(P in 1 hour) + (Q in 1 hour) = (together in 1 hour)
$$\frac{1}{3} \quad + \quad \frac{1}{4} \quad = \quad \frac{1}{x}.$$

Multiply each term of the equation by the LCM of 3, 4, and x, which is $12x$.

$$12x \cdot \frac{1}{3} + 12x \cdot \frac{1}{4} = 12x \cdot \frac{1}{x}$$
$$4x + 3x = 12$$
$$7x = 12$$
$$x = \frac{12}{7} \text{ or } 1\frac{5}{7} \text{ hours}$$

Working together, P and Q can produce the part in $1\frac{5}{7}$ hours.

Uniform motion problems

2. We use the relationship between distance traveled, rate of travel, and time traveled—distance (d) = rate (r) · time (t). An automobile can travel 200 miles in the same time that a truck can travel 150 miles. If the automobile travels at an average rate of 15 miles per hour faster than the truck, find the average rate of each.

The following table will be helpful in finding the equation necessary to solve this type of problem.

	d	r	t
Automobile			
Truck			

If we let r = the rate of the truck, then $r + 15$ = the rate of the automobile. We fill in the table.

	d	r	t
Automobile	200	$r + 15$	$\dfrac{200}{r + 15}$
Truck	150	r	$\dfrac{150}{r}$

Note
To find the time we used the relationship
$$\text{time} = \frac{\text{distance}}{\text{rate}}.$$

Since both vehicles traveled the same length of time, we obtain the equation
$$\frac{200}{r + 15} = \frac{150}{r}.$$

Multiply each term by the LCM of $r + 15$ and r, that is, $r(r + 15)$.

$$r(r + 15)\frac{200}{r + 15} = r(r + 15)\frac{150}{r}$$
$$200r = (r + 15)150$$
$$200r = 150r + 2{,}250$$
$$50r = 2{,}250$$
$$r = 45$$
$$r + 15 = 60$$

Thus the average rate of the truck is 45 miles per hour and of the automobile is 60 miles per hour.

3. John's boat travels at 20 miles per hour in still water. If the speed of the current is 5 miles per hour, how far can John travel downstream if it takes him 3 hours to go downstream and back?
Let x = the distance traveled downstream (and back upstream since they are the same distance).

Using $t = \dfrac{d}{r}$, we summarize the information in the table.

	d	r	t
Downstream	x	20 + 5 = 25	$\dfrac{x}{25}$
Upstream	x	20 − 5 = 15	$\dfrac{x}{15}$

Since total time downstream and back upstream is 3 hours, the equation is

$$\frac{x}{25} + \frac{x}{15} = 3.$$

Multiply each term by the LCM of 25 and 15, which is 75.

$$\frac{x}{25} \cdot 75 + \frac{x}{15} \cdot 75 = 3 \cdot 75$$

$$3x + 5x = 225$$

$$8x = 225$$

$$x = \frac{225}{8} = 28\frac{1}{8}$$

He traveled $28\frac{1}{8}$ miles downstream.

Number and reciprocal problem

4. One number is twice another number. The sum of their reciprocals is $\frac{9}{2}$. Find the numbers.

Let $x =$ one number, then $2x =$ the other number. Their reciprocals are then $\frac{1}{x}$ and $\frac{1}{2x}$. The sum of the reciprocals is $\frac{9}{2}$, so we have the equation

$$\frac{1}{x} + \frac{1}{2x} = \frac{9}{2}.$$

Multiply each term by the LCM of the denominators, that is, $2x$.

$$2x \cdot \frac{1}{x} + 2x \cdot \frac{1}{2x} = 2x \cdot \frac{9}{2}$$

$$2 + 1 = 9x$$

$$3 = 9x$$

$$x = \frac{3}{9} = \frac{1}{3}$$

$$2x = \frac{2}{3}$$

Since the sum of the reciprocals $3 + \frac{3}{2} = \frac{6}{2} + \frac{3}{2} = \frac{9}{2}$, the two numbers are $\frac{1}{3}$ and $\frac{2}{3}$.

Mastery points

Can you
- Set up and solve work problems?
- Set up and solve uniform motion problems?
- Set up and solve number and reciprocal problems?

Rational Expressions

Exercise 4–7

Directions Solve the following work problems. See example 4–7 A–1.

Work problems

1. Pete Hansen milks his herd of cattle in 2 hours, whereas his son John milks them in 3 hours. How many hours and minutes would it take them to do the same job working together?

2. Jake has two hay balers, *A* and *B*. If Jake can completely bale a given field in 3 hours using baler *A* and in $3\frac{3}{4}$ hours using baler *B,* how long would it take him to bale the field working both balers together?

3. During a "clean-up, paint-up, fix-up week" in Sault Sainte Marie, Jim, Toni, and Ken clean up a given vacant lot. If it takes each person $1\frac{1}{2}$ hours, 2 hours, and 3 hours, respectively, to do the job alone, how long would it take them working together to clean up the lot?

4. Tanya, Bill, and Neysa work in a bakery and can mix 12 loaves of bread individually in 36 minutes, 40 minutes, and 30 minutes, respectively. If they worked together, in how many minutes could they mix the 12 loaves?

5. If two water pipes can fill a swimming pool in 18 hours when both pipes are open and one of the pipes can fill the pool alone in 30 hours, how long would it take the other pipe to fill the pool alone?

6. In a machine shop, machine *A* can produce a gross of a given part in 1 hour and 20 minutes. If machines *A* and *B* working together can produce the parts in 50 minutes, how long would it take machine *B* to do the job alone?

7. Two inlet pipes can fill a water basin in 10 hours and 12 hours, respectively, when open individually. If an outlet pipe can empty the basin alone in 9 hours, how long would it take to fill the basin if all three pipes are open simultaneously?

8. An open sink drain can empty a sink full of water in 2 minutes. If the cold water and hot water faucets can, when open fully, fill the sink in $3\frac{1}{2}$ and 3 minutes, respectively, how long would it take to fill the sink if all three are open simultaneously?

9. It takes pump *A* twice as long to unload an oil tanker as it takes pump *B* to do the job. If the two pumps working together can unload the tanker in 12 hours, how long will it take each to do the job?

10. If one microprocessor can process a set of inputs in three-fifths of the time that a second microprocessor can do the job, and together the machines can do the job in 2 milliseconds, how long will it take each microprocessor to process the set of inputs individually?

11. It takes tug *A* one-half as long to push a series of barges up a river as it takes tug *B* to do the same job. If it takes tug *C* two times as long to do the job as it does tug *B* and all three tugs working together can do the job in 4 hours, how long would it take tug *A* to do the job alone?

12. A portable gasoline-powered generator, a solar cell, and a wind generator fully charge a dead storage battery in 9 hours when charging simultaneously. If the wind generator takes twice as long to fully charge the battery as the solar cell takes and the gasoline-powered generator takes three-fourths as long as the solar cell, how long would it take the solar cell to fully charge the battery alone?

Directions Solve the given uniform motion problems. See example 4–7 A–2 and 3.

Uniform motion problems

13. An excursion boat moves at 16 miles per hour in still water. If the boat travels 20 miles downstream in the same time it takes to travel 14 miles upstream, what is the speed of the current? (*Hint:* Let x = the speed of the current. Then $16 + x$ = the speed of the boat downstream and $16 - x$ = the speed of the boat upstream.)

14. An airplane can cruise at 300 miles per hour in still air. If the airplane takes the same time to fly 950 miles with the wind as it does to fly 650 miles against the wind, what is the speed of the wind?

15. A boat travels 40 kilometers upstream in the same time that it takes the same boat to travel 60 kilometers downstream. If the stream is flowing at 6 kilometers per hour, what is the speed of the boat in still water?

16. The speed of a wind is 15 miles per hour. Find the speed of an airplane in still air if it flies 200 miles against the wind in the same time that it flies 300 miles with the wind.

17. On a trip from Detroit to Los Angeles, a TWA plane takes one-third of an hour longer to make the trip than an American Airlines plane does. If the TWA plane flies at 300 miles per hour and the American Airlines plane flies at 320 miles per hour, how far is it from Detroit to Los Angeles?

18. It takes Bonnie 2 minutes longer to jog a certain distance than it does Maryann. What distance did they run if Bonnie can jog at 5 miles per hour and Maryann can jog at 7 miles per hour?

19. Mary can row her boat 3 miles per hour in still water. If the current of the river is 2 miles per hour, how far downstream can she row if it takes her 2 hours to go down and back?

20. Peter can average 10 miles per hour riding his bike to deliver his papers. By car he can average 30 miles per hour. If it takes him $\frac{1}{2}$ hour less time by car, how long is Peter's paper route?

Directions Solve the following number and reciprocal problems. See example 4–7 A–4.

Number and reciprocal problems

21. One number is three times another number. The sum of their reciprocals is 2. Find the numbers.

22. One number is twice another number. The sum of their reciprocals is $\frac{15}{8}$. Find the numbers.

23. If the same number is added to the numerator and the denominator of the fraction $\frac{3}{7}$, the result is $\frac{4}{5}$. Find the number.

24. If the same number is added to the numerator and the denominator of the fraction $\frac{1}{4}$, the result is $\frac{2}{3}$. Find the number.

25. What number must be added to the denominator of $\frac{6}{7}$ to obtain $\frac{3}{5}$?

26. What number must be added to the numerator and subtracted from the denominator of $\frac{5}{8}$ to obtain its reciprocal?

Chapter summary

1. A **rational expression** is any algebraic expression that can be written as a quotient of two polynomials.

2. The **domain** of a rational expression is the set of all replacement values of the variable for which the expression is defined.

3. The **fundamental principle of rational expressions** states that we obtain an equivalent expression when we multiply or divide the numerator and the denominator of a rational expression by the same nonzero polynomial.

4. Two polynomials are **relatively prime** if they have no common polynomial factors other than 1 or -1.

5. A rational expression is **reduced to its lowest terms** when the numerator and the denominator are relatively prime.

6. To **multiply** two, or more, rational expressions, multiply the numerators to determine the numerator of the product and multiply the denominators to determine the denominator of the product.

7. Given polynomials P, Q, R, and S,
$$\frac{P}{Q} \div \frac{R}{S} = \frac{P}{Q} \cdot \frac{S}{R} = \frac{P \cdot S}{Q \cdot R},$$
where $Q \neq 0$, $R \neq 0$, and $S \neq 0$.

8. Given polynomials P, Q, and R, $Q \neq 0$,
$$\frac{P}{Q} + \frac{R}{Q} = \frac{P + R}{Q}$$
and
$$\frac{P}{Q} - \frac{R}{Q} = \frac{P - R}{Q}.$$

9. Given polynomials P, Q, R, and S, $Q \neq 0$ and $S \neq 0$,
$$\frac{P}{Q} + \frac{R}{S} = \frac{P \cdot S + Q \cdot R}{Q \cdot S}$$
and
$$\frac{P}{Q} - \frac{R}{S} = \frac{P \cdot S - Q \cdot R}{Q \cdot S}.$$

10. To find the least common multiple (LCM) of a set of polynomials,
    a. completely factor the polynomials, then
    b. the LCM consists of the product of all distinct factors of the polynomials, each raised to the greatest power to which it appears in any of the factorizations.

11. Rational expressions can be added or subtracted by building each expression to an equivalent rational expression such that they have a common denominator—the least common denominator (LCD).

12. A **complex rational expression** is a rational expression whose numerator or denominator, or both, contain rational expressions.

13. In the complex rational expression
$$\frac{\dfrac{4}{5}}{\dfrac{7}{8}},$$
we call $\dfrac{4}{5}$ the **primary numerator**, $\dfrac{7}{8}$ the **primary denominator**, and 5 and 8 the **secondary denominators.**

14. To simplify a complex rational expression, multiply the primary numerator and the primary denominator by the least common multiple (LCM) of the secondary denominators.

15. To *divide* a multinomial by a monomial, simply divide each term of the multinomial by the monomial and add or subtract the resulting quotients.

16. A **rational equation** is an equation that contains at least one expression of the form $\dfrac{P}{Q}$, $Q \neq 0$, where P and Q are polynomials.

17. To obtain a simple equation, an equation containing no rational expression, multiply both members of the equation by the LCM of the denominators.

Chapter review

[4–1]
Directions State the domain of the given rational expression in set-builder notation.

1. $\dfrac{5}{x + 7}$

2. $\dfrac{2x + 1}{3x - 4}$

3. $\dfrac{4 - 2x}{x^2 - 10x + 25}$

4. $\dfrac{4a^2 + 1}{9a^2 - 16}$

5. $\dfrac{3z - 1}{6z^2 + 13z - 5}$

6. $\dfrac{4y - 11}{8y^3 + 27}$

Directions Reduce each rational expression to lowest terms. Assume all denominators are nonzero.

7. $\dfrac{a^3b^4c}{a^2bc^3}$

8. $\dfrac{-10m^3n^3p}{35m^4np^5}$

9. $\dfrac{5x - 10}{6x - 12}$

10. $\dfrac{24u}{8a^2 - 16a}$

11. $\dfrac{5x - 10y}{4y^2 - x^2}$

12. $\dfrac{y^3 - 64}{y^2 - 16}$

13. $\dfrac{a^2 - 24a + 144}{a^2 - 11a - 12}$

14. $\dfrac{4x^2 - 5x - 6}{5x^2 - 11x + 2}$

15. $\dfrac{10y^2 + 9y - 9}{12 - 2y - 30y^2}$

[4–2]
Directions Perform the indicated multiplication or division. Place necessary restrictions on the variables.

16. $\dfrac{14x}{3y} \cdot \dfrac{9y^2}{7x^2}$

17. $\dfrac{16a^2}{7y} \div \dfrac{4a}{21y^2}$

18. $\dfrac{p^2 - 16}{4p - 3} \cdot \dfrac{16p^2 - 9}{3p + 12}$

19. $\dfrac{z^2 + 1}{2z - 6} \div \dfrac{z^4 - 1}{z^2 - 6z + 9}$

20. $\dfrac{m^3 - 8}{m^2 - 3m} \cdot \dfrac{m^2 + 3m - 18}{m^2 + 3m - 10}$

21. $\dfrac{5a^2 + 17a + 6}{10a^2 - a - 2} \cdot \dfrac{2a^2 + 13a - 7}{a^2 + 10a + 21}$

22. $\dfrac{x^2 - 49}{8x^3 + 1} \div \dfrac{x^2 - 5x - 14}{4x^2 - 1}$

23. $\dfrac{4x^3 - 5x^2}{4x + 5} \div (16x^2 - 25)$

24. $\dfrac{8 - 10y - 7y^2}{y^2 + 4y + 4} \cdot \dfrac{y^2 + 6y + 9}{7y^2 + 17y - 12}$

25. $\dfrac{mq - mp - nq + np}{2mq - 2mp - nq + np} \div \dfrac{qm + qn + pm + pn}{2mp - np + 2qm - nq}$

[4–3]
Directions Find the least common multiple (LCM) of the set of polynomials.

26. $15xy^2, 20x^3y, 9x^2y^3$

27. $2x + 4, 6x^2, x^2 - 16$

28. $a^2 + 3a - 10, a^2 - 2a, 3a + 15$

29. $25 - p^2, p^2 + 10p + 25, p^2 - 5p$

Directions Perform the indicated additions and/or subtractions. Assume all denominators are nonzero.

30. $\dfrac{3x}{4y} + \dfrac{8x}{3y}$

31. $\dfrac{n - 7}{n + 4} - \dfrac{3n + 8}{n - 1}$

32. $\dfrac{10}{p^2 + 7p - 18} + \dfrac{9}{p^3 - 4p}$

33. $(2b + 7) - \dfrac{b + 9}{3b - 2}$

34. $\dfrac{3y - 4}{49 - y^2} + \dfrac{2y + 1}{y - 7}$

35. $\dfrac{2x}{48 + 4x - 2x^2} - \dfrac{5x + 1}{x^2 - 36}$

36. $\dfrac{7}{2a - 4} + \dfrac{9}{a - 2} - \dfrac{6}{2 - a}$

37. $\dfrac{x + 2}{x^2 - 3x - 28} - \dfrac{3x - 7}{4x + 16} - \dfrac{5 - x}{8x - 56}$

38. $\dfrac{a + b}{a^2 - ab - 2b^2} + \dfrac{3a - b}{a^2 - 4b^2}$

39. In electricity the total resistance R_t of any parallel circuit is given by

$$\dfrac{1}{R_t} = \dfrac{I_1}{E_1} + \dfrac{I_2}{E_2} + \dfrac{I_3}{E_3}.$$

Combine the expression in the right member.

[4–4]
Directions Simplify each complex rational expression. Assume all denominators are nonzero.

40. $\dfrac{\dfrac{4}{a^2}}{\dfrac{2}{a}}$

41. $\dfrac{\dfrac{5}{x - 3}}{\dfrac{4}{x}}$

42. $\dfrac{\dfrac{3x}{x - 5}}{\dfrac{x}{x + 2}}$

43. $\dfrac{7 - \dfrac{3}{b - 5}}{8 + \dfrac{6}{b - 5}}$

44. $\dfrac{\dfrac{x - y}{2}}{\dfrac{2}{x} + \dfrac{3}{y}}$

45. $\dfrac{(p - 3) + \dfrac{2}{p - 1}}{(p + 1) - \dfrac{3}{p - 1}}$

[4–5]
Directions Perform the indicated divisions. Assume that no divisor is equal to zero.

46. $\dfrac{25a^7 + 15a^3 + 10a}{5a}$

47. $\dfrac{36a^5b^5c^3 - 18a^2b^3c^4 + 6a^2b^3c}{6a^2b^3c}$

48. $(3x^3 + x + 4) \div (x + 1)$

49. $\dfrac{2x^4 + x^3 - 2x^2 + 2x - 1}{x^2 + x - 1}$

[4–6]
Directions Find the solution set S of each of the following rational equations.

50. $6 - \dfrac{10}{9a} = \dfrac{5}{12a}$

51. $\dfrac{4m - 3}{15} + \dfrac{2 - 5m}{20} = 1$

52. $\dfrac{4}{a + 6} - \dfrac{9}{2a + 12} = 3$

53. $\dfrac{9}{4a^2 - 1} - \dfrac{6}{2a - 1} = \dfrac{4}{2a + 1}$

54. $\dfrac{8}{6n^2 - 13n - 5} + \dfrac{6}{10n^2 - 25n} = 0$

Directions Solve the given equations for the indicated variable.

55. $\dfrac{2 + m}{6} - \dfrac{n - 3}{4} = \dfrac{p}{8}$ for p

56. $F = \dfrac{9}{5} C + 32$ for C (Temperature conversion)

57. $\dfrac{P_1 V_1}{T_1} = \dfrac{P_2 V_2}{T_2}$ for V_1 (Physics formula)

58. In electricity the total resistance R_t for two resistors connected in parallel is given by

$$R_t = \dfrac{R_1 R_2}{R_1 + R_2}.$$

Solve for R_2.

Directions Set up an equation and solve the following problems.

59. If Erin can do a job in 3 days and Sarah takes 6 days to do the same job, how long would it take the girls to do the job working together?

60. A train travels 120 miles in the same time that an automobile travels 80 miles. If the train travels 30 miles/hour faster than the automobile, what is the speed of each?

61. A boat travels 14 miles downstream in the same amount of time that it takes to travel 10 miles upstream. If the boat travels at 18 miles per hour in still water, what is the speed of the current?

Chapter 4 cumulative test

Directions Evaluate the following expressions when $x = 2$ and $y = -3$.

[1–5] **1.** $x^2 - 2xy + y^2$

[1–5] **2.** $3(2x - 1) - 2(3x + 2)$

[1–5] **3.** $\dfrac{3x^2}{5y^2}$

[1–5] **4.** $\dfrac{\dfrac{1}{x} - \dfrac{1}{y}}{\dfrac{1}{x} + \dfrac{1}{y}}$

[1–5] **5.** $\dfrac{2}{3x} - \dfrac{4}{5y} + \dfrac{1}{xy}$

Directions Perform the indicated operations and simplify.

[1–4] **6.** $3(2x - 1) + 4(x + 5) - (3x + 6)$

[3–3] **7.** $(4ab)(-2a^2b)(3a^3b^3)$

[3–3] **8.** $(x + 4)(x^2 - 2x + 5)$

[4–2] **9.** $\dfrac{10xy}{9x^2y} \cdot \dfrac{27x^3y^2}{15x^2y^2}$

[4–2] **10.** $\dfrac{x^2 + 5x - 24}{x^2 - 2x} \div \dfrac{x^2 - 9}{x^2 - 4x + 4}$

[4–3] **11.** $\dfrac{2y - 3}{8} - \dfrac{4y + 1}{6}$

[3–3] **12.** $(5x - 3)(2x + 9)$

[3–3] **13.** $(4x - 5)^2$

[3–3] **14.** $(3y - 5)(3y + 5)$

[4–5] **15.** $(2x^3 + 3x^2 - 1) \div (x - 3)$

[4–2] **16.** $\dfrac{6a^2 - a - 2}{a^2 + 4a - 21} \cdot \dfrac{2a^2 - a - 15}{8a^2 - 6a - 5}$

[4–3] **17.** $\dfrac{4}{2x^2 - x - 3} - \dfrac{5}{2x^2 + 3x + 1}$

Directions State the domain of the given rational expression.

[4–1] **18.** $\dfrac{4y + 1}{2y - 3}$

[4–1] **19.** $\dfrac{3x}{4x^2 - 1}$

[4–1] **20.** $\dfrac{5}{x^2 + 5}$

Directions Solve each equation.

[2–1] **21.** $4(y + 3) - 3(3y + 1) + (4y + 5) = 0$

[4–6] **22.** $\dfrac{3}{4x} - 2 = \dfrac{5}{3x}$

[4–6] **23.** $\dfrac{3}{y + 2} + \dfrac{2y + 1}{y^2 - 4} = \dfrac{-3}{y - 2}$

Directions Simplify each complex fraction.

[4–4] 24. $\dfrac{\dfrac{1}{4} + \dfrac{2}{5}}{\dfrac{3}{10} - 1}$

[4–4] 25. $\dfrac{4 - \dfrac{3}{x + 2}}{5 + \dfrac{2}{x - 1}}$

Directions Solve the following inequalities.

[2–4] 26. $5 - 3x \geq 8$

[2–4] 27. $3(4y - 1) < 2(5y + 7)$

[2–4] 28. $\dfrac{z + 1}{4} - \dfrac{z - 3}{6} > \dfrac{1}{3}$

[2–4] 29. $\dfrac{3}{4}(x + 1) - 2(x - 3) \leq \dfrac{2}{3}(x - 2)$

Directions Reduce the following rational expressions to lowest terms.

[4–1] 30. $\dfrac{36a^2b^3}{-24ab^5}$

[4–1] 31. $\dfrac{p^2 + p - 12}{p^2 - 9}$

[4–1] 32. $\dfrac{12 - 6y}{4y - 8}$

5

Exponents, Roots, and Radicals

5–1 Roots and rational exponents

The nth root

In chapter 3 we were concerned with raising some real number to a power. For example,

if $a = -4$, then $a^3 = (-4)^3 = (-4)(-4)(-4) = -64$ (read "-4 raised to the third power equals -64");

if $a = 2$, then $a^4 = (2)^4 = (2)(2)(2)(2) = 16$ (read "2 raised to the fourth power equals 16").

In this chapter we will reverse that process. That is, we will start with a power of a real number a and find a.

■ **Definition of nth root**
For every pair of real numbers a and b and every positive integer n greater than 1, if
$$a^n = b,$$
then a is called an nth root of b.

Concept
An nth root of a number is one of n equal factors that, when multiplied, equal the number.

Example 5–1   A

1. 2 is a fourth root of 16 since $2^4 = 16$.

2. -4 is a third root of -64 since $(-4)^3 = -64$.

3. 3 is a square root of 9 since $3^2 = 9$.

4. -3 is also a square root of 9 since $(-3)^2 = 9$.

We observe from the above examples that both 3 and -3 are second roots (square roots) of 9 since $3^2 = 9$ and $(-3)^2 = 9$. To avoid the ambiguity of two different answers for a problem, we now define the **principal nth root** of a number.

> ■ **Definition of principal nth root**
> If n is a positive integer greater than 1 and the nth root is a real number, then the principal nth root of a nonzero real number b, denoted by $\sqrt[n]{b}$, has the same sign as the number itself.

Since we most often talk about the principal nth root, we usually will eliminate the word "principal" and just say the "nth root," with the understanding that we are referring to the principal nth root.

> **Note**
> In the expression $\sqrt[n]{b}$, the $\sqrt{}$ is called a radical symbol,* b is called the radicand, n is called the index, and $\sqrt[n]{b}$ is called the radical. Also, the principal nth root of 0 is 0. That is, $\sqrt[n]{0} = 0$.

Table 5.1 lists the most common principal roots that we use in this book.

Square roots [a]		Cube roots		Fourth roots
$\sqrt{1} = 1$	$\sqrt{64} = 8$	$\sqrt[3]{1} = 1$	$\sqrt[3]{-1} = -1$	$\sqrt[4]{1} = 1$
$\sqrt{4} = 2$	$\sqrt{81} = 9$	$\sqrt[3]{8} = 2$	$\sqrt[3]{-8} = -2$	$\sqrt[4]{16} = 2$
$\sqrt{9} = 3$	$\sqrt{100} = 10$	$\sqrt[3]{27} = 3$	$\sqrt[3]{-27} = -3$	$\sqrt[4]{81} = 3$
$\sqrt{16} = 4$	$\sqrt{121} = 11$	$\sqrt[3]{64} = 4$	$\sqrt[3]{-64} = -4$	$\sqrt[4]{256} = 4$
$\sqrt{25} = 5$	$\sqrt{144} = 12$	$\sqrt[3]{125} = 5$	$\sqrt[3]{-125} = -5$	$\sqrt[4]{625} = 5$
$\sqrt{36} = 6$	$\sqrt{169} = 13$	$\sqrt[3]{216} = 6$	$\sqrt[3]{-216} = -6$	
$\sqrt{49} = 7$	$\sqrt{196} = 14$			

[a]When we write a square root, $\sqrt{}$, the index is understood to be 2.

Table 5.1

*The $\sqrt{}$ is the radical symbol and the $\overline{}$ means that what is under the radical symbol is a grouping.

Whenever we want the negative square root of a number, we indicate this by placing a minus sign in front of the square root symbol. That is, $-\sqrt{4} = -2$ and $-\sqrt{49} = -7$. Likewise, if we wanted the negative fourth root of 81, we would indicate it as $-\sqrt[4]{81} = -3$, or indicate the negative eighth root of 256 as $-\sqrt[8]{256} = -2$, and so forth.

In our definition of the principal nth root, we made the requirement that the nth root is a real number. This is because not all real numbers have a real nth root. Consider the example

$$\sqrt{-4} = \text{what?}$$

We know that all real numbers are positive, negative, or zero, and if we raise a real number to an even power (use it as a factor an even number of times), our answer will never be negative. Therefore the square root of a negative number does not exist in the set of real numbers, and, in general, *an even root of a negative number does not exist in the set of real numbers.*

In chapter 1 we defined the set of irrational numbers to be those numbers that cannot be represented by a terminating decimal or a nonterminating repeating decimal. Thus the nth root of b, $\sqrt[n]{b}$, is irrational if it cannot be expressed by a terminating decimal or a nonterminating repeating decimal. Some examples of irrational numbers are

$$\sqrt{8}, \quad \sqrt[5]{12}, \quad \sqrt[3]{-4}, \quad -\sqrt{10}, \quad \sqrt[4]{17}.$$

Whenever we are working with irrational numbers in a problem, we may have to approximate the number to as many decimal places as are required in the problem by using a calculator or an appropriate table.

Example 5–1 B

Find a decimal approximation to three decimal places by using a calculator.

1. $\sqrt{2} \approx 1.414$ **2.** $\sqrt{15} \approx 3.873$ **3.** $-\sqrt{11} \approx -3.317$

Note
"$\approx$" is read "is approximately equal to" and is used when our answer is not exact.

Rational exponents

In chapter 3 we developed a set of properties that guided our use of integers as exponents. We will now define rational exponents so that those same properties that apply to integer exponents will apply to rational exponents as well.

Consider the equation

$$b^{\frac{1}{3}} = a.$$

If we raise both sides of the equation to the third power, we have

$$\left(b^{\frac{1}{3}}\right)^3 = a^3.$$

Since the left member is a power to a power, we multiply the exponents to get

$$b^{\frac{1}{3} \cdot 3} = a^3$$
$$b^1 = a^3$$
$$b = a^3.$$

Since $a^3 = b$, then by the definition of the principal nth root, $a = \sqrt[3]{b}$. Also, we originally stated that $a = b^{\frac{1}{3}}$. Hence $b^{\frac{1}{3}}$ must be the same as $\sqrt[3]{b}$, that is,

$$b^{\frac{1}{3}} = \sqrt[3]{b}.$$

■ **Definition of $a^{\frac{1}{n}}$**

For every real number a and positive integer n greater than 1,

$$a^{\frac{1}{n}} = \sqrt[n]{a},$$

whenever the principal nth root of a is a real number.

Concept

The expression $a^{\frac{1}{n}}$ is equivalent to $\sqrt[n]{a}$ and represents the principal nth root of a.

Example 5–1 C

Rewrite the following in radical notation and use a calculator or table 5.1 to simplify. Round the answer to three decimal places when the value is irrational.

1. $6^{\frac{1}{3}} = \sqrt[3]{6} \approx 1.817$

2. $81^{\frac{1}{4}} = \sqrt[4]{81} = 3$

3. $19^{\frac{1}{2}} = \sqrt{19}$ (The index is understood to be 2) ≈ 4.359

4. $b^{\frac{1}{7}} = \sqrt[7]{b}$

Next, we need to decide how to define the expression $a^{\frac{m}{n}}$, where m and n are positive integers and a is a real number such that the nth root of a is also a real number. Before we determine a meaning for the expression $a^{\frac{m}{n}}$, we shall place a further restriction on the fraction $\frac{m}{n}$ such that $\frac{m}{n}$ is reduced to lowest terms. When the fraction $\frac{m}{n}$ is reduced to lowest terms, we say that m and n are **relatively prime.** That is, m and n contain no common positive integer factors other than 1. We can write the fraction $\frac{m}{n}$ as

$$\frac{m}{n} = \frac{1}{n} \cdot m.$$

Hence we have the following definition of $a^{\frac{m}{n}}$.

Note

When we have a rational exponent such as

$$\frac{m}{n},$$

the numerator (m) indicates the power to which the base is to be raised and the denominator (n) indicates the root to be taken.

In our previous definition the fractional exponent $\dfrac{m}{n}$ was rewritten as

$$\frac{1}{n} \cdot m.$$

If we apply the commutative property of multiplication, we can write $\dfrac{1}{n} \cdot m$ as

$$m \cdot \frac{1}{n}.$$

This fact leads us to the property of $\left(a^{\frac{1}{n}}\right)^m$.

Exponents, Roots, and Radicals

Example 5–1 D

Rewrite the following in radical form and use table 5.1 to simplify where possible. Assume that all variables represent nonnegative real numbers.

1. $x^{\frac{3}{4}} = (\sqrt[4]{x})^3$ or $\sqrt[4]{x^3}$

2. $27^{\frac{2}{3}} = (\sqrt[3]{27})^2 = (3)^2 = 9$ Alternate: $27^{\frac{2}{3}} = \sqrt[3]{27^2} = \sqrt[3]{729} = 9$

3. $(-8)^{\frac{2}{3}} = (\sqrt[3]{-8})^2 = (-2)^2 = 4$

4. $-8^{\frac{2}{3}} = -(\sqrt[3]{8})^2 = -(2)^2 = -4$

Example 5–1 E

Rewrite the following with rational exponents. Assume that all variables represent nonnegative real numbers.

1. $\sqrt[3]{a^2} = a^{\frac{2}{3}}$

2. $\sqrt[5]{x} = x^{\frac{1}{5}}$ (The power of x is understood to be 1.)

3. $\sqrt{5} = 5^{\frac{1}{2}}$ (The index of the radical is understood to be 2.)

In our previous definition of a rational exponent we required that the values of m and n be relatively prime. The following example illustrates what happens when m and n have a common factor of 2 and the radicand is negative. Consider the expression

$$[(-4)^2]^{\frac{1}{4}}.$$

Raising -4 to the second power

$$[(-4)^2]^{\frac{1}{4}} = (16)^{\frac{1}{4}},$$

and using table 5.1 to determine the fourth root, we have

$$(16)^{\frac{1}{4}} = \sqrt[4]{16} = 2.$$

If we take the original expression and write it as

$$[(-4)^2]^{\frac{1}{4}} = (-4)^{\frac{2}{4}} = (-4)^{\frac{1}{2}} = \sqrt{-4},$$

we see that the result, $\sqrt{-4}$, has no real solution. Therefore, depending on the method that we use, two different results are possible. For this reason, we have the following definition of $a^{\frac{m}{n}}$.

5–1 Roots and Rational Exponents

Definition of $a^{\frac{m}{n}}$, $a < 0$
If $a < 0$, and m and n are positive even integers,

$$(a^m)^{\frac{1}{n}} = |a|^{\frac{m}{n}}.$$

Note
In our example

$$[(-4)^2]^{\frac{1}{4}} = |-4|^{\frac{2}{4}} = 4^{\frac{1}{2}} = \sqrt{4} = 2.$$

If m and n are equal ($m = n$) and are even, the definition becomes

$$(a^n)^{\frac{1}{n}} = |a|^{\frac{n}{n}} = |a|^1 = |a|,$$

or, equivalently,

$$(a^n)^{\frac{1}{n}} = \sqrt[n]{a^n} = |a|,$$

when n is even. If $n = 2$, then

$$(a^2)^{\frac{1}{2}} = \sqrt{a^2} = |a|.$$

In general, we make the following definition for $(a^n)^{\frac{1}{n}}$.

Definition of $\sqrt[n]{a^n}$
For every real number a and positive integer n, where $n \geq 2$,

$$(a^n)^{\frac{1}{n}} = \sqrt[n]{a^n} = \begin{cases} |a| \text{ if } n \text{ is even} \\ a \text{ if } n \text{ is odd.} \end{cases}$$

Example 5–1 F

Simplify the following.

1. $b^{\frac{5}{5}} = \sqrt[5]{b^5} = b$

2. $\sqrt[3]{-8} = \sqrt[3]{(-2)^3} = -2$

3. $[(-2)^2]^{\frac{1}{2}} = \sqrt{(-2)^2} = |-2| = 2$

4. $\sqrt{(a+b)^2} = |a+b|$

We now make the following definition of $a^{-\frac{m}{n}}$.

Exponents, Roots, and Radicals

Example 5–1 G

Rewrite the following using positive exponents and use table 5.1 to simplify where possible.

1. $16^{-\frac{3}{4}} = \dfrac{1}{16^{\frac{3}{4}}} = \dfrac{1}{(\sqrt[4]{16})^3} = \dfrac{1}{(2)^3} = \dfrac{1}{8}$

2. $(-8)^{-\frac{2}{3}} = \dfrac{1}{(-8)^{\frac{2}{3}}} = \dfrac{1}{(\sqrt[3]{-8})^2} = \dfrac{1}{(-2)^2} = \dfrac{1}{4}$

Note
It is only the sign of the exponent that changes, not the sign of the base.

We have now developed the concept of a rational exponent so that the same properties that applied to integer exponents can now be extended to rational exponents.

Example 5–1 H

Perform the indicated operations and simplify. Assume that all variables represent positive real numbers.

1. $5^{\frac{1}{3}} \cdot 5^{\frac{1}{3}} = 5^{\frac{1}{3} + \frac{1}{3}} = 5^{\frac{2}{3}}$

2. $a^{\frac{2}{3}} \cdot a^{\frac{3}{4}} = a^{\frac{2}{3} + \frac{3}{4}} = a^{\frac{8}{12} + \frac{9}{12}} = a^{\frac{17}{12}}$

3. $\left(x^{\frac{4}{3}}\right)^{\frac{1}{2}} = x^{\frac{1}{2} \cdot \frac{4}{3}} = x^{\frac{2}{3}}$

4. $\dfrac{y^{\frac{1}{3}}}{y^{\frac{1}{4}}} = y^{\frac{1}{3} - \frac{1}{4}} = y^{\frac{4}{12} - \frac{3}{12}} = y^{\frac{1}{12}}$

5. $(2^3 a^{15} b^{21})^{\frac{1}{3}} = (2^3)^{\frac{1}{3}} (a^{15})^{\frac{1}{3}} (b^{21})^{\frac{1}{3}}$

$\qquad = 2^{3 \cdot \frac{1}{3}} a^{15 \cdot \frac{1}{3}} b^{21 \cdot \frac{1}{3}} = 2^1 a^5 b^7 = 2a^5 b^7$

6. $\dfrac{x^{-\frac{1}{4}} y^{\frac{2}{5}}}{x^{\frac{3}{4}} y^{-\frac{4}{5}}} = x^{-\frac{1}{4} - \frac{3}{4}} y^{\frac{2}{5} - \left(-\frac{4}{5}\right)} = x^{-\frac{4}{4}} y^{\frac{6}{5}} = x^{-1} y^{\frac{6}{5}} = \dfrac{y^{\frac{6}{5}}}{x^1} = \dfrac{y^{\frac{6}{5}}}{x}$

Exercise 5–1

Directions Find a decimal approximation to three decimal places by using a calculator. See example 5–1 B.

1. $\sqrt{18}$ **2.** $\sqrt{19}$ **3.** $-\sqrt{33}$ **4.** $-\sqrt{14}$

Directions Rewrite the following in radical notation and use table 5.1 to simplify wherever possible. Assume that all variables represent positive real numbers and no denominator is equal to zero. See examples 5–1 C, D, and G.

5. $6^{\frac{1}{3}}$ **6.** $2^{\frac{1}{2}}$ **7.** $x^{\frac{1}{2}}$ **8.** $a^{\frac{1}{3}}$ **9.** $b^{\frac{4}{5}}$ **10.** $a^{\frac{3}{4}}$

11. $64^{\frac{1}{3}}$ **12.** $(-8)^{\frac{1}{3}}$ **13.** $(-64)^{\frac{2}{3}}$ **14.** $16^{\frac{3}{4}}$ **15.** $81^{\frac{3}{4}}$ **16.** $64^{\frac{2}{3}}$

17. $16^{\frac{3}{2}}$ **18.** $(-27)^{-\frac{1}{3}}$ **19.** $8^{-\frac{1}{3}}$ **20.** $49^{-\frac{1}{2}}$ **21.** $16^{-\frac{1}{4}}$ **22.** $16^{-\frac{3}{4}}$

23. $27^{-\frac{2}{3}}$ **24.** $(-8)^{-\frac{2}{3}}$ **25.** $(-32)^{-\frac{3}{5}}$ **26.** $(-27)^{-\frac{2}{3}}$ **27.** $x^{-\frac{3}{4}}$ **28.** $a^{-\frac{2}{3}}$

Directions Rewrite the following with rational exponents. See example 5–1 E.

29. $\sqrt[7]{a^4}$ **30.** $\sqrt[6]{b}$ **31.** $\sqrt[5]{x}$

Directions Simplify the following. Variables represent *all* real numbers. See example 5–1 F.

32. $\sqrt{a^2}$ **33.** $\sqrt{b^2}$ **34.** $\sqrt{b^4}$ **35.** $\sqrt{x^6}$ **36.** $\sqrt{(ab)^2}$ **37.** $\sqrt{(x+y)^2}$

Directions Perform the indicated operations and simplify. Assume that all variables represent positive real numbers. Leave the answer with all exponents positive. See example 5–1 H.

38. $3^{\frac{1}{2}} \cdot 3^{\frac{3}{2}}$ **39.** $2^{\frac{1}{2}} \cdot 2^{\frac{1}{2}}$ **40.** $a^{\frac{1}{3}} \cdot a^{\frac{2}{3}}$ **41.** $b^{\frac{3}{4}} \cdot b^{\frac{2}{3}}$ **42.** $x^{\frac{1}{2}} \cdot x^{\frac{5}{4}}$ **43.** $\left(a^{\frac{2}{3}}\right)^{\frac{4}{5}}$

44. $\left(b^{\frac{2}{3}}\right)^{\frac{1}{2}}$ **45.** $\left(x^{\frac{3}{4}}\right)^4$ **46.** $\left(a^{\frac{1}{2}}\right)^{\frac{1}{2}}$ **47.** $(16y^4)^{\frac{3}{4}}$ **48.** $(a^3b^6)^{\frac{1}{3}}$ **49.** $(a^3b)^{\frac{2}{3}}$

50. $(8a^6b^{12})^{\frac{2}{3}}$ **51.** $(16a^{10}b^2)^{\frac{3}{4}}$ **52.** $\left(x^{-\frac{1}{4}}\right)^4$ **53.** $\left(y^{-\frac{2}{3}}\right)^3$ **54.** $\left(a^{-\frac{3}{4}}\right)^{-\frac{1}{3}}$ **55.** $\left(b^{-\frac{2}{3}}\right)^{-\frac{1}{2}}$

56. $\dfrac{y^{\frac{1}{4}}}{y^{\frac{1}{3}}}$
57. $\dfrac{a^{\frac{1}{3}}}{a^{\frac{5}{6}}}$
58. $\dfrac{b^{\frac{3}{4}}}{b}$
59. $\dfrac{x^{\frac{1}{3}}}{x}$
60. $\dfrac{a^{\frac{3}{4}}b^{\frac{1}{2}}}{a^{\frac{1}{4}}b^{\frac{1}{4}}}$
61. $\dfrac{xy^{\frac{3}{4}}}{x^{\frac{1}{2}}y^{\frac{1}{4}}}$

62. $\dfrac{ab}{a^{\frac{1}{2}}b^{\frac{1}{2}}}$
63. $\dfrac{x^{-\frac{1}{2}}x}{x^{\frac{1}{3}}}$
64. $\dfrac{b^{2}b^{\frac{1}{3}}}{b^{\frac{1}{2}}}$
65. $\dfrac{c^{\frac{2}{3}}c^{\frac{3}{4}}}{c^{-\frac{1}{3}}}$
66. $\dfrac{a^{-\frac{2}{3}}b^{\frac{1}{2}}}{a^{-\frac{1}{3}}b^{\frac{3}{4}}}$
67. $\dfrac{x^{-\frac{1}{2}}y^{\frac{5}{4}}}{x^{-\frac{2}{3}}y^{\frac{3}{4}}}$

Directions Solve the following verbal problems.

68. Find the number whose principal fourth root is 4.

69. Find the number whose principal third root is -3.

70. A square-shaped television picture tube has a surface area of 169 square inches. What is the length of the side of the tube? (*Hint:* Area of a square is found by squaring the length of a side, $A = s^2$.)

71. A garden in the shape of a square is 196 square feet. What is the length of a side? (See exercise 70.)

72. The electric-field intensity on the axis of a uniform charged ring is given by
$$E = \dfrac{T}{(x^2 + r^2)^{\frac{3}{2}}},$$
where T is the total charge on the ring and r is the radius of the ring. Express the fractional exponent in radical form.

73. A tank whose shape is a cube holds 216 cubic meters of water. What is the length of an edge of the cube? (*Hint:* The volume of a cube is found by raising the length of an edge to the third power, $V = e^3$.)

74. The formula for approximating the velocity V of a car based on the length of its skid marks S (in feet) on dry pavement is given by
$V = 2\sqrt{6S}$.
If the skid marks are 24 feet long, what was the velocity?

75. The formula for approximating the velocity V in miles per hour of a car based on the length of its skid marks S (in feet) on wet pavement is given by
$V = 2\sqrt{3S}$.
If the skid marks are 75 feet long, what was the velocity of the car?

76. To find the velocity of the center of mass of a rolling cylinder, we use the equation
$$v = \left(\dfrac{4}{3}gh\right)^{\frac{1}{2}}.$$
Write the expression in radical form.

77. At an altitude of h feet above the sea or level ground, the distance d in miles that a person can see an object is found by using the equation
$$d = \sqrt{\dfrac{3h}{2}}.$$
How far can someone see who is in a tower 216 feet above the ground?

78. When a gas is compressed with no gain or loss of heat, the pressure and volume of the gas are related by the formula
$$p = kv^{-\frac{7}{5}},$$
where p represents pressure, v represents volume, and k is a constant. Express the formula in radical form.

79. How can you find the fourth root of a number on a calculator using only the square root key?

80. How can you find the eighth root of a number on a calculator using only the square root key?

5-2 Simplifying radicals—I

Product property for radicals

In this section we will develop some properties for simplifying radicals. We will consider several forms of simplification that involve radicals. The first type of simplified radical is as follows:

The radicand (the quantity under the radical symbol) contains no factors that can be written to a power greater than or equal to the index.

The following property, called the **product property for radicals,** is useful for this type of simplification.

■ **Product property for radicals**
For all nonnegative real numbers a and b and positive integer n greater than 1,
$$\sqrt[n]{ab} = \sqrt[n]{a}\,\sqrt[n]{b}\,.$$

Concept
The nth root of a product is equal to the product of the nth roots of the factors.

To utilize this property in simplifying radicals, we look for factors that are perfect nth roots, that is, factors that are raised to the nth power. We have a perfect nth root when the value of a radical expression is a rational number. The following are examples of perfect nth roots:

$$\sqrt{25} = \sqrt{5^2} = 5, \quad \sqrt[3]{64} = \sqrt[3]{4^3} = 4, \quad \sqrt{\frac{4}{9}} = \sqrt{\left(\frac{2}{3}\right)^2} = \frac{2}{3},$$
$$\sqrt[4]{81} = \sqrt[4]{3^4} = 3, \quad \sqrt[5]{32} = \sqrt[5]{2^5} = 2.$$

Example 5-2 A

Simplify the following. Assume that all variables represent nonnegative real numbers.

1. $\sqrt{18}$
Since this is a square root, we are looking for factors that are raised to the second power. Since 9 is a factor of 18 and 9 is 3^2, we have
$$\sqrt{18} = \sqrt{9 \cdot 2} = \sqrt{9}\,\sqrt{2} = 3\sqrt{2}\,.$$
Hence $3\sqrt{2}$ is the simplified form of $\sqrt{18}$.

2. $\sqrt[3]{32}$
The index is 3, therefore we are looking for factors raised to the third power. Since 8 is a factor of 32 and $8 = 2^3$, our problem becomes
$$\sqrt[3]{32} = \sqrt[3]{8 \cdot 4} = \sqrt[3]{8}\,\sqrt[3]{4} = 2\sqrt[3]{4}\,,$$
and $2\sqrt[3]{4}$ is the simplified form of $\sqrt[3]{32}$.

3. $\sqrt[3]{a^5}$

We are looking for factors raised to the third power, and a^5 can be written as $a^3 \cdot a^2$. Hence

$$\sqrt[3]{a^5} = \sqrt[3]{a^3 a^2} = \sqrt[3]{a^3}\sqrt[3]{a^2} = a\sqrt[3]{a^2}$$

and is simplified.

4. $\sqrt[5]{a^{10}} = \sqrt[5]{a^5 \cdot a^5} = \sqrt[5]{a^5}\sqrt[5]{a^5} = a \cdot a = a^2$

Note

In example 4 the exponent, 10, was a multiple of the index, 5, and the radical was eliminated. When the exponent of a factor is a multiple of the index, that factor will no longer remain under the radical symbol.

The symmetric property from chapter 1 allows us to restate the product property for radicals.

■ **Product property for radicals**
For all nonnegative real numbers a and b and positive integer n greater than 1,

$$\sqrt[n]{a}\ \sqrt[n]{b} = \sqrt[n]{ab}\ .$$

Concept
The product of nth roots can be written as the nth root of the product.

Example 5–2 B

Multiply the following radicals and simplify where possible. Assume that all variables represent nonnegative real numbers.

1. $\sqrt{2}\ \sqrt{10}$

Since this is a square root times a square root, we can multiply the radicals together.

$$\sqrt{2}\ \sqrt{10} = \sqrt{2 \cdot 10} = \sqrt{20}$$

But $\sqrt{20}$ can be simplified.

$$\sqrt{20} = \sqrt{4 \cdot 5} = \sqrt{4}\ \sqrt{5} = 2\sqrt{5}$$

Therefore the simplified form of $\sqrt{2}\ \sqrt{10}$ is $2\sqrt{5}$.

2. $\sqrt[3]{9a^2b}\ \sqrt[3]{9ab}$ The indices are the same.

$= \sqrt[3]{9a^2b \cdot 9ab}$ Multiply the radicands.

$= \sqrt[3]{81a^3b^2}$ Simplify.

$= \sqrt[3]{27 \cdot 3a^3b^2} = \sqrt[3]{27}\ \sqrt[3]{3}\ \sqrt[3]{a^3}\ \sqrt[3]{b^2}$

$= 3\sqrt[3]{3} \cdot a\sqrt[3]{b^2} = 3a\sqrt[3]{3b^2}$

A second type of radical simplification can be seen in the following example. When we try to simplify the radical

$$\sqrt[6]{27} ,$$

we find that no factors of 27 can be written to a power greater than or equal to the index. Therefore it appears that no simplification is possible. But if we express the radical in fractional exponent form, we observe

$$\sqrt[6]{27} = \sqrt[6]{3^3} = 3^{\frac{3}{6}} = 3^{\frac{1}{2}} = \sqrt{3} .$$

We find that we can start with a radical whose index is 6 and reduce the index to 2 (square root). In reducing the index of the radical, it has been simplified. Therefore a second way that a radical is simplified is when the *exponent of the radicand and the index of the radical have no common factor other than 1*. That is, the exponent and the index are relatively prime.

■ **Property** $\sqrt[kn]{a^{km}}$
If a is a positive real number, m is an integer, and n and k are positive integers, then
$$\sqrt[kn]{a^{km}} = \sqrt[n]{a^m} .$$

Concept
We can divide out a common factor between the index and the exponent of the radicand.

Example 5–2 C

Simplify the following. Assume that all variables represent positive real numbers.

1. $\sqrt[6]{9} = \sqrt[6]{3^2}$
Since the exponent of the radicand and the index of the radical both have a common factor of 2, we can divide out (cancel) the common factor.

$$\sqrt[6]{3^2} = 3^{\frac{2}{6}} = 3^{\frac{1}{3}} = \sqrt[3]{3^1} = \sqrt[3]{3}$$

2. $\sqrt[9]{x^3 y^6}$
The radicand can be written as $(xy^2)^3$, and we see from this, that there is a common factor of 3 in the index and exponent.

$$\sqrt[9]{x^3 y^6} = \sqrt[9]{(xy^2)^3} = (xy^2)^{\frac{3}{9}} = (xy^2)^{\frac{1}{3}} = \sqrt[3]{(xy^2)^1} = \sqrt[3]{xy^2}$$

Note
When the radicand contains two or more different factors, we can reduce the index *only* if there is a common factor in the index and each of the exponents of the *different* factors. For example, $\sqrt[12]{a^4 b^3 c^6}$ is in simplest form because 1 is the only common factor between the index and the exponents.

Exercise 5–2

Directions Simplify the following. Assume that all variables represent nonnegative real numbers. See examples 5–2 A, B, and C.

1. $\sqrt{20}$
2. $\sqrt{63}$
3. $\sqrt[3]{24}$
4. $\sqrt[3]{32}$
5. $\sqrt[4]{a^5}$

6. $\sqrt[5]{a^7}$
7. $\sqrt[9]{a^{18}}$
8. $\sqrt[5]{b^{10}}$
9. $\sqrt[5]{c^5}$
10. $\sqrt{9x^2y^5}$

11. $\sqrt{25x^3y^9}$
12. $\sqrt{32a^4b^7}$
13. $\sqrt{50a^6bc^9}$
14. $\sqrt[3]{8x^5y^4}$
15. $\sqrt[3]{27a^3b^2c^{12}}$

16. $\sqrt[3]{16a^4b^5}$
17. $\sqrt[3]{81a^5b^{11}}$
18. $\sqrt[5]{64x^{10}y^{14}}$
19. $\sqrt[5]{32a^{10}b^4c^{12}}$
20. $\sqrt[5]{8a^7b^{15}c^3}$

21. $\sqrt{x^2 + 6x + 9}$
22. $\sqrt{a^2 + 10a + 25}$
23. $\sqrt{9a^2 + 6a + 1}$
24. $\sqrt{x^2 + y^2}$

25. $\sqrt{6}\,\sqrt{27}$
26. $\sqrt{10}\,\sqrt{20}$
27. $\sqrt{7}\,\sqrt{7}$
28. $\sqrt{12}\,\sqrt{12}$

29. $\sqrt{6}\,\sqrt{24}$
30. $\sqrt[3]{4}\,\sqrt[3]{4}$
31. $\sqrt[3]{6}\,\sqrt[3]{12}$
32. $\sqrt[3]{10}\,\sqrt[3]{4}$

33. $\sqrt{3a}\,\sqrt{15a}$
34. $\sqrt{7b}\,\sqrt{14b}$
35. $\sqrt[3]{a^2}\,\sqrt[3]{a^2}$
36. $\sqrt[5]{b^4}\,\sqrt[5]{b^3}$

37. $\sqrt[4]{x}\,\sqrt[4]{x^3}$
38. $\sqrt[3]{2x^2}\,\sqrt[3]{2x}$
39. $\sqrt[5]{8x^4}\,\sqrt[5]{4x^3}$
40. $\sqrt[3]{4a^2b}\,\sqrt[3]{4a^2b^2}$

41. $\sqrt[3]{5a^2b}\,\sqrt[3]{75a^2b^2}$
42. $\sqrt[3]{3ab^2}\,\sqrt[3]{18a^2b^2}$
43. $\sqrt[3]{25x^5y^7}\,\sqrt[3]{15xy^3}$
44. $\sqrt[3]{16a^{11}b^4}\,\sqrt[3]{12a^4b^6}$

45. $\sqrt[4]{8xy}\,\sqrt[4]{4x^3y^3}$
46. $\sqrt[6]{a^3}$
47. $\sqrt[10]{y^5}$
48. $\sqrt[6]{b^{10}}$

49. $\sqrt[8]{y^{14}}$
50. $\sqrt[4]{4x^7}$
51. $\sqrt[6]{8y^3}$
52. $\sqrt[9]{27x^6y^6}$

53. $\sqrt[9]{8a^3b^6}$

Directions Simplify the following. Variables represent *all* real numbers. See example 5–1 F.

Example
(a) $\sqrt{25a^2}$

(b) $\sqrt{x^2 + 2xy + y^2}$

Solution
(a) $= \sqrt{25}\,\sqrt{a^2}$
$= 5|a|$

(b) $= \sqrt{(x + y)^2}$
$= |x + y|$

54. $\sqrt{9a^2}$
55. $\sqrt{16x^2}$
56. $\sqrt{36a^2b^4}$
57. $\sqrt{49b^2c^2}$

58. $\sqrt{a^2 - 2ab + b^2}$
59. $\sqrt{4x^2 - 4xy + y^2}$
60. $\sqrt{a^2 + 2a + 1}$

Directions Solve the following verbal problems.

61. The moment of inertia for a rectangle is given by the formula

$$I = \frac{bh^3}{12}.$$

If we know the values of I and b, we can solve for h as follows:

$$h = \sqrt[3]{\frac{12I}{b}}.$$

Find h if $I = 27$ in.⁴ and $b = 4$ in.

62. Use exercise 61 to find h if $I = 2$ in.⁴ and $b = 3$ in.

63. The moment of inertia for a circle is given by the formula

$$I = \frac{\pi r^4}{4}.$$

If we know the value of I, we can solve for r as follows:

$$r = \sqrt[4]{\frac{4I}{\pi}}.$$

Find r if $I = 63.585$ in.⁴ and we use 3.14 for π.

64. Use exercise 63 to find r if $I = 12.56$ in.⁴.

65. The formula for finding the length of an edge e of a cube when the volume v is known is $e = \sqrt[3]{v}$. What is the length of the edge of a cube whose volume is 216 cubic units?

66. What is the length of the edge of a cube whose volume is 729 cubic units? (Refer to exercise 65.)

67. The current I (amperes) in a circuit is found by the formula

$$I = \sqrt{\frac{\text{watts}}{\text{ohms}}}.$$

What is the current of a circuit that has 3-ohms resistance and uses 450 watts?

68. What is the current of a circuit that has 2-ohms resistance and uses 1,728 watts? (Refer to exercise 67.)

Directions The following problems will make use of an important property of right triangles called the **Pythagorean Theorem.**

In a right triangle the square of the hypotenuse (the side opposite the right angle) is equal to the sum of the squares of the lengths of the legs (the sides that form the right angle). If c is the hypotenuse and a and b are the lengths of the legs, this property can be stated as:

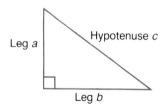

$$c^2 = a^2 + b^2 \text{ or } c = \sqrt{a^2 + b^2}$$

Also as $a^2 = c^2 - b^2$ or $a = \sqrt{c^2 - b^2}$

Also as $b^2 = c^2 - a^2$ or $b = \sqrt{c^2 - a^2}$

Example
Find the length of the hypotenuse of a right triangle whose legs are 6 cm and 8 cm.

Solution
We want to find c when $a = 6$ cm and $b = 8$ cm.

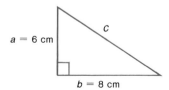

By the Pythagorean Theorem,

$$c = \sqrt{a^2 + b^2}$$
$$= \sqrt{6^2 + 8^2}$$
$$= \sqrt{36 + 64}$$
$$= \sqrt{100} = 10. \text{ Hence } c = 10 \text{ cm.}$$

Example

Find the second leg of a right triangle whose hypotenuse has length 13 in. and the first leg is 5 in. long.

Solution

We want to find b given $c = 13$ in. and $a = 5$ in. Using one of the forms of the theorem,

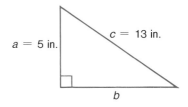

$$b = \sqrt{c^2 - a^2}$$
$$= \sqrt{13^2 - 5^2}$$
$$= \sqrt{169 - 25}$$
$$= \sqrt{144} = 12. \text{ Hence } b = 12 \text{ in.}$$

Directions Find the length of the unknown side in the following right triangles.

69. $a = 3$ m, $b = 4$ m

70. $a = 8$ ft, $c = 10$ ft

71. $a = 12$ in., $b = 5$ in.

72. $a = 15$ cm, $b = 8$ cm

73. $a = 5$ in., $b = 4$ in.

74. $a = 6$ yd, $c = 10$ yd

75. $a = 6$ ft, $b = 9$ ft

76. $b = 16$ m, $c = 20$ m

77. $a = 12$ mm, $b = 16$ mm

78. $a = 10$ in., $b = 24$ in.

79. $a = 4$ cm, $c = 10$ cm

Directions Solve the following verbal problems.

80. A 17-foot ladder is placed against the wall of a house. If the bottom of the ladder is 8 feet from the house, how far from the ground is the top of the ladder?

81. Find the width of a rectangle whose diagonal is 13 feet and length is 12 feet.

82. Find the diagonal of a rectangle whose length is 8 meters and whose width is 6 meters.

83. Under ideal conditions, the velocity v of an object falling freely from a height h is given by $v = \sqrt{2gh}$, where g is the accleration due to gravity. Use a calculator to find the velocity when h is 100 m. Round to two decimal places. Use $g = 9.8$ m/s^2.

5–3 Simplifying radicals—II

The quotient property for radicals

In this section we continue to develop some of the properties for simplifying radicals. A third way that a radical is simplified is when *the radicand contains no fractions*. The following property, called the **quotient property for radicals,** is useful for this type of simplification.

■ **Quotient property for radicals**

For all nonnegative real numbers a and b, where $b \neq 0$, and positive integer n greater than 1,

$$\sqrt[n]{\frac{a}{b}} = \frac{\sqrt[n]{a}}{\sqrt[n]{b}}.$$

Concept

The *n*th root of a quotient can be written as the *n*th root of the numerator divided by the *n*th root of the denominator.

When utilizing this property to simplify radicals, we should first determine if the fraction can be simplified (reduced).

Example 5–3 A

Simplify the following. Assume that all variables represent positive real numbers.

1. $\sqrt{\dfrac{a^3}{b^2}}$

There are no operations to be performed within the radical, so we apply the quotient property and simplify.

$$\sqrt{\frac{a^3}{b^2}} = \frac{\sqrt{a^3}}{\sqrt{b^2}} = \frac{a\sqrt{a}}{b}$$

2. $\sqrt[3]{\dfrac{8}{27}} = \dfrac{\sqrt[3]{8}}{\sqrt[3]{27}} = \dfrac{2}{3}$

3. $\sqrt[3]{\dfrac{x^5 y^2}{x^2}}$

Since there is a common factor of x^2, we reduce the fraction first.

$$\sqrt[3]{\frac{x^5 y^2}{x^2}} = \sqrt[3]{x^3 y^2} \qquad \text{Apply the product property.}$$

$$= \sqrt[3]{x^3}\,\sqrt[3]{y^2} = x\sqrt[3]{y^2}$$

Rationalizing a denominator that has a single term

When the problem has been simplified and a radical still remains in the denominator, the evaluation of the problem usually is an involved process. For this reason, the fourth way that a radical is simplified is when *no radicals appear in the denominator*. This procedure is called **rationalizing the denominator,** because it changes the denominator from a radical (irrational number) to a rational number.

Consider the example

$$\sqrt{\frac{4}{5}} = \frac{\sqrt{4}}{\sqrt{5}} = \frac{2}{\sqrt{5}}.$$

We would like to eliminate the radical symbol in the denominator. To remove this radical, we multiply the numerator and the denominator by a radical that yields a perfect square radicand and thereby allows us to eliminate the radical in the denominator.

$$\frac{2}{\sqrt{5}} \cdot \frac{\sqrt{5}}{\sqrt{5}} = \frac{2\sqrt{5}}{\sqrt{25}} = \frac{2\sqrt{5}}{5}$$

Since we multiplied the numerator and the denominator by the same number, using the fundamental principle of fractions, our answer is equivalent to the original fraction.

Example 5–3 B

Simplify the following. Assume that all variables represent positive real numbers.

1. $\sqrt{\dfrac{2}{3}}$ Apply the quotient property.

$= \dfrac{\sqrt{2}}{\sqrt{3}}$ Multiply by $\dfrac{\sqrt{3}}{\sqrt{3}}$ to rationalize the denominator.

$= \dfrac{\sqrt{2}}{\sqrt{3}} \cdot \dfrac{\sqrt{3}}{\sqrt{3}} = \dfrac{\sqrt{2 \cdot 3}}{\sqrt{3 \cdot 3}}$

$= \dfrac{\sqrt{6}}{\sqrt{9}}$ We now have a perfect square root and can eliminate the radical in the denominator.

$= \dfrac{\sqrt{6}}{3}$

2. $\sqrt{\dfrac{a^3}{b}}$ Apply the quotient rule.

$= \dfrac{\sqrt{a^3}}{\sqrt{b}}$ Simplify and multiply by $\dfrac{\sqrt{b}}{\sqrt{b}}$.

$= \dfrac{a\sqrt{a}}{\sqrt{b}} \cdot \dfrac{\sqrt{b}}{\sqrt{b}} = \dfrac{a\sqrt{ab}}{\sqrt{b \cdot b}}$

$= \dfrac{a\sqrt{ab}}{\sqrt{b^2}}$ Eliminate the radical in the denominator.

$= \dfrac{a\sqrt{ab}}{b}$

The following example will help us develop a general property for rationalizing a denominator that has a single term.

$$\sqrt[3]{\dfrac{1}{a}} = \dfrac{\sqrt[3]{1}}{\sqrt[3]{a}} = \dfrac{1}{\sqrt[3]{a}}$$

At this point, we have simplified the problem, but a radical still remains in the denominator. We must now determine what we can do to the fraction to remove the radical from the denominator.

Observations:

1. We can multiply the numerator and the denominator by the same number and form equivalent fractions.
2. If we multiply by a radical, the indices must be the same in order to carry out the multiplication.
3. To bring a factor out from under the radical symbol and not leave any of the factor behind, the exponent must be a multiple of the index.

With these observations in mind, we rationalize the fraction.

$$= \frac{1}{\sqrt[3]{a}} \cdot \frac{\sqrt[3]{a^2}}{\sqrt[3]{a^2}} \qquad \text{The indices are the same, and we multiply the numerator and the denominator by the same number.}$$

$$= \frac{\sqrt[3]{a^2}}{\sqrt[3]{a \cdot a^2}} = \frac{\sqrt[3]{a^2}}{\sqrt[3]{a^3}} \qquad \text{The sum of the exponents of } a \text{ in the denominator adds to the index, forming a perfect } n\text{th root (cube root).}$$

$$= \frac{\sqrt[3]{a^2}}{a} \qquad \text{The radical is eliminated from the denominator.}$$

We can summarize our rationalization procedure for a denominator of one term.

1. We multiply the numerator and the denominator by a radical with the same index as the radical that we wish to eliminate from the denominator.
2. The exponent of the factor under the radical must be such that when we add it to the original exponent of the factor under the radical in the denominator, the sum will be equal to, or a multiple of, the index of the radical.
3. We carry out the multiplication and reduce the fraction if possible.

Example 5–3 C

Simplify the following. Assume that all variables represent positive real numbers.

1. $\dfrac{1}{\sqrt[5]{x^2}}$

To eliminate the radical, we multiply by another 5th root, where the exponents of x will add up to 5 ($x^2 \cdot x^3 = x^{2+3} = x^5$).

$$\frac{1}{\sqrt[5]{x^2}} = \frac{1}{\sqrt[5]{x^2}} \cdot \frac{\sqrt[5]{x^3}}{\sqrt[5]{x^3}} = \frac{\sqrt[5]{x^3}}{\sqrt[5]{x^2 \cdot x^3}}$$

$$= \frac{\sqrt[5]{x^3}}{\sqrt[5]{x^5}} \qquad \text{The resulting denominator is a perfect } n\text{th root (5th root) and the radical symbol can be eliminated.}$$

$$= \frac{\sqrt[5]{x^3}}{x}$$

2. $\dfrac{x}{\sqrt[4]{x}} = \dfrac{x}{\sqrt[4]{x}} \cdot \dfrac{\sqrt[4]{x^3}}{\sqrt[4]{x^3}} = \dfrac{x\sqrt[4]{x^3}}{\sqrt[4]{x^4}} = \dfrac{x\sqrt[4]{x^3}}{x} = \sqrt[4]{x^3}$

3. $\dfrac{\sqrt[5]{a^3}}{\sqrt[5]{b^2}} = \dfrac{\sqrt[5]{a^3}}{\sqrt[5]{b^2}} \cdot \dfrac{\sqrt[5]{b^3}}{\sqrt[5]{b^3}} = \dfrac{\sqrt[5]{a^3 b^3}}{\sqrt[5]{b^5}} = \dfrac{\sqrt[5]{a^3 b^3}}{b}$

4. $\dfrac{1}{\sqrt[5]{a^2 b}}$

Since there are two different factors under the radical, the radicand of the radical that we multiply by must contain a and b with exponents such that the resulting radicand in the denominator is $a^5 b^5$. ($a^2 b \cdot a^3 b^4 = a^{2+3} b^{1+4} = a^5 b^5$). Therefore we multiply by $\sqrt[5]{a^3 b^4}$.

$$= \frac{1}{\sqrt[5]{a^2 b}} \cdot \frac{\sqrt[5]{a^3 b^4}}{\sqrt[5]{a^3 b^4}} = \frac{\sqrt[5]{a^3 b^4}}{\sqrt[5]{a^2 b a^3 b^4}} = \frac{\sqrt[5]{a^3 b^4}}{\sqrt[5]{a^{2+3} b^{1+4}}}$$

$$= \frac{\sqrt[5]{a^3 b^4}}{\sqrt[5]{a^5 b^5}} \qquad \text{Eliminate the radical in the denominator.}$$

$$= \frac{\sqrt[5]{a^3 b^4}}{ab}$$

The following is a summary of the conditions necessary for a radical expression to be in **simplest form,** also called **standard form.**

1. The radicand contains no factors that can be written to an exponent greater than or equal to the index. ($\sqrt[3]{a^4}$ violates this.)
2. The exponent of the radicand and the index of the radical have no common factor other than 1. ($\sqrt[9]{a^6}$ violates this.)
3. The radicand contains no fractions. $\left(\sqrt{\dfrac{a}{b}} \text{ violates this.} \right)$
4. No radicals appear in the denominator. $\left(\dfrac{1}{\sqrt{a}} \text{ violates this.} \right)$

Mastery points
Can you
- Simplify radicals containing fractions by using the quotient property for radicals?
- Rationalize fractions whose denominators are a single term?

Exercise 5–3

Directions Simplify the following expressions leaving no radicals in the denominator. Assume that all variables represent positive real numbers. See examples 5–3 A, B, and C.

1. $\sqrt{\dfrac{16}{25}}$
2. $\sqrt{\dfrac{25}{49}}$
3. $\sqrt{\dfrac{7}{9}}$
4. $\sqrt{\dfrac{3}{4}}$
5. $\sqrt[3]{\dfrac{8}{27}}$

6. $\sqrt[3]{\dfrac{27}{64}}$
7. $\sqrt{\dfrac{a^6}{9}}$
8. $\sqrt[3]{\dfrac{4x^9}{y^6}}$
9. $\sqrt[3]{\dfrac{2x}{y^{15}}}$
10. $\sqrt[5]{\dfrac{32a^3}{b^{15}}}$

11. $\sqrt[7]{\dfrac{x^{21}}{y^7 z^{14}}}$
12. $\sqrt[3]{\dfrac{a^7 b^2}{ab^5}}$
13. $\sqrt[3]{\dfrac{16x^4}{2xy^6}}$
14. $\sqrt[5]{\dfrac{2a^{12}b^4}{64b^9}}$
15. $\sqrt[5]{\dfrac{64x^{14}y^6}{x^4 y}}$

16. $\sqrt[4]{\dfrac{b^4 c^9}{a^{11}}}$
17. $\sqrt{\dfrac{1}{2}}$
18. $\sqrt{\dfrac{1}{3}}$
19. $\sqrt{\dfrac{9}{10}}$
20. $\sqrt{\dfrac{4}{11}}$

21. $\sqrt{\dfrac{1}{6}}$
22. $\sqrt{\dfrac{1}{10}}$
23. $\sqrt{\dfrac{9}{50}}$
24. $\sqrt{\dfrac{7}{12}}$
25. $\dfrac{2}{\sqrt{2}}$

26. $\dfrac{6}{\sqrt{3}}$
27. $\dfrac{12}{\sqrt{8}}$
28. $\dfrac{18}{\sqrt{27}}$
29. $\sqrt[3]{\dfrac{27}{4}}$
30. $\sqrt[3]{\dfrac{9}{25}}$

31. $\sqrt[5]{\dfrac{32}{81}}$
32. $\sqrt[3]{\dfrac{27}{16}}$
33. $\sqrt[4]{\dfrac{81}{64}}$
34. $\sqrt[4]{\dfrac{2}{9}}$
35. $\sqrt{\dfrac{x^2}{y}}$

36. $\sqrt{\dfrac{1}{b}}$ **37.** $\sqrt{\dfrac{1}{c}}$ **38.** $\sqrt{\dfrac{x^4}{y^3}}$ **39.** $\sqrt[3]{\dfrac{a^3}{b^2}}$ **40.** $\sqrt[3]{\dfrac{b^9}{c}}$

41. $\sqrt[3]{\dfrac{a}{b^2}}$ **42.** $\sqrt[3]{\dfrac{a}{b}}$ **43.** $\sqrt[5]{\dfrac{32x^5}{y^2}}$ **44.** $\dfrac{x^2}{\sqrt[3]{x^2}}$ **45.** $\dfrac{ab}{\sqrt[5]{b^4}}$

46. $\sqrt{\dfrac{a}{bc}}$ **47.** $\sqrt{\dfrac{2x}{yz}}$ **48.** $\sqrt[3]{\dfrac{x^3}{y^2z}}$ **49.** $\sqrt[3]{\dfrac{8x}{y^2z}}$ **50.** $\sqrt[3]{\dfrac{1}{2ab^2}}$

51. $\sqrt[7]{\dfrac{1}{16x^2y^3}}$ **52.** $\sqrt[4]{\dfrac{16}{a^3b^2}}$ **53.** $\dfrac{x}{\sqrt[5]{x^2y^4}}$ **54.** $\dfrac{a}{\sqrt[4]{ab^3}}$ **55.** $\dfrac{xy}{\sqrt[3]{xy^2}}$

56. $\dfrac{ab^2}{\sqrt[5]{a^4b^{12}}}$ **57.** $\dfrac{b^2c}{\sqrt[7]{b^4c^3}}$ **58.** $\dfrac{y^2z^3}{\sqrt[5]{y^7z^2}}$ **59.** $\sqrt{\dfrac{2y}{x}}\sqrt{\dfrac{x^2}{8}}$ **60.** $\sqrt{\dfrac{3a}{b^3}}\sqrt{\dfrac{ab}{27}}$

61. $\sqrt[3]{\dfrac{16a^7}{b^4}}\sqrt[3]{\dfrac{b}{2a}}$ **62.** $\sqrt[3]{\dfrac{3y}{x^7}}\sqrt[3]{\dfrac{x}{81y^4}}$ **63.** $\sqrt[5]{\dfrac{x^2y}{z^7}}\sqrt[5]{\dfrac{y^9z^3}{x^7}}$ **64.** $\sqrt[5]{\dfrac{a^3b}{c^8}}\sqrt[5]{\dfrac{b^4c^4}{a^8}}$

Directions Solve the following verbal problems.

65. If we wish to construct a sphere of specific volume, we can find the length of the radius necessary by the formula

$$r = \sqrt[3]{\dfrac{3V}{4\pi}}.$$

Find the radius necessary for a sphere to have a volume of 904.32 cubic units. (Use 3.14 for π.)

66. Use exercise 65 to find r if $V = 113.04$ cubic units. (Use 3.14 for π.)

67. To find the velocity of the center of mass of a rolling cylinder, we use the equation

$$v = \left(\dfrac{4}{3}gh\right)^{\frac{1}{2}}.$$

Write the expression in radical form and leave the answer in standard form.

68. When a gas is compressed with no gain or loss of heat, the pressure and volume of the gas are related by the formula

$$p = kv^{-\frac{7}{5}},$$

where p represents pressure, v represents volume, and k is a constant. Express the formula in radical form and leave the answer in standard form.

69. The formula below gives the length s of a side of an isosceles right triangle with hypotenuse c. Express the radical in standard form.

$$s = \sqrt{\dfrac{c^2}{2}}$$

70. If we know the length of the diagonal d of a square, we can find the length of the side s of the square with the formula

$$s = \sqrt{\dfrac{d^2}{2}}.$$

Express the radical in standard form.

71. The formula below gives the diagonal length d of a regular hexagon, where f is the distance across the flats. Express the radical in standard form.

$$d = \sqrt{\dfrac{4f^2}{3}}$$

72. The resonant frequency f of an AC series circuit is given by

$$f = \dfrac{1}{2\pi\sqrt{LC}}.$$

Express the radical in standard form.

73. The average speed v of a molecule of an ideal gas is given by

$$v = \sqrt{\dfrac{8kT}{\pi m}},$$

where m is the mass, T is the absolute temperature, and k is the Boltzmann constant. Express the radical in standard form.

74. The formula below is used to find the potential energy in a truss. Express the radical in standard form.

$$C = \dfrac{3KA}{\sqrt[3]{1,024L}}$$

5–4 Sums and differences of radicals

We have learned that we can only combine like terms in addition and subtraction. This same idea applies when we are dealing with radicals. *We can only add or subtract like radicals.*

To have like radicals, the following must be true:

1. the radicals must have the same index, and
2. the radicands must be the same.

For example, the expressions $-3\sqrt[5]{11}$ and $4\sqrt[5]{11}$ are like radicals since the indices, 5, are the same and the radicands, 11, are the same. The expressions $3\sqrt[5]{19}$ and $3\sqrt[4]{19}$ are not like radicals because the indices are different ($4 \neq 5$), and the expressions $7\sqrt[3]{13}$ and $7\sqrt[3]{14}$ are not like radicals because the radicands are different ($13 \neq 14$).

Addition and subtraction of radicals follow the same procedure as addition and subtraction of algebraic expressions. That is, *once we have determined that we have like radicals, the operations of addition and subtraction are performed only with the numerical coefficients using the distributive property.*

Example 5–4 A

Perform the indicated addition and subtraction. Assume that all variables represent nonnegative real numbers.

1. $3\sqrt{5} + 2\sqrt{5}$

Since we have like radicals, we can perform the addition by using the distributive property.

$$3\sqrt{5} + 2\sqrt{5} = (3 + 2)\sqrt{5} = 5\sqrt{5}$$

2. $8\sqrt[5]{x^2} - 2\sqrt[5]{x^2} + 3\sqrt[5]{x^2} = (8 - 2 + 3)\sqrt[5]{x^2} = 9\sqrt[5]{x^2}$

3. $4\sqrt{a} + \sqrt{b} - 2\sqrt{a} + 3\sqrt{b}$

Using the commutative and associative properties, we group the like radicals

$$- (4\sqrt{a} \quad 2\sqrt{a}) + (\sqrt{b} + 3\sqrt{b}),$$

and applying the distributive property, we perform the addition and subtraction.

$$= (4 - 2)\sqrt{a} + (1 + 3)\sqrt{b}$$
$$= 2\sqrt{a} + 4\sqrt{b}$$

Since the $\sqrt{a}$ and $\sqrt{b}$ are not like radicals, no further simplification can be performed.

Consider the problem

$$\sqrt{27} + 4\sqrt{3}.$$

It appears that the indicated addition cannot be performed since we do not have like radicals. However we should have observed that the $\sqrt{27}$ can be simplified.

$$\sqrt{27} = \sqrt{9 \cdot 3} = 3\sqrt{3}$$

Our problem then becomes

$$\sqrt{27} + 4\sqrt{3} = 3\sqrt{3} + 4\sqrt{3} = (3 + 4)\sqrt{3} = 7\sqrt{3}.$$

Therefore *whenever we are working with radicals, we must be certain that all radicals are in simplest form.*

Example 5–4 B

Perform the indicated addition and subtraction.

1. $5\sqrt{2} + \sqrt{18}$

Since $\sqrt{18}$ can be simplified, we have

$$= 5\sqrt{2} + \sqrt{9 \cdot 2} = 5\sqrt{2} + \sqrt{9}\sqrt{2} = 5\sqrt{2} + 3\sqrt{2},$$

and applying the distributive property,

$$= (5 + 3)\sqrt{2} = 8\sqrt{2}.$$

2. $\sqrt{27} + \sqrt{12}$ Simplify the radicals.

$$= \sqrt{9 \cdot 3} + \sqrt{4 \cdot 3}$$

$$= 3\sqrt{3} + 2\sqrt{3} \qquad \text{Add like radicals.}$$

$$= 5\sqrt{3}$$

3. $4\sqrt[3]{81} - \sqrt[3]{24}$ Simplify the radicals.

$$= 4\sqrt[3]{27 \cdot 3} - \sqrt[3]{8 \cdot 3}$$

$$= 4 \cdot 3\sqrt[3]{3} - 2\sqrt[3]{3}$$

$$= 12\sqrt[3]{3} - 2\sqrt[3]{3} \qquad \text{Subtract like radicals.}$$

$$= 10\sqrt[3]{3}$$

When we perform addition and subtraction of fractions that contain radicals, our first step is to simplify all radicals involved. We can then find the least common denominator and perform the indicated addition and subtraction.

Example 5–4 C

Perform the indicated addition and subtraction.

1. $\dfrac{2}{3} + \dfrac{1}{\sqrt{3}}$ Simplify the second fraction by rationalizing the denominator.

$$= \frac{2}{3} + \frac{1}{\sqrt{3}} \cdot \frac{\sqrt{3}}{\sqrt{3}} = \frac{2}{3} + \frac{\sqrt{3}}{\sqrt{9}} = \frac{2}{3} + \frac{\sqrt{3}}{3}$$

Since the denominators are the same, we can add the numerators and write the sum over the common denominator.

$$\frac{2}{3} + \frac{\sqrt{3}}{3} = \frac{2 + \sqrt{3}}{3}$$

2. $\dfrac{3}{\sqrt{2}} - \dfrac{2}{\sqrt{5}}$ Rationalize the denominators.

$$= \frac{3}{\sqrt{2}} \cdot \frac{\sqrt{2}}{\sqrt{2}} - \frac{2}{\sqrt{5}} \cdot \frac{\sqrt{5}}{\sqrt{5}} = \frac{3\sqrt{2}}{\sqrt{4}} - \frac{2\sqrt{5}}{\sqrt{25}} = \frac{3\sqrt{2}}{2} - \frac{2\sqrt{5}}{5}$$

The least common denominator is 10. Therefore we multiply the first fraction by $\dfrac{5}{5}$ and the second fraction by $\dfrac{2}{2}$.

$$\frac{3\sqrt{2}}{2} \cdot \frac{5}{5} - \frac{2\sqrt{5}}{5} \cdot \frac{2}{2} = \frac{15\sqrt{2}}{10} - \frac{4\sqrt{5}}{10}$$

We now have a common denominator and can finish the problem.

$$\frac{15\sqrt{2}}{10} - \frac{4\sqrt{5}}{10} = \frac{15\sqrt{2} - 4\sqrt{5}}{10}$$

Mastery points
Can you
- Identify like radicals?
- Add and subtract like radicals?
- Add and subtract fractions containing radicals?

Exercise 5–4

Directions Perform the indicated operations and simplify. Assume that all variables represent positive real numbers. See examples 5–4 A, B, and C.

1. $7\sqrt{5} + 4\sqrt{5}$

2. $9\sqrt{11} - 2\sqrt{11}$

3. $3\sqrt{3} + 4\sqrt{3}$

4. $10\sqrt{6} - 6\sqrt{6}$

5. $5\sqrt{5} + 7\sqrt{5} - 4\sqrt{5}$

6. $6\sqrt{2} - 4\sqrt{2} + 8\sqrt{2}$

7. $\sqrt{10} + 4\sqrt{10} - 6\sqrt{10}$

8. $8\sqrt{13} - 11\sqrt{13} - 4\sqrt{13}$

9. $5\sqrt[3]{4} + 2\sqrt[3]{4}$

10. $8\sqrt[5]{3} - 4\sqrt[5]{3} + 7\sqrt[5]{3}$

11. $10\sqrt[4]{3} + \sqrt[4]{3} - 5\sqrt[4]{3}$

12. $7\sqrt[3]{11} - 3\sqrt[3]{7} + 2\sqrt[3]{11}$

13. $\sqrt[5]{12} - \sqrt[5]{16} + 4\sqrt[5]{12}$

14. $\sqrt{2a} - 3\sqrt{a} + 4\sqrt{a}$

15. $2\sqrt{3x} - 4\sqrt{2x} + 2\sqrt{3x}$

16. $\sqrt{12} + 4\sqrt{3}$

17. $\sqrt{20} - 3\sqrt{5}$

18. $\sqrt{8} - 4\sqrt{2}$

19. $\sqrt{12} - \sqrt{75}$

20. $2\sqrt{48} - 3\sqrt{27}$

21. $5\sqrt{3} + 4\sqrt{12}$

22. $4\sqrt{7} - 5\sqrt{63}$

23. $2\sqrt{3} + \sqrt{27} - 2\sqrt{12}$

24. $2\sqrt{8} - \sqrt{50} + 5\sqrt{2}$

25. $\sqrt[3]{16} + \sqrt[3]{54}$

26. $\sqrt[3]{81} - \sqrt[3]{24}$

27. $3\sqrt[3]{16} + 5\sqrt[3]{24}$

28. $4\sqrt[3]{54} - 7\sqrt[3]{16}$

29. $\sqrt[3]{81} + 2\sqrt[3]{250}$

30. $\sqrt{50x} + \sqrt{8x}$

31. $\sqrt{32x} - \sqrt{18x}$

32. $4\sqrt{9x} - 5\sqrt{4x}$

33. $6\sqrt{4a^2b} + 5\sqrt{25a^2b}$

34. $-3\sqrt{8a} - 4\sqrt{50a} + 10\sqrt{2a}$

35. $7\sqrt{36a^2b} + 4\sqrt{49a^2b} - 11\sqrt{2b}$

36. $\sqrt[3]{8x^2} - \sqrt[3]{27x^2}$

37. $\sqrt[4]{16a} + \sqrt[4]{81a}$

38. $\sqrt[4]{256b^3} - \sqrt[4]{81b^3}$

39. $-5\sqrt[3]{27a^2} - 4\sqrt[3]{8a^2}$

40. $\sqrt[3]{64x^2y} - 2\sqrt[3]{27x^2y}$ **41.** $\sqrt[3]{a^6b} + 3a^2\sqrt[3]{b}$ **42.** $2\sqrt{x^3y} + 5x\sqrt{xy}$

43. $3a^2\sqrt{ab} - a\sqrt{a^3b}$ **44.** $4xy^2\sqrt{x^3y} + 2x^2y\sqrt{xy^3}$ **45.** $3a^2b\sqrt{ab^3} - ab^2\sqrt{a^3b}$

46. $\dfrac{1}{2} + \dfrac{1}{\sqrt{2}}$ **47.** $\dfrac{1}{5} + \dfrac{2}{\sqrt{5}}$ **48.** $\dfrac{3}{4} + \dfrac{5}{\sqrt{2}}$ **49.** $\dfrac{4}{9} - \dfrac{2}{\sqrt{3}}$ **50.** $\dfrac{1}{\sqrt{3}} + \dfrac{1}{\sqrt{5}}$

51. $\dfrac{2}{\sqrt{5}} + \dfrac{3}{\sqrt{6}}$ **52.** $\dfrac{4}{\sqrt{7}} - \dfrac{2}{\sqrt{14}}$ **53.** $\dfrac{6}{\sqrt{5}} + \dfrac{2}{\sqrt{10}}$ **54.** $\dfrac{5}{\sqrt{3}} + \dfrac{2}{\sqrt{6}}$ **55.** $\dfrac{4}{\sqrt{12}} - \dfrac{1}{\sqrt{48}}$

56. $\dfrac{2}{\sqrt{18}} - \dfrac{4}{\sqrt{50}}$ **57.** $\dfrac{7}{\sqrt{4x}} - \dfrac{3}{\sqrt{x}}$ **58.** $\dfrac{5}{\sqrt{9a}} + \dfrac{2}{\sqrt{a}}$ **59.** $\dfrac{5}{\sqrt{xy}} - \dfrac{4}{\sqrt{x}}$ **60.** $\dfrac{2}{\sqrt{a}} + \dfrac{3}{\sqrt{ab}}$

Directions Solve the following verbal problems.

61. We can find the height h of the given figure by finding b from the formula $b = \sqrt{c^2 - s^2}$. If $c = 13$ units and $s = 5$ units, find h.

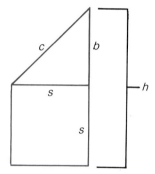

62. Use exercise 61 to find the height of the figure if $c = 10$ feet and $s = 6$ feet.

63. If two hallways of widths h_1 and h_2 meet at right angles, the longest board L that can be carried horizontally around the corner is given by the formula
$$L = \sqrt{(\sqrt[3]{h_1^2} + \sqrt[3]{h_2^2})^3}.$$
If h_1 is 8 ft and h_2 is 27 ft, find L. Leave the answer in radical form and also rounded to two decimal places.

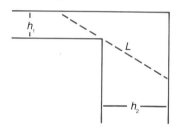

64. Use exercise 63 to find L if $h_1 = 6.859$ ft and $h_2 = 21.952$ ft. Leave the answer in radical form and also rounded to two decimal places.

65. The ideal keel length L (in feet) for a hang glider weighing 60 pounds with a pilot weighing P pounds is given by
$$L = \sqrt{\sqrt{2}\,(60 + P)}.$$
If the pilot weighs 175 pounds, find the ideal keel length. Round to two decimal places.

5–5 Further operations with radicals

Multiplying radicals

In section 5–2 we learned the procedure for multiplying two radicals. We now combine those ideas with the **distributive property,** $a(b + c) = ab + ac$, to perform multiplication of radical expressions that contain more than one term.

Example 5–5   A

Perform the indicated operations and simplify.

1. $\sqrt{2}(5 + \sqrt{2})$ Apply the distributive property.

$= \sqrt{2} \cdot 5 + \sqrt{2} \cdot \sqrt{2}$ Simplify.

$= 5\sqrt{2} + \sqrt{4}$

$= 5\sqrt{2} + 2$

2. $\sqrt{3}(\sqrt{15} - \sqrt{21})$ Apply the distributive property.

$= \sqrt{3}\sqrt{15} - \sqrt{3}\sqrt{21}$ Simplify.

$= \sqrt{45} - \sqrt{63}$

$= \sqrt{9 \cdot 5} - \sqrt{9 \cdot 7}$

$= 3\sqrt{5} - 3\sqrt{7}$

3. $(\sqrt{2} + \sqrt{3})(\sqrt{2} + 5\sqrt{3})$

> **Note**
>
> In this example we are multiplying quantities together.
> Therefore, as we did in chapter 3, we will *multiply each term in
> the first set of parentheses times each term in the second set
> of parentheses.*

$(\sqrt{2} + \sqrt{3})(\sqrt{2} + 5\sqrt{3})$

$= \sqrt{2}\sqrt{2} + \sqrt{2} \cdot 5\sqrt{3} + \sqrt{3}\sqrt{2} + \sqrt{3} \cdot 5\sqrt{3}$ Simplify.

$= \sqrt{4} + 5\sqrt{6} + \sqrt{6} + 5\sqrt{9}$

$= 2 + 5\sqrt{6} + \sqrt{6} + 5 \cdot 3$

Use the associative and commutative properties to add like terms and like radicals.

$= (2 + 15) + (5\sqrt{6} + \sqrt{6})$

$= 17 + 6\sqrt{6}$

4. $(3 - \sqrt{2})(3 + \sqrt{2})$ Apply the distributive property.

$= 3 \cdot 3 + 3\sqrt{2} - 3\sqrt{2} - \sqrt{2}\sqrt{2}$ Simplify.

$= 9 + 3\sqrt{2} - 3\sqrt{2} - \sqrt{4}$

$= 9 + 3\sqrt{2} - 3\sqrt{2} - 2$ Add and subtract like terms.

$= (9 - 2) + (3\sqrt{2} - 3\sqrt{2})$

$= 7 + 0$

$= 7$

We observe that when we added and subtracted the like terms, there were no longer any radicals in the problem.

Conjugate factors and rationalizing denominators

In example 4 we see that there are no radicals in the final answer. The type of factors that we are multiplying, called **conjugate factors,** are derived from the factorization of the special product, called the difference of two squares $[(a + b)(a - b) = a^2 - b^2]$. Conjugate factors are used to rationalize the denominator of a fraction when the denominator has two terms where one or both terms contain a square root. When multiplying conjugate factors, we can simply write our answer as the square of the second term subtracted from the square of the first term.

If we recognize that we are multiplying the factors of the difference of two squares in example 4, the multiplication can be performed as follows:

$$(3 - \sqrt{2})(3 + \sqrt{2}) = (3)^2 - (\sqrt{2})^2 = 9 - 2 = 7.$$

To determine the conjugate of a given factor, we write down the two terms of the factor and change the sign between them, that is, addition to subtraction, or subtraction to addition.

Example 5–5 B

Form the conjugates of the given expressions.

1. $5 - 2\sqrt{3}$ The conjugate is $5 + 2\sqrt{3}$.

2. $\sqrt{a} + \sqrt{b}$ The conjugate is $\sqrt{a} - \sqrt{b}$.

3. $-3 - \sqrt{7}$ The conjugate is $-3 + \sqrt{7}$.

If we wish to rationalize the denominator of the fraction

$$\frac{1}{3 - \sqrt{2}},$$

we recall from example 5–5 A, example 4 that when we multiplied $3 - \sqrt{2}$ by $3 + \sqrt{2}$, there were no radicals left in our product, and this is precisely what we want to occur in our denominator. Therefore to rationalize this fraction, we apply the fundamental principle of fractions and multiply the numerator and the denominator by $3 + \sqrt{2}$, which is the conjugate of the denominator.

$$\frac{1}{3 - \sqrt{2}} = \frac{1}{3 - \sqrt{2}} \cdot \frac{3 + \sqrt{2}}{3 + \sqrt{2}}$$

$$= \frac{1(3 + \sqrt{2})}{(3)^2 - (\sqrt{2})^2}$$

$$= \frac{3 + \sqrt{2}}{9 - 2}$$

$$= \frac{3 + \sqrt{2}}{7}$$

Example 5-5 C   Simplify the following by rationalizing the denominator. Assume that all variables represent positive real numbers and that no denominator is equal to zero.

1. $\dfrac{5}{\sqrt{13} - \sqrt{3}}$ The conjugate is $\sqrt{13} + \sqrt{3}$.

$= \dfrac{5}{\sqrt{13} - \sqrt{3}} \cdot \dfrac{\sqrt{13} + \sqrt{3}}{\sqrt{13} + \sqrt{3}}$ Use the special products.

$= \dfrac{5(\sqrt{13} + \sqrt{3})}{(\sqrt{13})^2 - (\sqrt{3})^2} = \dfrac{5(\sqrt{13} + \sqrt{3})}{13 - 3} = \dfrac{5(\sqrt{13} + \sqrt{3})}{10}$

There is a common factor of 5 and we can reduce the fraction.

$= \dfrac{5(\sqrt{13} + \sqrt{3})}{5 \cdot 2} = \dfrac{\sqrt{13} + \sqrt{3}}{2}$

2. $\dfrac{\sqrt{3}}{5 - 2\sqrt{3}}$ The conjugate is $5 + 2\sqrt{3}$.

$= \dfrac{\sqrt{3}}{5 - 2\sqrt{3}} \cdot \dfrac{5 + 2\sqrt{3}}{5 + 2\sqrt{3}}$ Use the special products.

$= \dfrac{\sqrt{3}(5 + 2\sqrt{3})}{(5)^2 - (2\sqrt{3})^2} = \dfrac{\sqrt{3}(5 + 2\sqrt{3})}{25 - 2^2(\sqrt{3})^2} = \dfrac{\sqrt{3}(5 + 2\sqrt{3})}{25 - 4 \cdot 3}$

$= \dfrac{\sqrt{3}(5 + 2\sqrt{3})}{25 - 12} = \dfrac{\sqrt{3}(5 + 2\sqrt{3})}{13}$ Multiply in the numerator.

$= \dfrac{5\sqrt{3} + 2\sqrt{9}}{13} = \dfrac{5\sqrt{3} + 2 \cdot 3}{13} = \dfrac{5\sqrt{3} + 6}{13}$

3. $\dfrac{\sqrt{ab}}{\sqrt{a} + \sqrt{b}}$ The conjugate is $\sqrt{a} - \sqrt{b}$.

$= \dfrac{\sqrt{ab}}{\sqrt{a} + \sqrt{b}} \cdot \dfrac{\sqrt{a} - \sqrt{b}}{\sqrt{a} - \sqrt{b}}$ Use the special products.

$= \dfrac{\sqrt{ab}(\sqrt{a} - \sqrt{b})}{(\sqrt{a})^2 - (\sqrt{b})^2} = \dfrac{\sqrt{ab}(\sqrt{a} - \sqrt{b})}{a - b}$ Multiply in the numerator and simplify.

$= \dfrac{\sqrt{a^2b} - \sqrt{ab^2}}{a - b} = \dfrac{a\sqrt{b} - b\sqrt{a}}{a - b}.$

Mastery points
Can you
- Multiply radical expressions containing more than one term?
- Form conjugate factors?
- Multiply conjugate factors?
- Rationalize a denominator that has two terms in which one or both terms contain a square root?

Exercise 5–5

Directions Perform the indicated operations and simplify. Assume that all variables represent positive real numbers and no denominator is equal to zero. See examples 5–5 A, B, and C.

1. $3(\sqrt{5} + \sqrt{3})$

2. $5(\sqrt{2} - \sqrt{3})$

3. $4(3\sqrt{7} + \sqrt{2})$

4. $\sqrt{5}(2\sqrt{11} - 3\sqrt{7})$

5. $\sqrt{3}(\sqrt{2} + \sqrt{5})$

6. $\sqrt{6}(\sqrt{7} - \sqrt{5})$

7. $\sqrt{2}(3\sqrt{3} - \sqrt{11})$

8. $2\sqrt{3}(3\sqrt{2} + 4\sqrt{5})$

9. $5\sqrt{5}(7\sqrt{2} - 4\sqrt{3})$

10. $\sqrt{6}(\sqrt{2} + \sqrt{3})$

11. $\sqrt{5}(\sqrt{15} - \sqrt{20})$

12. $\sqrt{14}(\sqrt{35} + \sqrt{10})$

13. $2\sqrt{7}(\sqrt{35} - 4\sqrt{14})$

14. $5\sqrt{3}(2\sqrt{3} - \sqrt{15})$

15. $\sqrt{x}(\sqrt{x} + \sqrt{y})$

16. $\sqrt{a}(\sqrt{ab} - \sqrt{b})$

17. $3\sqrt{x}(2\sqrt{xy} - 5\sqrt{x})$

18. $6\sqrt{ab}(2\sqrt{a} - 3\sqrt{b})$

19. $5\sqrt{xy}(\sqrt{x} + 4\sqrt{y})$

20. $(5 + \sqrt{2})(3 - \sqrt{2})$

21. $(3 + \sqrt{3})(2 + \sqrt{3})$

22. $(6 - \sqrt{6})(6 - \sqrt{6})$

23. $(4 + \sqrt{x})(5 + \sqrt{x})$

24. $(3 - 4\sqrt{x})(4 - 2\sqrt{x})$

25. $(1 + 2\sqrt{y})(3 - 4\sqrt{y})$

26. $(-3 + 5\sqrt{a})(-2 - \sqrt{a})$

27. $(3 - \sqrt{2})(3 + \sqrt{2})$

28. $(4 - \sqrt{5})(4 + \sqrt{5})$

29. $(3 - 2\sqrt{3})(3 + 2\sqrt{3})$

30. $(5 + 4\sqrt{5})(5 - 4\sqrt{5})$

31. $(\sqrt{5} - \sqrt{7})(\sqrt{5} + \sqrt{7})$

32. $(\sqrt{10} - \sqrt{5})(\sqrt{10} + \sqrt{5})$

33. $(\sqrt{a} - b)(\sqrt{a} + b)$

34. $(2\sqrt{a} - \sqrt{b})(2\sqrt{a} + \sqrt{b})$

35. $(3\sqrt{x} - 4\sqrt{y})(3\sqrt{x} + 4\sqrt{y})$

36. $(3 + \sqrt{5})^2$

37. $(2 - \sqrt{7})^2$

38. $(4\sqrt{3} + 2)^2$

39. $(5\sqrt{3} - 4)^2$

40. $(\sqrt{x} - 2y)^2$

41. $(2\sqrt{a} + b)^2$

42. $(3\sqrt{a} + 2\sqrt{b})(\sqrt{a} - 3\sqrt{b})$

43. $(4\sqrt{x} - \sqrt{y})(5\sqrt{x} + \sqrt{y})$

44. $(2\sqrt{a} + \sqrt{b})(3\sqrt{a} - \sqrt{b})$

45. $\dfrac{1}{\sqrt{3} + 2}$

46. $\dfrac{1}{\sqrt{5} - 2}$

47. $\dfrac{6}{4 + \sqrt{6}}$

48. $\dfrac{6}{3 - \sqrt{6}}$

49. $\dfrac{4}{\sqrt{10} - \sqrt{6}}$

50. $\dfrac{5}{\sqrt{11} - \sqrt{6}}$

51. $\dfrac{3}{4 - 2\sqrt{3}}$

52. $\dfrac{5}{6 - 2\sqrt{5}}$

53. $\dfrac{\sqrt{6}}{\sqrt{2} - 2\sqrt{3}}$

54. $\dfrac{\sqrt{10}}{\sqrt{5} + 2\sqrt{2}}$

55. $\dfrac{\sqrt{14}}{3\sqrt{7} - 2\sqrt{2}}$

56. $\dfrac{\sqrt{2}}{3\sqrt{6} - \sqrt{3}}$

57. $\dfrac{\sqrt{x}}{\sqrt{x} + \sqrt{y}}$

58. $\dfrac{\sqrt{a}}{\sqrt{a} + \sqrt{b}}$

59. $\dfrac{\sqrt{ab}}{\sqrt{ab} + \sqrt{b}}$

60. $\dfrac{\sqrt{a} + b}{\sqrt{a} - b}$

61. $\dfrac{\sqrt{x} - y}{\sqrt{x} + y}$

62. $\dfrac{a + \sqrt{b}}{a - \sqrt{b}}$

63. $\dfrac{a}{\sqrt{ab} - \sqrt{a}}$

64. $\dfrac{x^2y}{x\sqrt{x} - \sqrt{xy}}$

65. $\dfrac{2a}{\sqrt{a} - \sqrt{ab}}$

66. $\dfrac{3x}{\sqrt{x} - \sqrt{xy}}$

67. $\dfrac{\sqrt{a}}{\sqrt{ab} - \sqrt{a}}$

68. $\dfrac{2\sqrt{x}}{3\sqrt{x} - 2\sqrt{y}}$

69. The electric-field intensity on the axis of a uniform charged ring is given by

$$E = \frac{T}{(x^2 + r^2)^{\frac{3}{2}}},$$

where T is the total charge on the ring and r is the radius of the ring. Express the fractional exponent in radical form and leave the answer in standard form.

5-6 Complex numbers

Imaginary numbers

In this section we will examine what happens when we try to take the square root of a negative number. The expression $\sqrt{-4}$ has no meaning in the system of real numbers because there is no real number that when multiplied by itself equals a negative number, in this case, -4. However there are situations in which an answer for such a problem is required. For example, we need to find the square root of a negative number if we want to find the solution to the equation $x^2 + 4 = 0$. In electronics the impedance of a circuit, which is the total effective resistance to the flow of current caused by a combination of elements in the circuit, also requires that we solve for the square root of a negative number. We define a new kind of number to provide the required result.

> ■ **Definition of i**
> The number i is a number such that
> $$i = \sqrt{-1}$$
> and
> $$i^2 = -1.$$

We can use this fact to define the square root of any negative number. We now have the system of imaginary numbers that is the set of numbers which can be expressed in the form bi, where b is an element of the set of real numbers.

> ■ **Definition of $\sqrt{-b}$**
> For any positive real number b, we define
> $$\sqrt{-b} = i\sqrt{b}.$$

Example 5-6 A

1. $\sqrt{-4} = i\sqrt{4} = 2i$

2. $\sqrt{-2} = i\sqrt{2}$

If we wish to check our results, we can square the answer to get back the original radicand.

Example 5–6 B

1. $(2i)^2 = 2^2 \cdot i^2 = 4 \cdot (-1) = -4$

2. $(\sqrt{2}\,i)^2 = (\sqrt{2})^2 \cdot i^2 = 2(-1) = -2$

Consider the following examples; performing the same operation in two different ways, we have

(a) $(\sqrt{-4})^2 = (\sqrt{(4)(-1)})^2$

$= (\sqrt{4}\,\sqrt{-1})^2$

$= (2i)^2$

$= 2^2 i^2$

$= (4)(-1) = -4$

(b) $(\sqrt{-4})^2 = \sqrt{-4}\,\sqrt{-4}$

$= \sqrt{(-4)(-4)}$

$= \sqrt{16}$

$= 4$

We see that depending on the order in which we apply our properties of exponents, we get different results. Therefore *whenever we are dealing with the square root of a negative number, we must express our problem in terms of* i *before proceeding.* Hence $(\sqrt{-4})^2 = (2i)^2 = 2^2 i^2 = 4 \cdot (-1) = -4$.

Example 5–6 C

1. $\sqrt{-2}\,\sqrt{-8} = i\sqrt{2} \cdot i\sqrt{8} = i^2\sqrt{16} = (-1)(4) = -4$

2. $\sqrt{-6}\,\sqrt{-3} = i\sqrt{6} \cdot i\sqrt{3} = i^2\sqrt{18} = (-1) \cdot 3\sqrt{2} = -3\sqrt{2}$

If we apply the properties of exponents to i, we have

a
cycle $\begin{cases} i = i \\ i^2 = -1 \\ i^3 = i^2 i = (-1)i = -i \\ i^4 = i^2 i^2 = (-1)(-1) = 1 \end{cases}$

a
cycle $\begin{cases} i^5 = i^4 i = 1 \cdot i = i \\ i^6 = i^4 i^2 = 1(-1) = -1 \\ i^7 = i^4 i^3 = 1(-i) = -i \\ i^8 = i^4 i^4 = 1 \cdot 1 = 1 \end{cases}$

It can be seen that the powers of i go through the cycle of $i, -1, -i,$ and 1. Using this fact, it is possible to simplify i raised to any positive integer power.

Example 5–6 D

1. $i^{10} = i^4 \cdot i^4 \cdot i^2 = 1 \cdot 1 \cdot (-1) = -1$

2. $i^{15} = i^4 \cdot i^4 \cdot i^4 \cdot i^3 = 1 \cdot 1 \cdot 1 \cdot (-i) = -i$

3. $i^{20} = i^4 i^4 i^4 i^4 i^4 = 1 \cdot 1 \cdot 1 \cdot 1 \cdot 1 = 1$

From these examples, we see that *when we simplify* i *to a positive integer power, the resulting power of* i *is the remainder when we divide the original exponent by 4.*

Example 5–6 E

1. $i^{50} = i^2 = -1$ because $50 \div 4 = 12$ remainder 2

2. $i^{79} = i^3 = -i$ because $79 \div 4 = 19$ remainder 3

3. $i^{21} = i$ because $21 \div 4 = 5$ remainder 1

Complex numbers

Now let us define a new type of number that combines the system of real numbers and the system of imaginary numbers. These new numbers are called **complex numbers** and are composed of a real part denoted by *a* and an imaginary part denoted by *b*. *A complex number is any number that can be put in the form of* a + bi, *where* a *and* b *are real numbers and* i *represents* $\sqrt{-1}$. The number *a + bi* is called the **standard form** of a complex number.

Operations with complex numbers

The commutative, associative, and distributive properties for real numbers are also valid for complex numbers. If we wish to perform addition and subtraction of complex numbers, we do so by combining the real parts and combining the imaginary parts.

> ■ **Definition of addition of complex numbers**
> If *a, b, c,* and *d* are real numbers, then
> $$(a + bi) + (c + di) = (a + c) + (b + d)i,$$
> where $i = \sqrt{-1}$.

Example 5–6 F

Perform the indicated addition and subtraction.

1. $(4 + 5i) + (2 + 3i)$
To perform the addition, we add the real parts $(4 + 2)$ and the imaginary parts $(5 + 3)$.
$$(4 + 5i) + (2 + 3i) = (4 + 2) + (5 + 3)i$$
$$- 6 + 8i$$

2. $(6 + 11i) - (5 + 2i)$
To perform the subtraction, we subtract the real parts $(6 - 5)$ and the imaginary parts $(11 - 2)$.
$$(6 + 11i) - (5 + 2i) = (6 - 5) + (11 - 2)i$$
$$= 1 + 9i$$

When we are multiplying two complex numbers such as
$$(a + bi) (c + di),$$
we multiply each of the terms in the first parentheses times each of the terms in the second parentheses.

Example 5-6 G

Perform the indicated multiplication.

1. $(2 + 3i)(5 + 2i)$

   Multiply each of the terms of the first complex number times each of the terms of the second complex number and simplify.

   $$(2 + 3i)(5 + 2i) = 2 \cdot 5 + 2 \cdot 2i + 3i \cdot 5 + 3i \cdot 2i$$
   $$= 10 + 4i + 15i + 6i^2$$

   Since $i^2 = -1$, then $6i^2 = 6(-1) = -6$, and we have

   $$= 10 + 4i + 15i - 6$$
   $$= (10 - 6) + (4 + 15)i$$
   $$= 4 + 19i.$$

2. $(3 - 4i)(3 + 4i) = 3 \cdot 3 + 3 \cdot 4i - 4i \cdot 3 - 4i \cdot 4i$
   $$= 9 + 12i - 12i - 16i^2$$
   $$= 9 + 12i - 12i + 16$$
   $$= (9 + 16) + (12 - 12)i$$
   $$= 25 + 0i$$
   $$= 25$$

The factors $(3 - 4i)$ and $(3 + 4i)$ are called **complex conjugates** and their product will be a real number. We use complex conjugates to find the quotient of two complex numbers. Consider the quotient

$$(2 + 3i) \div (3 - 4i) = \frac{2 + 3i}{3 - 4i}.$$

We would like to perform the division of the two complex numbers and leave the answer in the standard form of $a + bi$. First of all, we will eliminate the i in the denominator, since this is just another form of the radical $\sqrt{-1}$. To rationalize the denominator, we multiply by the conjugate of $3 - 4i$, which is $3 + 4i$.

$$\frac{2 + 3i}{3 - 4i} \cdot \frac{3 + 4i}{3 + 4i} = \frac{2 \cdot 3 + 2 \cdot 4i + 3i \cdot 3 + 3i \cdot 4i}{(3)^2 - (4i)^2}$$
$$= \frac{6 + 8i + 9i + 12i^2}{9 - 16i^2}$$

Replacing i^2 with -1, $12i^2$ becomes -12 and $-16i^2$ becomes 16.

$$= \frac{6 + 8i + 9i - 12}{9 + 16}$$

Adding the like terms, we have

$$\frac{-6 + 17i}{25}.$$

Since the answer is to be stated in the form $a + bi$, we divide each term in the numerator by 25 to obtain

$$= \frac{-6}{25} + \frac{17}{25}i.$$

This is the quotient of the two complex numbers stated in standard form.

Example 5–6 H

Perform the indicated division.

1. $\dfrac{1 + \sqrt{-4}}{3 - \sqrt{-9}}$

We must first simplify the radicals before carrying out the division.

$$= \frac{1 + \sqrt{4(-1)}}{3 - \sqrt{9(-1)}} = \frac{1 + 2i}{3 - 3i} \qquad \begin{array}{l} \text{The conjugate of the denominator} \\ \text{is } 3 + 3i. \end{array}$$

$$= \frac{1 + 2i}{3 - 3i} \cdot \frac{3 + 3i}{3 + 3i} = \frac{3 + 3i + 6i + 6i^2}{(3)^2 - (3i)^2}$$

$$= \frac{3 + 3i + 6i - 6}{9 - 9i^2} = \frac{-3 + 9i}{9 + 9} = \frac{-3 + 9i}{18}$$

Divide and write the quotient in standard form.

$$= \frac{-3}{18} + \frac{9}{18}i = \frac{-1}{6} + \frac{1}{2}i$$

2. $\dfrac{3 - 2i}{i}$

Since $i = \sqrt{-1}$, we simply multiply by $\dfrac{i}{i}$.

$$= \frac{3 - 2i}{i} \cdot \frac{i}{i} = \frac{3i - 2i^2}{i^2} = \frac{3i + 2}{-1}$$

Divide and write in standard form.

$$= \frac{3i}{-1} + \frac{2}{-1} = -3i - 2 = -2 - 3i$$

Mastery points

Can you
- Simplify the square root of a negative number?
- Simplify i raised to a positive integer power?
- Add, subtract, multiply, and divide complex numbers?

Exercise 5–6

Directions Simplify the following. See examples 5–6 A, B, C, D, and E.

1. $\sqrt{-9}$ **2.** $\sqrt{-16}$ **3.** $\sqrt{-12}$ **4.** $\sqrt{-18}$ **5.** $(3i)^2$

6. $(4i)^2$ **7.** $(\sqrt{3}\,i)^2$ **8.** $(\sqrt{2}\,i)^2$ **9.** $\sqrt{-3}\sqrt{-5}$ **10.** $\sqrt{-7}\sqrt{-11}$

11. $\sqrt{-2}\sqrt{-2}$ **12.** $\sqrt{-6}\sqrt{-6}$ **13.** $(\sqrt{-5})^2$ **14.** $(\sqrt{-4})^2$ **15.** $\sqrt{-3}\sqrt{-12}$

16. $\sqrt{-2}\sqrt{-18}$ **17.** $\sqrt{-3}\sqrt{-15}$ **18.** $\sqrt{-7}\sqrt{-14}$ **19.** i^2 **20.** i^3

21. i^4 **22.** i^{15} **23.** i^{19} **24.** i^{60}

Directions Perform the indicated operations, and leave the answer in standard form. See examples 5–6 F, G, and H.

25. $(6 + 3i) + (2 + 4i)$

26. $(4 + 3i) + (1 + i)$

27. $(6 - 2i) - (8 - 4i)$

28. $(4 - 5i) - (3 - 7i)$

29. $(2 + \sqrt{-49}) - (1 - \sqrt{-1})$

30. $(9 + \sqrt{-64}) - (9 - \sqrt{-9})$

31. $(4 - \sqrt{-25}) - (4 - \sqrt{-36})$

32. $[(2 + 5i) + (3 - 2i)] + (3 - i)$

33. $[(-2 - i) + (3 + 2i)] + (4 - 3i)$

34. $[(8 - 5i) - (5 + 4i)] + (6 - 7i)$

35. $[(9 - i) - (6 - 4i)] + (5 + 5i)$

36. $(4 + 3i)(1 + i)$

37. $(3 - 2i)(3 + 2i)$

38. $(4 + 5i)(4 - 5i)$

39. $(3 + i)(5 - 4i)$

40. $(2 + \sqrt{-16})(3 - \sqrt{-25})$

41. $(7 + \sqrt{-1})(3 + \sqrt{-4})$

42. $(5 - \sqrt{-25})(4 + \sqrt{-16})$

43. $(5 - \sqrt{-9})(5 + \sqrt{-9})$

44. $(2 + i)^2$

45. $(4 - 3i)^2$

46. $(3 - \sqrt{-9})^2$

47. $(2 + \sqrt{-4})^2$

48. $\dfrac{3 - 2i}{i}$

49. $\dfrac{4 + 5i}{i}$

50. $\dfrac{6 - 2i}{3i}$

51. $\dfrac{2 + 4i}{2i}$

52. $\dfrac{4 - 9i}{\sqrt{-1}}$

53. $\dfrac{5 + 7i}{\sqrt{-9}}$

54. $\dfrac{4 - 3i}{1 + i}$

55. $\dfrac{5 - 2i}{5 - i}$

56. $\dfrac{4 - 5i}{2 + i}$

57. $\dfrac{3 - i}{3 + i}$

58. $\dfrac{5 - i}{5 + i}$

59. $\dfrac{2 + 5i}{3 - 2i}$

60. $\dfrac{4 + 3i}{3 - i}$

61. $\dfrac{6 + 3i}{3 + 4i}$

62. $\dfrac{5 - \sqrt{-4}}{3 + \sqrt{-9}}$

63. $\dfrac{2 + \sqrt{-16}}{4 - \sqrt{-1}}$

64. $\dfrac{7 - \sqrt{-25}}{3 + \sqrt{-36}}$

Directions Solve the following verbal problems.

65. The impedance of an electrical circuit is the measure of the total opposition to the flow of an electric current. The impedance Z in a series RCL circuit is given by
$Z = R + i(X_L - X_C)$.
Determine the impedance if $R = 30$ ohms, $X_L = 16$ ohms, and $X_C = 40$ ohms.

66. Use the formula in exercise 65 to find Z if $R = 28$ ohms, $X_L = 16$ ohms, and $X_C = 38$ ohms.

67. The impedance Z in an AC circuit is given by the formula
$Z = \dfrac{V}{I}$,
where V is the voltage and I is the current. Find Z when $V = 0.3 + 1.2i$ and $I = 2.1i$. Round all values to three decimal places.

68. Use the formula in exercise 67 to find Z if $V = 2.2 - 0.3i$ and $I = -1.1i$. Round all values to three decimal places.

69. The total impedance Z_T of an AC circuit containing impedances Z_1 and Z_2 in parallel is given by the formula
$Z_T = \dfrac{Z_1 Z_2}{Z_1 + Z_2}$.
Find Z_T when $Z_1 = 3 - i$ and $Z_2 = 2 + i$.

70. Use the formula in exercise 69 to find Z_T if $Z_1 = 4 - i$ and $Z_2 = 3 + i$.

71. If three resistors in an AC circuit are connected in parallel, the total impedance Z_T is given by the formula
$Z_T = \dfrac{Z_1 Z_2 Z_3}{Z_1 Z_2 + Z_1 Z_3 + Z_2 Z_3}$.
Find Z_T when $Z_1 = 3 - i$, $Z_2 = 3 + i$, and $Z_3 = 2i$.

72. Use the formula in exercise 71 to find Z_T if $Z_1 = 2 - i$, $Z_2 = 2 + i$, and $Z_3 = 2i$.

73. For what values of x does the expression $\sqrt{5 - x}$ represent a real number?

74. For what values of x does the expression $\sqrt{x + 4}$ represent a real number?

75. For what values of x does the expression $\sqrt{x + 11}$ represent an imaginary number?

76. For what values of x does the expression $\sqrt{8 - x}$ represent an imaginary number?

Chapter summary

1. $a^{\frac{1}{n}} = \sqrt[n]{a}$, whenever the principal n^{th} root of a is a real number.

2. $a^{\frac{m}{n}} = (\sqrt[n]{a})^m = \sqrt[n]{a^m}$, if the principal n^{th} root of a is a real number.

3. For all nonnegative real numbers a and b and positive integer n greater than 1,
$$\sqrt[n]{ab} = \sqrt[n]{a}\,\sqrt[n]{b},$$
and
$$\sqrt[n]{\frac{a}{b}} = \frac{\sqrt[n]{a}}{\sqrt[n]{b}}, \qquad b \neq 0.$$

4. We eliminate radicals from the denominator of a fraction by **rationalizing** the denominator.

5. We can only add or subtract *like* radicals.

6. **Conjugate factors** are used to rationalize the denominator of a fraction when the denominator has two terms where one or both terms contain a square root.

7. We define $i = \sqrt{-1}$.

8. A complex number is any number that can be put in the form $a + bi$, where a and b are real numbers and i represents $\sqrt{-1}$.

Chapter review

Assume that all variables represent positive real numbers and no denominator is equal to zero.

[5–1]

Directions Rewrite the following in radical notation and use table 5.1 to simplify.

1. $36^{\frac{1}{2}}$

2. $16^{-\frac{3}{4}}$

3. $(-27)^{\frac{2}{3}}$

Directions Perform the indicated operations and simplify. Leave the answer with all exponents positive.

4. $a^{\frac{2}{3}} \cdot a^{\frac{1}{4}}$

5. $\left(c^{\frac{3}{4}}\right)^{\frac{1}{3}}$

6. $(27x^3)^{\frac{2}{3}}$

7. $\left(b^{-\frac{2}{3}}\right)^{-3}$

8. $\dfrac{a^{\frac{1}{2}}}{a^{\frac{1}{3}}}$

9. $(8x^{12}y^6)^{\frac{2}{3}}$

10. $\dfrac{x^3 x^{\frac{2}{3}}}{x^{\frac{1}{2}}}$

11. $\dfrac{a^{\frac{2}{3}}}{a^{-\frac{1}{2}}}$

Directions Simplify the following.

12. $\sqrt{12}$ **13.** $\sqrt{10}\ \sqrt{15}$ **14.** $\sqrt[5]{x^4}\ \sqrt[5]{x^3}$

15. $\sqrt[3]{6ab^2}\ \sqrt[3]{4a^2b^2}$ **16.** $\sqrt[4]{a^6}$ **17.** $\sqrt[9]{8a^3b^6}$

18. If the hypotenuse of a right triangle is 10 inches long and one of the legs is 6 inches long, find the length of the other leg.

Directions Simplify the following expressions leaving no radicals in the denominator.

19. $\sqrt{\dfrac{49}{64}}$ **20.** $\sqrt{\dfrac{16}{27}}$ **21.** $\sqrt[3]{\dfrac{16x^3y^2}{z^6}}$ **22.** $\sqrt{\dfrac{1}{8}}$ **23.** $\dfrac{6}{\sqrt{2}}$

24. $\sqrt[3]{\dfrac{8}{25}}$ **25.** $\sqrt{\dfrac{x^2}{y}}$ **26.** $\sqrt[3]{\dfrac{a^2}{b}}$ **27.** $\dfrac{x}{\sqrt[5]{x^2}}$ **28.** $\sqrt[3]{\dfrac{a}{bc^2}}$

29. $\dfrac{a}{\sqrt[4]{ab^2}}$ **30.** $\sqrt{\dfrac{3x}{y^2}}\ \sqrt{\dfrac{xy}{3}}$

Directions Perform the indicated operations and simplify.

31. $4\sqrt{3}\ -\ 3\sqrt{3}\ +\ 7\sqrt{3}$ **32.** $\sqrt{18}\ +\ \sqrt{50}$ **33.** $\sqrt{8a}\ +\ 9\sqrt{18a}$

34. $5x^2\sqrt{xy}\ -\ 2x\sqrt{x^3y}$ **35.** $\dfrac{5}{\sqrt{6}}\ -\ \dfrac{1}{\sqrt{3}}$ **36.** $\dfrac{2}{\sqrt{ab}}\ -\ \dfrac{1}{\sqrt{a}}$

Directions Simplify the following expressions leaving no radicals in the denominator.

37. $\sqrt{2}\,(\sqrt{6}\ -\ \sqrt{10}\,)$ **38.** $2\sqrt{a}\,(\sqrt{ab}\ +\ 2\sqrt{a}\,)$ **39.** $(5\ -\ \sqrt{5}\,)^2$

40. $(\sqrt{10}\ -\ \sqrt{7}\,)(\sqrt{10}\ +\ \sqrt{7}\,)$ **41.** $(2\sqrt{a}\ +\ 3\sqrt{b}\,)(2\sqrt{a}\ -\ 3\sqrt{b}\,)$ **42.** $(3\sqrt{x}\ +\ y)^2$

43. $\dfrac{1}{\sqrt{6}\ +\ 2}$ **44.** $\dfrac{10}{4\ +\ \sqrt{6}}$ **45.** $\dfrac{\sqrt{3}}{\sqrt{6}\ -\ \sqrt{3}}$

46. $\dfrac{a^2b}{a\sqrt{a}\ -\ \sqrt{ab}}$

Directions Simplify the following.

47. $\sqrt{-49}$ **48.** $\sqrt{-28}$ **49.** $(2i)^2$ **50.** $(\sqrt{7}\,i)^2$ **51.** $\sqrt{-3}\ \sqrt{-12}$

52. $(\sqrt{-3}\,)^2$ **53.** $\sqrt{-2}\ \sqrt{-3}$ **54.** i^{37}

Directions Simplify the following and leave the answer in standard form.

55. $(4\ +\ 2i)\ +\ (3\ +\ 5i)$ **56.** $(2\ -\ \sqrt{-36}\,)\ -\ (3\ +\ \sqrt{-25}\,)$

57. $(2\ -\ \sqrt{-9}\,)(3\ +\ \sqrt{-16}\,)$

58. $(2 + 5i)^2$ **59.** $\dfrac{3 + 4i}{i}$ **60.** $\dfrac{7 - 6i}{\sqrt{-9}}$ **61.** $\dfrac{4 - i}{2 + i}$

62. $\dfrac{9 - \sqrt{-4}}{7 - \sqrt{-9}}$

Chapter 5 cumulative test

Directions Factor completely.

[3–7] **1.** $a^2 - 7a - 8$ [3–5] **2.** $4x^2 - 3x$ [3–8] **3.** $9x^2 - 36$

[3–6] **4.** $2x^2 + 11x + 12$ [3–6] **5.** $3a^2 - 11a - 20$ [3–6] **6.** $6x^2 + 17x + 12$

[1–5] **7.** Evaluate the expression $b^2 - 4ac$ at
(a) $a = 1$, $b = 4$, and $c = -3$;
(b) $a = 2$, $b = -4$, and $c = 3$.

Directions Solve for x.

[2–1] **8.** $3(2x - 1) + 4 = x + 3$ [2–4] **9.** $5x + 7 > 2x - 4$ [2–2] **10.** $3x - 2y = 4(x + y)$

[2–5] **11.** $|3x - 1| = 4$ [2–5] **12.** $|2x + 3| > 8$ [2–1] **13.** $x(x + 1) - (x + 3)^2 = 4$

[2–5] **14.** $|1 - 4x| \le 7$

Directions Simplify the following and leave in standard form. Assume that all variables represent positive real numbers.

[5–2] **15.** $\sqrt[5]{64a^{10}b^7}$ [5–6] **16.** $(3 - 4i)(2 + i)$ [5–2] **17.** $\sqrt{48}$ [5–1] **18.** $a^{\frac{1}{3}} \cdot a^{\frac{1}{4}}$

[5–2] **19.** $\sqrt[3]{8a^4b^6}$ [3–2] **20.** $(2a^3b^4c)^3$ [5–6] **21.** $\sqrt{-18}$ [5–5] **22.** $\dfrac{4}{\sqrt{10} + \sqrt{6}}$

[5–6] **23.** $\dfrac{1 - 3i}{2 + 3i}$ [5–1] **24.** $(4a^6)^{\frac{1}{2}}$ [3–4] **25.** $\dfrac{3x^{-2}y^4}{9x^{-5}y^2}$ [5–3] **26.** $\sqrt[3]{\dfrac{a^2}{4bc^2}}$

Directions Solve the following verbal problems.

[2–3] **27.** When the length of a side of a square is increased by 4 inches, the area is increased by 72 square inches. Find the original length of a side.

[2–3] **28.** A metallurgist wishes to form 1,000 kg of an alloy that is 62% copper. This alloy is to be obtained by fusing some alloy that is 80% copper and some that is 50% copper. How many kilograms of each alloy must be used?

[5–1] **29.** In the theory of ballistics the ballistic limit v of a material is approximated by the formula

$$v = kT^{\frac{6}{5}},$$

where T is the thickness of a sheet of material and k is a constant that is dependent on the material being used. Compute the ballistic limit if $k = 24{,}000$ and $T = 0.03125$.

[5–1] **30.** Use the formula in exercise 29 to find v if $k = 25{,}000$ and $T = 0.07776$.

Chapter

6

Quadratic Equations and Inequalities

6–1 Solution by factoring and extracting the roots

A quadratic equation

In chapter 3 we found the solution set S of first-degree (linear) equations in one variable. We now consider the solution set of a second-degree equation in one variable, called a **quadratic equation** in one variable. Such an equation contains the second, but no higher, power of the variable. For example, in electrical circuits the flow of current varies with time. In a certain circuit this relationship is expressed as

$$I = 12 - 12t^2,$$

where I is the current in amperes and t is the time in seconds. When will the current flow be 0? To answer this question, it is necessary to set $I = 0$ in the above equation. This yields

$$0 = 12 - 12t^2,$$

which is a quadratic equation in the variable t.

> ■ **Definition of a quadratic equation in one variable**
> A **quadratic equation in one variable** is any second-degree equation that can be written in the form
> $$ax^2 + bx + c = 0,$$
> where a, b, and c are real numbers, $a \neq 0$. We call this the **standard form** of a quadratic equation in one variable, x.

Note
It is necessary to restrict a so that it is different from zero. If $a = 0$, then we have
$$0 \cdot x^2 + bx + c = 0 \quad \text{or} \quad bx + c = 0,$$
and we have a first-degree (linear) equation.

If $b = 0$ or $c = 0$, then the equation is of the form
$$ax^2 + c = 0 \quad \text{or} \quad ax^2 + bx = 0,$$
and we still have a second-degree (quadratic) equation.

In this chapter we will discuss four basic methods for solving (finding the solution set of) a quadratic equation. The first of these methods involves factoring the quadratic expression when possible and using the following property, called the **zero product property**.

> ■ **Zero product property**
> If P and Q are polynomials, then
> $$P \cdot Q = 0,$$
> if and only if $P = 0$ or $Q = 0$.
>
> **Concept**
> The product of two polynomials is zero provided one or both of the polynomials is equal to zero.

To solve a quadratic equation by factoring when using this property, we

1. write the equation in standard form, if it is not in this form;
2. completely factor the quadratic expression; and
3. set each factor equal to 0 and solve the resulting linear equations.

Example 6–1 A

Find the solution set S of the following equations.

1. $x^2 + 6 = 7x$
Since the equation is not in standard form, we subtract $7x$ from each member to obtain the standard form
$$x^2 - 7x + 6 = 0.$$

Factoring $x^2 - 7x + 6$, we obtain the equation

$(x - 6)(x - 1) = 0$.

Then $x - 6 = 0$ or $x - 1 = 0$.
Since $x = 6$ when $x - 6 = 0$ and $x = 1$ when $x - 1 = 0$, the solution set is

$S = \{6,1\}$.

To check our solutions, as we should always do, substitute 6 for x and then 1 for x in the original equation.

(1) When $x = 6$,
$$x^2 + 6 = 7x$$
$$(6)^2 + 6 = 7(6)$$
$$36 + 6 = 42$$
$$42 = 42 \text{ (true)}.$$

(2) When $x = 1$,
$$x^2 + 6 = 7x$$
$$(1)^2 + 6 = 7(1)$$
$$1 + 6 = 7$$
$$7 = 7 \text{ (true)}.$$

We will not show a check in the remaining examples, but we should *always* check our solutions.

2. $6p^2 - 5p = 0$

The equation is already in standard form, so we factor the common factor p.

$p(6p - 5) = 0$
$p = 0$ or $6p - 5 = 0$ Set each factor equal to 0.
$$6p = 5$$
$$p = \frac{5}{6}$$
$$S = \left\{0, \frac{5}{6}\right\}$$

Note
A common error is to forget the factor p. That is, the solution $p = 0$ is sometimes omitted by students.

3. $3q^2 + 4q + \dfrac{4}{3} = 0$

Since we have a rational equation (an equation containing at least one rational term), we clear the denominator by multiplying all terms in both members of the equation by 3.

$9q^2 + 12q + 4 = 0$
$\quad (3q + 2)^2 = 0$ Factor $9q^2 + 12q + 4$.
$\quad\quad 3q + 2 = 0$ Set each factor equal to 0.
$\quad\quad\quad 3q = -2$
$$q = -\frac{2}{3}$$
$$S = \left\{-\frac{2}{3}\right\}$$

The linear factor $3q + 2$ appears twice. Hence $-\dfrac{2}{3}$ is said to be a solution of *multiplicity two.*

Consider now the quadratic equation $x^2 - 16 = 0$. Factoring, we obtain

$$(x - 4)(x + 4) = 0.$$

Since $x = 4$ when $x - 4 = 0$ and $x = -4$ when $x + 4 = 0$, the solution set is

$$S = \{4, -4\}.$$

We can also solve this equation by writing it in the form $x^2 = 16$. Since 16 is greater than or equal to zero, x is a number that, when squared, yields 16. This can be accomplished when

$$x = \sqrt{16} \quad \text{or} \quad x = -\sqrt{16},$$

since $(\sqrt{16})^2 = 16$ and $(-\sqrt{16})^2 = 16$. Thus the solutions of the quadratic equation $x^2 = 16$ are

$$x = \sqrt{16} = 4 \quad \text{or} \quad x = -\sqrt{16} = -4,$$

which are the same values that we determined by factoring.

This example illustrates the square root property.

> ■ **Square root property**
> Given real number p and $x^2 = p$, then
> $$x = \sqrt{p} \quad \text{or} \quad x = -\sqrt{p}.$$

We use this property to solve certain types of quadratic equations by **extracting the roots.**

> **Note**
> We can write $x = \sqrt{p}$ or $x = -\sqrt{p}$ as $x = \pm \sqrt{p}$, which we read "x equals plus or minus square root of p."

Example 6–1 B

Find the solution set by extracting the roots.

1. $x^2 = 13$

Extract the roots.

$$x = \sqrt{13} \quad \text{or} \quad x = -\sqrt{13}$$

So the solution set is

$$S = \{\sqrt{13}, -\sqrt{13}\}.$$

2. $(3x + 2)^2 = 7$

Extract the roots.

$$3x + 2 = \sqrt{7} \quad \text{or} \quad 3x + 2 = -\sqrt{7}$$

Solve each equation for x.

$$3x = -2 + \sqrt{7} \quad \text{or} \quad 3x = -2 - \sqrt{7}$$

$$x = \frac{-2 + \sqrt{7}}{3} \quad \text{or} \quad x = \frac{-2 - \sqrt{7}}{3}$$

$$S = \left\{ \frac{-2 + \sqrt{7}}{3}, \frac{-2 - \sqrt{7}}{3} \right\}$$

3. $(x - 3)^2 = -4$

Extract the roots.

$$x - 3 = \sqrt{-4} \quad \text{or} \quad x - 3 = -\sqrt{-4}$$

However $\sqrt{-4}$ is not a real number and the equation has *no real number solution*. We have learned that $\sqrt{-4}$ is equal to $2i$, where $i = \sqrt{-1}$. Then

$$x - 3 = 2i \quad \text{or} \quad x - 3 = -2i$$
$$x = 3 + 2i \quad \text{or} \quad x = 3 - 2i$$
$$S = \{3 + 2i, 3 - 2i\}.$$

4. The power output P of an 80-volt electric generator is defined by $P = 80I - 5I^2$, where I is the current in amperes. What current I is necessary for an output power of 140 watts?

Given $P = 140$, we substitute to obtain the equation

$$140 = 80I - 5I^2.$$

Add $5I^2 - 80I$ to both members to write the equation in standard form.

$$5I^2 - 80I + 140 = 0$$
$$5(I^2 - 16I + 28) = 0 \qquad \text{Factor the left member.}$$
$$5(I - 14)(I - 2) = 0$$
$$I - 14 = 0 \quad \text{or} \quad I - 2 = 0 \quad (\textit{Note:} \ 5 \neq 0)$$
$$I = 14 \text{ or} \qquad I = 2$$

The generator will produce 140 watts when $I = 14$ amperes or when $I = 2$ amperes.

Many times when we translate a verbal problem into mathematical language, we obtain a quadratic equation. Consider the following examples.

Example 6–1 C

Set up an equation and solve the following problems.

1. Find the length of each side of a square if the diagonal is 12 centimeters long.

Let $s =$ the length of the side of the square. Using the Pythagorean Theorem for right triangles, we obtain the equation

$$s^2 + s^2 = (12)^2$$
$$2s^2 = 144$$
$$s^2 = 72.$$

Extract the roots.

$$s = \sqrt{72} = 6\sqrt{2} \quad \text{or} \quad s = -\sqrt{72} = -6\sqrt{2}$$

We reject the negative solution since we are finding lengths of sides.

Thus $s = 6\sqrt{2}$ centimeters.

2. The lengths of the three sides of a right triangle are three consecutive integers. Find the lengths of the three sides.

Let $x =$ the length of the shortest side, then
$x + 1 =$ the length of the next longest side, and
$x + 2 =$ the length of the longest side.
Using the Pythagorean Theorem,

$$x^2 + (x + 1)^2 = (x + 2)^2$$
$$x^2 + x^2 + 2x + 1 = x^2 + 4x + 4 \qquad \text{Square each term.}$$
$$2x^2 + 2x + 1 = x^2 + 4x + 4 \qquad \text{Combine like terms.}$$
$$x^2 - 2x - 3 = 0 \qquad \text{Write in standard form.}$$
$$(x - 3)(x + 1) = 0 \qquad \text{Factor the left member.}$$
$$x - 3 = 0 \text{ or } x + 1 = 0 \qquad \text{Set each factor equal to 0.}$$
$$x = 3 \text{ or } \qquad x = -1.$$

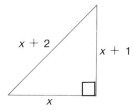

Reject -1 as an answer since we are finding lengths of sides. Then

$x = 3$, $x + 1 = 4$, and $x + 2 = 5$

and the lengths of the sides of the triangle are 3, 4, and 5 units.

3. The sum of the squares of two consecutive even integers is 52. Find the integers.

Let $x =$ the first integer, then
$x + 2 -$ the next consecutive even integer.

$$x^2 + (x + 2)^2 = 52$$
$$x^2 + x^2 + 4x + 4 = 52$$
$$2x^2 + 4x + 4 = 52$$
$$2x^2 + 4x - 48 = 0 \qquad \text{Write in standard form.}$$
$$2(x^2 + 2x - 24) = 0 \qquad \text{Factor the left member.}$$
$$2(x + 6)(x - 4) = 0$$
$$x + 6 = 0 \quad \text{ or } x - 4 = 0 \qquad (\textit{Note: } 2 \neq 0)$$
$$x = -6 \text{ or } \qquad x = 4$$

When $x = -6$, then $x + 2 = -4$ and when $x = 4$, then $x + 2 = 6$. The sets of even integers are -6 and -4 or 4 and 6.

Exercise 6–1

Directions Find the solution set S of the following equations.

Example

$(x + 2)(3x - 1) = 0$

Solution

By the square root property, $(x + 2)(3x - 1) = 0$ if and only if $x + 2 = 0$ or $3x - 1 = 0$. Solving these equations, $x = -2$ when $x + 2 = 0$ and $x = \dfrac{1}{3}$ when $3x - 1 = 0$. Thus $S = \left\{-2, \dfrac{1}{3}\right\}$.

1. $(x - 3)(x + 4) = 0$ **2.** $(x - 7)(x + 1) = 0$ **3.** $(3y - 1)(2y + 5) = 0$

4. $(2z - 3)(3z - 4) = 0$ **5.** $x(3x + 4)(x + 10) = 0$ **6.** $4p(2p - 1)(5p + 6) = 0$

Directions Find the solution set S of the following equations by factoring. See example 6–1 A.

7. $x^2 - 5x + 6 = 0$ **8.** $x^2 - 7x - 8 = 0$ **9.** $y^2 - 10y + 25 = 0$ **10.** $y^2 - 18y + 81 = 0$

11. $p^2 - 5p = 24$ **12.** $z^2 - 12 = 4z$ **13.** $m^2 - m = 0$ **14.** $q^2 + 3q = 0$

15. $-3y^2 + 27 = 0$ **16.** $-9z^2 + 144 = 0$ **17.** $2x^2 - 3x - 2 = 0$ **18.** $2n^2 + 11n - 6 = 0$

19. $4y^2 + 5y = 6$ **20.** $3x^2 - 8 = 10x$ **21.** $x - 1 - \dfrac{x^2}{4} = 0$ **22.** $\dfrac{x^2}{6} - \dfrac{x}{3} - \dfrac{1}{2} = 0$

23. $\dfrac{x}{2} + \dfrac{7}{2} = \dfrac{4}{x}$ **24.** $x + \dfrac{1}{3} = \dfrac{4}{3x}$

25. $(y + 6)(y - 2) = -7$ **26.** $(p + 4)(p - 6) = -16$

27. $(3m + 2)(m - 1) = 4m$ **28.** $(3x + 2)(x - 1) = -(7x - 7)$

29. $3x(x - 3) = (x - 5)(x - 3)$ **30.** $(y - 1)(y + 4) = -2y(y + 4)$

31. $(x - 1)^2 = (2x + 5)(x - 1)$

Directions Solve the following equations for x.

Example

$x^2 - 3ax - 4a^2 = 0$

Solution

Factoring the left member, we have $(x - 4a)(x + a) = 0$. This is true when $x - 4a = 0$ or $x + a = 0$.
Solving each equation for x, we obtain
$x = 4a$ or $x = -a$.

32. $x^2 + 4ax + 3a^2 = 0$

33. $x^2 - 10bx - 24b^2 = 0$

34. $3x^2 - 13xy + 4y^2 = 0$

35. $4x^2 - ax - 14a^2 = 0$

36. $12x^2 = 8ax + 15a^2$

37. $12xy - 6y^2 = 6x^2$

38. $5x^2 - 6y^2 = 7xy$

39. $6x^2 + 5xy = 4y^2$

Directions Find the solution set of the following equations by extracting the roots. Express radicals in simplest form. See example 6–1 B.

40. $x^2 = 81$

41. $x^2 = 121$

42. $3y^2 = 27$

43. $5z^2 = 245$

44. $p^2 = 20$

45. $q^2 = 32$

46. $m^2 - 40 = 0$

47. $y^2 - 72 = 0$

48. $16p^2 - 400 = 0$

49. $7y^2 - 56 = 0$

50. $9x^2 - 162 = 0$

51. $2n^2 - 100 = 0$

52. $(x + 3)^2 = 16$

53. $(y + 7)^2 = 36$

54. $(x - 9)^2 = -144$

55. $(x - 12)^2 = -121$

56. $(x - 6)^2 = 12$

57. $(y + 10)^2 = 48$

58. $(3x - 2)^2 = 25$

59. $(4x + 1)^2 = 81$

60. $(9y + 1)^2 = -24$

61. $(10p - 3)^2 = -84$

62. $(x - 7)^2 = a^2, a > 0$ **63.** $(y + 8)^2 = b^2, b > 0$

Directions Solve the following verbal problems. See examples 6–1 B–4 and C.

64. Given $P = 100I - 5I^2$, find I in amperes, $I > 0$, when (a) $P = 420$, (b) $P = 0$.

65. A ball rolls down a slope and travels a distance d defined by the equation $d = 6t + \dfrac{t^2}{2}$ feet in t seconds. How long does it take the ball to roll

(a) $d = 32$ feet, (b) $d = 14$ feet?

66. The height h that an object will reach in t seconds if it is propelled vertically upward with an initial velocity V_0 feet per second is defined by the equation
$h = -16t^2 + V_0 t$.
When will the object hit the ground?

67. An object with an initial velocity V_0 accelerates at rate a in time t. The displacement of the object for this time is given by the equation

$s = V_0 t + \dfrac{1}{2} at^2$.

If $V_0 = 3$ and $a = 6$, when will the displacement s be 6 feet?

68. The sum S of the first n even positive integers is given by $S = n(n + 1)$. Find n when $S = 30$.

69. The sum S of the first n of the numbers 5, 8, 11,

$\cdots$, $3n + 2$ is given by $S = \dfrac{1}{2}n(3n + 7)$.

Find n when $S = 98$.

70. If the diagonal of a square is 32 feet, find the length of each side of the square.

71. If the diagonal of a square is $7\sqrt{2}$ meters long, find the length of each side of the square.

72. The three sides of a right triangle are three consecutive even integers. Find the lengths of the three sides.

73. One leg of a right triangle is 7 inches longer than the other leg. If the hypotenuse is 9 inches longer than the shortest leg, find the lengths of the three sides of the triangle.

74. The longest side of a right triangle is 1 yard longer than twice the length of the shortest side. If the third side measures 15 yards, find the lengths of the other two sides.

75. The square of the sum of two consecutive even integers is 100. Find the integers.

76. The square of the sum of two consecutive odd integers is 144. Find the integers.

77. The sum of the squares of two consecutive odd integers is 130. Find the integers.

78. One integer is 1 more than twice the other integer. Their product is 105. Find the integers.

6–2 Solutions of quadratic equations by completing the square

Completing the square

Finding the solution set of quadratic equations by factoring and by extracting the roots required special cases of the quadratic equation. Now we develop a method that can be applied to any quadratic equation. The method, called **completing the square,** involves transforming the standard quadratic equation

$$ax^2 + bx + c = 0, a \neq 0,$$

into the form

$$(x + k)^2 = d,$$

where k and d are constants. This latter equation can then be solved by extracting the roots as we did in section 6–1.

Consider the identities

$$(x + 8)^2 = x^2 + 16x + 64$$
$$(x - 7)^2 = x^2 - 14x + 49.$$

Observe first that the coefficient of x^2 in each case is 1. This is necessary for what we do next. We then consider the relationship between the second term (the linear term) and the third term (the constant term) of the trinomial. Notice that the constant term in each case is the *square of one-half of the coefficient of the linear term, x.*

1. In $x^2 + 16x + 64$, the constant term, 64, is the square of one-half of the coefficient of the middle (linear) term, 16.

$$\left[\frac{1}{2}(16)\right]^2 = 8^2 = 64$$

2. In $x^2 - 14x + 49$, the constant term, 49, is the square of one-half of the coefficient of the middle (linear) term, -14.

$$\left[\frac{1}{2}(-14)\right]^2 = (-7)^2 = 49$$

Further, we see that the constant term in the binomial is the number that is one-half of the coefficient of the middle (linear) term of the perfect-square trinomial.

1. Given $x^2 + 16x + 64 = (x + 8)^2$,

$$\left[\frac{1}{2}(16)\right] = 8.$$

2. Given $x^2 - 14x + 49 = [x + (-7)]^2 = (x - 7)^2$,

$$\left[\frac{1}{2}(-14)\right] = -7.$$

We now use these observations to "build" perfect-square trinomials by *completing the square* and thereby obtain their equivalent binomial squares.

Example 6–2 A	Complete the square in each of the following expressions and state the equivalent perfect square.

1. $x^2 + 10x$

The coefficient of the linear term, $10x$, is 10. Now we square one-half of 10 to obtain

$$\left[\frac{1}{2}(10)\right]^2 = 5^2 = 25.$$

Adding 25 to the given expression, we have

$$x^2 + 10x + 25 = (x + 5)^2.$$

2. $x^2 - 7x$

The coefficient of the linear term is -7. The square of one-half of -7 is

$$\left[\frac{1}{2}(-7)\right]^2 = \left(-\frac{7}{2}\right)^2 = \frac{49}{4}.$$

Then

$$x^2 - 7x + \frac{49}{4} = \left(x - \frac{7}{2}\right)^2.$$

Solving by completing the square

We now use this procedure to obtain the solution set of a quadratic equation by completing the square.

Example 6–2 B	Find the solution set S by completing the square.

1. $x^2 - 6x + 5 - 0$

Isolate $x^2 - 6x$ by subtracting 5 from each member.

$$x^2 - 6x = -5$$

Complete the square in the left member.

$$\left[\frac{1}{2}(-6)\right]^2 = (-3)^2 = 9$$

$x^2 - 6x + 9 = -5 + 9$ Add 9 to each member.

$(x - 3)^2 = 4$ $x^2 - 6x + 9 = (x - 3)^2$ and $-5 + 9 = 4$.

Extract the roots.

$x - 3 = 2$ or $x - 3 = -2$

$x = 5$ or $x = 1$

Thus $S = \{5,1\}$.

Note
A very common error when solving by this method is *to **fail** to add the same number to each member* when completing the square. Failure to do so changes the equation.

2. $\frac{1}{2}x^2 - x = -\frac{5}{4}$

Clear the denominators by multiplying each term by the LCM of the denominators, 4.

$$4 \cdot \frac{1}{2}x^2 - 4 \cdot x = 4 \cdot -\frac{5}{4}$$

$$2x^2 - 4x = -5$$

$$x^2 - 2x = -\frac{5}{2} \qquad \text{Divide each term by 2.}$$

Complete the square by adding $\left[\frac{1}{2}(-2)\right]^2 = (-1)^2 = 1$ to each member.

$$x^2 - 2x + 1 = -\frac{5}{2} + 1$$

$$(x - 1)^2 = -\frac{3}{2}$$

$$x - 1 = \sqrt{-\frac{3}{2}} = i\sqrt{\frac{3}{2}} \quad \text{or} \quad x - 1 = -\sqrt{-\frac{3}{2}} = -i\sqrt{\frac{3}{2}}$$

But $\sqrt{\frac{3}{2}} = \frac{\sqrt{3}}{\sqrt{2}} = \frac{\sqrt{3}}{\sqrt{2}} \cdot \frac{\sqrt{2}}{\sqrt{2}} = \frac{\sqrt{6}}{2}$.

Thus $x - 1 = i\frac{\sqrt{6}}{2} \quad \text{or} \quad x - 1 = -i\frac{\sqrt{6}}{2}$

$$x = 1 + i\frac{\sqrt{6}}{2} = \frac{2 + i\sqrt{6}}{2} \quad \text{or} \quad x = 1 - i\frac{\sqrt{6}}{2} = \frac{2 - i\sqrt{6}}{2}$$

$$S = \left\{\frac{2 + i\sqrt{6}}{2}, \frac{2 - i\sqrt{6}}{2}\right\}.$$

Note
The solutions are complex numbers.

In summary, to find the solution set of the quadratic equation $ax^2 + bx + c = 0$, $a \neq 0$, by completing the square, we follow this procedure.

1. If $a = 1$, proceed to number 2. If $a \neq 1$, divide each term of the equation by a.
2. Write the equation with the variable terms in the left member and the constant in the right member.
3. Add to each member of the equation the square of one-half of the coefficient of the linear term.
4. Write the left member as a perfect square and combine in the right member.
5. Extract the root and solve the resulting linear equations.

Exercise 6–2

Directions Complete the square of each of the following binomials and state their equivalent binomial square. See example 6–2 A.

1. $x^2 + 4x$ **2.** $x^2 + 8x$ **3.** $y^2 - 18y$ **4.** $z^2 - 24z$ **5.** $p^2 + 2p$ **6.** $m^2 - 30m$

7. $x^2 + 3x$ **8.** $y^2 + y$ **9.** $w^2 - 11w$ **10.** $q^2 - 5q$ **11.** $x^2 + 13x$ **12.** $y^2 - 15y$

Directions Find the solution set by completing the square. See example 6–2 B.

13. $x^2 + 12x + 11 = 0$ **14.** $x^2 + 5x + 4 = 0$ **15.** $y^2 - 11y + 10 = 0$ **16.** $p^2 - 4p = 8$

17. $n^2 + 8n = -25$ **18.** $x^2 + 6x = -10$ **19.** $x^2 - 8x = 0$ **20.** $x^2 + 4x = 0$

21. $z^2 + z = 0$ **22.** $m^2 - 5m = 0$ **23.** $y^2 = 3 - y$ **24.** $x^2 + 2 = -3x$

25. $2 - p = 6p^2$ **26.** $2n = 4 - n^2$ **27.** $-2x^2 + 4 = -7x$ **28.** $1 - z^2 = 3z$

29. $2x^2 + 3x - 2 = 0$ **30.** $2y^2 - 5y - 3 = 0$ **31.** $3x^2 - 10x + 3 = 0$ **32.** $2z^2 + 7z + 3 = 0$

33. $4u^2 - 4u = 3$ **34.** $4x^2 + 9x + 4 = 0$ **35.** $5m^2 - 3m + 1 = 0$ **36.** $5q^2 + 4q + 1 = 0$

37. $(x + 2)(x - 3) = 1$ **38.** $(2y - 1)(y + 5) = -3$ **39.** $(3x + 1)^2 = (x - 2)^2$

40. $(2y - 3)^2 = (y + 4)^2$ **41.** $5p(2p - 3) = p + 1$ **42.** $8p(p + 3) = 2p - 5$

43. $\frac{1}{2}x^2 - \frac{3}{4}x - 1$ **44.** $y^2 - \frac{1}{3}y = \frac{2}{3}$ **45.** $2x^2 - \frac{2}{3}x + \frac{1}{2} = 0$

46. $3z^2 + \frac{1}{2}z - \frac{3}{4} = 0$ **47.** $\frac{3}{4}x = x^2 - 2$ **48.** $\frac{1}{5}n^2 = 1 - 2n$

49. $\frac{5}{x} - 2x + 3 = 0$ **50.** $\frac{3}{x} - 2 = 3x$ **51.** $5 - \frac{2}{t} = \frac{3}{t^2}$

52. $3 + \frac{5}{p} = \frac{1}{p^2}$

Directions Solve the following verbal problems by completing the square.

53. When an object is thrown downward with an initial velocity v_0 of 9 feet per second, the relationship between the distance s it travels in time t is given by
$s = 9t + 16t^2$.
How long does it take for the object to fall 100 feet?

54. When an object is dropped from rest, the distance s that the object falls is given by the equation
$s = 16t^2$.
How long does it take the object to hit the bottom of a gorge when it is dropped a distance of 205 feet from a bridge across the gorge?

55. The supply equation for a specific product is given by

$$S = 32p + p^2,$$

where p cents is the price per unit of the product. What should the price to the nearest cent be when the supply S is 500 units?

56. The demand equation for a specific product is given by

$$D = 36p + p^2,$$

where p is the price per 1,000 units. What should the price to the nearest cent per 1,000 units be when the demand D is 100,000 units?

57. The radius r of a circular arch having height h and span b is given by

$$r = \frac{(b^2 + 4h^2)}{8h}.$$

Find h when $b = 10$ and $r = 13$.

58. The square of a number added to three times the number is 6. Find the number.

59. The square of twice a number less the number is 4. Find the number.

60. The square of the difference between four times a number and 5 is 24. Find the number.

61. The formula for the volume of a cylinder with height h and radius r is

$$V = \pi r^2 h.$$

Find r when V is 235 cubic meters and $h = 10$ meters.

6–3 Solutions of quadratic equations by the quadratic formula

The quadratic formula

In previous sections we have found the solution set of quadratic equations by factoring, extracting the roots, and completing the square. Even though the solution set of any quadratic equation can be found by completing the square, this can be a time-consuming chore. In this section we will use the method of completing the square to develop a general formula that will find the solution set. We call this formula the **quadratic formula.**

Given the quadratic equation in standard form,

$$ax^2 + bx + c = 0 \qquad \text{(assume } a > 0\text{)},$$

where a, b, and c are real numbers, we first divide each term by a to change the coefficient of x^2 to 1.

$$x^2 + \frac{b}{a}x + \frac{c}{a} = 0$$

Subtract $\frac{c}{a}$ from both members of the equation to get the terms containing the variable in one member. We have

$$x^2 + \frac{b}{a}x = -\frac{c}{a}.$$

To complete the square in the left member, take one-half of $\frac{b}{a}$ and square the result.

$$\left[\frac{1}{2}\left(\frac{b}{a}\right)\right]^2 = \left(\frac{b}{2a}\right)^2 = \frac{b^2}{4a^2}$$

Add this quantity to both members of the equation.

$$x^2 + \frac{b}{a}x + \frac{b^2}{4a^2} = \frac{b^2}{4a^2} + \left(-\frac{c}{a}\right)$$

$$\left(x + \frac{b}{2a}\right)^2 = \frac{b^2 - 4ac}{4a^2} \qquad \text{Add in the right member.}$$

Extract the roots.

$$x + \frac{b}{2a} = \sqrt{\frac{b^2 - 4ac}{4a^2}} \quad \text{or} \quad x + \frac{b}{2a} = -\sqrt{\frac{b^2 - 4ac}{4a^2}}$$

$$x + \frac{b}{2a} = \frac{\sqrt{b^2 - 4ac}}{\sqrt{4a^2}} \quad \text{or} \quad x + \frac{b}{2a} = -\frac{\sqrt{b^2 - 4ac}}{\sqrt{4a^2}}$$

$$x + \frac{b}{2a} = \frac{\sqrt{b^2 - 4ac}}{2a} \quad \text{or} \quad x + \frac{b}{2a} = -\frac{\sqrt{b^2 - 4ac}}{2a}$$

$$x = -\frac{b}{2a} + \frac{\sqrt{b^2 - 4ac}}{2a} = \frac{-b + \sqrt{b^2 - 4ac}}{2a}$$

$$\text{or } x = -\frac{b}{2a} - \frac{\sqrt{b^2 - 4ac}}{2a} = \frac{-b - \sqrt{b^2 - 4ac}}{2a}$$

$$S = \left\{ \frac{-b + \sqrt{b^2 - 4ac}}{2a}, \frac{-b - \sqrt{b^2 - 4ac}}{2a} \right\}$$

This is often abbreviated, using the $\pm$ symbol we previously introduced, to

$$S = \left\{ \frac{-b \pm \sqrt{b^2 - 4ac}}{2a} \right\}.$$

This is called the **quadratic formula.**

To use the quadratic formula, the equation must be written in the standard form

$$ax^2 + bx + c = 0,$$

and we note that a is the coefficient of the squared term, b is the coefficient of the linear term, and c is the constant.

Solving by quadratic formula

To find the solution set of any quadratic equation by using the quadratic formula, we only need to substitute the numerical values for a, b, and c into the formula and to simplify the result. Consider the equation

$$x^2 + 4x - 12 = 0.$$

Since $a = 1$, $b = 4$, and $c = -12$, we substitute in the quadratic formula to obtain

$$x = \frac{-4 \pm \sqrt{(4)^2 - 4(1)(-12)}}{2(1)}$$

$$= \frac{-4 \pm \sqrt{16 + 48}}{2} = \frac{-4 \pm \sqrt{64}}{2} = \frac{-4 \pm 8}{2}.$$

Then

$$x = \frac{-4 + 8}{2} = \frac{4}{2} = 2 \quad \text{or} \quad x = \frac{-4 - 8}{2} = \frac{-12}{2} = -6.$$

$$S = \{2, -6\}.$$

We will now summarize the procedure for solving a quadratic equation by using the quadratic formula.

1. Write the equation in standard form (if necessary).
2. Identify the numerical values of a, b, and c.
3. Substitute these values into the formula.
4. Simplify the resulting expression.

Example 6–3 A

Find the solution set using the quadratic formula.

1. $x^2 = 5 - 3x$

Write the equation in standard form by adding $3x - 5$ to both members.

$x^2 + 3x - 5 = 0$

Then $a = 1$, $b = 3$, $c = -5$. Substitute in the quadratic formula.

$$x = \frac{-3 \pm \sqrt{3^2 - 4(1)(-5)}}{2(1)}$$

$$= \frac{-3 \pm \sqrt{9 + 20}}{2} = \frac{-3 \pm \sqrt{29}}{2}$$

Then $x = \dfrac{-3 + \sqrt{29}}{2}$ or $x = \dfrac{-3 - \sqrt{29}}{2}$.

$$S = \left\{ \frac{-3 + \sqrt{29}}{2}, \frac{-3 - \sqrt{29}}{2} \right\}$$

2. $2y^2 - 4y + 1 = 0$

Now $a = 2$, $b = -4$, and $c = 1$, so

$$y = \frac{-(-4) \pm \sqrt{(-4)^2 - 4(2)(1)}}{2(2)}$$

$$= \frac{4 \pm \sqrt{16 - 8}}{4} = \frac{4 \pm \sqrt{8}}{4}.$$

Since $\sqrt{8} = \sqrt{4 \cdot 2} = 2\sqrt{2}$, then

$$y = \frac{4 \pm 2\sqrt{2}}{4} = \frac{2(2 \pm \sqrt{2})}{4} = \frac{2 \pm \sqrt{2}}{2}.$$

$$y = \frac{2 + \sqrt{2}}{2} \quad \text{or} \quad y = \frac{2 - \sqrt{2}}{2}.$$

$$S = \left\{ \frac{2 + \sqrt{2}}{2}, \frac{2 - \sqrt{2}}{2} \right\}$$

3. $4 - \dfrac{3}{x} + \dfrac{4}{x^2} = 0$

Multiply each term by the LCM of x and x^2, that is, x^2, to clear the denominators.

$$x^2 \cdot 4 - x^2 \cdot \dfrac{3}{x} + x^2 \cdot \dfrac{4}{x^2} = x^2 \cdot 0$$

$$4x^2 - 3x + 4 = 0$$

Since $a = 4$, $b = -3$, and $c = 4$, then

$$x = \dfrac{-(-3) \pm \sqrt{(-3)^2 - 4(4)(4)}}{2(4)} = \dfrac{3 \pm \sqrt{9 - 64}}{8}$$

$$= \dfrac{3 \pm \sqrt{-55}}{8} = \dfrac{3 \pm i\sqrt{55}}{8}.$$

Thus $S = \left\{ \dfrac{3 + i\sqrt{55}}{8}, \dfrac{3 - i\sqrt{55}}{8} \right\}.$

Nature of the solutions

The general quadratic equation $ax^2 + bx + c = 0$ has the solutions x_1 and x_2 such that, using the quadratic formula,

$$x_1 = \dfrac{-b + \sqrt{b^2 - 4ac}}{2a} \quad \text{and} \quad x_2 = \dfrac{-b - \sqrt{b^2 - 4ac}}{2a}.$$

We can determine the nature of the solutions x_1 and x_2 (that is, rational, irrational, or complex) by using the expression under the radical sign, $b^2 - 4ac$, where a, b, and c are **rational** numbers. For this reason, we call $b^2 - 4ac$ the **discriminant.** The type of solutions can be determined as follows:
When $b^2 - 4ac$ is

1. 0, then there is only *one* unique *rational* solution and that solution is $\dfrac{-b}{2a}$;

2. *positive* and a *perfect square,* then there are *two* distinct *rational* solutions;

3. *positive* but *not a perfect square,* then $\sqrt{b^2 - 4ac}$ is irrational and there are *two* distinct *irrational* solutions;

4. *negative,* then there are *two* distinct *complex* (nonreal) solutions.

Example 6–3 B

Use the discriminant to decide the number and the nature of the solutions of the given quadratic equation.

1. $4x^2 + 12x + 9 = 0$

Since $a = 4$, $b = 12$, and $c = 9$, then

$$b^2 - 4ac = (12)^2 - 4(4)(9) = 144 - 144 = 0.$$

There is only *one rational* solution and that solution is

$$\dfrac{-b}{2a} = \dfrac{-12}{2(4)} = \dfrac{-3}{2} = -\dfrac{3}{2}.$$

2. $3x^2 - 4 = 4x$

Write the equation in standard form, $3x^2 - 4x - 4 = 0$. Thus $a = 3$, $b = -4$, and $c = -4$. Then

$$b^2 - 4ac = (-4)^2 - 4(3)(-4) = 16 + 48 = 64.$$

Since $64 = 8^2$, the discriminant is positive and a perfect square, there are *two* distinct *rational* solutions.

3. $x^2 = 4x + 6$

We must first write the equation in standard form, $x^2 - 4x - 6 = 0$. Then $a = 1$, $b = -4$, and $c = -6$, and

$$b^2 - 4ac = (-4)^2 - 4(1)(-6) = 16 + 24 = 40.$$

Since 40 is not a perfect square, but is positive, there are *two* distinct *irrational* solutions.

4. $2y^2 - 3y + 5 = 0$

Since $a = 2$, $b = -3$, and $c = 5$, then

$$b^2 - 4ac = (-3)^2 - 4(2)(5) = 9 - 40 = -31.$$

The discriminant is negative and the equation has *two* distinct *complex* (nonreal) solutions.

Mastery points
Can you
- Identify the numerical values of a, b, and c in a quadratic equation?
- Solve a quadratic equation by using the quadratic formula?
- Use the discriminant to determine the nature of solutions?

Exercise 6–3

Directions Find the solution set of each quadratic equation using the quadratic formula. See example 6–3 A.

1. $y^2 + 6y - 16 = 0$ **2.** $z^2 + 6z + 9 = 0$ **3.** $p^2 - 14p + 49 = 0$ **4.** $x^2 - 28 = 0$

5. $3y^2 = 20$ **6.** $4z^2 - 3z = 0$ **7.** $2x = 5x^2$ **8.** $m^2 - 2m = 4$

9. $p^2 = -5p - 7$ **10.** $y^2 + 6 = 2y$ **11.** $-3x - 5 = x^2$ **12.** $18 = 10x - x^2$

13. $2x^2 - 7x + 6 = 0$ **14.** $3y^2 - 5y - 6 = 0$ **15.** $4z^2 - 8z + 1 = 0$ **16.** $9p^2 - 18p + 7 = 0$

17. $9x^2 - 12x + 4 = 0$ **18.** $4y^2 + 20y + 25 = 0$ **19.** $3q^2 - 2q + 7 = 0$ **20.** $3z^2 - 4z = -3$

21. $9y^2 - 12y = -5$ **22.** $2t^2 = 6t - 5$ **23.** $11 - 18m = 9m^2$ **24.** $2v = 5v^2 - 2$

25. $x = 2x^2 + 7$ **26.** $3x - \dfrac{2}{x} + 5 = 0$ **27.** $2y^2 - \dfrac{7}{2} = \dfrac{y}{2}$ **28.** $\dfrac{2}{3}x^2 - \dfrac{4}{3} = x$

29. $\frac{2}{3}y - \frac{1}{3} = \frac{4}{9}y^2$

30. $\frac{2p}{3} - \frac{p^2}{4} = 1$

31. $\frac{3}{4}q^2 = \frac{1}{2}q + 4$

32. $\frac{1}{x + 2} + \frac{1}{x - 3} - 2 = 0$

33. $\frac{3}{y - 5} - \frac{2}{y + 1} + 3 = 0$

34. $\frac{1}{2} - \frac{3}{2x + 3} = \frac{3}{x - 4}$

35. $(z - 3)(z + 2) = 2z - 3$

36. $(x + 6)(x - 5) = 10 - x$

37. $(2x + 1)^2 = (x - 3)^2$

Directions Solve the following equations for x in terms of the other variables or constants. Assume that all other variables are positive real numbers.

Example
$2x^2 - 3xy - 5y^2 = 0$

Solution
Here $a = 2$, $b = -3y$ (coefficient of x), and $c = -5y^2$ (term not containing x). Then

$$x = \frac{-(-3y) \pm \sqrt{(-3y)^2 - 4(2)(-5y^2)}}{2(2)}$$

$$= \frac{3y \pm \sqrt{9y^2 + 40y^2}}{4}$$

$$= \frac{3y \pm \sqrt{49y^2}}{4}$$

$$= \frac{3y \pm 7y}{4}.$$

Thus

$$x = \frac{3y + 7y}{4} = \frac{10y}{4} = \frac{5}{2}y \text{ or } x = \frac{3y - 7y}{4} = \frac{-4y}{4} = -y.$$

Therefore $x = \frac{5}{2}y$ or $x = -y$.

38. $x^2 - xy - 2y^2 = 0$

39. $x^2 - 3xy - 18y^2 = 0$

40. $2x^2 - 3ax + 5a^2 = 0$

41. $4x^2 + 2x - 3y = 0$

42. $2x^2 - 3x + 4a = 0$

43. $x^2 - 4ax + 3a = 0$

Directions Determine the number and the nature of the solutions of each quadratic equation. See example 6–3 B.

44. $x^2 + 4x + 1 = 0$

45. $x^2 - 3x - 5 = 0$

46. $2y^2 - y + 1 = 0$

47. $3x^2 + 3x + 4 = 0$

48. $4y^2 - 4y + 1 = 0$

49. $9x^2 - 30x + 25 = 0$

50. $4y^2 - y = 10$

51. $3m^2 - 5m = 2$

52. $y^2 + 6 = 0$

53. $x^2 = -4$

54. $y^2 = 20$

55. $m^2 - 18 = 0$

56. $5t^2 - 3t = 0$

57. $7p^2 = 8p$

58. $x^2 - \frac{1}{2}x + \frac{3}{5} = 0$

59. $x^2 - 4x + \frac{9}{4} = 0$

60. $\frac{y^2}{2} + 5y = 1$

61. $\frac{m^2}{4} + \frac{3m}{2} = 5$

Directions Solve the following using the quadratic formula.

62. The distance s through which an object will fall in t seconds is given by
$$s = \frac{1}{2}gt^2,$$
where $g = 32$ feet per second per second (32 ft/sec²). Find t to the nearest tenth of a second when (a) $s = 96$ feet, (b) $s = 60$ feet.

63. A metal strip is shaped into a right triangle as shown in the diagram. If $a = x$, $b = x + 3$, and $c = x + 6$, find x.
(*Hint:* Use the Pythagorean Theorem previously discussed.)

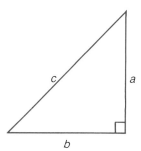

64. In a given electric circuit the relationship between I (in amperes), E (in volts), and R (in ohms) is given by
$$I^2R + IE = 8,000.$$
Find I ($I > 0$) when $R = 4$ and $E = 100$.

65. Using the formula
$$s = v_0t + \frac{1}{2}at^2,$$
find t when (a) $s = 8$, $v_0 = 3$, $a = 4$; (b) $s = 80$, $v_0 = 36$, $a = 32$. ($t > 0$)

66. In chemistry an equation in connection with equilibrium in liquid flow is given by
$$k = \frac{x^2}{(a - x)(b - x)}.$$
Solve for x if $a = 1$, $b = 2$, and $k = 3$.

67. If P dollars are invested at $r\%$ interest compounded annually, after two years the worth A in dollars is given by
$$A = P(1 + 0.01r)^2.$$
If the amount $A = \$1,200$ after two years when $P = \$1,000$ is invested, find the rate of interest r, to the nearest tenth of a percent.

6–4 Applications of quadratic equations

A number of physical situations yield mathematical models that are quadratic equations. Because of this, there may be two answers to the problem. Sometimes, because of the nature of the situation, only one of the answers is legitimate. For example, it would not be feasible to accept -25 as the measurement of a dimension of a room or $\frac{7}{6}$ as the number of books on a shelf. We now consider some such physical situations.

Example 6–4 A

1. When a ball is thrown straight upward into the air, the equation
$$s = -16t^2 + 80t + 44$$
gives the distance s in feet that the ball is above the ground t seconds after it is thrown. How long does it take for the ball to hit the ground?

The ball hits the ground when $s = 0$. Thus we have
$$0 = -16t^2 + 80t + 44.$$

Multiply each member by -1 to obtain

$16t^2 - 80t - 44 = 0$
$4(4t^2 - 20t - 11) = 0.$

Since $4t^2 - 20t - 11 = (2t - 11)(2t + 1)$, we have

$4(2t - 11)(2t + 1) = 0$
$2t - 11 = 0$ or $2t + 1 = 0$

$t = \dfrac{11}{2}$ or $t = -\dfrac{1}{2}$.

We reject $t = -\dfrac{1}{2}$, so $t = \dfrac{11}{2}$ or $5\dfrac{1}{2}$.

The ball hits the ground in $5\dfrac{1}{2}$ seconds.

Right triangle problem

2. Find the length of side b of the given right triangle if $c = 13$ units and side b is 7 units longer than side a.

Use the Pythagorean Theorem, $a^2 + b^2 = c^2$. Since b is 7 units longer than a, $b = a + 7$. Substitute $a + 7$ for b and 13 for c.

$a^2 + (a + 7)^2 = (13)^2$
$a^2 + a^2 + 14a + 49 = 169$
$2a^2 + 14a - 120 = 0$
$2(a^2 + 7a - 60) = 0$
$2(a + 12)(a - 5) = 0$
$a + 12 = 0$ or $a - 5 = 0$
Then $a = -12$ or $a = 5$.

We reject $a = -12$, so

$a = 5$ and $b = a + 7 = 5 + 7 = 12$.

Side b is 12 units long.

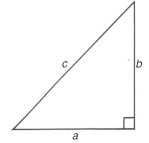

Geometric figure

3. A rectangle has an area, A, of 65 square centimeters. If the length ℓ of the rectangle is 3 centimeters more than twice the width w, find the dimensions of the rectangle ($A = \ell w$).

Let w be the width of the rectangle, then $\ell = 2w + 3$. Replace A by 65 and ℓ by $2w + 3$ in $A = \ell w$.

$65 = (2w + 3)w$
$65 = 2w^2 + 3w$
$2w^2 + 3w - 65 = 0$ Write in standard form.
$(2w + 13)(w - 5) = 0$
Then $2w + 13 = 0$ or $w - 5 = 0$.

Since $w = -\dfrac{13}{2}$ when $2w + 13 = 0$ and $w = 5$ when $w - 5 = 0$,

then $w = -\dfrac{13}{2}$ or $w = 5$.

Reject $w = -\dfrac{13}{2}$ since we cannot have a negative width, then $w = 5$ and $\ell = 2w + 3 = 2(5) + 3 = 13$. The rectangle is 5 centimeters wide and 13 centimeters long.

4. Two pipes when opened can fill a tank in 5 hours. If one pipe takes 2 hours less to fill the tank than the other one does, how long will it take each pipe to fill the tank alone? (Round off to the nearest tenth.)

Let $x =$ time in hours that the larger pipe takes to fill the tank.
Then $x + 2 =$ time in hours that the smaller pipe takes to fill the tank.

Now working together, the pipes fill $\dfrac{1}{5}$ of the tank per hour, the larger pipe

fills $\dfrac{1}{x}$ of the tank per hour, and the smaller pipe fills $\dfrac{1}{x + 2}$ of the tank per

hour.

Since the sum of the individual rates equals the rate working together, then

$$\frac{1}{x} + \frac{1}{x + 2} = \frac{1}{5}$$

Clear the denominators by multiplying each term by the LCM of x, $x + 2$, and 5, that is, $5x(x + 2)$.

$$5x(x + 2) \cdot \frac{1}{x} + 5x(x + 2) \cdot \frac{1}{x + 2} = 5x(x + 2) \cdot \frac{1}{5}$$

$$5(x + 2) + 5x = x(x + 2)$$
$$5x + 10 + 5x = x^2 + 2x$$
$$10x + 10 = x^2 + 2x$$
$$x^2 - 8x - 10 = 0 \qquad \text{Write in standard form.}$$

Since $x^2 - 8x - 10$ is not factorable, we use the quadratic formula.

$$x = \frac{-(-8) \pm \sqrt{(-8)^2 - 4(1)(-10)}}{2(1)}$$

$$= \frac{8 \pm \sqrt{64 + 40}}{2} = \frac{8 \pm \sqrt{104}}{2} = \frac{8 \pm 2\sqrt{26}}{2} = \frac{2(4 \pm \sqrt{26})}{2}$$

$$= 4 \pm \sqrt{26}$$

Since $4 - \sqrt{26}$ is negative, we reject that solution, so

$$x = 4 + \sqrt{26} \approx 4 + 5.1 \approx 9.1$$
$$x + 2 = (4 + \sqrt{26}) + 2 \approx 6 + 5.1 \approx 11.1.$$

It would take the larger pipe approximately 9.1 hours and the smaller pipe approximately 11.1 hours to fill the tank alone.

Mastery points
Can you
- Substitute and solve physical formulas that are quadratic?
- Solve verbal problems involving the use of a right triangle and the relationship $a^2 + b^2 = c^2$?
- Solve verbal problems involving the areas of geometric figures?
- Solve verbal problems involving economic relationships?
- Solve verbal work problems?

Exercise 6–4

Directions Solve the following using quadratic equations. Compute all answers to the nearest tenth where necessary. See example 6–4 A–1.

1. The distance s a body falls when air resistance is neglected is given by
$$s = v_0 t + 16t^2,$$
where s is in feet, t is in seconds, and v_0 is the initial velocity of the body. How long will it take a body to fall 80 feet if it has an initial velocity $v_0 = 64$ feet per second?

2. Using the information in exercise 1, how long will it take the body in a vacuum to fall 80 feet if the body starts from rest? (*Hint:* $v_0 = 0$ feet per second.)

3. Using the formula in exercise 1, how long will it take a body in a vacuum to fall 240 feet if the initial velocity is $v_0 = 32$ feet per second?

4. An object fired vertically into the air with an initial velocity of v_0 feet per second will be at a distance h in feet, t seconds after launching, determined by the equation
$$h = v_0 t - 16t^2.$$
If the initial velocity is 96 feet per second, how long will it take the object to reach a height of 80 feet?

5. Using the formula in exercise 4, how long will it take for the object to return to the ground? (*Hint:* $h = 0$ when the object is on the ground.)

6. Using exercise 4, find the times when the object will be 12 feet off the ground.

7. An object is dropped from the top of the Washington Monument (555 feet tall). How long will it take the object to strike the ground? (*Hint:* Use exercise 1.)

8. The current in an electric circuit flows according to the equation
$$I = 18t - 12t^2,$$
where I is the current in amperes and t is the time in seconds. In how many seconds will there be 6 amperes of current?

9. Using the formula in exercise 8, in how many seconds t will there be no current?

10. In a given circuit the relationship between I (in amperes), E (in volts), and R (in ohms) is given by
$$I^2 R + IE = 6,000.$$
Find I when $E = 60$ volts and $R = 3$ ohms.

11. Using exercise 10, find I when $E = 50$ volts and $R = 5$ ohms.

12. The formula for rating engine horsepower, hp, based on an average effective pressure on the piston of 67 pounds per square inch at a piston speed of 1,000 feet per minute is given by
$$hp = 0.4D^2 \times N,$$
where D is the diameter of the piston bore and N is the number of pistons. Find D when an 8-cylinder engine has 36 horsepower.

13. A polygon is a closed geometric plane figure that has n sides, where $n \geq 3$. A diagonal of a polygon is a line segment connecting any two nonadjacent vertices of the polygon. The number of diagonals D of a polygon of n sides is given by
$$D = \frac{n(n - 3)}{2}.$$
How many sides does the polygon have that has 405 diagonals?

14. Using exercise 13, how many sides does the polygon have that has 189 diagonals?

Directions Use the Pythagorean Theorem. See example 6–4 A–2.

Right triangle problems

15. Find the length of the shortest side of a right triangle whose sides are y inches, $y + 1$ inches, and $y + 8$ inches.

16. The hypotenuse (longest side) of a right triangle is 10 millimeters long. One leg is 2 millimeters longer than the other. What are the lengths of the two legs?

17. The hypotenuse of a right triangle is 1 inch longer than the longer of the two legs. The shorter leg is 7 inches. Find the length of the hypotenuse.

18. A plot of ground has the shape of a right triangle. If the longer of the two legs is 9 dekameters longer than the shorter leg, and the hypotenuse is 8 dekameters longer than the longer leg, find the lengths of the three sides.

Directions See example 6–4 A–3.

Geometric figures

19. Given a rectangular plot of land whose length is 30 feet longer than its width, what are the dimensions of the plot if a diagonal path across the plot is 240 feet long? (*Hint:* The diagonal is the hypotenuse of a right triangle.)

20. If a rectangular-shaped playground has a length 5 feet less than three times the width and the distance from one corner to the opposite corner is 100 feet, find the dimensions of the playground.

21. The diagonal of a square flower bed is $\sqrt{50}$ feet. Find the length of each side.

22. The diagonal of a square piece of metal is 50 centimeters long. Find the length of each side of the piece of metal.

23. A triangular-shaped plate has an altitude that is 5 inches longer than its base. If the area of the plate is 52 square inches, what is the altitude of the triangle? ($A = \frac{1}{2}bh$.)

24. The area of a rectangular floor is 12 square meters. Find the dimensions of the floor if the width is 2 meters less than the length. ($A = \ell w$.)

25. A rectangular field is twice as long as it is wide. Find the dimensions of the field if it contains 5,000 square yards.

26. A rectangle that is 6 centimeters long and 3 centimeters wide has its dimensions increased by the same amount. The area of this new rectangle is three times that of the old rectangle. What are the dimensions of the new rectangle? (*Hint:* Let x be the increase in the length and width.)

27. The base and altitude of a triangle are 3 inches and 5 inches, respectively. If the base is increased by twice as much as the altitude, the area of the new triangle is twice that of the old triangle. What are the lengths of the base and the altitude of the new triangle?

28. Find the length of the radius r of a circular disk whose face area A is 154 square feet. (*Note:* $A = \pi r^2$ and use $\pi = \frac{22}{7}$ as an approximation of π.)

29. Find the length of the diameter D of a circular gear whose face area A is 12.56 square feet. (*Note:* $A = \pi r^2$ and $D = 2r$. Use $\pi = 3.14$ as an approximation of π.)

Business problems

30. The demand equation for a specific commodity is given by
$$D = \frac{2,000}{p},$$
where D is the demand for the specific commodity at price p dollars per unit. If the supply equation is given by
$$S = 300p - 400,$$
where S is the quantity of the commodity that the supplier is willing to supply at p dollars per unit, find the equilibrium price.
(*Note:* Equilibrium price occurs when $D = S$.)

31. Suppose that a manufacturer of ballpoint pens finds the demand equation to be
$$D = 24 - p^2$$
and the supply equation to be
$$S = p^2 + 2p,$$
where p is the price of each pen in dollars. What is the equilibrium price?

32. In exercise 31, at what price, to the nearest cent where necessary, (a) is there no demand, (b) will the quantity that the supplier is willing to sell be zero?

33. In exercise 31, (a) how many pens will be supplied at the equilibrium price, (b) what is the demand at that price?

34. A small manufacturer finds the total cost C for a solar energy device to be
$$C = 50x^2 - 24,000,$$
and the total revenue R at a price $100 per unit to be
$$R = 100x,$$
where x is the number of units manufactured and sold. What is the break-even point (that is, where total cost = total revenue) to the nearest unit?

35. As in exercise 34, a steel producer's annual cost and revenue are given by
$$C = 20 - 0.4x^2 \text{ and } R = 0.8x,$$
where x is the number of units produced and sold. To the nearest unit, what is the break-even point?

36. The profit P in dollars in the manufacture and sale of a product is given by
$$P = \frac{1}{100}n^2 - 20n,$$
where n is the number of units manufactured. How many units of the product must be manufactured to make a profit of $20,000?

37. A baker makes a profit P in cents according to the equation
$$P = -n^2 + 240n,$$
where n is the number of cakes baked and sold. How many cakes should be made to realize a profit of $144?

38. What is the profit if the baker makes (a) 100 cakes, (b) 120 cakes, (c) 200 cakes? Can we draw any conclusion from these results?

39. A rug company determines that its marginal profit MP in thousands of rugs is given by
$$MP = -n^2 + 40n - 80,$$
where n is the number of rugs produced in thousands. The maximum profit is where MP is zero. Find n, to the nearest tenth.

Directions See example 6–4 A–4.

40. Working together, Tom Roggenbeck and Amy Miyazaki can mow a lawn in 1 hour. Working alone, it would take Amy 90 minutes longer than it would take Tom. How long would it take Amy to mow the lawn alone?

41. It takes Debbie 39 minutes longer to do a job than it takes Lisa to do the same job. Working together, they can do the job in 40 minutes. How long would it take each girl to do the job working alone?

42. Tim can paint a room in 6 hours less time than Tom. If they can paint the room in 4 hours working together, how long would it take each boy to paint the room working alone?

43. A water tank has an inlet pipe (to fill the tank) and an outlet pipe (to empty the tank). The tank will fill in 8 hours when both pipes are open. If it takes 2 hours longer to empty the tank than it does to fill the tank, how long does it take the outlet pipe to empty the tank?
[*Hint:* (rate of inlet) − (rate of outlet) = (rate to fill).]

6–5 Equations involving radicals

Equations in which at least one member contains a radical expression that has a variable in the radicand are called **radical equations**. When the radicals are of the second-order (involving square root), we use the methods of solving equations together with the following property to find the solution set.

■ **Property of *n*th power**
Given a natural number *n* and real algebraic expressions *P* and *Q*, all of the solutions of the equation
$$P = Q$$
are contained in the solution set of the equation
$$P^n = Q^n.$$

Concept
If each member of an equation is raised to some natural number power, the solution set of the original equation is a subset of the solution set of the resulting equation.

From this property, we can see that the equation $P^n = Q^n$ *may* contain solutions that are not solutions of the equation $P = Q$. Such solutions are called **extraneous solutions.** To illustrate, consider the equation $x - 3 = 5$, whose solution is 8. If we square each member of the equation, we obtain

$$(x - 3)^2 = 5^2$$
$$x^2 - 6x + 9 = 25$$
$$x^2 - 6x - 16 = 0 \qquad \text{Write in standard form.}$$
$$(x - 8)(x + 2) = 0$$
$$x - 8 = 0 \quad \text{or} \quad x + 2 = 0$$
$$x = 8 \quad \text{or} \qquad x = -2$$
$$S = \{8, -2\}.$$

Since −2 is not a solution of the original equation, $x - 3 = 5$, we call −2 an extraneous solution.

Since the application of the previous property may produce extraneous solutions, then *each solution* obtained through the use of the property *must be checked* into the original equation. Now consider some radical equations where we apply this property.

Quadratic Equations and Inequalities

Example 6–5   A   Find the solution set of each equation. Identify extraneous solutions, if they exist.

1. $\sqrt{x} - 4 = 5$

To isolate the radical, we add 4 to each member and obtain the equation

$\sqrt{x} = 9.$

To remove the radical sign, square each member.

$(\sqrt{x})^2 = (9)^2$
$x = 81$

Check our solution in the *original* equation.

$\sqrt{x} - 4 = 5$
$\sqrt{81} - 4 = 5$
$9 - 4 = 5$
$5 = 5 \quad \text{(true)}$

No extraneous solutions exist and the solution set is $S = \{81\}$.

2. $\sqrt{2x - 5} = 3$

$\sqrt{2x - 5} = 3$
$(\sqrt{2x - 5})^2 = 3^2 \qquad$ Square each member to remove the radical.
$2x - 5 = 9$
$2x = 14$
$x = 7$

We must check our possible solution.

$\sqrt{2x - 5} = 3$
$\sqrt{2(7) - 5} = 3$
$\sqrt{9} = 3$
$3 = 3 \quad \text{(true)}$

There is no extraneous solution and $S = \{7\}$.

3. $\sqrt{x} = x - 12$

To remove the radical, we square both members of the equation.

$(\sqrt{x})^2 = (x - 12)^2$
$x = x^2 - 24x + 144$

> **Note**
> Remember, $(x - 12)^2 = x^2 - 24x + 144$. A common error we
> can make is to say $(x - 12)^2 = x^2 - 12^2 = x^2 - 144$.
> $(x - 12)^2 \neq x^2 - 144$.

Subtract x from both members.

$0 = x^2 - 25x + 144$
$0 = (x - 16)(x - 9)$
$x - 16 = 0 \quad \text{or} \quad x - 9 = 0$
$x = 16 \quad \text{or} \qquad x = 9$

Check: Let

(1) $x = 16$, then   (2) $x = 9$, then

$$\sqrt{16} = 16 - 12 \qquad\qquad \sqrt{9} = 9 - 12$$
$$4 = 16 - 12 \qquad\qquad\quad 3 = 9 - 12$$
$$4 = 4 \quad \text{(true)} \qquad\qquad 3 = -3 \quad \text{(false)}$$

Therefore 9 is an extraneous solution and the solution set of the equation $\sqrt{x} = x - 12$ is $S = \{16\}$.

4. $\sqrt{x + 4} - \sqrt{x - 3} = 1$

When two terms contain radical expressions, we must rewrite the equation with a radical expression in each member. Add $\sqrt{x - 3}$ to each member.

$$\sqrt{x + 4} - \sqrt{x - 3} + 1$$

Square each member.

$$(\sqrt{x + 4})^2 = (\sqrt{x - 3} + 1)^2$$

$$x + 4 = x - 3 + 2\sqrt{x - 3} + 1 \qquad \text{Watch this step carefully.}$$

Then $x + 4 = x - 2 + 2\sqrt{x - 3}$.

Isolate the remaining radical expression in one member by adding $-x + 2$ to both members. Then

$$x + 4 - x + 2 = 2\sqrt{x - 3}$$
$$6 = 2\sqrt{x - 3} \qquad \text{Divide each member by 2.}$$
$$3 = \sqrt{x - 3}$$
$$9 = x - 3 \qquad\quad \text{Square each member.}$$
$$x = 12.$$

Check: Let $x = 12$, then

$$\sqrt{12 + 4} - \sqrt{12 - 3} = 1$$

$$\sqrt{16} - \sqrt{9} = 1$$

$$4 - 3 = 1$$

$$1 = 1 \quad \text{(true)}$$

There is no extraneous solution and the solution set is $S = \{12\}$.

5. $\sqrt[3]{2x - 3} = 2$

Cube both members of the equation.

$$(\sqrt[3]{2x - 3})^3 = 2^3$$
$$2x - 3 = 8$$
$$2x = 11$$
$$x = \frac{11}{2}$$

Check: Let $x = \dfrac{11}{2}$.

$$\sqrt[3]{2\left(\dfrac{11}{2}\right) - 3} = 2$$
$$\sqrt[3]{11 - 3} = 2$$
$$\sqrt[3]{8} = 2$$
$$2 = 2 \quad \text{(true)}$$

There is no extraneous solution, so the solution set is

$$S = \left\{\dfrac{11}{2}\right\}.$$

6. $\sqrt{x} + 1 = 0$

Subtract 1 from each member to obtain

$$\sqrt{x} = -1$$
$$x = 1. \qquad \text{Square each member.}$$

Check: Let $x = 1$, then

$$\sqrt{1} + 1 = 0$$
$$1 + 1 = 0$$
$$2 = 0 \quad \text{(false)}$$

There is no solution and

$$S = \emptyset.$$

In general, we use this procedure to solve a radical equation.

1. If only one term is a radical expression, isolate that term in one member of the equation.
2. If two terms contain radical expressions, separate them so that they are in opposite members of the equation.
3. Raise each member of the equation to the necessary power (the same as the index) to eliminate the radical symbols. Continue to do so until all radicals are eliminated.
4. Solve the resulting linear, or quadratic, equation for the variable.
5. Check for extraneous solutions.

Mastery points
Can you
- Identify extraneous solutions of an equation?
- Find the solution set of a radical equation?

Exercise 6–5

Directions Find the solution set of each equation. Identify extraneous solutions, if they exist. See example 6–5 A.

1. $\sqrt{x} - 6 = 3$

2. $\sqrt{x} + 9 = 13$

3. $\sqrt{x} + 5 = 2$

4. $\sqrt{x + 2} = 5$

5. $\sqrt{y - 4} = 7$

6. $\sqrt{w^2 - 6w} = 4$

7. $\sqrt{t^2 + 10t} = 3$

8. $\sqrt{3k - 1} - 5 = 0$

9. $\sqrt{5z + 1} - 11 = 0$

10. $\sqrt{9a + 5} = \sqrt{3a - 1}$

11. $\sqrt{2p + 5} = \sqrt{3p + 4}$

12. $\sqrt{r + 6} = \sqrt{3r + 2}$

13. $2\sqrt{3z} = \sqrt{5z + 7}$

14. $3\sqrt{x - 3} = \sqrt{2x - 5}$

15. $2\sqrt{2z - 1} = \sqrt{4z}$

16. $\sqrt{p} \ \sqrt{p - 8} = 3$

17. $\sqrt{y} \ \sqrt{y - 5} = 6$

18. $\sqrt{m} \ \sqrt{2m + 4} = 4$

19. $\sqrt{2z} \ \sqrt{z - 3} = 6$

20. $\sqrt{z} + 12 = z$

21. $\sqrt{2x} = x - 4$

22. $p = \sqrt{6 - p}$

23. $q = \sqrt{5q - 4}$

24. $\sqrt{x - 2} = x - 2$

25. $\sqrt{u - 1} = u - 3$

26. $\sqrt{3x + 10} - 3x = 4$

27. $2t - \sqrt{5 - 2t} = 5$

28. $1 + \sqrt{5x + 9} = x$

29. $\sqrt{5x + 1} - 1 = \sqrt{3x}$

30. $\sqrt{v + 5} = 5 - \sqrt{v}$

31. $\sqrt{2n + 3} - \sqrt{n - 2} = 2$

32. $3 - \sqrt{y + 4} = \sqrt{y + 7}$

33. $\sqrt{p + 1} = \sqrt{2p + 9} - 2$

34. $(2y + 3)^{\frac{1}{2}} - (4y - 1)^{\frac{1}{2}} = 0$

    [*Hint:* $(2y + 3)^{\frac{1}{2}} = \sqrt{2y + 3}$.]

35. $(2x^2 + 3x - 5)^{\frac{1}{2}} = (2x^2 - x - 2)^{\frac{1}{2}}$

36. $(4x + 2)^{\frac{1}{2}} - (2x)^{\frac{1}{2}} = 0$

37. $(1 - 2y)^{\frac{1}{2}} + (y + 5)^{\frac{1}{2}} = 4$

38. $(x - 2)^{\frac{1}{2}} = (5x + 1)^{\frac{1}{2}} - 3$

39. $\sqrt[3]{x - 7} = 3$

40. $\sqrt[4]{2y - 3} = 1$

41. $\sqrt[3]{x^2 - 6x - 8} = 2$

42. $\sqrt[4]{2x^2 - 3x - 8} = -1$

43. $(4p - 3)^{\frac{1}{3}} = -2$

44. $(2q - 5)^{\frac{1}{5}} = -1$

Directions Solve the following equations and formulas for the indicated variable. Assume all denominators are nonzero.

Example

$3x\sqrt{x + y} = 2$ for y

Solution

Square both members.

    $(3x\sqrt{x + y})^2 = 2^2$

Then $9x^2(x + y) = 4$.

Perform the indicated multiplication.

$9x^3 + 9x^2y = 4$

Subtract $9x^3$ from both members.

$9x^2y = 4 - 9x^3$

Divide each member by $9x^3$.

$y = \dfrac{4 - 9x^3}{9x^2}$

45. $4x\sqrt{xy} = 3$ for y **46.** $\sqrt{s - t} = 3 + t$ for s **47.** $\sqrt{p^2 - q^2} = r$ for p

48. $\sqrt{\dfrac{y^2 - x^2}{3}} = y - 3$ for x **49.** $r = \sqrt{\dfrac{A}{4\pi h}}$ for A **50.** $t = \sqrt{\dfrac{2s}{g}}$ for s

51. $r = \sqrt{\dfrac{A}{\pi} - R^2}$ for A **52.** $r = \sqrt{\dfrac{V}{\pi h}}$ for h **53.** $D = \sqrt[3]{\dfrac{6A}{\pi}}$ for A

54. $v = \sqrt{\dfrac{2gKE}{W}}$ for W

55. At an altitude of h ft above the sea or level ground, the distance d in miles that a person can see an object is given by
$$d = \sqrt{\dfrac{3h}{2}}.$$
How tall must a person be to see an object 3 miles away?

56. The formula for approximating the velocity V of a car based on the length of its skid marks S (in feet) is given by
$$V = 2\sqrt{16S}.$$
If the velocity is 48 mph, how long will the skid marks be?

57. On wet pavement, the formula in exercise 56 is given by
$$V = 2\sqrt{3S}.$$
How long will the skid marks be if the car is traveling at 30 mph on wet pavement?

58. Find the number whose principal square root is $3i$. (*Hint:* Let $\sqrt{x} = 3i$.)

59. Find the number whose principal third root is -3.

60. Find the number whose principal fourth root is 4.

6–6 Quadratic-type equations

There are a number of equations that are not quadratic equations but they can, nevertheless, be written in quadratic form
$$au^2 + bu + c = 0, \qquad a \neq 0$$
and solved as we solved quadratic equations. The variable u in the above equation represents some expression in another variable.

The following equations are examples of quadratic-type equations.

1. $x^4 + 4x^2 - 12 = 0$
2. $y + 3\sqrt{y} - 4 = 0$
3. $3x^{-2} + 7x^{-1} - 6 = 0$

To recognize each of these as being the quadratic type, we must observe the characteristic that these equations have in common. Notice that the square of the variable factor of the middle term yields the variable factor of the first term. That is,

1. $(x^2)^2 = x^4$
2. $(\sqrt{y})^2 = y$
3. $(x^{-1})^2 = x^{-2}$

The equations that we are going to discuss will have this characteristic. We let u represent this variable factor of the middle term so that we express the given equation in quadratic form.

Example 6–6   A

Find the solution set of each equation.

1. $x^4 + 4x^2 - 12 = 0$

We let $u = x^2$ and then $u^2 = (x^2)^2 = x^4$. Substitute u for x^2 and u^2 for x^4.

$u^2 + 4u - 12 = 0$

$(u + 6)(u - 2) = 0$ Factor the left member.

Then $u = -6$ when $u + 6 = 0$ and $u = 2$ when $u - 2 = 0$.
Replace u by x^2 to get back to the original variable.

$x^2 = -6$ or $x^2 = 2$

Extract the roots.

When $x^2 = -6$, $x = \sqrt{-6} = i\sqrt{6}$ or $x = -\sqrt{-6} = -i\sqrt{6}$.
When $x^2 = 2$, $x = \sqrt{2}$ or $x = -\sqrt{2}$.

$S = \{i\sqrt{6}, -i\sqrt{6}, \sqrt{2}, -\sqrt{2}\}$

Note

The equation is of fourth degree and we obtained four
solutions. We should expect *at most* four solutions.

2. $y + 3\sqrt{y} - 4 = 0$

Let $u = \sqrt{y}$, then $u^2 = (\sqrt{y})^2 = y$ and we substitute to obtain the
quadratic equation in u.

$u^2 + 3u - 4 = 0$

$(u + 4)(u - 1) = 0$

$u = -4$ when $u + 4 = 0$ and $u = 1$ when $u - 1 = 0$.

Replace u by $\sqrt{y}$ and square each member.

$\sqrt{y} = -4$ or $\sqrt{y} = 1$

$y = 16$ or $y = 1$

Since we squared both members of each equation to get these *possible*
solutions, we must check the original equation $y + 3\sqrt{y} - 4 = 0$ for
extraneous solutions.

Check:

(1) Let $y = 16$, then

$16 + 3\sqrt{16} - 4 = 0$

$16 + 3(4) - 4 = 0$

$16 + 12 - 4 = 0$

$28 - 4 = 0$

$24 = 0$ (false)

(2) Let $y = 1$, then

$1 + 3\sqrt{1} - 4 = 0$

$1 + 3 - 4 = 0$

$4 - 4 = 0$

$0 = 0$ (true)

Thus 16 is an extraneous solution and the solution set is $S = \{1\}$.

3. $3x^{-2} + 7x^{-1} - 6 = 0$

Let $u = x^{-1}$ and then $u^2 = (x^{-1})^2 = x^{-2}$. Substitute to get a quadratic
equation in u.

$3u^2 + 7u - 6 = 0$

$(3u - 2)(u + 3) = 0$

Now $u = \dfrac{2}{3}$ when $3u - 2 = 0$ and $u = -3$ when $u + 3 = 0$.

Replace u with x^{-1} or $\dfrac{1}{x}$.

$$\frac{1}{x} = \frac{2}{3} \quad \text{or} \quad \frac{1}{x} = -3$$
$$2x = 3 \quad \text{or} \quad -3x = 1$$
$$x = \frac{3}{2} \quad \text{or} \quad x = -\frac{1}{3}$$

The solution set S is $S = \left\{ \dfrac{3}{2}, -\dfrac{1}{3} \right\}$.

Mastery points
Can you
- Identify a quadratic-type equation?
- Solve any quadratic-type equation?

Exercise 6–6

Directions Find the solution set of each equation. Identify extraneous solutions, if they exist. See example 6–6 A.

1. $x^4 - 6x^2 + 5 = 0$ **2.** $y^4 + 3y^2 - 28 = 0$ **3.** $3z^4 + z^2 = 2$ **4.** $4p^4 = 25p^2 - 6$

5. $(x - 2)^4 + 9(x - 2)^2 + 8 = 0$ **6.** $(m + 5)^4 - 4(m + 5)^2 + 4 = 0$

7. $q - 6\sqrt{q} - 27 = 0$ **8.** $x + \sqrt{x} = 12$ **9.** $2y + 3\sqrt{y} + 1 = 0$

10. $3t - 4\sqrt{t} = 4$ **11.** $p - 5p^{\frac{1}{2}} = -4$ **12.** $2x = 9 - 3x^{\frac{1}{2}}$

13. $(x^2 - 3) - 3\sqrt{x^2 - 3} - 28 = 0$ **14.** $(m^2 + 1) + \sqrt{m^2 + 1} = 20$

15. $y^{\frac{2}{3}} + 3y^{\frac{1}{3}} - 10 = 0$ **16.** $2y^{\frac{2}{3}} - 3y^{\frac{1}{3}} = 2$ **17.** $5p^{\frac{2}{3}} + 11p^{\frac{1}{3}} + 2 = 0$

18. $u^{\frac{3}{2}} - 7u^{\frac{3}{4}} - 8 = 0$ **19.** $x^{\frac{3}{2}} - 2x^{\frac{3}{4}} + 1 = 0$ **20.** $p^{\frac{3}{4}} - 3p^{\frac{3}{8}} + 2 = 0$

21. $z^{-2} - z^{-1} - 12 = 0$ **22.** $r^{-2} - 7r^{-1} + 6 = 0$ **23.** $2y^{-2} = 9y^{-1} - 4$

24. $x^{-4} = 5x^{-2} - 4$ **25.** $4y^{-4} + 4 = 17y^{-2}$

26. $(t^2 - t)^2 - 4(t^2 - t) - 12 = 0$ **27.** $(x^2 + 3x)^2 - 8(x^2 + 3x) = 20$

28. Given the equation $x - 2\sqrt{x} - 15 = 0$, find the solution set by the method in (a) section 6–5 and (b) section 6–6.

6-7 Quadratic and rational inequalities

A quadratic inequality

In section 4–4 we discussed the solution set of a linear inequality in one variable of the form

$$ax + b < 0; \quad ax + b \le 0; \quad ax + b > 0; \quad ax + b \ge 0.$$

We now extend the methods used there together with the methods for solving quadratic equations to find the solution set of a *quadratic inequality*—an inequality of the form

$$ax^2 + bx + c < 0; \quad ax^2 + bx + c \le 0;$$
$$ax^2 + bx + c > 0; \quad ax^2 + bx + c \ge 0; \, (a \ne 0).$$

We will use the number line to help in solving quadratic inequalities. To illustrate, consider the inequality

$$x^2 + 2x - 3 > 0.$$

If we factor the left member, we have

$$(x + 3)(x - 1) > 0.$$

We are not interested in the values of $x^2 + 2x - 3$ at this point. Rather, the *signs* of the values of $x^2 + 2x - 3$ are our concern.

Since $x + 3 = 0$ when $x = -3$ and $x - 1 = 0$ when $x = 1$, we consider what happens when values of x are taken around -3 and 1. The numbers -3 and 1 divide the number into three intervals as shown in figure 6.1.

Figure 6.1

It can be shown that if one number in a given interval makes the product, $(x + 3)(x - 1)$, positive (as we want it to be), then all numbers in that interval will make the product positive.

We now choose a number within each interval as a *test number* to see if it satisfies the inequality. Suppose we choose the numbers

$$-4 \text{ in } x < -3; \quad 0 \text{ in } -3 < x < 1; \quad 2 \text{ in } x > 1.$$

Substitute these numbers into the inequality.

When $x = -4$, then $(x + 3)(x - 1) = (-4 + 3)(-4 - 1)$
$$= (-1)(-5) = 5 \quad \text{(positive)}$$
When $x = 0$, then $(x + 3)(x - 1) \quad = (0 + 3)(0 - 1)$
$$= (3)(-1) = -3 \quad \text{(negative)}$$
When $x = 2$, then $(x + 3)(x - 1) \quad = (2 + 3)(2 - 1)$
$$= (5)(1) = 5 \quad \text{(positive)}$$

Thus the solution set includes all points in the regions $x < -3$ or $x > 1$. The solution set of the inequality $x^2 + 2x - 3 > 0$ is

$$S = \{x \mid x < -3 \text{ or } x > 1\}.$$

The solution set is graphed on the number line in figure 6.2

Figure 6.2

The numbers -3 and 1 are called *critical numbers*. Critical numbers divide the number line into the intervals that we must consider when solving the inequality. The critical numbers are determined by setting each factor equal to 0.

To solve a quadratic inequality we use these steps.

Step 1 Write the inequality in the form

$$ax^2 + bx + c < 0 \text{ or } ax^2 + bx + c > 0 \text{ or}$$
$$ax^2 + bx + c \leq 0 \text{ or } ax^2 + bx + c \geq 0.$$

Step 2 Factor the quadratic trinomial $ax^2 + bx + c$.

Step 3 Find the critical numbers by setting each factor equal to 0 and solving for x.

Step 4 Divide the number line into intervals using the critical numbers. Write the intervals excluding the critical numbers.

Step 5 Choose a test number within each interval obtained and check the *sign* of the product for the test number.

Step 6 Choose the interval(s) that satisfy the conditions of the original inequality.

Example 6-7 A

Find the solution set of the following quadratic inequalities. Graph the solution set on the number line.

1. $x^2 - x - 6 \leq 0$

Since the inequality is in the standard form, we factor $x^2 - x - 6$.

$$x^2 - x - 6 \leq 0$$
$$(x - 3)(x + 2) \leq 0$$

Since $x - 3 = 0$ when $x = 3$ and $x + 2 = 0$ when $x = -2$, the critical numbers are -2 and 3, which yield the following intervals.

$$x < -2 \qquad -2 < x < 3 \qquad x > 3$$
$$\overline{} \quad -2 \quad \overline{} \quad 3$$

Choose the test numbers

-3 in $x < -2$; 0 in $-2 < x < 3$; 4 in $x > 3$.

Substitute the test numbers into the factors of the inequality $(x - 3)(x + 2) < 0$ and check the signs.

If $x = -3$, then $(x - 3)(x + 2) > 0$.
If $x = 0$, then $(x - 3)(x + 2) < 0$.
If $x = 4$, then $(x - 3)(x + 2) > 0$.

6-7 Quadratic and Rational Inequalities   265

The numbers in the interval $-2 \le x \le 3$ satisfy the inequality. Since we have the weak inequality $\le$, meaning *less than or equal to*, the critical numbers are included in the set.

The solution set is

$S = \{x \mid -2 \le x \le 3\}.$

2. $3x^2 - 4x - 4 \ge 0$

$3x^2 - 4x - 4 \ge 0$ Factor the left member.

$(3x + 2)(x - 2) \ge 0$

Since $3x + 2 = 0$ when $x = -\dfrac{2}{3}$ and $x - 2 = 0$ when $x = 2$, the critical

numbers are $-\dfrac{2}{3}$ and 2. Choose test numbers

-1 in $x < -\dfrac{2}{3}$;   0 in $-\dfrac{2}{3} < x < 2$;   3 in $x > 2$.

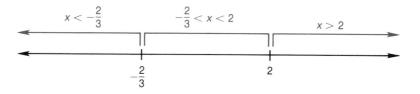

Substitute these test numbers in the inequality $(3x + 2)(x - 2) \ge 0$ and check the signs.

If $x = -1$, then $(3x + 2)(x - 2) > 0$.
If $x = 0$, then $(3x + 2)(x - 2) < 0$.
If $x = 3$, then $(3x + 2)(x - 2) > 0$.

The solution set is $S = \left\{ x \mid x \le -\dfrac{2}{3} \text{ or } x \ge 2 \right\}$.

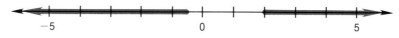

We can apply this same method in solving a *rational* inequality such as

$$\frac{z}{z - 5} > 3.$$

Our first inclination is to multiply each member by the denominator to clear the denominator. However we do not know whether the denominator represents a *positive* or a *negative* number. Recall that this affects the order symbol involved. We could solve this problem by cases. However an easier approach is to use the same method we use in solving quadratic inequalities.

3. $\dfrac{z}{z-5} > 3$

We must first change the form of the inequality by subtracting 3 from each member to isolate 0 in the right member. Then subtract in the left member.

$$\frac{z}{z-5} - 3 > 0$$

$$\frac{z}{z-5} - \frac{3(z-5)}{z-5} > 0$$

$$\frac{z - 3z + 15}{z-5} > 0$$

$$\frac{-2z + 15}{z-5} > 0$$

Since $-2z + 15 = 0$ when $z = \dfrac{15}{2}$ and $z - 5 = 0$ when $z = 5$, the critical numbers are 5 and $\dfrac{15}{2}$. We have the following intervals.

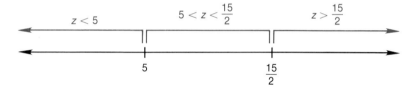

Choose the test numbers

0 in $z < 5$; 6 in $5 < z < \dfrac{15}{2}$; 8 in $z > \dfrac{15}{2}$.

If $z = 0$, then $\dfrac{-2z + 15}{z-5} < 0$.

If $z = 6$, then $\dfrac{-2z + 15}{z-5} > 0$.

If $z = 8$, then $\dfrac{-2z + 15}{z-5} < 0$.

The solution set is $S = \left\{ z \mid 5 < z < \dfrac{15}{2} \right\}$.

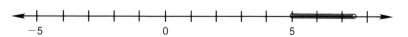

Exercise 6–7

Directions Find the solution set and graph the solution set S for each of the following inequalities. Write the answer in set-builder notation. See example 6–7 A.

1. $(x + 3)(x - 1) > 0$ **2.** $(y - 4)(y - 5) \geq 0$ **3.** $(p + 7)(p + 2) \leq 0$ **4.** $(z - 6)(z + 8) < 0$

5. $r(r - 1) \geq 0$ **6.** $q(3q + 4) < 0$ **7.** $x^2 - 5x + 4 < 0$ **8.** $m^2 + 6m + 5 \geq 0$

9. $q^2 - 3q \leq 18$ **10.** $t^2 + t > 30$ **11.** $x^2 + 2 < 3x$ **12.** $w^2 - 8 \leq 2w$

13. $u^2 \geq 12 - 4u$ **14.** $x^2 - 1 < 0$ **15.** $y^2 - 4 > 0$ **16.** $p^2 \geq 5$

17. $5x^2 - 15 < 0$ **18.** $6y^2 \geq 42$ **19.** $x^2 - 5x < 0$ **20.** $2m^2 - 3m < 0$

21. $2w^2 \leq -6w$ **22.** $2x^2 - 7x - 4 > 0$ **23.** $2y^2 + y - 6 < 0$ **24.** $4z^2 + 7z + 3 \leq 0$

25. $3w^2 + 16w + 5 \geq 0$ **26.** $2p^2 - 3p < 9$ **27.** $9v^2 - 8 < 6v$ **28.** $6w^2 - 7 \leq -11w$

29. $x^2 + 2x - 3 \geq 0$ **30.** $(x - 3)(x + 1)(x - 2) > 0$ **31.** $(y + 4)(y - 5)(y + 6) < 0$

32. $r(3r + 5)(r - 6) \leq 0$ **33.** $t(t + 5)(3t - 8) \geq 0$

Directions Find the solution set of the following rational inequalities. Graph the solution set. Write the answer in set-builder notation. See example 6–7 A–3.

34. $\dfrac{x - 4}{x - 2} \leq 0$ **35.** $\dfrac{p + 3}{p + 1} \geq 0$ **36.** $\dfrac{3q - 2}{2q + 5} > 0$ **37.** $\dfrac{5t + 4}{7t - 3} < 0$

38. $\dfrac{1}{x} \geq 2$ **39.** $\dfrac{5}{y} \leq 3$ **40.** $\dfrac{3}{u + 4} \geq 2$ **41.** $\dfrac{-5}{v - 6} \leq 1$

42. $\dfrac{1}{2r - 3} < -3$ **43.** $\dfrac{-4}{3t + 8} > 1$ **44.** $\dfrac{-1}{2x - 7} \leq -2$ **45.** $\dfrac{x}{x + 5} \geq 3$

46. $\dfrac{y}{y - 3} > 4$ **47.** $\dfrac{x + 6}{x - 7} \leq 4$ **48.** $\dfrac{y - 4}{2y + 5} < -1$ **49.** $\dfrac{2z - 5}{z + 2} > 1$

50. $\dfrac{2p}{p - 2} \leq p$ **51.** $\dfrac{q}{q + 4} \geq 2q$ **52.** $\dfrac{3r}{2r + 1} > -2r$ **53.** $\dfrac{5t}{3t - 2} < -3t$

Chapter summary

1. The **standard form** of a **quadratic equation** in one variable is given by
$ax^2 + bx + c = 0, a \neq 0$.

2. To solve a quadratic equation by factoring, we use the property $P \cdot Q = 0$, if and only if $P = 0$ or $Q = 0$, where P and Q are polynomials.

3. If $x^2 = p$, then $x = \pm \sqrt{p}$.

4. The quadratic equation $ax^2 + bx + c = 0$ can be solved by **completing the square** by writing the equation in the form $(x + k)^2 = d$.

5. The quadratic equation $ax^2 + bx + c = 0$ can be solved using the **quadratic formula**
$$x = \frac{-b \pm \sqrt{b^2 - 4ac}}{2a}.$$

6. Given the equation $P = Q$, where P and Q are polynomials, the solution set of the equation $P = Q$ is a subset of the solution set of the equation $P^n = Q^n$, where n is a positive integer.

7. Solutions of the equation $P^n = Q^n$, where n is a positive integer, that *are not* solutions of the equation $P = Q$ are called **extraneous solutions.**

8. A quadratic inequality is any inequality written in the form
$ax^2 + bx + c < 0, ax^2 + bx + c > 0,$
$ax^2 + bx + c \leq 0$, or $ax^2 + bx + c \geq 0$,
where $a \neq 0$.

Chapter review

[6–1]

Directions Find the solution set of the following equations by factoring or extracting the roots.

1. $x^2 - 9x - 10 = 0$

2. $y^2 - 4y = 32$

3. $4p^2 = 7p$

4. $9x^2 - 36 = 0$

5. $4z^2 - 12z + 5 = 0$

6. $2n^2 = 3n + 5$

7. $\frac{y}{2} + \frac{1}{3y} = \frac{7}{6}$

8. $(5x - 1)(x + 3) - 2x(2x + 6) = 0$

9. $(m + 7)^2 = 64$

10. $(3z - 4)^2 = 16$

11. A projectile is fired vertically upward with an initial velocity of 640 ft/sec. The distance S (in feet) above the ground after t seconds is given by
$S = -16t^2 + 640t.$
When will the projectile reach a height of 1,600 feet? When will the projectile hit the ground?

[6–2]

Directions Find the solution set of the given quadratic equations by completing the square.

12. $y^2 + 3y - 8 = 0$

13. $2p^2 - 3p + 1 = 0$

14. $5p^2 + 3p = -1$

15. $(4p + 3)(2p - 1) = p(p + 3)$

16. $\frac{2}{3}x^2 - \frac{1}{4}x + 1 = 0$

17. Solve the equation $2x^2 + 3ax - a^2 = 0$ for x in terms of a by completing the square, where $a > 0$.

Directions Find the solution set of the given quadratic equation by using the quadratic formula.

18. $x^2 - 11x + 10 = 0$ **19.** $3y^2 + 3y - 2 = 0$ **20.** $2z^2 - 3 = 0$ **21.** $p^2 + 7p = 0$

22. $2z^2 = -z - 4$ **23.** $3x - 2 = \dfrac{5}{x}$ **24.** $(2y - 3)^2 = 3y - 2$

Directions Solve the following verbal problems.

25. Solve the equation $4x^2 + xy - 2y^2 = 0$ for x in terms of y using the quadratic formula, where y is positive.

26. A beam of length L has a maximum displacement at a distance x from the end. This is expressed by the equation $2x^2 - 3xL + L^2 - 0$. Solve the equation for x by using the quadratic formula.

Directions Solve the following by using quadratic equations.

27. An object fired vertically into the air with an initial velocity v_0 feet per second will be at a distance h feet in the air at t seconds after launching, according to the equation $h = v_0 t - 16t^2$.
How long will it take the object to reach a height of 75 feet if the initial velocity is 120 feet per second (to the nearest tenth)?

28. A rectangular solar panel has an area of 1.0 square meter. If the length is 150 centimeters more than the width, find the dimensions of the panel. (*Hint:* Use $A = \ell w$ and 1 meter = 100 centimeters.)

29. The shortest leg of a right triangular brace is 9.0 feet shorter than the hypotenuse and 7.0 feet shorter than the other leg. What are the lengths of the sides of the brace?

30. Mary can paint the exterior of a house in 2 hours less time than Dick can. Working together, they can paint the house in 12 hours. How long would it take each person to paint the house working alone?

Directions Find the solution set of each radical equation. Indicate any extraneous solutions.

31. $y = \sqrt{y + 20}$ **32.** $\sqrt{3x + 5} + 1 = 3x$ **33.** $\sqrt{y + 2} + 1 = \sqrt{y + 6}$

34. $\sqrt[4]{x^2 - x - 5} = 1$ **35.** $\sqrt{5x} + 2x = \sqrt{20x} + 5$

36. The radius r of a sphere is determined by the equation
$$r = \sqrt[3]{\frac{3V}{4\pi}}.$$
Solve the equation for V.

Directions Find the solution set of each equation. Identify extraneous solutions, if they exist.

37. $x^4 - 5x^2 - 14 = 0$ **38.** $(y + 1)^2 - 6(y + 1) + 5 = 0$ **39.** $q + 9\sqrt{q} + 8 = 0$

40. $3p - 5p^{\frac{1}{2}} - 2 = 0$ **41.** $2x^{\frac{2}{3}} + x^{\frac{1}{3}} - 6 = 0$ **42.** $z^{-2} = 10z^{-1} + 11$

43. $5y^{-4} - 8y^{-2} = 4$

[6–7]

Directions Find the solution set and graph each inequality. Write the answer in set-builder and interval notation.

44. $(x - 3)(x + 7) > 0$ **45.** $(2x - 1)(3x + 4) \leq 0$ **46.** $2y^2 - 3y \geq 0$

47. $9z^2 - 4 < 0$ **48.** $m^2 + 7 > 6m$ **49.** $4p^2 - 5p < 6$

50. $5x^2 \geq 40$ **51.** $y^2 - 14y \leq 49$ **52.** $(y - 2)(y + 4)(y - 5) < 0$

53. $(2z + 1)(3z - 2)(z + 1) \geq 0$ **54.** $\dfrac{m + 3}{m - 1} \geq 0$ **55.** $\dfrac{x - 3}{x + 7} \leq 2$

Chapter 6 cumulative test

Directions Perform the indicated operations and simplify. Assume all variables are nonzero real numbers. Leave all answers with positive exponents.

[1–2] **1.** $\dfrac{(-18)(-4)}{-6}$ [1–3] **2.** $16 - 8[3 - 4(12 - 8) + 14] - 6$

[1–3] **3.** $56 - 28 \div 7 - 5 + 3^2$ [3–1] **4.** $(4xy^2 - 5xy + 2x^2y) - (-3xy^2 + xy - 4x^2y)$

[3–3] **5.** $(3x + 2y)^2$ [3–3] **6.** $(4y + 1)(4y - 1)$

[3–3] **7.** $(x - 2)(9x^2 + 2x + 4)$ [3–2] **8.** $(5xy^2)^3(-3xy)^3$

[3–4] **9.** $\left(\dfrac{a^2b^0}{a^{-3}}\right)^4$ [3–4] **10.** $\dfrac{-24a^{-3}b^2}{12a^2b^{-3}}$

[3–1] **11.** Given $P(x) = 3x^2 - 2x + 1$,
find (a) $P(-1)$, (b) $P(0)$,
(c) $P(4)$.

Directions Reduce the following rational expressions to lowest terms.

[4–1] **12.** $\dfrac{12a - 12b}{b - a}$ [4–1] **13.** $\dfrac{6a^2 - 6}{3a^2 - 15a + 12}$ [4–1] **14.** $\dfrac{3b^2 - 12}{4b^2 - 16}$

Directions Perform the indicated operations and simplify. All denominators are nonzero.

[4–2] **15.** $\dfrac{x + 2}{x - 1} \cdot \dfrac{x^2 - 1}{x^2 - x - 6}$

[4–2] **16.** $\dfrac{x^2 - 2x + 1}{49 - x^2} \div \dfrac{x^2 - 7x + 6}{2x^2 - 15x + 7}$

[4–3] **17.** $\dfrac{15}{p^2 - 7p - 18} - \dfrac{3}{p^2 - 4}$

[4–3] **18.** $\dfrac{4a + 1}{a - 6} + \dfrac{3a - 4}{6 - a}$

Directions Solve the following equations and inequalities.

[2–5] **19.** $|4x + 7| = 5$

[2–5] **20.** $|2x - 5| \le 5$

[2–5] **21.** $|7 - 2x| > 4$

[2–5] **22.** $4(2x + 1) - 3(x - 1) = 4x$

[2–5] **23.** $4x - 3 \le 5(x + 6)$

[2–5] **24.** $\dfrac{3}{x} - \dfrac{4}{x} - \dfrac{6}{10}$

[2–5] **25.** $P = 2\ell + 2w$ for w

[4–4] **26.** Simplify the complex fraction $\dfrac{5 - \dfrac{6}{3y}}{2 + \dfrac{5}{2y}}$.

Directions Perform the indicated operations and simplify.

[5–2] **27.** $\sqrt{3}\,(6 - \sqrt{3}\,)$

[5–2] **28.** $3\sqrt{75} + \sqrt{27} - \sqrt{12}$

[5–2] **29.** $\sqrt[3]{81} - 3\sqrt[3]{24}$

[5–6] **30.** $(3 - 2i)(3 + 2i)$

[5–2] **31.** $(2\sqrt{5} - 3\sqrt{3}\,)^2$

Directions Rationalize the denominator.

[5–3] **32.** $\sqrt{\dfrac{16}{5}}$

[5–3] **33.** $\dfrac{\sqrt{2} - \sqrt{3}}{\sqrt{2} + \sqrt{3}}$

[5–6] **34.** $\dfrac{i}{2 - 5i}$

Directions Find the solution sets of the following equations and inequalities.

[6–1] **35.** $x^2 - 15x + 14 = 0$

[6–3] **36.** $5y^2 - y + 3 = 0$

[6–2] **37.** $3y^2 + 1 = y - 2$

[6–7] **38.** $z^2 - 2z \le 3$

[6–6] **39.** $p^4 - 5p^2 - 50 = 0$

[6–7] **40.** $\dfrac{2y - 3}{y + 7} < 1$

Chapter 7

Graphing Linear Equations and Inequalities in Two Variables

7–1 The rectangular coordinate system

In chapter 3 we considered the solution set of linear equations (first-degree equations) in one variable. That is, equations of the form

$$ax + b = 0,$$

where a and b are real numbers, $a \neq 0$. In chapter 6 we studied the solution set of quadratic equations (second-degree equations) in one variable of the form

$$ax^2 + bx + c = 0,$$

where a, b, and c are real numbers, $a \neq 0$. The solution sets of these equations are sets of real numbers or complex numbers.

In this chapter we expand our work with these equations to consider **linear equations in two variables,** x and y. Such equations will be of the form

$$ax + by = c,$$

where a, b, and c are real numbers. The equations

$$3x + y = 4, \quad 4y - x = 0, \quad y = 2x - 1, \quad \text{and} \quad x = y - 4$$

are examples of linear equations in two variables.

In any equation in two variables, x and y, any *pair* of values, one for x and one for y, that satisfies the equation (makes the equation a true statement) is a solution of the equation. Consider the linear equation $3x - y = 5$. Let $x = 2$ and $y = 1$. If we substitute 2 for x and 1 for y in the equation, we have

$$3(2) - (1) = 5$$
$$6 - 1 = 5$$
$$5 = 5. \quad \text{(true)}$$

Therefore the values 2 for x and 1 for y form a solution of the equation $3x - y = 5$. The pair of numbers $x = 2$ and $y = 1$ that form this solution are usually written in the form (2,1), called an **ordered pair of real numbers** because the numbers are written in a specific order.

To graph an ordered pair, we use two real number lines that intersect at right angles with each other at their zero points. The point of intersection is called the **origin.** The two lines, one *horizontal* and the other *vertical,* are called **axes.** The horizontal line, called the **x-axis,** is associated with the first number of the ordered pair and the vertical line, called the **y-axis,** is associated with the second number of the ordered pair. The x-axis and the y-axis form the **rectangular coordinate system** and partition the plane into four equal regions called **quadrants.** The quadrants are numbered I, II, III, and IV in a counterclockwise direction. See figure 7.1.

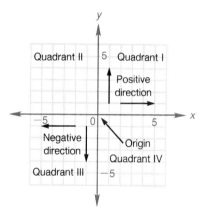

Figure 7.1

Ordered pairs

To locate the point in the plane that corresponds to the ordered pair (3,5), we start at the origin and move 3 units to the *right* (the positive direction) along the x-axis, and then we move 5 units *up* (the positive direction) parallel to the y-axis. See figure 7.2(a). To locate the point in the plane that corresponds to the ordered pair $(-5,-4)$, we start at the origin and move 5 units to the *left* (the negative direction) along the x-axis, and then we move 4 units *down* (the negative direction) parallel to the y-axis. See figure 7.2(b).

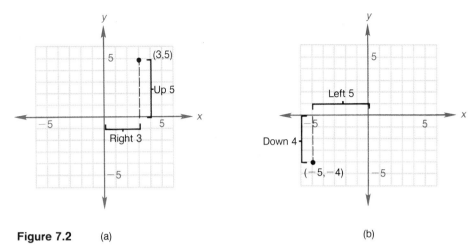

Figure 7.2   (a)   (b)

In the ordered pair $(-5,-4)$, we call the numbers -5 and -4 the **coordinates of the point,** where the first number, -5, is called the **abscissa,** or x-coordinate, of the point and the second number, -4, is called the **ordinate,** or y-coordinate, of the point.

> **Note**
> -5 is called the **first component** of the ordered pair and -4 is called the **second component** of the ordered pair.

Points are usually named by capital letters. When we use the notation $P(x,y)$, we mean the point P whose coordinates are x and y. That is, $P(x,y)$ represents the point P whose abscissa is x and whose ordinate is y. For example, in figure 7.3 the points $A(4,4)$, $B(0,3)$, $C(-4,2)$, $D(-3,0)$, $E(-6,-2)$, $F(0,-5)$, $G(2,-3)$, $H(7,0)$, and $I(0,0)$ have been located in the plane. Point A lies in quadrant I, point C lies in quadrant II, point E lies in quadrant III, and point G lies in quadrant IV.

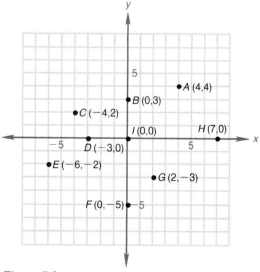

Figure 7.3

275

Graphing an equation

Now we consider the graph of the solution set of an equation in two variables, called the **graph of the equation.** The solution set can be described using the given equation and set-builder notation. To illustrate, the solution set of the equation $2x + 3y = 6$ can be expressed as

$$\{(x,y) \mid 2x + 3y = 6\},$$

which we read "the set of all ordered pairs (x,y) such that $2x + 3y = 6$." To obtain ordered pairs that belong to this set, we choose arbitrary real number values for one variable, substitute for that variable, and solve the resulting equation for the remaining variable. Let x take on the values -6, -3, 0, 3, and 6. When

(1) $x = -6$

$2(-6) + 3y = 6$

$-12 + 3y = 6$

$3y = 18$

$y = 6$

(2) $x = -3$

$2(-3) + 3y = 6$

$-6 + 3y = 6$

$3y = 12$

$y = 4$

(3) $x = 0$

$2(0) + 3y = 6$

$0 + 3y = 6$

$3y = 6$

$y = 2$

(4) $x = 3$

$2(3) + 3y = 6$

$6 + 3y = 6$

$3y = 0$

$y = 0$

(5) $x = 6$

$2(6) + 3y = 6$

$12 + 3y = 6$

$3y = -6$

$y = -2$

We have determined that the ordered pairs

$$(-6,6), (-3,4), (0,2), (3,0), \text{ and } (6,-2)$$

belong to the solution set of the equation $2x + 3y = 6$. We can now draw what we call a *sketch of the graph* of $\{(x,y) \mid 2x + 3y = 6\}$ by plotting the points associated with these ordered pairs and connecting the dots that represent these points. See figure 7.4.

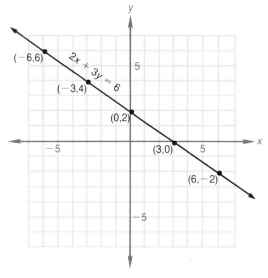

Figure 7.4

Note
It is impossible to draw the complete graph since x can take on numerically large and numerically small values. Arrowheads have been placed at each end of our graph to indicate that the graph continues indefinitely in both directions.

The graph of the linear equation $2x + 3y = 6$ turns out to be a straight line. In fact, **the graph of any linear (first-degree) equation in two variables,**

$$Ax + By = C,$$

where A and B are not both zero, is a straight line. Since any two points in the plane determine a unique straight line, we only need to find two solutions, plot the corresponding points, and draw the line through the points. However a third point should be found that lies on the line as a check. If a straight line cannot be drawn through all three points, a mistake has been made in determining the points.

x- and *y*-intercepts

Consider again the graph of the equation $2x + 3y = 6$ in figure 7.4. The graph passes through the points $(0,2)$, where the graph crosses the y-axis, and $(3,0)$, where the graph crosses the x-axis. These solutions, in which one of the components is zero, are generally the easiest to obtain. For this reason, these are the two that we will use to graph the equation.

> ■ **Definition of *x*- and *y*-intercepts**
> The *abscissa* (x-coordinate) of the point at which a line intersects the x-axis is called the **x-intercept** of the line. The *ordinate* (y-coordinate) of the point at which a line intersects the y-axis is called the **y-intercept** of the line.

Thus in the graph of $2x + 3y = 6$, the x-intercept is 3 and the y-intercept is 2. Since any point on the x-axis has y-coordinate of 0, we find the x-intercept by letting $y = 0$ in the equation and solving for x. Similarly, we can find the y-intercept by letting $x = 0$ and solving for y. We then use the intercepts to graph linear equations in two variables.

Example 7–1 A

Find the x- and y-intercepts and sketch the graph of each equation.

1. $3x + 5y = 15$
   (a) Let $y = 0$, then $3x + 5(0) = 15$
$$3x + 0 = 15$$
$$3x = 15$$
$$x = 5.$$
   The x-intercept is 5.

(b) Let $x = 0$, then $3(0) + 5y = 15$
$$0 + 5y = 15$$
$$5y = 15$$
$$y = 3.$$

The y-intercept is 3.

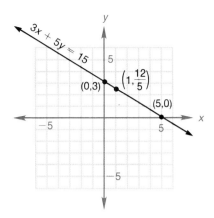

If we let $x = 1$ and substitute this into the equation, we obtain

$$3(1) + 5y = 15$$
$$3 + 5y = 15$$
$$5y = 12$$
$$y = \frac{12}{5}.$$

Therefore the ordered pair $\left(1, \frac{12}{5}\right)$ is a solution of the equation and a check shows that its graph lies on our line. We can feel certain that our graph is correct.

Note
In future examples we will not show the checkpoint; however we should always make this check.

2. $4x = y$

When $y = 0$, we have $4x = 0$ which is true only when $x = 0$. Thus we have found the x- and y-intercepts are at the same point $(0,0)$, which is the origin. We must find a second distinct point. Let $x = 1$, then $4(1) = y$ and $y = 4$, so $(1,4)$ is a solution. We graph the ordered pairs $(0,0)$ and $(1,4)$ to complete our graph.

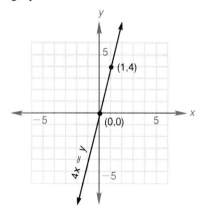

3. $y = -3$

We can write this equation as

$$0 \cdot x + y = -3,$$

and for any value of x that we might choose, y is *always* -3. That is,

$$(-5,-3), \ (-2,-3), \ (0,-3), \ (1,-3), \text{ and } (6,-3)$$

are all solutions of the equation. Plotting these points and drawing a straight line through them, the graph is a horizontal line (parallel to the x-axis) having a y-intercept of -3 and no x-intercept.

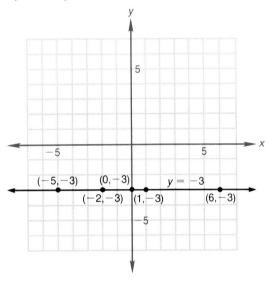

4. $x = 2$

We can write the equation as

$$x + 0 \cdot y = 2,$$

and for any value of y that we choose, x is always 2. If we choose two solutions, say $(2,-3)$ and $(2,0)$, and draw a straight line through these points, we have the graph of $x = 2$. The graph is a vertical line (parallel to the y-axis) having an x-intercept of 2 and no y-intercept.

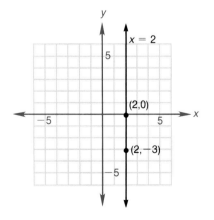

In general, from examples 3 and 4, we see that if k is some real number, $k \neq 0$, the graph of

1. $x = k$ is a vertical line with an x-intercept equal to k and no y-intercept.
2. $y = k$ is a horizontal line with a y-intercept equal to k and no x-intercept.

Mastery points

Can you
- Plot the graph of an ordered pair of real numbers?
- Determine in what quadrant a point lies?
- Find the x- and y-intercepts of a linear equation in two variables?
- Express y in terms of x in a linear equation in two variables?
- Sketch the graph of a linear equation in two variables?

Exercise 7–1

Directions Plot the graph of the following ordered pairs of real numbers. State the quadrant in which the point lies.

1. $(2,4)$ **2.** $(5,2)$ **3.** $(-4,3)$ **4.** $(-6,5)$ **5.** $(-1,-3)$

6. $(-4,-1)$ **7.** $(4,0)$ **8.** $(-6,0)$ **9.** $(0,-1)$ **10.** $(0,5)$

11. $\left(\dfrac{1}{2},3\right)$ **12.** $\left(-2,\dfrac{3}{2}\right)$ **13.** $\left(-\dfrac{7}{2},-\dfrac{5}{2}\right)$

Directions Plot the x- and y-intercepts on the graph of each equation, if they exist. Sketch the lines. See example 7–1 A.

14. $x + 2y = 4$ **15.** $x - 3y = -6$ **16.** $4x - 5y = 20$ **17.** $5x + 2y = 20$

18. $4x + y = 8$ **19.** $5x - y = -10$ **20.** $x = 3y$ **21.** $x = -2y$

22. $y - 3x = 0$ **23.** $y + 2x = 0$ **24.** $4y - x = 0$ **25.** $5y - 3x = 6$

26. $3x + 2y = 8$ **27.** $y = -2$ **28.** $y = 6$ **29.** $x = -1$

30. $x = 8$ **31.** $x = 0$ **32.** $y = 0$

Directions For each equation, express y in terms of x and find the missing value of y in the ordered pairs. Sketch the graph of the equation using these ordered pairs.

Example
$y - 4x = 5; (-3, \quad), (-1, \quad), (0, \quad), (1, \quad)$

Solution
To express y in terms of x, we solve the equation for y by adding $4x$ to each member. Then $y = 4x + 5$.

When
(1) $x = -3, y = 4(-3) + 5 = -12 + 5 = -7; \quad (-3,-7)$
(2) $x = -1, y = 4(-1) + 5 = -4 + 5 = 1; \quad (-1,1)$
(3) $x = 0, y = 4(0) + 5 = 0 + 5 = 5; \quad (0,5)$
(4) $x = 1, y = 4(1) + 5 = 4 + 5 = 9; \quad (1,9)$

The ordered pairs
$(-3,-7), (-1,1), (0,5)$, and $(1,9)$
are solutions of the equation $y - 4x = 5$.

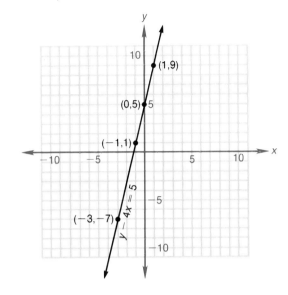

33. $2x + y = 3; (-2, \quad), (0, \quad), (2, \quad), (4, \quad)$

34. $3x + y = -1; (-3, \quad), (-1, \quad), (0, \quad), (3, \quad)$

35. $x + y = 4; (-5, \quad), (-3, \quad), (0, \quad), (4, \quad)$

36. $x - y = 2; (-2, \quad), (0, \quad), (2, \quad), (3, \quad)$

37. $y - x = -2; (-3, \quad), (0, \quad), (2, \quad), (4, \quad)$

38. $x + 2y = 4; (-2, \quad), (0, \quad), (2, \quad), (3, \quad)$

39. $x - 3y = 1; (-1, \quad), (0, \quad), (1, \quad), (3, \quad)$

40. $2x + 5y = 20; (-5, \quad), (0, \quad), (5, \quad), (10, \quad)$

41. $3x + 2y = 8; (-2, \quad), (0, \quad), (2, \quad), (4, \quad)$

42. $4x - 3y = 6; (-6, \quad), (-3, \quad), (0, \quad), (3, \quad)$

Directions Translate each of the following statements into an equation and graph the equation.

43. If 4 is added to x, the result is y.

44. Two times the value of x less 3 is equal to y.

45. Three times x less two times y is 12.

46. If 3 is added to the y-value, the result is four times the x-value.

47. Temperature measured in Fahrenheit degrees can be converted to Celsius degrees using the equation

$$C = \frac{5}{9}(F - 32).$$

Let the horizontal axis represent F and the vertical axis represent C. Graph the equation.

48. Graph the lines $y = x + 2$ and $y = x - 1$ on the same rectangular coordinate system. What appears to be true of the lines? From this, the graph of $y = x + 6$ will be where in the plane?

49. Graph the lines $y = x + 2$ and $y = 2x - 3$ on the same rectangular coordinate system. At what point do the lines appear to cross?

7–2 The distance formula and the slope of a line

Distance formula

In section 7–1 we studied the graph of a linear equation that is a straight line. If we choose any two points on that line, the portion of the line between the two points is called a **line segment.** We cannot determine the length of a line since it continues indefinitely in both directions, but we can determine the length of a line segment. The length of the line segment is defined as the **distance** between the two points. See figure 7.5.

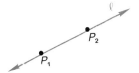

Figure 7.5

Given line ℓ containing points P_1 and P_2, the length of the line segment from P_1 to P_2 is then called the *distance* from P_1 to P_2.

Now consider the distance between arbitrary points $P_1(x_1,y_1)$ and $P_2(x_2,y_2)$, denoted by $d(P_1P_2)$. See figure 7.6.

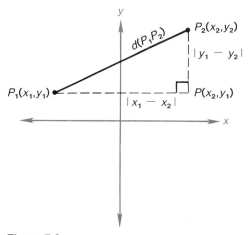

Figure 7.6

We have drawn a horizontal dashed line segment from P_1 and a vertical dashed line segment from P_2 so that these segments meet at point $P(x_2,y_1)$, thus forming a right triangle. The lengths of the dashed line segments by definition are $d(P_1P) = |x_1 - x_2|$ and $d(P_2P) = |y_1 - y_2|$. To find the distance from P_1 to P_2, denoted by $d(P_1P_2)$, we use a well-known property of right triangles called the **Pythagorean Theorem.**

> ■ **Pythagorean Theorem**
> The square of the length of the longest side of a right triangle (called the hypotenuse) is equal to the sum of the squares of the lengths of the other two sides (called the legs).

Graphing Linear Equations and Inequalities in Two Variables

Since the longest side has length $d(P_1P_2)$ and the lengths of the legs are $d(P_1P) = |x_1 - x_2|$ and $d(P_2P) = |y_1 - y_2|$, then

$$[d(P_1P_2)]^2 = [d(P_1P)]^2 + [d(P_2P)]^2,$$

and substituting we have

$$[d(P_1P_2)]^2 = [|x_1 - x_2|]^2 + [|y_1 - y_2|]^2.$$

Considering only the positive square root of the right member, we obtain the **distance formula.**

■ **Distance formula**

$$d(P_1P_2) = \sqrt{(x_1 - x_2)^2 + (y_1 - y_2)^2}\,.$$

Note

Since the square of any number other than 0 is positive, the absolute value symbols are unnecessary in the formula.

Example 7–2 A

Find the distance from $P_1(3,2)$ to $P_2(6,6)$.

Using the formula $d(P_1P_2) = \sqrt{(x_1 - x_2)^2 + (y_1 - y_2)^2}$, we substitute 3 for x_1, 6 for x_2, 2 for y_1, and 6 for y_2 to obtain

$$\begin{aligned}
d(P_1P_2) &= \sqrt{(3 - 6)^2 + (2 - 6)^2} \\
&= \sqrt{(-3)^2 + (-4)^2} \\
&= \sqrt{9 + 16} \\
&= \sqrt{25} = 5.
\end{aligned}$$

Thus $d(P_1P_2) = 5$ units.

The midpoint of a line segment

From time to time we must find the midpoint of a line segment. We now give the formula for finding the coordinates of this point.

■ **Midpoint of a line segment**

The **midpoint** of the line segment joining points $P_1(x_1,y_1)$ and $P_2(x_2,y_2)$ has coordinates

$$\left(\frac{x_1 + x_2}{2}, \frac{y_1 + y_2}{2}\right).$$

Example 7–2 B

Find the midpoint of the line segment whose endpoints are $(6, -4)$ and $(4, 2)$.
The midpoint has coordinates

$$\left(\frac{6 + 4}{2}, \frac{-4 + 2}{2}\right) = \left(\frac{10}{2}, \frac{-2}{2}\right) = (5, -1).$$

The slope of a line

Now consider the portions of two inclines as one moves from point P_1 to point P_2 on each incline. See figure 7.7.

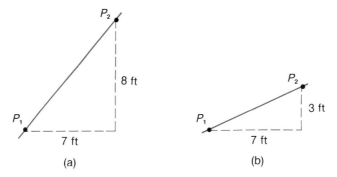

(a)   (b)

Figure 7.7

The incline in figure 7.7(a) is "steeper" than the incline in figure 7.7(b). That is, the inclination in figure 7.7(a) is greater than the inclination in figure 7.7(b). In moving from point P_1 to P_2, the horizontal change is 7 feet in both cases, but the vertical change in (a) is 8 feet and the vertical change in (b) is 3 feet. If we measure this "steepness," or inclination, by the quotient

$$\frac{\text{vertical change}}{\text{horizontal change}},$$

the "steepness" of

$$\text{(a) is } \frac{8 \text{ feet}}{7 \text{ feet}} = \frac{8}{7}$$

and of

$$\text{(b) is } \frac{3 \text{ feet}}{7 \text{ feet}} = \frac{3}{7}.$$

Note
$\frac{8}{7}$ is greater than $\frac{3}{7}$, so the line in (a) is "steeper" than the line in (b).

When applying this concept to any straight line, this "steepness" is called the **slope** of the line. Therefore the slope of a line ℓ is given by

$$\text{slope} = \frac{\text{vertical change}}{\text{horizontal change}}.$$

See figure 7.8.

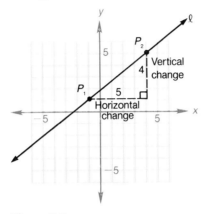

Figure 7.8

The vertical change is 4 units and the horizontal change is 5 units. Then the slope of line ℓ is given by

$$\text{slope} = \frac{\text{vertical change}}{\text{horizontal change}} = \frac{4}{5}.$$

Note
This means that for every 5 units moved to the right from a point on line ℓ, there must be a rise of 4 units to get back to the line.

Referring to figure 7.8 again, we see that the coordinates of point P_1 are $(-1,1)$ and of P_2 are $(4,5)$. The vertical change of 4 units can be determined by the difference in the ordinates, $5 - 1 = 4$. The horizontal change of 5 units can be determined by the difference in the abscissas, $4 - (-1) = 4 + 1 = 5$. We now formally define the slope of a straight line.

> ■ **Definition of the slope of a straight line**
> Given the points $P_1(x_1,y_1)$ and $P_2(x_2,y_2)$ on line ℓ, the *slope,* denoted by m, of line ℓ is given by
>
> $$m = \frac{y_2 - y_1}{x_2 - x_1} \quad \text{or} \quad m = \frac{y_1 - y_2}{x_1 - x_2} \qquad (x_1 \neq x_2).$$
>
> **Concept**
> The slope of a line is obtained by dividing the change in *y*-values by the change in *x*-values.

Note

It is a common mistake to write $m = \dfrac{y_1 - y_2}{x_2 - x_1}$

or $m = \dfrac{y_2 - y_1}{x_1 - x_2}$. The order in which the x-values are subtracted *must be the same* as the order in which the y-values are subtracted.

The slope of a line can alternately be defined by

$$m = \frac{\text{rise}}{\text{run}},$$

where the vertical change is the **rise** and the horizontal change is the **run.** See figure 7.9.

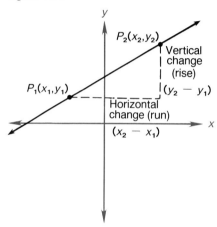

Figure 7.9

Example 7–2 C

Find the slope of a line passing through the given points. Sketch the line.

1. $P_1(4,1)$ and $P_2(8,3)$

Using the formula slope

$$m = \frac{y_2 - y_1}{x_2 - x_1},$$

we substitute to obtain slope

$$m = \frac{3 - 1}{8 - 4} = \frac{2}{4} = \frac{1}{2}.$$

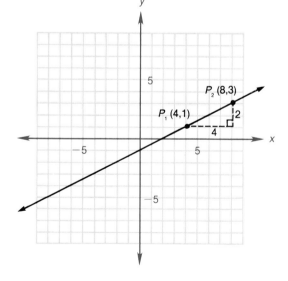

Thus the slope of the line is $\frac{1}{2}$. This means that for every move 2 units to the right, there is a rise of 1 unit.

Note

The x- and y-values may be subtracted in any order so long as the coordinates of each point are in the same position in the numerator and the denominator. That is, we could use

$$m = \frac{y_1 - y_2}{x_1 - x_2}$$

$$= \frac{1 - 3}{4 - 8} = \frac{-2}{-4} = \frac{1}{2}.$$

2. $P_1(-5,4)$ and $P_2(2,1)$

Using $m = \frac{y_2 - y_1}{x_2 - x_1}$, we substitute to obtain

$$m = \frac{1 - 4}{2 - (-5)} = \frac{-3}{7} = -\frac{3}{7}.$$

Thus the slope of the line is $-\frac{3}{7}$. This means that *for every move 7 units to the right, there is a "negative rise" (fall) of 3 units.*

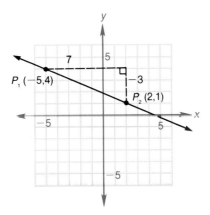

We see that the graph of a line having a positive slope (example 1) "slants" *up* from left to right, and the graph of a line having a negative slope (example 2) "slants" *down* from left to right. This will always be the case.

3. $P_1(-6,2)$ and $P_2(1,2)$

The slope $m = \dfrac{y_2 - y_1}{x_2 - x_1}$

$= \dfrac{2 - 2}{1 - (-6)} = \dfrac{0}{7} = 0.$

Thus the slope of the line is 0.

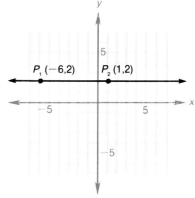

P_1 and P_2 lie on a horizontal line and the slope $m = 0$.

> ■ **Slope of a horizontal line**
> The slope m of any horizontal line is **0**.

4. $P_1(-3,-1)$ and $P_2(-3,4)$

Using $m = \dfrac{y_2 - y_1}{x_2 - x_1}$,

$m = \dfrac{4 - (-1)}{-3 - (-3)}$

$= \dfrac{5}{-3 + 3}$

$= \dfrac{5}{0}$.　　(undefined)

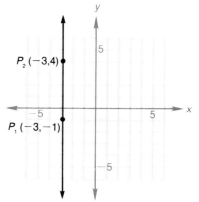

P_1 and P_2 lie on a vertical line and the slope is undefined.

> ■ **Slope of a vertical line**
> The slope of a vertical line is undefined.

Parallel lines

Now we consider a special condition that can exist between two nonvertical lines ℓ_1 and ℓ_2 that have slopes m_1 and m_2, respectively. We can show that ℓ_1 is parallel to ℓ_2 if and only if $m_1 = m_2$.

> **■ Definition of parallel lines**
> If ℓ_1 is parallel to ℓ_2, then $m_1 = m_2$, and
> if $m_1 = m_2$, then lines ℓ_1 and ℓ_2 are parallel.
>
> **Concept**
> Two nonvertical lines are parallel if and only if their slopes are
> the same.

Example 7–2 D

Determine if lines ℓ_1 and ℓ_2 are parallel given ℓ_1 contains points $(-3,2)$ and $(5,1)$ and ℓ_2 contains points $(1,5)$ and $(-7,6)$.

Using

$$m = \frac{y_2 - y_1}{x_2 - x_1},$$

we substitute to find that line ℓ_1 has slope

$$m_1 = \frac{1 - 2}{5 - (-3)} = \frac{-1}{8} = -\frac{1}{8}$$

and line ℓ_2 has slope

$$m_2 = \frac{6 - 5}{-7 - 1} = \frac{1}{-8} = -\frac{1}{8}.$$

Therefore $m_1 = m_2$ and ℓ_1 is parallel to ℓ_2.

Perpendicular lines

There is another special condition that can exist between two nonvertical lines ℓ_1 and ℓ_2 that have slopes m_1 and m_2, respectively. We can show that ℓ_1 is perpendicular to ℓ_2 if and only if $m_1 \cdot m_2 = -1$.

> **■ Definition of perpendicular lines**
> If ℓ_1 is perpendicular to ℓ_2, then $m_1 \cdot m_2 = -1$, and
> if $m_1 \cdot m_2 = -1$, then ℓ_1 is perpendicular to ℓ_2.
>
> **Concept**
> Two nonvertical lines are perpendicular provided the product
> of their slopes is -1. That is, their slopes are negative
> reciprocals.

Example 7–2 E

1. Consider lines ℓ_1 and ℓ_2 such that
   (a) $(-2,3)$ and $(1,5)$ lie on ℓ_1 and (b) $(5,-3)$ and $(3,0)$ lie on ℓ_2.

   The slope

   $$m_1 \text{ of } \ell_1 = \frac{3-5}{-2-1} = \frac{-2}{-3} = \frac{2}{3}$$

   and the slope

   $$m_2 \text{ of } \ell_2 = \frac{-3-0}{5-3} = \frac{-3}{2} = -\frac{3}{2}.$$

   Since $\dfrac{2}{3}\left(-\dfrac{3}{2}\right) = -1$, lines ℓ_1 and ℓ_2 are perpendicular lines.

   Note

   The slope of ℓ_2, that is, $m_2 = -\dfrac{3}{2}$, is the **negative reciprocal** of the slope of ℓ_1, $m_1 = \dfrac{2}{3}$, and vice versa.

   Perpendicular lines form a right angle.

2. Determine if the graphs of the equations $2x - 4y = 4$ and $4x + 2y = -5$ are perpendicular.

   If we choose two ordered pairs that are in the solution set of each equation, we can determine the slopes of each line. Find the intercepts of each graph.

   (a) For $2x - 4y = 4$, the x-intercept is 2 and the y-intercept is -1. Then $(2,0)$ and $(0,-1)$ are solutions and

   $$m_1 = \frac{0-(-1)}{2-0} = \frac{1}{2}.$$

   (b) For $4x + 2y = -5$, the x-intercept is $-\dfrac{5}{4}$ and the y-intercept is $-\dfrac{5}{2}$. Therefore $\left(-\dfrac{5}{4},0\right)$ and $\left(0,-\dfrac{5}{2}\right)$ are solutions and

   $$m_2 = \frac{0-\left(-\dfrac{5}{2}\right)}{-\dfrac{5}{4}-0} = \frac{\dfrac{5}{2}}{-\dfrac{5}{4}} = \frac{5}{2}\cdot\left(-\frac{4}{5}\right) = -2.$$

   Thus the graphs of the equations $2x - 4y = 4$ and $4x + 2y = -5$ are perpendicular lines since $m_1 m_2 = \dfrac{1}{2}(-2) = -1$.

Exercise 7–2

Directions Find the distance between the given pairs of points P_1 and P_2. Find the slope of the line containing the points. Find the midpoint in exercises 1–10. See examples 7–2 A, B, and C.

1. (2,2) and (6,7)

2. (4,1) and (1,4)

3. (−1,5) and (−3,0)

4. (−4,6) and (−1,2)

5. (−1,4) and (−1,9)

6. (2,−6) and (2,1)

7. (3,6) and (−4,6)

8. (−1,−3) and (5,−3)

9. (−3,−1) and (−4,0)

10. (3,5) and (−2,6)

11. (7,−3) and (−4,4)

12. (−2,−2) and (6,−3)

13. (4,−5) and (2,4)

14. (0,8) and (0,−1)

15. (0,−6) and (0,4)

Directions Determine if lines ℓ_1 and ℓ_2 are parallel. If parallel, draw the lines. See example 7–2 D.

16. ℓ_1 contains (4,2) and (1,−1)
ℓ_2 contains (1,1) and (0,0)

17. ℓ_1 contains (5,−2) and (4,1)
ℓ_2 contains (0,6) and (2,0)

18. ℓ_1 contains (5,1) and (−4,2)
ℓ_2 contains (4,−3) and (2,1)

19. ℓ_1 contains (−6,3) and (−2,−5)
ℓ_2 contains (−4,1) and (7,−6)

Directions Determine whether lines ℓ_1 and ℓ_2 are perpendicular. If perpendicular, draw the lines. See example 7–2 E–1.

20. ℓ_1 contains (2,−3) and (4,3)
ℓ_2 contains (1,0) and (−2,1)

21. ℓ_1 contains (5,−2) and (−3,−3)
ℓ_2 contains (4,6) and (5,−2)

22. ℓ_1 contains (1,1) and (4,4)
ℓ_2 contains (−2,2) and (3,−3)

23. ℓ_1 contains (−6,−6) and (−1,−1)
ℓ_2 contains (4,−4) and (−4,4)

Directions Determine if the graphs of the given pairs of equations are parallel, perpendicular, or neither. If parallel or perpendicular, graph the equations. See example 7–2 E–2.

24. $x + y = 1$ and $3x + 3y = −6$

25. $2x + y = −4$ and $6x + 3y = 12$

26. $y − x = −5$ and $4x + 4y = 8$

27. $4y − 3x = 12$ and $4x + 3y = 24$

28. $2y − 3x = 1$ and $3y + 2x = 5$

29. $x + 4y = 5$ and $2x − 5y = 2$

30. $3y − 4x = 1$ and $8y + 3x = 6$

31. $2y − x = 1$ and $6x + 3y = 0$

32. $x + 5y = 0$ and $3x + 15y = 2$

33. $3x − 7y = 0$ and $3y + 5x = 0$

34. (1,3) and (2,4); (7,2) and (8,3)

35. (−2,0) and (1,4); (5,2) and (9,5)

36. (1,5) and (−2,−3); (0,1) and (3,2)

37. (5,6) and (1,3); (−1,6) and (2,1)

38. (7,2) and (−3,4); (−1,−2) and (5,2)

Directions Solve the following verbal problems.

39. The *pitch* of a roof is the slope of a roof. If the roof of a house rises vertically a distance of 9 feet through a horizontal distance of 15 feet, what is the pitch of the roof?

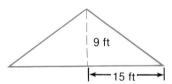

40. The roof of a school building rises 10 feet through a horizontal run of 35 feet. What is the pitch of the roof? (Refer to exercise 39.)

41. The guy wire of a telephone pole is attached to the pole 5 meters above the ground and attached to the ground 7.5 meters from the base of the pole. What is the slope of the guy wire?

42. A ladder leaning against a house touches the building at a point 14 feet above the ground. If the foot of the ladder is 9 feet from the base of the house, what is the slope of the ladder?

43. If a company's profits (*P*) are related to the number of items produced (*x*) by a linear equation, what is the slope of the graph of the equation if the profits rise by $25,000 for every 175 items produced?

44. A company's profits (*P*) are related to increases in the workers' average pay (*x*) by a linear equation. If the company's profits drop by $25,000 per year for every increase of $550 per year in the workers' average pay, what is the slope of the graph of the equation?

45. The increase in a jogger's heartbeat in beats per minute is related to her increase in speed in feet per second by a linear equation. What is the slope of the graph of the equation if an increase in speed of 2 feet per second causes an increase of 10 heartbeats per minute and an increase in speed of 4 feet per second causes an increase of 25 heartbeats per minute? [*Hint:* Use ordered pairs (2,10) and (4,25).]

46. The bacteria count in a culture is related to the hours it exists by a linear equation. What is the slope of the graph if after $1\frac{1}{2}$ hours the bacteria count is 1,000,000 and after 2 hours the bacteria count is 4,500,000?

47. The decay of a substance is related by a linear equation to the time in years that the substance is left in the open air. If after 10 years there are 75 grams remaining and after 25 years there are 40 grams remaining, what is the slope of the equation?

48. The vertices of a triangle in the plane are at the points (−2,4), (5,2), and (0,−4). Find the perimeter (distance around) of the triangle.

49. Show that the points (4,2), (3,0), (−1,0), and (0,2) are the vertices of a parallelogram. Find the perimeter of the parallelogram. (*Hint:* A parallelogram is a four-sided figure whose opposite sides are parallel.)

50. Show that the points (−2,1), (5,3), (3,4), and (0,0) are the vertices of a parallelogram. Find the perimeter of the parallelogram. (Refer to exercise 49.)

51. Show that the points (4,2), (−2,−3), and (4,−3) are the vertices of a right triangle. (*Hint:* Show that two line segments are perpendicular.)

52. Show that the points $(2,-3)$, $(5,1)$, and $(-2,0)$ are the vertices of a right triangle by (a) using the Pythagorean Theorem and (b) showing two sides are perpendicular.

53. A trapezoid is a four-sided figure with one pair of opposite sides that are parallel. Show that the points $(-3,2)$, $(-1,-4)$, $(5,2)$, and $(9,-4)$ are the vertices of a trapezoid.

54. Three points that lie on the same straight line are said to be **collinear.** Three points are collinear if the sum of the distances between two pairs of points is equal to the distance between the third pair of points and if the slopes between all pairs of points are the same. Show that the points $(6,7)$, $(5,2)$, and $(3,-8)$ are collinear using (a) slopes and (b) the distance formula.

55. Show that the points $(1,2)$, $(-3,-4)$, and $(-5,-7)$ are collinear using (a) slopes and (b) the distance formula. (Refer to exercise 54.)

56. Find the abscissa of the points whose ordinate is -6 if the points are at a distance of $5\sqrt{5}$ from the point $(4,5)$.

57. Find the ordinate of the points whose abscissa is 4 if the points are at a distance of $2\sqrt{13}$ from the point $(-2,-7)$.

58. Given one endpoint of a line segment is $(-2,3)$ and the midpoint of the line segment is $(2,-3)$, what are the coordinates of the other endpoint?

59. Given the midpoint of a line segment is the point $(1,5)$, find the other endpoint of the line segment if one endpoint is $(4,-2)$.

7–3 The equations of a line

In section 7–1 we discussed the graph of a linear equation in two variables of the form $ax + by = c$ ($a \neq 0$ or $b \neq 0$). The graph of the equation is a straight line. In this section we will determine the **equation of the graph** of a straight line. That is, we want an equation that is satisfied only by the coordinates of the points on the line. Thus the equation must be such that, for any arbitrary point P,

1. if P is on the graph, then its coordinates satisfy the equation, and
2. if P is not on the graph, then its coordinates do not satisfy the equation.

Slope of a line

Consider a line in the plane having slope m that passes through a given point $P_1(x_1,y_1)$. Let $P(x,y)$ be any other arbitrary point on the line. See figure 7.10.

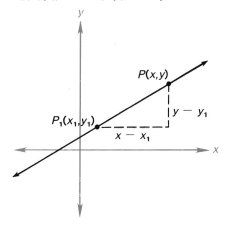

Figure 7.10

Then by the definition of the slope of a line, provided that $x \neq x_1$, we have

$$m = \frac{y - y_1}{x - x_1}.$$

Then multiply each member by $x - x_1$,

$$y - y_1 = m(x - x_1).$$

We call this the **point-slope** equation of a line.

> ■ **Point-slope form**
> The point-slope form of the equation of a line having slope m and passing through the known point (x_1, y_1) is given by
> $$y - y_1 = m(x - x_1).$$

Example 7–3 A

1. Find the equation of the line having a slope of $-\frac{1}{2}$ and passing through the point $(-1,4)$.

Using the point-slope form $y - y_1 = m(x - x_1)$, we are given $m = -\frac{1}{2}$ and $(x_1, y_1) = (-1,4)$. Replacing m by $-\frac{1}{2}$, x_1 by -1, and y_1 by 4, we obtain

$$y - 4 = -\frac{1}{2}[x - (-1)]$$

$$y - 4 = -\frac{1}{2}(x + 1)$$

$2y - 8 = -1(x + 1)$ Multiply each member by 2.
$2y - 8 = -x - 1$
$x + 2y = 7.$ Add $x + 8$ to each member.

We say that the equation $x + 2y = 7$ is written in **standard form.**

> ■ **Standard form**
> The standard form of the equation of a line is written as
> $$Ax + By = C,$$
> where A, B, and C are real numbers.

---

2. Find the equation of the line passing through the points $(2,4)$ and $(-5,-2)$. Write the equation in standard form.

To use the point-slope form $y - y_1 = m(x - x_1)$, we need the slope m, which is not given. Using the formula $m = \dfrac{y_2 - y_1}{x_2 - x_1}$,

$$m = \frac{-2 - 4}{-5 - 2} = \frac{-6}{-7} = \frac{6}{7}.$$

Replacing m by $\dfrac{6}{7}$ and using either of the given points to replace x_1 and y_1, say 2 for x_1 and 4 for y_1, we have

$$y - 4 = \frac{6}{7}(x - 2)$$

$$
\begin{array}{ll}
7y - 28 = 6(x - 2) & \text{Multiply each member by 7.} \\
7y - 28 = 6x - 12 & \\
-6x + 7y = 16 & \text{Add } -6x + 28 \text{ to each member.} \\
6x - 7y = -16. & \text{Multiply each member by } -1.
\end{array}
$$

The equation in standard form is $6x - 7y = -16$.

Consider again the point-slope form of the equation of a line. Given

$$y - y_1 = m(x - x_1),$$

suppose we let the known point be $(0,b)$, the y-intercept. Substituting 0 for x_1 and b for y_1, we have

$$y - b = m(x - 0)$$
$$y - b = mx.$$
$$y = mx + b.$$

We call this the **slope-intercept** form of the equation of a line.

■ **Slope-intercept form**
The slope-intercept form of the equation of a line is written as
$$y = mx + b,$$
where m is the slope and b is the y-intercept.

Example 7–3 B

1. Find the slope and y-intercept of the line whose equation is $2y + 3x = 4$.

To find the slope and y-intercept, we will write the equation in slope-intercept form, which is accomplished by solving the equation for y. Thus

$$2y + 3x = 4$$
$$2y = -3x + 4$$
$$y = -\frac{3}{2}x + 2.$$

Then the slope is $-\dfrac{3}{2}$ (the coefficient of x) and the y-intercept is 2.

7–3 The Equations of a Line

2. Find the equation of the line whose slope is $-\dfrac{5}{3}$ and whose y-intercept is 3. Write the equation in standard form.

Using the slope-intercept form of the equation, $y = mx + b$, substitute $-\dfrac{5}{3}$ for m and 3 for b to obtain

$$y = -\frac{5}{3}x + 3.$$

Since we want the equation in standard form, we multiply both members by 3. Then

$$3y = -5x + 9 \quad \text{or} \quad 5x + 3y = 9.$$

We saw in the previous section that the slopes of two nonvertical *parallel* lines are the same and the slopes of two nonvertical *perpendicular* lines are negative reciprocals. Now we use these facts in the next two examples.

---

3. Find the equation of the line through the point $(3,1)$ that is parallel to the line whose equation is $3x - 2y = 4$.

In order for the lines to be parallel, the line we want must have the same slope as the equation $3x - 2y = 4$. We solve this equation for y to obtain the slope-intercept form.

$$3x - 2y = 4$$
$$-2y = -3x + 4$$
$$y = \frac{3}{2}x - 2$$

The slope we want is the coefficient of x, so $m = \dfrac{3}{2}$. Using the point-slope form, the slope $\dfrac{3}{2}$, and the coordinates of the given point $(3,1)$,

$$y - y_1 = m(x - x_1)$$
$$y - 1 = \frac{3}{2}(x - 3)$$
$$2y - 2 = 3(x - 3)$$
$$2y - 2 = 3x - 9$$
$$2y - 3x = -7 \qquad \text{Multiply each member by } -1.$$
$$3x - 2y = 7. \qquad \text{Standard form}$$

The equation of the line is $3x - 2y = 7$.

4. Find the equation of the line that is perpendicular to the line $4x - 2y = 5$ and passes through the point $(-3,4)$.

We write the equation $4x - 2y = 5$ in slope-intercept form to find the slope.

$$4x - 2y = 5$$
$$-2y = -4x + 5 \qquad \text{Subtract } 4x \text{ from each member.}$$
$$y = 2x - \frac{5}{2} \qquad \text{Divide each member by } -2.$$

The slope of the given line is 2. The slope of the line we desire is the negative reciprocal of 2, which is $-\frac{1}{2}$. We want the equation with slope $m = -\frac{1}{2}$ and passing through $(-3,4)$. Use the point-slope form.

$$y - 4 = -\frac{1}{2}[x - (-3)]$$

$$y - 4 = -\frac{1}{2}(x + 3)$$

$$2y - 8 = -1(x + 3)$$
$$2y - 8 = -x - 3$$
$$x + 2y = 5$$

The equation of the line is $x + 2y = 5$.

The slope-intercept form of the equation of a line can be used to graph a linear equation in two variables.

Example 7–3 C

Graph the following equations using the slope and the y-intercept.

1. $3x + 4y = 4$

Write the equation in slope-intercept form.

$$3x + 4y = 4$$
$$4y = -3x + 4$$
$$y = -\frac{3}{4}x + 1$$

The slope is $m = -\frac{3}{4}$ and the y-intercept is 1.

Recall that a definition of the slope of a line is

$$m = \frac{\text{rise}}{\text{run}}.$$

If we write the slope $m = \dfrac{-3}{4}$, then the rise (in this case, negative rise or fall) is -3 units and the run is 4 units. Then from the point $(0,1)$, the y-intercept, we move 4 units to the *right* (the run) and 3 units *down* (the fall) to find another point on the graph. We can then draw a line through this point and the y-intercept to sketch the graph of the equation $3x + 4y = 4$.

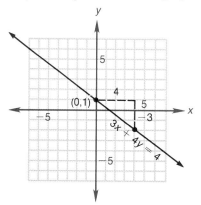

2. $y - 3x = 3$

Write the equation in slope-intercept form.

$y = 3x + 3$

The y-intercept is 3 and the slope is $m = 3 = \dfrac{3}{1}$. The rise is then 3 units and the run is 1 unit. From the point $(0,3)$, move 1 unit to the right (the run) and 3 units up (the rise) to obtain a second point on the graph. We can then draw a line through the y-intercept and this second point.

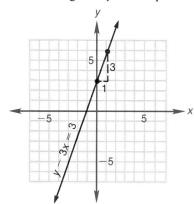

We now summarize the different forms of a linear equation.

$$Ax + By = C \qquad \text{Standard form}$$
$$y - y_1 = m(x - x_1) \qquad \text{Point-slope form}$$
$$y = mx + b \qquad \text{Slope-intercept form}$$
$$x = k \qquad \text{Equation of a vertical line}$$
$$y = k \qquad \text{Equation of a horizontal line}$$

Mastery points
Can you
- Find the equation of a line using the point-slope form $y - y_1 = m(x - x_1)$?
- Write the equation of a line in standard form $Ax + By = C$?
- Write the equation of a line in slope-intercept form $y = mx + b$?
- Find the slope, m, and the y-intercept, b, of a line given its equation?
- Sketch the graph of a linear equation in two variables using the slope and y-intercept?

Exercise 7–3

Directions Find the equation of the line that satisfies the given conditions. Write the equation in standard form. See example 7–3 A–1.

1. Slope $m = \dfrac{1}{2}$ and passing through (3,4)

2. Slope $m = -\dfrac{5}{4}$ and passing through $(-1,-6)$

3. Slope $m = 5$ and passing through (0,7)

4. Slope $m = -\dfrac{3}{2}$ and having x-intercept -5

5. Slope $m = -\dfrac{5}{6}$ and passing through (0,0)

6. Slope $m = -6$ and having y-intercept 2

7. Slope $m = 1$ and having y-intercept -3

8. Horizontal line through $(5,-3)$

9. Horizontal line through (1,4)

10. Horizontal line through (0,0)

11. Slope is undefined and passing through $(-5,6)$

12. Slope is undefined and passing through (1,2)

13. Vertical line passing through $(5,-4)$

14. Vertical line passing through $(-6,2)$

15. Vertical line passing through (0,0)

Directions Find the equation of the line through the given points. Write the equation in standard form. See example 7–3 A–2.

16. (1,2) and (5,4)

17. (3,6) and (1,1)

18. $(-3,4)$ and $(5,-1)$

19. (2,6) and $(-1,-4)$

20. (5,0) and $(-2,-3)$

21. (4,0) and $(5,-2)$

22. $(3,-6)$ and $(2,-6)$

23. (5,4) and $(-2,4)$

24. $(-2,-2)$ and (4,4)

25. $(-4,4)$ and $(3,-3)$

26. $(-5,5)$ and $(1,-1)$

27. $(4,-3)$ and $(4,-7)$

28. $(-2,6)$ and $(-2,-5)$

Directions Write each equation in slope-intercept form $y = mx + b$ and identify the slope m and the y-intercept b. Sketch the graph using the slope and y-intercept. See examples 7–3 B and C.

29. $2x + y = 5$ **30.** $3x + y = -3$ **31.** $4x - y = 6$ **32.** $3x - 7y = 0$

33. $5y - 2x = 0$ **34.** $2y + 9x = 0$ **35.** $8x + 3y = 0$ **36.** $x - y = 0$

37. $3y + 5 = 0$ **38.** $2y - 7 = 0$

Directions Find the equation of the line satisfying the given conditions. Write each equation in standard form. See example 7–3 B–3 and 4.

39. Parallel to $3x + y = 6$ and passing through $(1,2)$

40. Perpendicular to $2x + 5y = 3$ and passing through $(1,7)$

41. Passing through $(-3,-1)$ and parallel to $3x - 2y = 9$

42. Passing through $(-7,4)$ and perpendicular to $7y - 2x = 0$

43. Passing through $(0,0)$ and perpendicular to $9x - 2y = 1$

44. Passing through $(5,-3)$ and parallel to $y - 2 = 0$

45. Passing through $(4,0)$ and perpendicular to $y = 5$

Directions Using the slope-intercept form of the equation of a line, determine if the given pairs of equations represent parallel lines, perpendicular lines, or neither.

46. $x + 2y = 5$ and $6x - 3y = 4$

47. $3y + 2x = 5$ and $2y + 3x = -1$

48. $2y - 5x = -3$ and $y + 3x = 4$

49. $4y - 5x = -1$ and $8x + 10y = 4$

50. $3x + 8y = 2$ and $6x - 16y = 5$

51. $8x - 5y = 1$ and $16x - 10y = 7$

Directions Solve the following verbal problems.

52. Given the point-slope equation of a line,
$$y - y_1 = m(x - x_1),$$
if the points $P_1(x_1,y_1)$ and $P_2(x_2,y_2)$ lie on the graph, then
$$m = \frac{y_2 - y_1}{x_2 - x_1},$$
and if we substitute, we obtain the **two-point** form of the equation of a line given by
$$y - y_1 = \frac{y_2 - y_1}{x_2 - x_1}(x - x_1).$$
Using this form, find the equation in standard form for the line passing through the given points (a) $(3,2)$ and $(4,1)$; (b) $(5,-1)$ and $(-3,0)$.

53. Given the two-point form found in exercise 52 of the equation of a line, use the points $(a,0)$ and $(0,b)$ to show that
$$\frac{x}{a} + \frac{y}{b} = 1$$
is an equation of the line with x-intercept a and y-intercept b. We call this the **intercept** form of the equation of a line.

54. Write each equation in **intercept** form and determine the x- and y-intercepts. (Refer to exercise 53.)
a. $4x + 3y = 12$ b. $2x + 5y = 10$
c. $3x - y = 3$ d. $3x - 5y = 6$
e. $5y - 2x = 8$ f. $6x + 5y = 12$

55. Find the slope and y-intercept of the equation $Ax + By = C$ in terms of A, B, and C.

56. What is the slope of a line that is parallel to the line with the equation $Ax + By = C$?

57. What is the slope of a line that is perpendicular to the line with the equation $Ax + By = C$?

Directions There are a number of real-life situations that can be described using linear equations in two variables. If two pairs of values are known, we can use the *two-point* form demonstrated in exercise 52. Express each equation in standard form.

58. A company produces 300 boxes of cereal for $150 and 600 boxes of the same cereal for $250. Let x be the number of boxes of cereal and y be the total cost.

59. John Doe makes $150 profit on 4 waterfront paintings that he does and $400 profit on 7 paintings. Let x be the number of paintings and y be the profit on the paintings.

60. A company found its total sales were $35,000 in the third year of operation and $105,000 in the fifth year of operation. Let x be the year of operation and y the total sales that year.

61. Use the resulting equation of exercise 60 to predict the total sales in the sixth year.

62. Use the resulting equation of exercise 58 to determine the approximate cost of producing 1,000 boxes.

7–4 Graphs of linear inequalities

Linear inequalities in two variables

In chapter 2 we discussed the solution set of linear inequalities in one variable such as

$$x + 5 < 3, \quad 2x - 3 \le 1, \quad 4x - 3 > 0, \quad \text{and} \quad -4 \le 2x + 1 < 3.$$

In this chapter we have discussed the graph of the solution set of linear equations in two variables such as

$$2x - 3y = 5, \quad x - 2y = 6, \quad y = -4, \quad \text{and} \quad x = 5.$$

Now we extend these ideas to consider the solution set and the graph of **linear inequalities in two variables** such as

$$y \le x, \quad 3x - 2y > 4, \quad \text{and} \quad y < 2x + 3.$$

> ■ **Linear inequalities**
> $Ax + By < C$, $Ax + By \le C$, $Ax + By > C$, and $Ax + By \ge C$, where A, B, and C are real numbers and A and B are not both 0, are called linear inequalities in two variables.

The graph of any linear inequality in two variables will be a *half-plane*, as illustrated in the following examples.

Example 7–4 A

Graph the following linear inequalities.

1. $2x + y \leq 4$

To graph the inequality $2x + y \leq 4$, we first graph the straight line $2x + y = 4$.

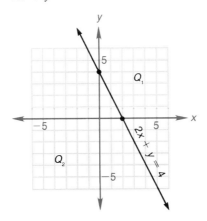

The graph of the inequality includes the points of this line together with all points of the plane either *above* or *below* this line. To decide which half-plane is the solution set, choose *any point not on the line* $2x + y = 4$ and substitute these values into the inequality $2x + y \leq 4$. The origin, (0,0), is most often a good choice. Replacing x and y with 0 in the inequality $2x + y \leq 4$,

$$2(0) + 0 \leq 4$$
$$0 \leq 4. \quad \text{(true)}$$

Since the result is true, (0,0) does satisfy the inequality and the solution set includes all points on the side of the line where (0,0) lies. We then shade that region.

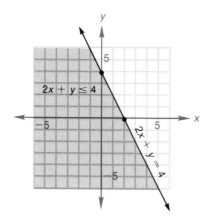

Note

If our inequality had been the *strict* inequality $2x + y < 4$, instead of the *weak* inequality $2x + y \leq 4$, the line $2x + y = 4$ would be *dashed* since the points of the line are not in the solution set anymore.

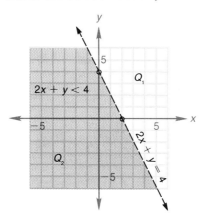

2. $3x + y > 6$

Graph the equation $3x + y = 6$ and make the line dashed since the order symbol, $>$, does not include equality. Since $(0,0)$ does not lie on the line, choose this as a test point and substitute 0 for x and 0 for y in the inequality.

$$3x + y > 6$$
$$3(0) + (0) > 6$$
$$0 + 0 > 6 \quad \text{(false)}$$

Since $(0,0)$ does not satisfy the inequality, shade the half-plane that *does not* contain the origin.

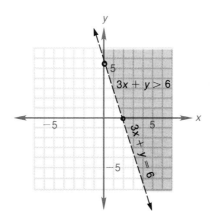

To graph a linear inequality in two variables we use the following steps:

Step 1 Replace the inequality symbol by the equality symbol.

Step 2 Graph this line making it (1) a solid line if the inequality symbol is $\leq$ or $\geq$ (which includes the line in the solution set) or (2) a dashed line if the inequality symbol is $<$ or $>$ (which does *not* include the line in the solution set).

Step 3 Choose some test point that is *not* on the line [if possible, the origin (0,0) since the arithmetic is easiest for this point].

Step 4 Substitute the coordinates of the test point in the inequality.

Step 5 If the test point's coordinates satisfy the inequality, shade the half-plane containing that point for the solution set; if the test point's coordinates *do not* satisfy the inequality, shade the other half-plane for the solution set.

Example 7–4 B

Graph the following inequalities.

1. $x > -3$

Graph the line $x = -3$ with a dashed line since the order symbol is $>$ and does not include equality. Choose test point (0,0) and substitute 0 for x in the inequality.

$$x > -3$$
$$0 > -3 \quad \text{(true)}$$

Shade the half-plane containing the origin.

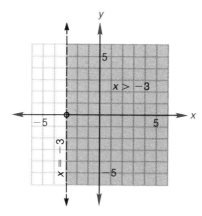

2. $5x - 3y \leq 0$

Graph the line $5x - 3y = 0$ and make it a solid line. The point (0,0) is on the line since the graph passes through the origin. Therefore we cannot use (0,0) as a test point and we arbitrarily choose some point a fair distance from the line that does not lie on the line. Suppose we choose (3,0). We substitute 3 for x and 0 for y in the inequality.

$$5x - 3y \leq 0$$
$$5(3) - 3(0) \leq 0$$
$$15 - 0 \leq 0$$
$$15 \leq 0 \quad \text{(false)}$$

Since the statement is false, shade the half-plane that *does not* contain (3,0).

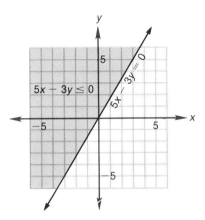

Now consider the following graphs.

Example 7–4 C

Graph the given sets of ordered pairs.

1. $\{(x,y) \mid 4x + y \leq 2\} \cap \{(x,y) \mid x \leq 2\}$

Since the intersection of two sets includes all elements that belong to both sets, the graph of the intersection of these two sets must contain all points that belong to both inequalities. The graph of $\{(x,y) \mid 4x + y \leq 2\}$ is shown in diagram (a) and the graph of $\{(x,y) \mid x \leq 2\}$ is shown in diagram (b). We put these two graphs together to get the graph of their intersection, shown in diagram (c).

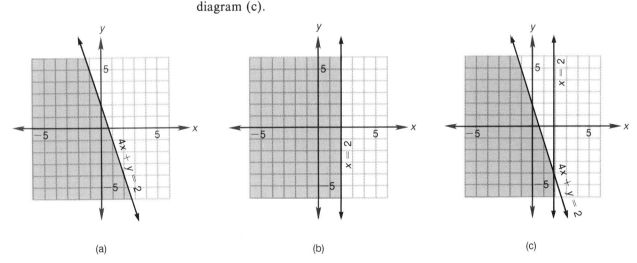

(a)   (b)   (c)

2. $\{(x,y) \mid 4x + y \leq 2\} \cup \{(x,y) \mid x \leq 2\}$

Since the union of two sets includes all elements that belong to either set, the graph of the union of these two sets must contain those points found in either inequality. The graphs of $4x + y \leq 2$ and $x \leq 2$ are shown in parts (a) and (b) of the previous illustration. Include all points in the two graphs.

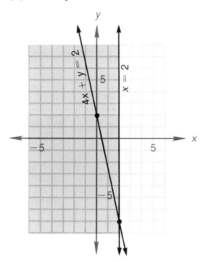

3. $|x| < 1$

If $|x| < 1$, then $-1 < x < 1$ and we graph $\{(x,y) \mid -1 < x < 1\}$, that is, all points whose first components are between -1 and 1. We shade the portion of the plane that lies between the vertical dashed lines $x = -1$ and $x = 1$.

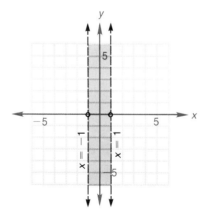

4. $|y - 2| \geq 3$

If $|y - 2| \geq 3$, then $y - 2 \geq 3$ or $y - 2 \leq -3$, and adding 2 to all members, we obtain the inequalities $y \geq 5$ or $y \leq -1$. We then graph $\{(x,y) \,|\, y \geq 5\} \cup \{(x,y) \,|\, y \leq -1\}$. We shade the portion of the plane above the horizontal solid line $y = 5$ and below the horizontal solid line $y = -1$.

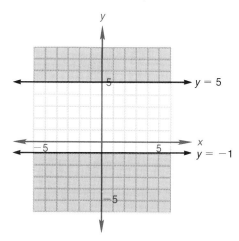

Mastery points

Can you

- Determine when the boundary line is solid and when it is dashed?
- Graph the solution set of a linear inequality in two variables?
- Graph the intersection or union of two linear inequalities?
- Graph an absolute value inequality?

Exercise 7–4

Directions Graph the solution set of each linear inequality. See examples 7–4 A and B.

1. $y < 3$ **2.** $y \leq -4$ **3.** $x \geq 1$ **4.** $x > -5$ **5.** $y > 2$

6. $y \geq -1$ **7.** $x - 7 \geq -3$ **8.** $x + y > 5$ **9.** $x + y < -1$ **10.** $y < 3x - 1$

11. $y > 4 - 3x$ **12.** $x \leq 2y$ **13.** $3y > 4x$ **14.** $2y \geq -x$ **15.** $2x - 3y \leq 0$

16. $y - x > 0$ **17.** $5x + 2y \geq -10$ **18.** $3y - 5x > -15$

Directions Graph the following sets of ordered pairs. See example 7–4 C–1 and 2.

19. $\{(x,y) \,|\, y - x \leq 3\} \cup \{(x,y) \,|\, x \leq 2\}$ **20.** $\{(x,y) \,|\, y + x > 1\} \cup \{(x,y) \,|\, y > 0\}$

21. $\{(x,y) \,|\, 4x + y > 4\} \cup \{(x,y) \,|\, x > -1\}$ **22.** $\{(x,y) \,|\, x + 2y > 3\} \cap \{(x,y) \,|\, y > 2\}$

23. $\{(x,y) \,|\, 2x - 4y > 8\} \cap \{(x,y) \,|\, x > 0\}$ **24.** $\{(x,y) \,|\, 3x + 4y \leq 12\} \cap \{(x,y) \,|\, y \leq 1\}$

Directions Graph the following absolute value inequalities in two variables. See example 7–4 C–3 and 4.

25. $|x| < 2$ **26.** $|x| > 3$ **27.** $|x| \geq 1$ **28.** $|x| \leq 4$ **29.** $|y| < 1$

30. $|y| > 2$ **31.** $|y| \geq 3$ **32.** $|y| \leq 5$ **33.** $|x - 1| \leq 3$ **34.** $|x + 2| < 1$

35. $|x - 5| > 2$ **36.** $|3 - x| \geq 4$ **37.** $|y - 4| \leq 3$ **38.** $|y + 6| < 2$ **39.** $|y + 7| > 1$

40. $|5 - y| \geq 3$ **41.** $|2x - 3| \leq 3$ **42.** $|3y + 1| < 5$

Chapter summary

1. An **ordered pair** of real numbers is written as (x,y), where x and y are real numbers.

2. The **rectangular coordinate system** is formed by two real number lines, called **axes,** drawn in the plane—one horizontal (x-axis) and the other vertical (y-axis)—that intersect at the **origin** 0 of each line.

3. The axes divide the plane into four equal regions called **quadrants.**

4. Each point in the rectangular coordinate system is associated with only one ordered pair of real numbers, called the **coordinates of the point.** The point is called the **graph** of the ordered pair.

5. In the ordered pair (x,y), x is called the **first component** and y is called the **second component** of the ordered pair.

6. The **abscissa** of a point in the plane is the first component of the ordered pair and the **ordinate** of a point is the second component of the ordered pair.

7. The **graph of an equation** is the set of points in the plane associated with the solutions of the equation.

8. The graph of the linear equation in two variables $Ax + By = C$ is a **straight line.**

9. The abscissa of the point at which a line intersects the x-axis is called the **x-intercept** of the line.

10. The **y-intercept** of a line is the **ordinate** of the point at which the line intersects the y-axis.

11. The distance between two points in the plane, $P_1(x_1,y_1)$ and $P_2(x_2,y_2)$, denoted by $d(P_1P_2)$, is given by
$$d(P_1P_2) = \sqrt{(x_1 - x_2)^2 + (y_1 - y_2)^2}\,.$$
We call this the **distance formula.**

12. The **slope** m of the line containing $P_1(x_1,y_1)$ and $P_2(x_2,y_2)$ is given by
$$m = \frac{y_2 - y_1}{x_2 - x_1} \quad \text{or} \quad m = \frac{y_1 - y_2}{x_1 - x_2}\ (x_1 \neq x_2).$$

13. Two nonvertical lines are **parallel** if and only if they have the same slopes.

14. Two lines are **perpendicular** if and only if their slopes are negative reciprocals.

15. The **point-slope** form of the equation of a line having slope m and passing through (x_1,y_1) is given by $y - y_1 = m(x - x_1)$.

16. The **slope-intercept** form of the equation of a line is written as $y = mx + b$, where m is the slope and b is the y-intercept.

17. The graph of a linear inequality in two variables, $Ax + By > C$, $Ax + By \geq C$, $Ax + By < C$, or $Ax + By \leq C$, is the **half-plane** on one side of the line $Ax + By = C$. The graph may or may not include the line.

Chapter review

[7-1]

Directions Plot the graph of each ordered pair of real numbers and state the quadrant in which the point lies, if it lies in a quadrant.

1. $(-3,5)$ **2.** $(-2,-7)$ **3.** $(0,-1)$ **4.** $(5,0)$

5. $\left(1,-\dfrac{3}{2}\right)$ **6.** $\left(\dfrac{1}{2},\dfrac{11}{2}\right)$ **7.** $(0,4)$ **8.** $\left(-\dfrac{5}{2},-4\right)$

Directions Find the x- and y-intercepts of the graph for each equation (if they exist).

9. $x - 3y = 6$ **10.** $2x - y = 8$ **11.** $5x - 2y = 10$ **12.** $x = -6$

13. $y = 5$ **14.** $4y + 2x = 7$

Directions Sketch the graph of each equation using the x- and y-intercepts (if they exist).

15. $y = 2x - 3$ **16.** $y = -5x + 1$ **17.** $y = 3x$ **18.** $x - 2y = 4$

19. $3y - 2x = 9$ **20.** $y = -6$ **21.** $x = 1$

[7-2]

Directions Find the distance between each pair of points.

22. $(3,2)$ and $(0,4)$ **23.** $(-2,1)$ and $(-2,4)$ **24.** $(1,5)$ and $(-3,5)$

Directions Determine if the lines ℓ_1 and ℓ_2 containing the given points are parallel, perpendicular, or neither.

25. ℓ_1 contains $(-3,2)$ and $(1,3)$
ℓ_2 contains $(5,-1)$ and $(3,2)$

26. ℓ_1 contains $(0,-1)$ and $(2,3)$
ℓ_2 contains $(3,2)$ and $(1,3)$

27. ℓ_1 contains $(4,0)$ and $(-3,-4)$
ℓ_2 contains $(5,6)$ and $(-2,2)$

28. ℓ_1 contains $(0,0)$ and $(9,-3)$
ℓ_2 contains $(-1,-3)$ and $(8,2)$

Directions Determine if the graphs of each pair of equations are parallel lines, perpendicular lines, or neither.

29. $2x + y = 1$ and $4x + 2y = -6$

30. $x - 3y = -2$ and $2x + 3y = 0$

31. $2x - 5y = 3$ and $5x + 2y = 1$

32. $3x - 2y = 1$ and $6x + 9y = 3$

33. A brace for a wall shelf is attached to the wall 9 inches below where the shelf meets the wall and to the bottom of the shelf $6\dfrac{1}{2}$ inches from the wall. What is the slope of the brace?

34. Show that the points $(-3,1)$, $(4,1)$, and $(-3,4)$ are the vertices of a right triangle.

[7-3]

Directions Find the equation of the line satisfying the given conditions. Write the equation in standard form $Ax + By = C$.

35. Slope $m = \dfrac{2}{3}$ and passing through $(-1,5)$

36. Horizontal line through $(-2,4)$

37. Undefined slope and passing through $(1,-9)$

38. Passing through points $(2,-5)$ and $(3,3)$

Directions Write each equation in slope-intercept form $y = mx + b$, identify the slope m and y-intercept b, and sketch the graph of the equation using the slope and y-intercept.

39. $3x - y = 4$ **40.** $2x + 3y = 9$ **41.** $3x - 2y = 0$ **42.** $2y + 3 = 0$

43. Find the equation of the line (in standard form) that passes through the point $(-7,2)$ and is parallel to the line $2y - x = 3$.

44. Find the equation of the line (in standard form) that is perpendicular to the line $4x + 5y = -2$ and passing through the point $(0,3)$.

[7–4]
Directions Sketch the graph of each linear inequality in two variables.

45. $x \geq 0$ **46.** $y < -1$ **47.** $x + 3 < 0$ **48.** $y - 4 \geq 0$

49. $x + y \geq 4$ **50.** $3x - y < 6$ **51.** $x + 4y \geq 8$ **52.** $2x + 7y \leq 14$

Directions Sketch the graph of each absolute value inequality.

53. $|x| \leq 5$ **54.** $|y| > 1$ **55.** $|x + 3| < 2$ **56.** $|x - 4| \geq 1$

57. $|y + 3| \leq 1$ **58.** $|2x - 3| \geq 1$

Chapter 7 cumulative test

Directions Perform the indicated operations and simplify. Assume all variables are nonzero real numbers.

[3–2]   **1.** $(3a^2b^3)(-6ab^2)$

[3–4]   **2.** $\dfrac{x^{-3}y^2}{x^2y^{-4}}$

[3–4]   **3.** $(4a^{-3}b^2c^{-1})^{-2}$

[3–1]   **4.** $5y^2 - \{5y - [4y^2 + 2y - 3]\}$

[3–1]   **5.** $(x - 2y)^2 - (2x + y)^2$

[3–3]   **6.** $6xy(2x^2 - 3y^2 + x^2y - x^3y^3)$

Directions Completely factor the following expressions.

[3–5]   **7.** $3a^2 - 12b^2$

[3–6]   **8.** $8x^2 - 55x - 7$

[3–8]   **9.** $8x^2 + 27y^3$

[3–7]   **10.** $16x^2 - 40xy + 25y^2$

Directions Find the solution set of the following equations and inequalities.

[2–1]   **11.** $3(2a - 5) - 5a = 5(a + 6)$

[2–5]   **12.** $|2x + 3| = 5$

[2–5]   **13.** $|2y - 1| < 4$

[2–5]   **14.** $|y + 3| \geq 1$

[6–1]   **15.** $4x^2 - x = 0$

[6–1]   **16.** $x^2 - 7x - 18 = 0$

[6–2]   **17.** $4y^2 = 2y + 6$

[2–5]   **18.** $-5 < 4 - 3x \leq 5$

[6–3]   **19.** $2x^2 - 3x = 7$

Directions Perform the indicated operations and simplify. Assume all denominators are nonzero.

[4-2] **20.** $\dfrac{y + 2}{y^2 - y - 20} \cdot \dfrac{y^2 - 16}{y^2 - 2y - 8}$

[4-3] **21.** $\dfrac{7y - 3}{2y^2 - 14y} - \dfrac{5y}{y^2 - 49}$

[4-4] **22.** Simplify the complex fraction $\dfrac{\dfrac{4}{y} - \dfrac{3}{x}}{\dfrac{4x - 3y}{xy}}$.

[4-6] **23.** Solve the equation $\dfrac{3}{8x - 1} = \dfrac{2}{2x + 3}$.

[2-3] **24.** Solve the equation $\dfrac{3a - 1}{4} - 2b = \dfrac{5a + 1}{6}$ for b.

Directions Perform the indicated operations, simplify, and rationalize all denominators.

[5-2] **25.** $\sqrt{18} + \sqrt{50} - \sqrt{8}$

[5-5] **26.** $(4 + \sqrt{5})^2$

[5-3] **27.** $\sqrt{\dfrac{9}{5}}$

[5-3] **28.** $\dfrac{3}{2 - \sqrt{3}}$

[5-6] **29.** $(3 + 2i)(3 - 2i)$

[5-6] **30.** $(4 - 3i) - (2 + 4i)$

[6-5] **31.** Solve the equation $\sqrt{x - 3} - 1 = \sqrt{x + 2}$. Name any extraneous solutions.

Directions Find the equation of the line satisfying the following conditions.

[7-3] **32.** Passing through points $(1, -3)$ and $(2,4)$

[7-3] **33.** Passing through $(-1,2)$ and parallel to $2x - 3y = 1$

[7-3] **34.** A vertical line passing through $(5, -3)$

[7-3] **35.** Find the slope and y-intercept of the line $3x - 5y = 10$. Sketch the graph using these facts.

[7-3] **36.** Are the lines $2x - y = 6$ and $4x + 8y = 1$ parallel, perpendicular, or neither?

[7-3] **37.** Find the distance between the points $(1, -3)$ and $(2,5)$. Find the coordinates of the midpoint of the line segment.

Chapter *8*

Systems of Linear Equations

8–1 Systems of linear equations in two variables

In chapter 7 we learned that the graph of a linear equation in two variables is a straight line. In this chapter we will consider the relationship that exists between two or more linear equations involving the same variables. These equations form what we call a **system of linear equations.**

There are a number of physical situations where two equations may be used to find two unknown quantities. For example, suppose we wish to cut a 12-foot-long board into two pieces such that one piece is 6 feet longer than the other. If we let

> x be the length of the shorter piece and
> y be the length of the longer piece,

we can form the two equations $x + y = 12$, since the two pieces make up the 12-foot board, and $y = x + 6$, since y (the longer piece) is 6 feet longer than x (the shorter piece).

These two equations form the linear system

$$x + y = 12$$
$$y = x + 6.$$

We want the ordered pairs, if there are any, that the solution sets of these equations have in common. We can show that the ordered pair (3,9) satisfies both equations and we say that (3,9) is the **simultaneous solution** of the system of equations.

Example 8–1 A

Determine if the given ordered pair is a simultaneous solution of the system of equations.

1. $(x,y) = (2,5)$ and the system
$$x + 2y = 12$$
$$3x - y = 1.$$

If we substitute 2 for x and 5 for y into each equation, we find

$x + 2y = 12$	$3x - y = 1$
$(2) + 2(5) = 12$	$3(2) - (5) = 1$
$2 + 10 = 12$	$6 - 5 = 1$
$12 = 12$ (true)	$1 = 1.$ (true)

Therefore (2,5) is a simultaneous solution of the system of linear equations since it satisfies both of the original equations.

2. $(x,y) = (-5,-2)$ and the system
$$3x - 2y = -11$$
$$2x + y = 10.$$

If we substitute -5 for x and -2 for y into each equation, we find

$3x - 2y = -11$	$2x + y = 10$
$3(-5) - 2(-2) = -11$	$2(-5) + (-2) = 10$
$-15 + 4 = -11$	$-10 + (-2) = 10$
$-11 = -11$ (true)	$-12 = 10.$ (false)

The ordered pair $(-5,-2)$ is not a simultaneous solution of the system of linear equations since it does not satisfy both equations.

If we wish to geometrically represent a system of linear equations, the graph will be a pair of straight lines. Two straight lines, ℓ_1 and ℓ_2, in the plane can be related in one of three ways.

1. They can intersect in only one point. See figure 8.1(a).
2. They can be the same line, in which case they intersect in all points. See figure 8.1(b).
3. They do not intersect, in which case the lines are parallel. See figure 8.1(c).

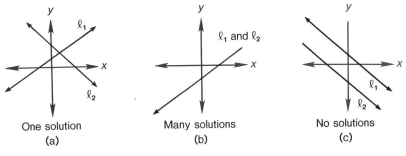

One solution (a)   Many solutions (b)   No solutions (c)

Figure 8.1

313

The three possibilities lead us to conclude that their simultaneous solution(s) must be one of the following:

1. The intersection of the two solution sets contains *only one* ordered pair. See figure 8.1(a). The system is called **consistent and independent.**
2. The intersection of the two solution sets contains all ordered pairs found in either one of the solution sets. See figure 8.1(b). The system is called **dependent.**
3. The intersection of the solution sets contains no ordered pair and is the null (empty) set. See figure 8.1(c). The system is called **inconsistent.**

Example 8–1 B

Solve the system of equations $4x - y = 1$ by graphing.
$$x + y = 4$$
Using the intercepts, we graph each equation.

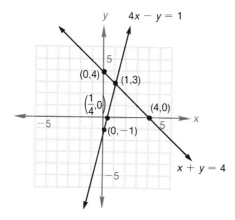

The lines appear to intersect at the point (1,3), so the system is consistent and independent and the intersection of the solution sets is

$$\{(x,y) \,|\, 4x - y = 1\} \cap \{(x,y) \,|\, x + y = 4\} = \{(1,3)\}.$$

Solution by elimination

Since it is time-consuming to graph and it can be difficult to read exact solutions from the graph, we usually use other, algebraic, methods to solve a system. One such method is called the **elimination** (or **addition**) method. The following examples demonstrate this method.

Example 8–1   C   Find the solution set of the following systems by elimination.

1. $4x - 3y = 5$ (1)
  $2x + 3y = 7$ (2)
The elimination method involves adding the respective members of the two equations so that one of the variables is eliminated.

$$4x - 3y = 5$$
$$\underline{2x + 3y = 7}$$
$6x \quad\quad = 12$ Add left and right members.
$\quad\quad x = 2$ Divide each member by 6.

We now replace x by 2 in either equation and solve for y. Using $4x - 3y = 5$,

$4(2) - 3y = 5$
$\quad 8 - 3y = 5$
$\quad\quad -3y = -3$ Subtract 8 from each member.
$\quad\quad\quad y = 1.$ Divide each member by -3.

The simultaneous solution of the system is $x = 2$ and $y = 1$, which we write as the ordered pair (2,1). The solution set is then

$$S = \{(2,1)\},$$

which can be checked by substituting into each equation.

(1) $4(2) - 3(1) = 5$ (2) $2(2) + 3(1) = 7$
$\quad\quad 8 - 3 = 5$ $\quad\quad 4 + 3 = 7$
$\quad\quad 5 = 5$ (true) $\quad\quad 7 = 7$ (true)

Note
We will not check our solution in future examples, but this is something we should get into the habit of doing automatically.

2. $3x - 5y = -1$ (1)
  $2x - 3y = 8$ (2)
One way to eliminate one variable (and there are other ways) is to multiply equation (1) by 2 and equation (2) by -3.

$3x - 5y = -1$ Multiply by 2. $\rightarrow$ $6x - 10y = -2$
$2x - 3y = 8$ Multiply by -3. $\rightarrow$ $\underline{-6x + 9y = -24}$
$\quad\quad\quad\quad\quad\quad\quad\quad\quad\quad$ Add. $\quad\quad -y = -26$
$\quad\quad\quad\quad\quad\quad\quad\quad\quad\quad\quad\quad\quad\quad\quad y = 26$

Substitute 26 for y in equation (2) and solve for x.

$2x - 3(26) = 8$
$\quad 2x - 78 = 8$
$\quad\quad\quad 2x = 86$
$\quad\quad\quad\quad x = 43$

The solution set is $S = \{(43,26)\}$.

3. $3x - 4y = 1$ (1)
$\quad\ 6x - 8y = -5$ (2)

If we multiply the members of $3x - 4y = 1$ by -2, we obtain the system

$$-6x + 8y = -2$$
$$\underline{\ \ 6x - 8y = -5}$$
$$\quad\quad\quad 0 = -7. \text{ (false)}$$

This indicates the system has no solution and the solution set is $S = \emptyset$. The system is *inconsistent*.

4. $3x - 2y = 1$ (1)
$\quad\ 9x - 6y = 3$ (2)

If we multiply each member of equation (1) by -3, we obtain the system

$$-9x + 6y = -3$$
$$\underline{\ \ 9x - 6y = 3}$$
$$\quad\quad\quad 0 = 0. \text{ (true)}$$

By adding the resulting system we obtain a true statement. This indicates every solution for one equation is also a solution for the other. The system is a *dependent* system and the solution set is

$$S = \{(x,y) \,|\, 3x - 2y = 1\}.$$

    To summarize the elimination method of solving a linear system of two equations in two variables, we use the following procedure:

Step 1 Write the system of equations in standard form $Ax + By = C$.

Step 2 Multiply one or both equations by appropriate constants so that one of the variables is eliminated by adding corresponding members.

Step 3 Add the corresponding members of the resulting equations.

Step 4 Solve the resulting equation in *one variable*.

Step 5 Substitute the value of the variable into one of the *original* equations and solve for the other variable.

Step 6 Check the solution into each of the original equations.

Step 7 If we get $0 = a$, then $S = \emptyset$ (inconsistent) and if we get $0 = 0$, then $S =$ the solution set of either equation (dependent).

Solution by substitution

A second algebraic method that is used to solve a system of linear equations involves solving one of the equations for one of the variables and substituting for that variable into the other equation to obtain an equation in one variable. We call this the **substitution** method for solving a system of linear equations.

Example 8–1 D Solve the following systems by substitution.

1. $4x + 3y = 14$ (1)
   $y = 3x - 4$ (2)

Notice that equation (2) is solved for y. Substitute $3x - 4$ for y in equation (1).

$4x + 3(3x - 4) = 14$
$4x + 9x - 12 = 14$
$13x - 12 = 14$ Combine like terms.
$13x = 26$ Add 12 to each member.
$x = 2$ Divide each member by 13.

Since $y = 3x - 4$, substitute 2 for x in this equation.

$y = 3(2) - 4$
$ = 6 - 4$
$y = 2$

The solution set is $S = \{(2,2)\}$.
Check the solution by substituting 2 for x and 2 for y in equation (1).

2. $x - 3y = 4$ (1)
   $3x + 4y = 1$ (2)

Solving the equation $x - 3y = 4$ for x, we have the system

$x = 3y + 4$
$3x + 4y = 1.$

Substituting $3y + 4$ for x into the equation $3x + 4y = 1$, we obtain

$3(3y + 4) + 4y = 1$
$9y + 12 + 4y = 1$
$13y + 12 = 1$
$13y = -11$
$y = -\dfrac{11}{13}.$

To find x, substitute $-\dfrac{11}{13}$ for y into one of the original equations, say $x - 3y = 4$, then

$x - 3\left(-\dfrac{11}{13}\right) = 4$

$x + \dfrac{33}{13} = 4$

$x = 4 - \dfrac{33}{13} = \dfrac{52}{13} - \dfrac{33}{13}$

$x = \dfrac{19}{13}.$

The solution set of the system is $S = \left\{\left(\dfrac{19}{13}, -\dfrac{11}{13}\right)\right\}$.

To summarize the substitution method of solving a linear system of two equations in two variables, we use the following procedures:

Step 1 Solve one of the equations for one of the variables.

Step 2 Substitute the expression obtained in step 1 for that variable in the other equation.

Step 3 Solve the resulting equation in one variable.

Step 4 Substitute the value obtained for the variable into the equation obtained from step 1 and solve for the other variable.

Step 5 Check the solution in the original equations.

Step 6 If we get $0 = a$, then $S = \emptyset$ (inconsistent) and if we get $0 = 0$, then $S =$ the solution set of either equation (dependent).

Being able to solve a system of linear equations lets us solve many verbal problems using two variables as opposed to the one-variable approach we used in chapter 7.

Example 8–1 E

1. The perimeter of a rectangular plot of ground is 238 meters. If the length of the rectangle is 11 meters longer than twice the width, what are the dimensions of the rectangle? [Use perimeter = 2 (length) + 2 (width).]

Let w be the width of the rectangle. From the phrase "the length of the rectangle is 11 meters longer than twice the width," we determine that the length is given by the equation $\ell = 2w + 11$ meters. Since $P = 2\ell + 2w$ and $P = 238$, we substitute 238 for P to obtain the second equation.

$$238 = 2\ell + 2w$$

We form the system of equations

$$\ell = 2w + 11$$
$$2\ell + 2w = 238.$$

Since one equation is already solved for ℓ, we use the substitution method of solving the system. Substitute $2w + 11$ for ℓ in the equation $2\ell + 2w = 238$. Then

$$2(2w + 11) + 2w = 238$$
$$4w + 22 + 2w = 238$$
$$6w + 22 = 238$$
$$6w = 216$$
$$w = 36.$$

Since $\ell = 2w + 11$, then $\ell = 2(36) + 11 = 72 + 11 = 83$. The rectangle is 83 meters long and 36 meters wide.

2. Arlene wishes to invest $5,000. If she invests part at 7% simple interest, part at 6% simple interest, and receives a total interest of $332 after one year, how much does she invest at each rate?

> **Note**
> We must use the formula $i = Prt$, where i is the simple interest received when principal P is invested at rate r for t years. Time will be one year in these problems.

If we let
x = the amount invested at 7% and
y = the amount invested at 6%,
then since the total amount invested is $5,000, we have the equation

$x + y = 5,000$.

The interest i received at 7% is given by $i = 0.07x$ and the interest i received at 6% is given by $i = 0.06y$. Then the second equation of our system is

$0.07x + 0.06y = 332$.

To eliminate decimal numbers, we multiply the terms of this equation by 100 to obtain the equation $7x + 6y = 33,200$. Thus we have the system of equations

$$x + y = 5,000 \qquad \text{Multiply by } -7. \qquad -7x - 7y = -35,000$$
$$7x + 6y = 33,200. \qquad\qquad\qquad\qquad \underline{7x + 6y = 33,200}$$
$$-y = -1,800$$
$$y = 1,800$$

Since $x + y = 5,000$, then

$x + 1,800 = 5,000$
$\qquad\quad x = 3,200$.

Arlene invests $1,800 at 6% interest and $3,200 at 7% interest.

> **Mastery points**
> **Can you**
> - Determine whether an ordered pair is a solution of a system?
> - Solve a system of linear equations in two variables using elimination?
> - Solve a system of linear equations in two variables using substitution?
> - Solve a verbal problem using systems of linear equations in two variables?

Exercise 8–1

Directions Determine whether the given ordered pair is a simultaneous solution of the system of linear equations. See example 8–1 A.

1. $x + y = 6$
   $x - y = -2$; (2,4)

2. $3x - y = 1$
   $x + y = -5$; $(-1,-4)$

3. $2x + 3y = 6$
   $x - 2y = 3$; (3,0)

4. $x - 6y = 0$
   $2x + 3y = 8$; (4,1)

Directions Find the solution set of each system of equations by elimination. If the equations are inconsistent or dependent, so state. See example 8–1 C.

5. $x - y = 3$
   $x + y = -6$

6. $3x + y = 1$
   $2x - y = 9$

7. $x + 4y = 10$
   $x + 2y = 4$

8. $x - y = 3$
   $3x - y = -7$

9. $5x - 3y = 1$
   $2x - 3y = -5$

10. $-5x - y = 4$
    $-5x + 2y = 7$

11. $3x + 2y = 11$
    $x - y = 2$

12. $3x + 4y = 18$
    $2x - y = 1$

13. $4x + y = 11$
    $2x + 3y = 6$

14. $4x - 3y = 7$
    $3x - 2y = 6$

15. $8x - 3y = -4$
    $4x + 7y = 11$

16. $3x - y = 10$
    $6x - 2y = 5$

17. $3x + y = 2$
    $9x + 3y = 6$

18. $-x + 2y = -7$
    $-2x + 6y = 1$

19. $6x - 5y = 0$
    $12x - 10y = -3$

20. $x - y = 9$
    $-4x + 4y = -36$

21. $5x - 2y = 4$
    $3x + 3y = 5$

22. $-2x + y = 6$
    $4x + y = 1$

23. $\frac{1}{2}x + \frac{1}{3}y = 1$
    $\frac{1}{4}x - \frac{2}{3}y = \frac{1}{12}$

24. $\frac{2}{3}x - \frac{1}{4}y = 4$
    $x + \frac{1}{3}y = 2$

25. $\frac{3}{2}x + \frac{2}{5}y = \frac{9}{10}$
    $\frac{1}{2}x + \frac{6}{5}y = \frac{3}{10}$

26. $\frac{5}{7}x - \frac{4}{5}y = \frac{9}{10}$
    $\frac{2}{7}x - \frac{2}{5}y = \frac{3}{10}$

27. $\frac{5}{2}x - y = \frac{-17}{2}$
    $\frac{2}{3}x + \frac{4}{3}y = 1$

28. $(0.3)x - (0.8)y = 0.9$
    $(0.1)x + (0.4)y = 1.2$

29. $(1.3)x + (2.1)y = -0.3$
    $(0.5)x + (1.3)y = 3.2$

30. $x - (0.5)y = 3$
    $(0.9)x + y = -1$

Directions Find the solution set of the following systems of two equations by substitution. If the system is dependent or inconsistent, so state. See example 8–1 D.

31. $2x + y = 10$
    $y = -x + 3$

32. $3x + 2y = 9$
    $y = 2x - 3$

33. $x = y + 5$
    $x + 5y = -4$

34. $5x + y = 10$
    $x = -2 + 3y$

35. $y = 3 - 6x$
    $2x + 5y = 5$

36. $3x - 5y = 4$
    $x + 2y = -2$

37. $x - y = 3$
    $2x + 2y = 11$

38. $3x + 3y = 0$
    $x + y = 5$

39. $5x - y = 8$
    $2x - y = -2$

40. $-x + 3y = 4$
    $-x - 8y = 1$

41. $4x - 3y = 7$
    $y = -5$

42. $2x + 7y = 8$
    $y = 1$

43. $5x - 6y = -7$
$x = 6$

44. $7x + 6y = -9$
$x = -8$

45. $-4x - 3y = 4$
$y = 0$

46. $3x + 2y = -1$
$2x - 3y = 1$

47. $-4x + 7y = 3$
$3x + 2y = 0$

Directions Solve each system of equations by either elimination or substitution. Try to choose the most suitable method. See examples 8–1 C and D.

48. $-6x + 3y = 4$
$-12x - 6y = 1$

49. $3x + 2y = -6$
$-6x - 4y = 1$

50. $5x - 10y = 5$
$x - 2y = 1$

51. $2x - y = 7$
$6x - 3y = 21$

52. $2x - 3y = 5$
$3x + 4y = 1$

53. $4x - 3y = 4$
$-2x + 4y = 3$

54. $7x - 8y = 14$
$5x + 3y = -4$

55. $\dfrac{-1}{3}x + y = 4$

$x = \dfrac{1}{3}y - 1$

56. $\dfrac{2}{3}x - \dfrac{1}{3}y = 4$

$x = \dfrac{1}{3}y - 1$

57. $\dfrac{2}{5}x - \dfrac{3}{5}y = 3$

$y = \dfrac{5}{2}x - 3$

58. $\dfrac{1}{6}x + y = -7$

$y = \dfrac{-2}{3}x + 2$

Directions Find the solution set of the following systems. Let $p = \dfrac{1}{x}$ and $q = \dfrac{1}{y}$. Substitute, solve for p and q, then solve for x and y.

59. $\dfrac{1}{x} + \dfrac{1}{y} = 3$

$\dfrac{1}{x} - \dfrac{1}{y} = -5$

60. $\dfrac{2}{x} - \dfrac{3}{y} = 1$

$\dfrac{1}{x} + \dfrac{2}{y} = 2$

61. $\dfrac{4}{x} + \dfrac{5}{y} = 0$

$\dfrac{3}{x} + \dfrac{2}{y} = 1$

62. $\dfrac{-3}{x} + \dfrac{1}{y} = 5$

$\dfrac{2}{x} - \dfrac{4}{y} = -1$

Directions Find equations for two lines, write them in standard form, and solve the system.

63. Find the equation of the line passing through points $(-1,-2)$ and $(3,4)$ and the equation of the line through points $(4,1)$ and $(2,-4)$. Find their point of intersection by solving the resulting system of linear equations.

64. Do the same as in exercise 63 for the line passing through points $(0,5)$ and $(-6,2)$ and another line passing through points $(1,1)$ and $(5,-7)$.

65. Find the point of intersection of line ℓ_1 having slope 2 and y-intercept -6 and line ℓ_2 having slope -3 and passing through the point $(1,2)$.

66. Find the point of intersection of line ℓ_1 having slope 0 and passing through the point $(-4,-3)$ and line ℓ_2 having slope 4 and y-intercept 2.

67. Find the point of intersection of line ℓ_1 having x-intercept 3 and y-intercept -1 and line ℓ_2 having slope 5 and passing through the point $(2,2)$.

Directions Set up a system of two linear equations and solve. See example 8–1 E–1.

68. The distance around a rectangular flower garden is 64 feet. If the length is three times the width, what are the dimensions?

69. If twice the length of a rectangular floor is increased by three times the width, the sum is 48 feet. The perimeter of the room is 40 feet. What are the dimensions of the floor?

70. The perimeter of a rectangular plot of ground is 30 meters. Three times the length minus four times the width is 3 meters. Find the length and width of the plot.

See example 8–1 E–2.

71. Phoebe wishes to invest $20,000, part at 7% interest and the rest at $6\frac{1}{2}$ %. If her total income from the two investments for one year is $1,370, how much should she invest at each rate?

72. The income from two investments for one year is $1,485. If $19,000 is invested, part at 8% and the rest at $7\frac{1}{2}$ %, how much is invested at each rate?

73. The income from an 8% investment is $300 more than the income from a 6% investment. How much is invested at each rate if a total of $30,000 is invested?

74. Jamie invests a total of $16,000, part at 7% interest and part at 9% interest. If the income from each investment is the same, how much money does Jamie invest at each rate?

75. The income from two investments is the same. If $36,000 is invested, part at 7% and the rest at 8%, how much is invested at each rate?

76. Juanita invests a total of $22,000. She suffers a net loss of $160 on her two investments. If one investment made her a 12% profit and the other investment caused an 8% loss, how much was each investment?

77. Simone makes a 15% profit on one investment but takes a 12% loss on a second investment. If she invests a total of $25,000 and realizes a net gain of $1,320 on her investments, how much is each investment?

78. A 20-foot board must be cut into two pieces so that one piece is 4 feet longer than the other piece. How long is each piece?

79. A 21-foot piece of pipe must be cut into two pieces so that one piece is 9 feet longer than the other piece. How long is each piece of pipe?

80. The sum of two electric currents is 96 amps. If one current is 22 amps less than the other, how many amps are there in each current?

81. The sum of two voltages in an electric circuit is 47 volts and their difference is 25 volts. Find the voltages.

82. The difference in the number of teeth in two gears is 14 and their sum is 72. How many teeth are there in each gear?

Example

Sam's Pizza Emporium sells two kinds of large pizzas at $6.50 and $8.00 per pizza. If Sam sells 43 large pizzas for $306.50, how many of each kind of pizza does he sell?

Solution

Let x = number of $6.50 pizzas sold and y = number of $8.00 pizzas sold. Since 43 pizzas were sold, $x + y = 43$. At $6.50 per pizza, Sam took in $6.50x$ dollars and at $8.00 per pizza, he took in $8.00y$ dollars. Then
$6.50x + 8.00y = \$306.50$.
Multiplying by 10 to clear the decimal point, we have the system

$$x + y = 43 \qquad \text{Multiply by } -65. \qquad -65x - 65y = -2{,}795$$
$$65x + 80y = 3{,}065. \qquad\qquad \underline{65x + 80y = 3{,}065}$$
$$15y = 270$$
$$y = 18$$

Replacing y by 18 in the equation $x + y = 43$, we have $x + 18 = 43$ so $x = 25$. Sam sold 25 pizzas at $6.50 per pizza and 18 pizzas at $8.00 per pizza.

83. A clothing store sells men's suits at $152 and $205 per suit. The store sells 32 suits and takes in $5,659. How many suits at each price does the store sell?

84. A hardware supply company sells two types of doorknobs. The chromium-plated knob sells at $8 per knob and the solid brass knob sells for $11.50 per knob. The company sold 420 doorknobs for $2,867.50. How many of each type were sold?

85. A road construction crew consists of cat operators working at $90 per day and laborers working at $50 per day. The total payroll per day is $1,600. If there are 3 laborers doing odd jobs and 4 laborers are assigned to work with each cat operator, how many laborers are there in the crew?

86. A keypunch operator at a local firm works for $9 per hour and an entry-level typist works for $6.50 per hour. The total pay for an 8-hour day is $476, and there are two more typists than keypunch operators. How many keypunch operators does the firm employ?

87. Skilled and unskilled workers are employed by a construction firm. If 5 skilled workers and 8 unskilled workers are employed, the total wages per day are $948. When 3 skilled and 5 unskilled workers are employed, the total wages per day are $580. What is the daily rate of pay of each type of worker?

Example

An auto mechanic has two bottles of battery acid. One bottle contains a 10% acid solution and the other contains a 4% acid solution. How many cubic centimeters (cm³) of each solution are needed to make 120 cubic centimeters of a 6% acid solution?

Solution

We let $x =$ the number of cubic centimeters of 10% acid solution and $y =$ the number of cubic centimeters of 4% acid solution. Then, since the total mixture will be 120 cm³, we have
$x + y = 120$.
The amount of acid in each bottle is given by $0.10x$ of the 10% acid solution and $0.04y$ of the 4% acid solution, whereas there is $0.06(120)$ acid in the final solution. We then obtain
$0.10x + 0.04y = 0.06(120)$,
and multiplying each term by 100, we have
$10x + 4y = 6(120)$ or $10x + 4y = 720$.
We then must find the simultaneous solution of the system of equations.

$$
\begin{array}{lll}
x + y = 120 & \text{Multiply by } -4. & -4x - 4y = -480 \\
10x + 4y = 720 & & \underline{10x + 4y = 720} \\
& & 6x = 240 \\
& & x = 40
\end{array}
$$

Since $x + y = 120$, then $40 + y = 120$ and $y = 80$.
Thus 40 cm³ of 10% acid solution must be mixed with 80 cm³ of 4% acid solution to obtain 120 cm³ of 6% acid solution.

88. How many grams of silver that is 60% pure must be mixed together with silver that is 35% pure to obtain a mixture of 90 grams of silver that is 45% pure?

89. A metallurgist wishes to form 2,000 kilograms of an alloy that is 80% copper. He is to obtain this alloy by fusing together some alloy that is 60% copper with some alloy that is 85% copper. How many kilograms of each alloy must be used?

90. How much of each substance must be mixed together if a jeweler wishes to form 16 ounces of 65% pure gold from sources that are 50% and 70% pure gold?

91. How many liters of a 3.5% solution and a 6% solution of acid must a chemist mix together to form 800 liters of a 4.5% acid solution?

Example

A boat can travel 24 miles downstream in 2 hours and 16 miles upstream in the same amount of time. What is the speed of the boat in still water and what is the speed of the current? [*Hint:* Use distance (d) = rate (r) × time (t).]

Solution

Let x = speed of the boat in still water and y = speed of the current. Going with the current, the speed of the boat is $x + y$ and going against the current, the boat travels at a speed of $x - y$. Using the formula $t = \dfrac{d}{r}$, the time traveled

(1) downstream (2 hours) $= \dfrac{24}{x + y}$; (2) upstream (2 hours) $= \dfrac{16}{x - y}$.

Since,

(1) $2 = \dfrac{24}{x + y}$, then $2(x + y) = 24$ and $x + y = 12$;

(2) $2 = \dfrac{16}{x - y}$, then $2(x - y) = 16$ and $x - y = 8$.

We then must find the simultaneous solution of the system of equations.

$$
\begin{aligned}
x + y &= 12 \\
\underline{x - y} &= \underline{8} \\
2x &= 20 \\
x &= 10
\end{aligned}
$$

Using $x + y = 12$, then $10 + y = 12$ and $y = 2$.
Thus the boat travels at 10 mph in still water and the current has a speed of 2 mph.

92. An airplane can fly at 268 miles per hour against the wind and 380 miles per hour with the wind. What is the speed of the airplane in still air and what is the speed of the wind?

93. Terry travels upstream in his motorboat at top speed to town, a distance of 24 miles, in $1\frac{1}{2}$ hours. If the return trip downstream takes 1 hour, what is the top speed of Terry's motorboat and what is the speed of the stream?

94. Two canoeists make a 30-mile trip in 7 hours. If they paddle at a rate of 4.5 miles per hour part of the time and 4 miles per hour for the remaining time, how many hours did they travel at each rate?

95. A jogger runs a given distance and then catches a ride back to his home by car. If the round trip of 10 miles takes 1 hour, the car travels at 40 miles per hour, and he jogs at 5 miles per hour, how long did the jogger run?

96. Two trains leave the same city at 2:00 P.M., traveling in opposite directions. If one train travels at 48 mph and the other at 60 mph, at what time will they be 594 miles apart?

97. A mother and her daughter set out at the same time from their home, jogging in opposite directions. Maintaining their normal rate, the two women are 12 miles apart after 2 hours. What is the rate of each if the daughter jogs twice as fast as her mother?

98. Two automobiles start from towns 450 miles apart and travel toward each other. They meet after 5 hours. If one automobile travels 12 miles per hour faster than the other, what is the average speed of each automobile?

99. A cyclist and a pedestrian are 40 miles apart. If they travel toward each other, they will meet in $2\frac{1}{4}$ hours. But if they travel in the same direction, the cyclist will overtake the pedestrian in 5 hours. At what rate is each traveling?

100. Two automobiles are 150 miles apart. If they drive toward each other, they will meet in $1\frac{1}{2}$ hours; if they drive in the same direction, they will meet in 3 hours. What are their speeds?

101. The tickets for a puppet show cost $3.50 for adults and $1.25 for children. If $853.75 in tickets are sold to an audience of 503, how many children's tickets were sold?

102. A movie theater sold 323 tickets for $831.50. If adult tickets cost $3 and children's tickets cost $1.75, how many tickets of each type were sold?

103. Fernando has saved 43 coins in dimes and quarters. If he has saved a total of $7.15, how many dimes and how many quarters does he have?

104. Pam has a collection of fifty-two 13-cent and 20-cent stamps. If the face value of her collection is $9.28, how many 20-cent stamps does she have?

8–2 Systems of linear equations in three variables

Consider the equation $x + 2y - 4z = 12$, which involves three variables x, y, and z. Such an equation is called a **linear** (first-degree) **equation in three variables.** A solution of this equation is an **ordered triple** of real numbers, (x, y, z), if the resulting statement is true when the variables x, y, and z are replaced by real numbers. Then the set of all ordered triples of real numbers that satisfy the equation is the **solution set,** S, of the equation.

To illustrate, the ordered triple $(4,0,-2)$ is a solution of the equation $x + 2y - 4z = 12$ since when we replace x by 4, y by 0, and z by -2, we obtain

$$4 + 2(0) - 4(-2) = 12$$
$$4 + 0 + 8 = 12$$
$$12 = 12. \text{ (true)}$$

Other ordered triples that satisfy the equation are

$$(2,1,-2), (10,1,0), (-4,0,-4), \text{ and } (-8,4,-3).$$

In this section we will discuss the solution(s) of a system of three linear equations in three variables such as

$$3x + 2y - z = 4$$
$$x - 3y + z = -1$$
$$2x + y - z = 0.$$

The graph of a linear equation in three variables is a **plane.** Graphing this requires three-dimensional graphing, which is not very practical in this course. However, as with linear equations in two variables, there are a number of possible solutions.

1. The planes can intersect in one point. See figure 8.2(a).
2. The planes can have no common solutions. Two or more of the planes are parallel. See figure 8.2(b).
3. The planes can intersect in a common line. See figure 8.2(c).
4. The three planes can coincide so that the solution is all points of the plane.

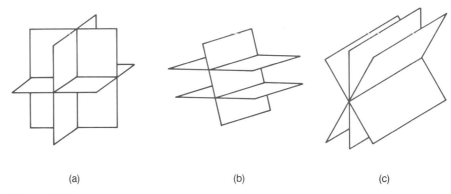

(a)　　　　　　　　(b)　　　　　　　　(c)

Figure 8.2

Since solving a system of three linear equations in three variables by graphing is difficult and impractical, we can find the simultaneous solution by *eliminating* variables as we did with two linear equations in two variables.

Example 8–2　A

Find the solution set of each system of linear equations.

1.　$\begin{aligned} 2x - 5y - z &= -8 \quad (1) \\ -x + 2y + 3z &= 13 \quad (2) \\ x + 3y - z &= 5 \quad (3) \end{aligned}$

The first objective is to obtain an equivalent system of two equations in two variables by eliminating one of the variables. Let us eliminate x.

To begin, use equations (2) and (3). We can eliminate x by adding the respective members of the two equations. Thus

$$\begin{aligned} -x + 2y + 3z &= 13 \\ \underline{x + 3y - z} &= \underline{5} \\ 5y + 2z &= 18, \quad (4) \end{aligned}$$

and we obtain equation (4) involving the variables y and z. We must find another equation involving the variables y and z. Using equation (1) (the equation that has not been involved thus far) and one of the other equations, say (2), we must again eliminate x. Multiply the terms of equation (2) by 2 and add the respective members of the system.

$$\begin{array}{r} 2x - 5y - z = -8 \\ -2x + 4y + 6z = 26 \\ \hline -y + 5z = 18 \qquad (5) \end{array}$$

Now solve the system of linear equations in two variables formed by equations (4) and (5).

$$5y + 2z = 18 \qquad (4)$$
$$-y + 5z = 18 \qquad (5)$$

To do this, use one of the methods that we learned in section 8–1. Multiply the terms of equation (5) by 5 to obtain the system

$$\begin{array}{r} 5y + 2z = 18 \\ -5y + 25z = 90 \\ \hline 27z = 108 \\ z = 4. \end{array}$$

Substitute 4 for z in equation (5), or in equation (4).

$$\begin{array}{r} -y + 5(4) = 18 \\ -y + 20 = 18 \\ -y = -2 \\ y = 2 \end{array}$$

To find the third variable, x, we substitute 2 for y and 4 for z into equation (1), (2), or (3). Using equation (3),

$$\begin{array}{r} x + 3y - z = 5 \\ x + 3(2) - 4 = 5 \\ x + 6 - 4 = 5 \\ x + 2 = 5 \\ x = 3. \end{array}$$

We have found the ordered triple (3,2,4), to be the only simultaneous solution of the system. The solution set is $S = \{(3,2,4)\}$.

In general, to find the solution set of a system of three linear equations in three variables we use the following procedure:

Step 1 Eliminate any one variable from any two of the given three equations to obtain an equation in two variables.

Step 2 Eliminate the *same* variable using the equation not yet involved and one of the other two equations. The result is another equation in the same two variables as in step 1.

Step 3 Solve the resulting system of linear equations in the two variables found in steps 1 and 2. (If this system is dependent or inconsistent, our given system is also dependent or inconsistent.)

Step 4 Substitute the values of the variables found in step 3 into any one of the three original equations to find the value of the third variable.

2. $2x + z = 7$ (1)
$x + y = 2$ (2)
$y - z = -2$ (3)

Note that one variable is missing in each equation and it is not the same variable. Suppose we use equations (1) and (2). Since equation (3) does not contain an x, we must eliminate x when working with equations (1) and (2) to obtain a system of two equations in two variables. Multiply the members of equation (2) by -2 and add.

$$\begin{array}{r} 2x + z = 7 \\ -2x - 2y = -4 \\ \hline z - 2y = 3 \qquad (4) \end{array}$$

Now solve the system involving equations (3) and (4) for y and z.

$$\begin{array}{r} y - z = -2 \qquad (3) \\ -2y + z = 3 \qquad (4) \\ \hline \text{(add)} \qquad -y = 1 \\ y = -1 \end{array}$$

Replace y by -1 in equation (3).

$$\begin{array}{r} y - z = -2 \\ (-1) - z = -2 \\ -z = -1 \\ z = 1 \end{array}$$

Finally, replace z by 1 in equation (1).

$$\begin{array}{r} 2x + z = 7 \\ 2x + 1 = 7 \\ 2x = 6 \\ x = 3 \end{array}$$

Thus the ordered triple $(3, -1, 1)$ is the only simultaneous solution and the solution set is $S = \{(3, -1, 1)\}$.

3. $2x - 11y + 3z = 2$ (1)
$x - 3y + z = 1$ (2)
$x - 8y + 2z = -1$ (3)

Suppose we choose to eliminate z. Using equations (1) and (2), we multiply the members of equation (2) by -3 to obtain the system

$$\begin{array}{r} 2x - 11y + 3z = 2 \\ -3x + 9y - 3z = -3 \\ \hline \text{(add)} \qquad -x - 2y = -1. \qquad (4) \end{array}$$

Using equations (2) and (3), we again must eliminate z. Multiply equation (2) by -2 to obtain the system

$$\begin{array}{r} -2x + 6y - 2z = -2 \\ x - 8y + 2z = -1 \\ \hline \text{(add)} \qquad -x - 2y = -3. \qquad (5) \end{array}$$

We now solve the system of equations (4) and (5).

$$-x - 2y = -1 \qquad (4)$$
$$-x - 2y = -3 \qquad (5)$$

Multiply by -1.

$$x + 2y = 1$$
$$\underline{-x - 2y = -3}$$
$$0 = -2 \text{ (false)}$$

Thus the system is inconsistent and so the solution set is $S = \emptyset$.

4.
$$3x - 2y - z = 3 \qquad (1)$$
$$-x + \frac{2}{3}y + \frac{1}{3}z = -1 \qquad (2)$$
$$6x - 4y - 2z = 6 \qquad (3)$$

If we multiply the terms of equation (1) by 2, we obtain equation (3),

$$6x - 4y - 2z = 6,$$

and if we multiply equation (2) by -6, we again obtain equation (3). Because of this the system is *dependent*, and all three equations have the same graph. The solution set is $\{(x,y,z) \mid 3x - 2y - z = 3\}$, or use either of the other equations.

Now we will consider verbal problems with three unknown quantities and set up a system of three equations in three variables.

Example 8–2 B

The sum of the measure of the three angles of a triangle is 180°. The middle-sized angle has measure 8° less than twice the measure of the smallest angle, and the largest angle has measure 20° less than the sum of the measure of the other two angles. Find the measure of the three angles of the triangle.

Let $x =$ the measure of the smallest angle
$\quad y =$ the measure of the middle-sized angle
$\quad z =$ the measure of the largest angle.

From "the sum of the measure of the three angles of a triangle is 180°,"

$$x + y + z = 180.$$

From "the middle-sized angle (y) has measure 8° less than twice the measure of the smallest angle (x),"

$$y = 2x - 8.$$

From "the largest angle has measure 20° less than the sum of the measure of the other two angles,"

$$z = x + y - 20.$$

Thus the answer to our problem is the simultaneous solution of the system of equations

$$x + y + z = 180$$
$$y = 2x - 8$$
$$z = x + y - 20.$$

Rewrite the system in the standard form.

$$\begin{aligned} x + y + z &= 180 \quad (1) \\ -2x + y &= -8 \quad (2) \\ -x - y + z &= -20 \quad (3) \end{aligned}$$

We solve the system by elimination. Since the variable z is missing from equation (2), we use equations (1) and (3) and eliminate z. Multiplying equation (3) by -1, we have the system

$$\begin{aligned} x + y + z &= 180 \\ \underline{x + y - z} &= \underline{20} \\ 2x + 2y &= 200. \end{aligned}$$

We now solve the system

$$\begin{aligned} -2x + y &= -8 \\ \underline{2x + 2y} &= \underline{200} \\ 3y &= 192 \\ y &= 64. \end{aligned}$$

Using $-2x + y = -8$, replace y by 64 and solve for x.

$$\begin{aligned} -2x + 64 &= -8 \\ -2x &= -72 \\ x &= 36 \end{aligned}$$

Replacing x by 36 and y by 64 in equation (1),

$$\begin{aligned} 36 + 64 + z &= 180 \\ z &= 80. \end{aligned}$$

The three angles of the triangle measure $36°$, $64°$, and $80°$.

Mastery points

Can you
- Find the solution set of a system of three linear equations in three variables?
- Determine when a system of three linear equations in three variables is dependent or inconsistent?
- Solve a verbal problem by setting up a system of three equations in three variables and solving the problem?

Exercise 8–2

Directions   Find the solution set of the given system of equations. If the system is dependent or inconsistent, so state. See example 8–2 A.

1. $\begin{aligned} x + y + z &= 6 \\ x - 2y - z &= -1 \\ x + y - z &= 2 \end{aligned}$

2. $\begin{aligned} x + y - z &= 9 \\ x + y + z &= 5 \\ x - y - z &= 1 \end{aligned}$

3. $\begin{aligned} 2x + 3y - z &= 7 \\ x + y + z &= 6 \\ 3x - y - z &= 6 \end{aligned}$

4. $\begin{aligned} x + y + z &= 1 \\ 2x - y + 3z &= 2 \\ 2x - y - z &= 2 \end{aligned}$

5. $-2x + y + 4z = 3$
$x + y - 3z = 2$
$x + y - 2z = 1$

6. $3x - y + z = -8$
$4x - 2y - 3z = 3$
$2x + 3y - 2z = -1$

7. $-x + y - z = -6$
$2x + 3y - z = 1$
$x + 2y + 2z = 5$

8. $x + 2y + 3z = 5$
$-x + y - z = -6$
$2x + y + 4z = 4$

9. $-x + y + z = -3$
$3x + 9y + 5z = 5$
$x + 3y + 2z = 4$

10. $3x - 2y + 3z = 11$
$2x + 3y - 2z = -5$
$x + 4y - z = -5$

11. $6x + 4y + 2z = -1$
$3x + 2y + z = 3$
$x - 3y + z = 4$

12. $x - 2y + 3z = 4$
$3x - 3y + 4z = 5$
$2x - y + z = 1$

13. $x - 4y + z = -5$
$3x - 12y + 3z = -15$
$-2x + 8y - 2z = 10$

14. $5x - 2y + z = 6$
$-2x - 3y + 4z = -2$
$4x + 6y - 8z = 4$

15. $-5x + 3y - 2z = -13$
$4x - 2y + 5z = 13$
$2x + 4y - 3z = -9$

16. $4x + 6y + 5z = 2$
$-2x + 3y - 2z = -10$
$5x + 2y + 3z = -2$

17. $7x + 8y - 2z = -5$
$-2x + 5y + z = -3$
$5x + 14y - z = -11$

18. $x - y = -1$
$x + z = -2$
$y - z = 2$

19. $x + 2y = 10$
$-x + 3z = -23$
$4y - z = 9$

20. $3y + 4z = 12$
$2x - 5y = 2$
$-4x + 3z = 5$

21. $2x + z = 0$
$-4x + y = 1$
$3y + z = -7$

22. $2y + z = -4$
$y = -3$
$x - 3y + 2z = 9$

23. $4x + y - 2z = 1$
$x - y = 5$
$z = -2$

24. $3x + 2y = -3$
$-2x + 5z = 38$
$4y + 3z = 12$

25. $2x - 5y = 2$
$-4x + 3z = 5$
$3y + 4z = 12$

26. $2x - y + 3z = -5$
$x + y = -4$
$2x - y + 2z = 6$

27. $x + y - z = 8$
$-x + y + z = -3$
$y + z = 5$

28. $3x + y + 5z = 3$
$5x + y - z = -3$
$8x + 2y - z = -5$

29. $2x + 8y - 2z = 6$
$3x + 12y - 3z = 9$
$-x - 4y + z = -3$

30. $x + y - 3z = 4$
$-3x + y - z = 2$
$2x + 2y - 6z = 5$

Directions In each of the following exercises, set up a system of three equations in three variables and solve. See example 8–2 B.

31. The sum of the measures of the three angles of a triangle is 180°. If the largest angle is 20° more than the sum of the other two angles and the smallest angle is 67° less than the larger angle, find the measure of the three angles of the triangle.

32. The sum of the measures of the three angles of a triangle is 180°. The sum of the smallest angle and the largest angle is 120°. If the middle-sized angle has measure 30° more than the smallest angle, what is the measure of the three angles of the triangle?

33. The perimeter of a triangular-shaped garden is 122 meters. If the length of the longest side is equal to the sum of the lengths of the other two sides, and twice the length of the shortest side is 11 meters less than the length of the longest side, find the lengths of the three sides. (*Note:* Perimeter is the distance around the triangle.)

34. The longest side of a triangle is twice the length of the shortest side, and the middle-length side is 9 inches longer than the shortest side. If the perimeter is 65 inches, what are the lengths of the three sides?

35. Tickets for a Harry Belafonte concert have three prices, "expensive," "middle-priced," and "cheapest." The "middle-priced" tickets cost $4 more than the "cheapest," and the "expensive" tickets cost $6 more than the "middle-priced" tickets. If the "expensive" tickets cost $1 less than twice the "cheapest" tickets, find the price of each kind of ticket.

36. A used-car salesman must sell a quota of cars before receiving a bonus. The cars are placed in three different price categories—A, B, and C. He must sell two more at price B than at price A and three times as many at price C as at price A. If the number at price C is one more than twice the number at price B, find the number of each car that he must sell.

37. Bill has a special stamp collection that is worth approximately $151,000 on the market. The stamps are separated into three different approximate price categories—$750, $1,500, and $25,000 per stamp. If the number of $750 stamps is four times the number of $25,000 stamps and the number of $1,500 stamps is ten more than the number of $750 stamps, how many of each kind does Bill's collection contain?

38. Find the values of a, b, and c so that the points (0,5), (−1,2), and (2,17) lie on the graph of $y = ax^2 + bx + c$. (*Hint:* Substitute the coordinates of each point into the equation to obtain three linear equations in variables a, b, and c.)

39. Find the values of a, b, and c so that the points (1,−2), (0,2), and (−2,13) lie on the graph of $y = ax^2 + bx + c$.

40. Find the values of a, b, and c so that the points (2, 8), (1, 2), and (3, 22) lie on the graph of $y = ax^2 + bx + c$.

41. Jay has 33 bills in denominations of fives, tens, and twenties. If he has $360 total and the number of five-dollar bills is two more than the number of twenty-dollar bills, how many of each denomination does he have?

42. Erin has a collection of pennies, nickels, and dimes in her piggy bank. She has twice as many pennies as dimes and eight more nickels than dimes. If she has $2.44 altogether, how many of each coin does she have?

8–3 Solutions of systems of equations by determinants

In sections 8–1 and 8–2 we solved systems of equations in two and three variables by using algebraic methods involving the elimination of a variable and by substituting an expression for one of the variables. We now consider another method for solving these systems by using **determinants.** A determinant is the number associated with an array of numbers called a **matrix.**

> ■ **Definition of a matrix**
> A **matrix** is an ordered array of rows and columns of numbers.

The array is enclosed in brackets, as illustrated by the following examples.

I

$$\begin{bmatrix} 1 & 2 \\ 3 & 4 \end{bmatrix}$$

II

$$\begin{bmatrix} 4 & -7 & 5 \\ 2 & 7 & 4 \\ 0 & -1 & 6 \end{bmatrix}$$

III

$$\begin{bmatrix} -3 & 5 & 9 \\ 8 & 0 & -5 \end{bmatrix}$$

IV

$$\begin{bmatrix} -9 \\ 3 \\ 1 \end{bmatrix}$$

The numbers that make up the matrix are called the **elements** of the matrix. The elements that appear next to one another horizontally form a **row** of the matrix, and the elements that appear vertically form a **column** of the matrix.

Matrices are named according to the number of rows and the number of columns they contain. Our examples are

   I. 2×2 (read "two-by-two")
  II. 3×3 (read "three-by-three")
 III. 2×3 (read "two-by-three")
 IV. 3×1 (read "three-by-one").

A **square matrix** is one that has the same number of rows as columns. The 2×2 and 3×3 matrices in our examples are square matrices. Associated with each square matrix having real number entries is a real number that is called a *determinant*. Thus for the 2×2 matrix

$$\begin{bmatrix} a_1 & b_1 \\ a_2 & b_2 \end{bmatrix},$$

the determinant is written as

$$\begin{vmatrix} a_1 & b_1 \\ a_2 & b_2 \end{vmatrix}.$$

  ■ **Definition of a 2 × 2 determinant**

$$\begin{vmatrix} a_1 & b_1 \\ a_2 & b_2 \end{vmatrix} = a_1 b_2 - a_2 b_1.$$

The value of this 2×2 determinant can be indicated as

$$\begin{vmatrix} a_1 & b_1 \\ a_2 & b_2 \end{vmatrix} = a_1 b_2 - a_2 b_1.$$

Example 8–3 A

Find the value of each determinant.

1. $\begin{vmatrix} 1 & -2 \\ 3 & 4 \end{vmatrix} = (1)(4) - (3)(-2) = 4 + 6 = 10$

2. $\begin{vmatrix} 0 & 3 \\ -4 & -1 \end{vmatrix} = (0)(-1) - (-4)(3) = 0 + 12 = 12$

A 3 × 3 determinant can be evaluated in a similar way.

■ **Definition of a 3 × 3 determinant**

$$\begin{vmatrix} a_1 & b_1 & c_1 \\ a_2 & b_2 & c_2 \\ a_3 & b_3 & c_3 \end{vmatrix} = a_1b_2c_3 - a_1b_3c_2 + a_3b_1c_2 - a_2b_1c_3 + a_2b_3c_1 - a_3b_2c_1$$

Since it would be difficult to remember the indicated products in the definition, a method for calculating this determinant has been developed.

We rewrite the definition as

$$\begin{vmatrix} a_1 & b_1 & c_1 \\ a_2 & b_2 & c_2 \\ a_3 & b_3 & c_3 \end{vmatrix} = a_1b_2c_3 - a_1b_3c_2 - a_2b_1c_3 + a_2b_3c_1 + a_3b_1c_2 - a_3b_2c_1,$$

and factoring by pairs of terms, we obtain

$$(1) \quad \begin{vmatrix} a_1 & b_1 & c_1 \\ a_2 & b_2 & c_2 \\ a_3 & b_3 & c_3 \end{vmatrix} = a_1(b_2c_3 - b_3c_2) - a_2(b_1c_3 - b_3c_1) + a_3(b_1c_2 - b_2c_1).$$

Now we define the minor of an element of a determinant.

■ **Definition of a minor**

The **minor** of an element of a 3 × 3 determinant is defined to be the 2 × 2 determinant that remains after the row and column in which the element appears have been deleted.

Thus given the 3 × 3 determinant
$$\begin{vmatrix} a_1 & b_1 & c_1 \\ a_2 & b_2 & c_2 \\ a_3 & b_3 & c_3 \end{vmatrix},$$
the minor of a_1 is

$$\begin{vmatrix} a_1 & b_1 & c_1 \\ a_2 & b_2 & c_2 \\ a_3 & b_3 & c_3 \end{vmatrix} \rightarrow \begin{vmatrix} b_2 & c_2 \\ b_3 & c_3 \end{vmatrix} = b_2c_3 - b_3c_2,$$

the minor of a_2 is

$$\begin{vmatrix} a_1 & b_1 & c_1 \\ a_2 & b_2 & c_2 \\ a_3 & b_3 & c_3 \end{vmatrix} \rightarrow \begin{vmatrix} b_1 & c_1 \\ b_3 & c_3 \end{vmatrix} = b_1c_3 - b_3c_1,$$

and the minor of a_3 is

$$\begin{vmatrix} a_1 & b_1 & c_1 \\ a_2 & b_2 & c_2 \\ a_3 & b_3 & c_3 \end{vmatrix} \rightarrow \begin{vmatrix} b_1 & c_1 \\ b_2 & c_2 \end{vmatrix} = b_1c_2 - b_2c_1.$$

Thus the determinant (1) can be expanded by

$$(2) \quad \begin{vmatrix} a_1 & b_1 & c_1 \\ a_2 & b_2 & c_2 \\ a_3 & b_3 & c_3 \end{vmatrix} = a_1 \begin{vmatrix} b_2 & c_2 \\ b_3 & c_3 \end{vmatrix} - a_2 \begin{vmatrix} b_1 & c_1 \\ b_3 & c_3 \end{vmatrix} + a_3 \begin{vmatrix} b_1 & c_1 \\ b_2 & c_2 \end{vmatrix}.$$

The right member of (2) is called the **expansion of the determinant by minors about the first column.**

Example 8–3 B

Expand the determinant by minors about the first column.

$$\begin{vmatrix} 2 & -3 & -1 \\ 1 & 2 & 3 \\ -4 & -2 & 1 \end{vmatrix} = 2 \begin{vmatrix} 2 & 3 \\ -2 & 1 \end{vmatrix} - 1 \begin{vmatrix} -3 & -1 \\ -2 & 1 \end{vmatrix} + (-4) \begin{vmatrix} -3 & -1 \\ 2 & 3 \end{vmatrix}$$

$$= 2[(2)(1) - (-2)(3)] - 1[(-3)(1) - (-2)(-1)]$$
$$\quad + (-4)[(-3)(3) - (2)(-1)]$$
$$= 2(2 + 6) - 1(-3 - 2) + (-4)(-9 + 2)$$
$$= 2(8) - 1(-5) + (-4)(-7)$$
$$= 16 + 5 + 28$$
$$= 49$$

With the proper use of signs we could expand this determinant by minors about *any row* or *any column* and obtain the same numerical value. To determine the correct signs in the expansion of a 3 × 3 determinant by minors about other rows and columns, we use the following **sign array** of alternating signs.

$$
\begin{array}{ccc}
+ & - & + \\
- & + & - \\
+ & - & +
\end{array}
$$

Note

These are signs of *position only.*

Example 8–3 C

Expand the determinant by minors about *row two.*

$$
\begin{vmatrix} 2 & -3 & -1 \\ 1 & 2 & 3 \\ -4 & -2 & 1 \end{vmatrix} = -1 \begin{vmatrix} -3 & -1 \\ -2 & 1 \end{vmatrix} + 2 \begin{vmatrix} 2 & -1 \\ -4 & 1 \end{vmatrix} - 3 \begin{vmatrix} 2 & -3 \\ -4 & -2 \end{vmatrix}
$$

$$
\begin{aligned}
&= -1[(-3)(1) - (-2)(-1)] + 2[(2)(1) - (-4)(-1)] \\
&\quad - 3[(2)(-2) - (-4)(-3)] \\
&= -1(-3 - 2) + 2(2 - 4) - 3(-4 - 12) \\
&= -1(-5) + 2(-2) - 3(-16) \\
&= 5 - 4 + 48 \\
&= 49
\end{aligned}
$$

The real number value of the determinant is the same.

The expansion of higher-order determinants by minors can be accomplished in the same way as with third-order determinants. The pattern of alternating signs in the sign array extends to higher-order determinants. The determinants that are the minors in each expansion will be of order one less than the order of the original determinant. The following is the sign array of a 4 × 4 determinant.

$$
\begin{array}{cccc}
+ & - & + & - \\
- & + & - & + \\
+ & - & + & - \\
- & + & - & +
\end{array}
$$

Determinants can be used to solve systems of linear equations. Consider the system of two linear equations in two variables

$$
a_1x + b_1y = c_1 \\
a_2x + b_2y = c_2.
$$

Suppose we solve the system of equations by elimination. To eliminate x multiplying each term of the first equation by $-a_2$ and each term of the second equation by a_1, we obtain the equivalent system

$$-a_1a_2x - a_2b_1y = -a_2c_1$$
$$a_1a_2x + a_1b_2y = a_1c_2.$$

Adding the terms, the resulting equation in y is given by

$$a_1b_2y - a_2b_1y = a_1c_2 - a_2c_1.$$

Factoring y from the terms in the left member, we obtain

$$(a_1b_2 - a_2b_1)y = a_1c_2 - a_2c_1,$$

and dividing each member by $a_1b_2 - a_2b_1$, we have

$$y = \frac{a_1c_2 - a_2c_1}{a_1b_2 - a_2b_1} \qquad (a_1b_2 - a_2b_1 \neq 0).$$

In like fashion, if we solve the system for x, we obtain

$$x = \frac{c_1b_2 - c_2b_1}{a_1b_2 - a_2b_1} \qquad (a_1b_2 - a_2b_1 \neq 0).$$

Now by definition,

$$\begin{vmatrix} a_1 & b_1 \\ a_2 & b_2 \end{vmatrix} = a_1b_2 - a_2b_1 \text{ (which we denote by } D),$$

$$\begin{vmatrix} c_1 & b_1 \\ c_2 & b_2 \end{vmatrix} = c_1b_2 - c_2b_1 \text{ (which we denote by } D_x),$$

$$\begin{vmatrix} a_1 & c_1 \\ a_2 & c_2 \end{vmatrix} = a_1c_2 - a_2c_1 \text{ (which we denote by } D_y).$$

Note

The elements in D are the coefficients of the variables in the two equations, whereas the elements in
1. D_x contain the coefficients of the variables with the constants c_1 and c_2 replacing the coefficients of x and
2. D_y contain the coefficients of the variables with the constants c_1 and c_2 replacing the coefficients of y.

Solving a system of linear equations using determinants is known as **Cramer's Rule.**

> **■ Cramer's Rule for 2 × 2 systems**
> Given the system of linear equations $a_1x + b_1y = c_1$
> $a_2x + b_2y = c_2,$
> then
> $$x = \frac{\begin{vmatrix} c_1 & b_1 \\ c_2 & b_2 \\ a_1 & b_1 \\ a_2 & b_2 \end{vmatrix}}{} = \frac{D_x}{D} \quad \text{and} \quad y = \frac{\begin{vmatrix} a_1 & c_1 \\ a_2 & c_2 \\ a_1 & b_1 \\ a_2 & b_2 \end{vmatrix}}{} = \frac{D_y}{D},$$
> where $D = a_1b_2 - a_2b_1 \neq 0.$

Example 8–3 D

Use Cramer's Rule to find the solution set of each system of linear equations.
$2x - 3y = 4$
$x + 4y = -1$

By Cramer's Rule, $x = \dfrac{D_x}{D}$ and $y = \dfrac{D_y}{D}$. Consider D first since if $D = 0$,

the rule does not work and the system is then inconsistent or dependent. To find D, we use the coefficients of the variables, so

$$D = \begin{vmatrix} 2 & -3 \\ 1 & 4 \end{vmatrix} = 2(4) - 1(-3) = 8 + 3 = 11.$$

To find D_x, replace the column $\begin{smallmatrix} 2 \\ 1 \end{smallmatrix}$ of D by the constants $\begin{smallmatrix} 4 \\ -1 \end{smallmatrix}$. Thus

$$D_x = \begin{vmatrix} 4 & -3 \\ -1 & 4 \end{vmatrix} = 4(4) - (-1)(-3) = 16 - 3 = 13.$$

To find D_y, replace the column $\begin{smallmatrix} -3 \\ 4 \end{smallmatrix}$ of D by the constants $\begin{smallmatrix} 4 \\ -1 \end{smallmatrix}$. Thus

$$D_y = \begin{vmatrix} 2 & 4 \\ 1 & -1 \end{vmatrix} = 2(-1) - 1(4) = -2 - 4 = -6.$$

Then

$$x = \frac{D_x}{D} = \frac{13}{11} \text{ and } y = \frac{D_y}{D} = \frac{-6}{11},$$

and the solution set of the system is $S = \left\{ \left(\dfrac{13}{11}, -\dfrac{6}{11} \right) \right\}.$

If $D = 0$, the system of equations is then either inconsistent or dependent. When

$$D = 0 \text{ and } D_y \neq 0, D_x \neq 0,$$

the system of equations is inconsistent. If

$$D = 0 \text{ and } D_y = D_x = 0,$$

the system of equations is dependent.

Now we consider a system of three linear equations in three variables. The procedure is similar to that used to solve a system of two linear equations in two variables. That is, given the system of equations

$$a_1x + b_1y + c_1z = d_1$$
$$a_2x + b_2y + c_2z = d_2$$
$$a_3x + b_3y + c_3z = d_3,$$

the solution to the system is given by

$$x = \frac{D_x}{D}, y = \frac{D_y}{D}, z = \frac{D_z}{D}, D \neq 0,$$

where

$$D = \begin{vmatrix} a_1 & b_1 & c_1 \\ a_2 & b_2 & c_2 \\ a_3 & b_3 & c_3 \end{vmatrix},$$

$$D_x = \begin{vmatrix} d_1 & b_1 & c_1 \\ d_2 & b_2 & c_2 \\ d_3 & b_3 & c_3 \end{vmatrix},$$

$$D_y = \begin{vmatrix} a_1 & d_1 & c_1 \\ a_2 & d_2 & c_2 \\ a_3 & d_3 & c_3 \end{vmatrix},$$

$$D_z = \begin{vmatrix} a_1 & b_1 & d_1 \\ a_2 & b_2 & d_2 \\ a_3 & b_3 & d_3 \end{vmatrix}.$$

Example 8–3 E

Use Cramer's Rule to find the solution set of the system of linear equations

$2x - 2y + z = 0$
$x + 5y - 7z = 3$
$x - y - 3z = -7.$

We first evaluate the denominator D by minors about the first column.

$$D = \begin{vmatrix} 2 & -2 & 1 \\ 1 & 5 & -7 \\ 1 & -1 & -3 \end{vmatrix} = 2 \begin{vmatrix} 5 & -7 \\ -1 & -3 \end{vmatrix} - 1 \begin{vmatrix} -2 & 1 \\ -1 & -3 \end{vmatrix} + 1 \begin{vmatrix} -2 & 1 \\ 5 & -7 \end{vmatrix}$$

$$= 2[(5)(-3) - (-1)(-7)] - 1[(-2)(-3) - (-1)(1)]$$
$$+ 1[(-2)(-7) - (5)(1)]$$
$$= 2(-15 - 7) - 1(6 + 1) + 1(14 - 5)$$
$$= 2(-22) - 1(7) + 1(9) = -44 - 7 + 9$$
$$= -42$$

Evaluating each numerator by minors about the first column, we have

$$D_x = \begin{vmatrix} 0 & -2 & 1 \\ 3 & 5 & -7 \\ -7 & -1 & -3 \end{vmatrix} = 0 \begin{vmatrix} 5 & -7 \\ -1 & -3 \end{vmatrix} - 3 \begin{vmatrix} -2 & 1 \\ -1 & -3 \end{vmatrix} + (-7) \begin{vmatrix} -2 & 1 \\ 5 & -7 \end{vmatrix}$$

$$= 0[(5)(-3) - (-1)(-7)] - 3[(-2)(-3) - (-1)(1)]$$
$$- 7[(-2)(-7) - (5)(1)]$$
$$= 0(-15 - 7) - 3(6 + 1) - 7(14 - 5)$$
$$= 0 - 3(7) - 7(9) = 0 - 21 - 63 = -84,$$

$$D_y = \begin{vmatrix} 2 & 0 & 1 \\ 1 & 3 & -7 \\ 1 & -7 & -3 \end{vmatrix} = 2 \begin{vmatrix} 3 & -7 \\ -7 & -3 \end{vmatrix} - 1 \begin{vmatrix} 0 & 1 \\ -7 & -3 \end{vmatrix} + 1 \begin{vmatrix} 0 & 1 \\ 3 & -7 \end{vmatrix}$$

$$= 2[(3)(-3) - (-7)(-7)] - 1[(0)(-3) - (-7)(1)]$$
$$+ 1[(0)(-7) - (3)(1)]$$
$$= 2(-9 - 49) - 1(0 + 7) + 1(0 - 3)$$
$$= 2(-58) - 1(7) + 1(-3) = -116 - 7 - 3 = -126,$$

$$D_z = \begin{vmatrix} 2 & -2 & 0 \\ 1 & 5 & 3 \\ 1 & -1 & -7 \end{vmatrix} = 2\begin{vmatrix} 5 & 3 \\ -1 & -7 \end{vmatrix} - 1\begin{vmatrix} -2 & 0 \\ -1 & -7 \end{vmatrix} + 1\begin{vmatrix} -2 & 0 \\ 5 & 3 \end{vmatrix}$$

$$= 2[(5)(-7) - (-1)(3)] - 1[(-2)(-7) - (-1)(0)]$$
$$+ 1[(-2)(3) - (5)(0)]$$
$$= 2(-35 + 3) - 1(14 - 0) + 1(-6 - 0)$$
$$= 2(-32) - 1(14) + 1(-6) = -64 - 14 - 6 = -84.$$

Therefore $x = \dfrac{-84}{-42} = 2$, $y = \dfrac{-126}{-42} = 3$, $z = \dfrac{-84}{-42} = 2$, and the

solution set is given by $S = \{(2,3,2)\}$.

Again, when $D = 0$, the system is either dependent or inconsistent depending on whether D_x, D_y, and D_z are zero.

Note
When a variable (or variables) is missing in any equation, a zero must be inserted in the determinant.

Mastery points
Can you
• Evaluate 2 × 2 and 3 × 3 determinants?
• Solve a system of two linear equations in two variables using determinants?
• Solve a system of three linear equations in three variables using determinants?

Exercise 8-3

Dircotiono Evaluate the following determinants. See example 8 3 A.

1. $\begin{vmatrix} 1 & -3 \\ 2 & 5 \end{vmatrix}$

2. $\begin{vmatrix} 4 & 1 \\ 2 & 7 \end{vmatrix}$

3. $\begin{vmatrix} 4 & -2 \\ -6 & -3 \end{vmatrix}$

4. $\begin{vmatrix} 2 & 5 \\ 4 & 10 \end{vmatrix}$

5. $\begin{vmatrix} 0 & -5 \\ 1 & 2 \end{vmatrix}$

6. $\begin{vmatrix} 4 & 0 \\ 2 & -7 \end{vmatrix}$

7. $\begin{vmatrix} 5 & 4 \\ 0 & -7 \end{vmatrix}$

8. $\begin{vmatrix} 1 & 2 \\ 3 & 0 \end{vmatrix}$

Directions Evaluate the following determinants using expansion by minors about the first column. See example 8–3 B.

9. $\begin{vmatrix} 1 & 2 & 3 \\ 3 & 2 & 1 \\ 2 & 1 & 3 \end{vmatrix}$

10. $\begin{vmatrix} -2 & 1 & 3 \\ 1 & 0 & 4 \\ -1 & 4 & 3 \end{vmatrix}$

11. $\begin{vmatrix} -3 & 2 & 1 \\ -1 & 3 & 2 \\ 1 & 4 & 5 \end{vmatrix}$

12. $\begin{vmatrix} 3 & 0 & 1 \\ 2 & 0 & 4 \\ 3 & 0 & -1 \end{vmatrix}$

13. $\begin{vmatrix} 3 & -1 & 2 \\ 0 & 0 & 0 \\ 4 & 3 & 1 \end{vmatrix}$

14. $\begin{vmatrix} 2 & 0 & 0 \\ 0 & 2 & 0 \\ 0 & 0 & 2 \end{vmatrix}$

Directions Evaluate each determinant by expanding about any row or column. See example 8–3 C.

15. $\begin{vmatrix} 2 & 3 & 1 \\ 1 & 2 & 3 \\ 0 & 1 & 1 \end{vmatrix}$

16. $\begin{vmatrix} -1 & -1 & -1 \\ 2 & 2 & 2 \\ 3 & -3 & 3 \end{vmatrix}$

17. $\begin{vmatrix} -1 & 4 & 0 \\ -2 & 1 & -3 \\ -3 & 7 & 2 \end{vmatrix}$

18. $\begin{vmatrix} 0 & 1 & 7 \\ 3 & 0 & 2 \\ 4 & 0 & -1 \end{vmatrix}$

19. $\begin{vmatrix} -4 & 0 & 1 \\ 2 & 0 & 3 \\ -5 & -1 & -2 \end{vmatrix}$

20. $\begin{vmatrix} 5 & 10 & 15 \\ 1 & -1 & 0 \\ 2 & 1 & 0 \end{vmatrix}$

21. $\begin{vmatrix} 2 & 2 & -2 \\ 3 & -3 & 3 \\ -1 & 1 & 1 \end{vmatrix}$

22. $\begin{vmatrix} 0 & -1 & 2 \\ 1 & 0 & 2 \\ -5 & 6 & 0 \end{vmatrix}$

23. $\begin{vmatrix} a & 0 & a \\ 0 & a & 0 \\ 0 & 0 & a \end{vmatrix}$

24. $\begin{vmatrix} 0 & x & 0 \\ 0 & 0 & x \\ x & 0 & 0 \end{vmatrix}$

25. $\begin{vmatrix} x & y & 0 \\ 0 & x & y \\ y & x & 0 \end{vmatrix}$

26. $\begin{vmatrix} a & b & 0 \\ b & 0 & a \\ 0 & a & b \end{vmatrix}$

Directions Expand and evaluate each determinant about row one using the sign array $+ - + -$.

27.
$$\begin{vmatrix} 1 & 2 & 3 & -1 \\ 2 & 0 & 1 & 3 \\ -2 & 1 & 0 & -1 \\ 0 & 3 & 2 & 0 \end{vmatrix}$$

28.
$$\begin{vmatrix} 0 & 5 & 3 & 1 \\ 2 & 0 & 4 & -2 \\ -1 & -1 & 0 & 1 \\ 2 & 1 & 3 & 0 \end{vmatrix}$$

29.
$$\begin{vmatrix} 4 & 0 & 1 & 0 \\ -5 & -1 & 0 & -1 \\ 0 & 1 & 0 & -2 \\ 2 & 0 & -3 & 1 \end{vmatrix}$$

30.
$$\begin{vmatrix} 0 & 0 & 1 & 4 \\ -5 & 6 & 0 & -3 \\ 2 & 0 & 1 & 0 \\ 4 & 2 & 0 & 7 \end{vmatrix}$$

Directions Using Cramer's Rule, find the solution set of the following systems of linear equations. See example 8–3 D.

31. $x - y = 2$
$2x + y = 3$

32. $3x + y = 8$
$x - 2y = -1$

33. $3x + 5y = 6$
$2x - 3y = -4$

34. $4x - y = 3$
$8x - 2y = 1$

35. $10x - 2y = -3$
$5x - y = 0$

36. $x + 2y = 7$
$3x + 6y = 21$

37. $-4x - 6y = 2$
$2x + 3y = -1$

38. $6x - 2y = 7$
$x - 6y = 11$

39. $-3x - y = 1$
$4x + 5y = -5$

40. $8x - 5y = 0$
$-3x + 4y = -3$

41. $6x - 2y = 7$
$2y = 8$

42. $x - 7y = -3$
$-y = 8$

See example 8–3 E.

43. $x - y + 3z = -1$
$2x + y - z = 0$
$-3x - 4y + z = 1$

44. $2x + 2y - z = 3$
$x - 4y + 5z = 0$
$-x + y + z = 2$

45. $x - 2y - z = 0$
$2x - y + z = 0$
$4x + 2y - 2z = 0$

46. $2x + y - z = 0$
$x - 3y + 3z = 0$
$-3x + 2y - 2z = 0$

47. $3x - y - 6z = 5$
$-6x + 2y - 2z = -4$
$3x - y + z = 2$

48. $6x - 9y + 12z = 24$
$-4x + 6y - 8z = -16$
$2x - 3y + 4z = 8$

49. $2x - y + 3z = 5$
$3x - y + 2z = 10$
$3x - 2y + 7z = 3$

50. $4x + 2y - 3z = 2$
$6x + y - 2z = 0$
$2x + 3y - z = 0$

51. $3x + 2y + 4z = 3$
$3x - y - 2z = 0$
$3x + y + 2z = 2$

52. $x - 2y = 3$
$y + 3z = -2$
$2x - z = 4$

53. $x + 2z = 5$
$2x - y = 4$
$2y - z = 5$

54. $2x + 3y - z = 10$
$2y + 2z = -1$
$x - 4z = 0$

55. $x - 2y + z = -2$
$3x + y = 7$
$2x - z = 0$

56. $y + 4z = 6$
$4x + z = 0$
$5y - z = 9$

8–4 Solving systems of equations by the matrix method (optional)

We have learned several methods for solving linear systems of equations. Now we will consider a method for solving these systems using the *matrix* that was mentioned in section 8–3. This method is easily adapted to a computer.

With every system of equations, we can associate a matrix that consists of the coefficients and constant terms. To illustrate, consider the system

$$6x + 5y = -3$$
$$2x + y = -7,$$

with which we can associate the matrix

$$\begin{bmatrix} 6 & 5 & | & -3 \\ 2 & 1 & | & -7 \end{bmatrix}.$$

We call this the **augmented matrix** of the system. The coefficients of the system are placed to the left of the vertical bar and the constants to the right. We now operate on the rows of the augmented matrix just as we did with the equations of the system. Using the following **elementary row operations,** we produce new matrices that lead to systems containing solutions of the original system.

Elementary row operations
1. Any two rows of the augmented matrix can be interchanged.
2. Any row can be multiplied by a nonzero constant.
3. Any row of the augmented matrix can be changed by adding a nonzero multiple of another row to that row.

Use row operations to rewrite the matrix until it is a system whose solution is easy to find. The object is to obtain the first column $\begin{smallmatrix}1\\0\end{smallmatrix}$ and the second column $\begin{smallmatrix}\text{constant}\\1\end{smallmatrix}$ that now represent the coefficients of the variables x and y. We can then easily solve the resulting system.

Example 8–4 A

Find the solution set of the following system using row operations.
$$6x + 5y = -3$$
$$2x + y = -7$$

The augmented matrix of the system is $\begin{bmatrix} 6 & 5 & | & -3 \\ 2 & 1 & | & -7 \end{bmatrix}.$

To start with, we want to make sure that there is a 1 in the first row, first column position. We multiply the first row by $\frac{1}{6}$ to obtain the matrix

$$\begin{bmatrix} 1 & \frac{5}{6} & \bigg| & -\frac{1}{2} \\ 2 & 1 & \bigg| & -7 \end{bmatrix}.$$

Next, we must get 0s in every position below the first. To get 0 in row two, column one, use the third row operation: add to the numbers in row two the results of multiplying the numbers in row one by -2. We obtain

$$\begin{bmatrix} 1 & \frac{5}{6} & \bigg| & -\frac{1}{2} \\ 2 + 1(-2) & 1 + \frac{5}{6}(-2) & \bigg| & -7 + \left(-\frac{1}{2}\right)(-2) \end{bmatrix}$$

Original + -2 times
number   number from row one

$$\begin{bmatrix} 1 & \frac{5}{6} & \bigg| & -\frac{1}{2} \\ 0 & -\frac{2}{3} & \bigg| & -6 \end{bmatrix}.$$

We now have 1 in the first row, first column position and 0s in every position below that. Now we go to the second column and obtain 1 in the second row, second column position. To get this, use the second row operation and multiply row two by $-\frac{3}{2}$.

$$\begin{bmatrix} 1 & \frac{5}{6} & \bigg| & -\frac{1}{2} \\ 0 & 1 & \bigg| & 9 \end{bmatrix}$$

This augmented matrix yields the system
$$x + \frac{5}{6}y = -\frac{1}{2}$$
$$0x + 1y = 9 \text{ (or } y = 9).$$

Thus $y = 9$ and substituting 9 for y in the equation $x + \frac{5}{6}y = -\frac{1}{2}$, we obtain

$$x + \frac{5}{6}(9) = -\frac{1}{2}$$
$$x + \frac{15}{2} = -\frac{1}{2}$$
$$x = -\frac{1}{2} - \frac{15}{2}$$
$$x = -8.$$

The solution set of the system is $S = \{(-8,9)\}$.

To solve systems with three equations use row operations to get 1s down the diagonal from left to right and 0s below each 1. The following example demonstrates this method.

Example 8–4 B

Find the solution set of the following system using row operations.

$$2x - y + 2z = -8$$
$$x + 2y - 3z = 9$$
$$3x - y - 4z = 3$$

Write the augmented matrix of the system.

$$\begin{bmatrix} 2 & -1 & 2 & | & -8 \\ 1 & 2 & -3 & | & 9 \\ 3 & -1 & -4 & | & 3 \end{bmatrix}$$

To obtain 1 in row one, column one, we interchange rows one and two.

$$\begin{bmatrix} 1 & 2 & -3 & | & 9 \\ 2 & -1 & 2 & | & -8 \\ 3 & -1 & -4 & | & 3 \end{bmatrix}$$

We must now get 0s in column one below the first row. Add to row two the results of multiplying row one by -2.

$$\begin{bmatrix} 1 & 2 & -3 & | & 9 \\ 0 & -5 & 8 & | & -26 \\ 3 & -1 & -4 & | & 3 \end{bmatrix}$$

Add to row three the results of multiplying row one by -3.

$$\begin{bmatrix} 1 & 2 & -3 & | & 9 \\ 0 & -5 & 8 & | & -26 \\ 0 & -7 & 5 & | & -24 \end{bmatrix}$$

We must now obtain a 0 in row three, column two. Add to row three the results of multiplying row two by $-\dfrac{7}{5}$.

$$\begin{bmatrix} 1 & 2 & -3 & | & 9 \\ 0 & -5 & 8 & | & -26 \\ 0 & 0 & -\dfrac{31}{5} & | & \dfrac{62}{5} \end{bmatrix}$$

Multiply row three by $-\frac{5}{31}$ $\left(\text{the reciprocal of } -\frac{31}{5}\right)$ to get 1 in the row three, column three position.

$$\begin{bmatrix} 1 & 2 & -3 & | & 9 \\ 0 & -5 & 8 & | & -26 \\ 0 & 0 & 1 & | & -2 \end{bmatrix}$$

Multiply each member of row two by $-\frac{1}{5}$.

$$\begin{bmatrix} 1 & 2 & -3 & | & 9 \\ 0 & 1 & -\frac{8}{5} & | & \frac{26}{5} \\ 0 & 0 & 1 & | & -2 \end{bmatrix}$$

We now have the system
$$x + 2y - 3z = 9$$
$$y - \frac{8}{5}z = \frac{26}{5}$$
$$z = -2.$$

Replace z by -2 in $y - \frac{8}{5}z = \frac{26}{5}$ and solve for y.

$$y - \frac{8}{5}(-2) = \frac{26}{5}$$
$$y + \frac{16}{5} = \frac{26}{5}$$
$$y = \frac{26}{5} - \frac{16}{5} = \frac{10}{5} = 2$$

Replace y by 2 and z by -2 in the equation $x + 2y - 3z = 9$.
$$x + 2(2) - 3(-2) = 9$$
$$x + 4 + 6 = 9$$
$$x + 10 = 9$$
$$x = -1$$

The solution set of the system is $S = \{(-1,2,-2)\}$.

Mastery points
Can you
- Solve a system of two equations in two variables using the augmented matrix?
- Solve a system of three equations in three variables using the augmented matrix?

Exercise 8–4

Directions Solve the following systems of equations using the augmented matrix. See examples 8–4 A and B.

1. $x + 3y = 11$
   $2x - y = 1$

2. $x - 5y = 11$
   $2x + 3y = -4$

3. $x - 4y = -6$
   $3x + y = -5$

4. $x + 6y = -14$
   $5x - 3y = -4$

5. $2x + y = 5$
   $3x - 5y = 14$

6. $3x - 2y = 16$
   $4x + 2y = 12$

7. $5x - y = 0$
   $2x + 3y = -1$

8. $-3x + 2y = 1$
   $x - y = 4$

9. $-x - 2y = 4$
   $x + 6y = -2$

10. $x + 3y - z = 5$
    $3x - y + 2z = 5$
    $x + y + 2z = 7$

11. $x - 2y + 3z = -11$
    $2x + 3y - z = 6$
    $3x - y - z = 3$

12. $2x - y + z = 8$
    $x - 2y - 3z = 4$
    $3x + 3y - z = -4$

13. $x - 2y - 2z = 4$
    $2x + y - 3z = 7$
    $x - y - z = 3$

14. $2x - y - z = -4$
    $x + 3y - 4z = 12$
    $x + y + z = -5$

15. $2x + 3y + z = 11$
    $3x - y - z = 11$
    $x - 2y - 5z = 2$

16. $2x - y = -1$
    $2y - z = 6$
    $x + z = 1$

17. $x - y = 1$
    $2x - z = 0$
    $2y - z = -2$

Chapter summary

1. Two or more linear equations that involve the same variables are called a **system of linear equations.**

2. A system of equations is **consistent and independent** if the system has only one solution.

3. A system of equations is **dependent** if all solutions of one equation are also solutions of the other equation(s).

4. A system of equations is **inconsistent** if the system has no solution.

5. A system of linear equations can be solved by **elimination, substitution,** or by **determinants** using Cramer's Rule.

6. A **matrix** is an ordered array of rows and columns of numbers.

7. The 2×2 determinant $\begin{vmatrix} a_1 & b_1 \\ a_2 & b_2 \end{vmatrix}$ is defined by

$a_1b_2 - a_2b_1$.

8. By definition, the determinant

$$\begin{vmatrix} a_1 & b_1 & c_1 \\ a_2 & b_2 & c_2 \\ a_3 & b_3 & c_3 \end{vmatrix} = a_1 \begin{vmatrix} b_2 & c_2 \\ b_3 & c_3 \end{vmatrix} - a_2 \begin{vmatrix} b_1 & c_1 \\ b_3 & c_3 \end{vmatrix} + a_3 \begin{vmatrix} b_1 & c_1 \\ b_2 & c_2 \end{vmatrix}.$$

9. By Cramer's Rule, given the system of linear
   equations $a_1x + b_2y = c_1$
   $\qquad\qquad a_2x + b_2y = c_2,$
   $x = \dfrac{D_x}{D}$ and $y = \dfrac{D_y}{D}$,

   where $D = \begin{vmatrix} a_1 & b_1 \\ a_2 & b_2 \end{vmatrix}$ $(D \neq 0)$, $D_x = \begin{vmatrix} c_1 & b_1 \\ c_2 & b_2 \end{vmatrix}$, and $D_y = \begin{vmatrix} a_1 & c_1 \\ a_2 & c_2 \end{vmatrix}$.

10. The augmented matrix of the system of linear
    equations $a_1x + b_1y = c_1$ is given by
    $\qquad\qquad a_2x + b_2y = c_2$

    $$\begin{bmatrix} a_1 & b_1 & | & c_1 \\ a_2 & b_2 & | & c_2 \end{bmatrix}.$$

Chapter review

[8–1]

Directions Find the solution set of each system of linear equations by elimination. If the system is inconsistent or dependent, so state.

1. $x + y = 4$
   $x - y = 2$

2. $2x + 3y = 2$
   $2x - 3y = 0$

3. $2x + 3y = 4$
   $4x + 6y = 8$

4. $x - 4y = 5$
   $3x - 12y = 0$

5. $x + y = 1$
   $4x - 4y = 6$

6. $\dfrac{1}{2}x + y = 4$
   $x - \dfrac{1}{2}y = -5$

7. Forces F_1 and F_2 on a structure yield the system of equations
   $$\dfrac{1}{3}F_1 + \dfrac{2}{3}F_2 = 3$$
   $$\dfrac{2}{3}F_1 - \dfrac{1}{3}F_2 = 5.$$
   Find the forces F_1 and F_2.

Directions Find the solution set of each system of linear equations by substitution. If the system is inconsistent or dependent, so state.

8. $3x + 2y = 1$
   $y = -4$

9. $4x - y = 6$
   $x = -1$

10. $5y - x = 1$
    $y = 3x + 1$

11. $6x + 2y = -3$
    $x = 4 - y$

12. $x - 4y = 1$
    $2y - 2x = 3$

Directions Set up a system of linear equations and solve each problem.

13. Find the point of intersection of the two lines ℓ_1 and ℓ_2 if ℓ_1 contains the points $(-3,2)$ and $(1,0)$ and ℓ_2 contains the points $(5,1)$ and $(-2,-3)$.

14. Noel Doe wishes to enclose her rectangular yard with 180 feet of fencing. If she wishes to make the enclosed yard twice as long as it is wide, what are the dimensions of the yard? (*Hint:* The perimeter $P = 2\ell + 2w$. Set up two equations in ℓ and w and solve simultaneously.)

15. The total number of fire alarms in Detroit on a given day is 30. If the number of real alarms is two more than six times the number of false alarms, how many of each alarm are sounded?

16. A woman has $3,000 to invest. If she invests part at 7% simple interest and the rest at $6\frac{1}{2}\%$ simple interest, and the total income for the year is $201, how much does she invest at each rate?

17. Solder made of 20% tin is to be melted with solder made of 5% tin to produce 50 grams of solder containing 10% tin. How many grams of each should be used?

18. George bought a suit and a topcoat for $177. If one-fifth of the cost of the suit is $9 more than one-sixth of the cost of the topcoat, what is the price of each article of clothing?

19. Two airplanes leave from Detroit, one flying East and the other flying West. If one plane flies at 250 mph and the other flies at 305 mph, in how many hours will they be 2,886 miles apart?

[8–2]
Directions Find the solution set of each system of three linear equations. If the system is inconsistent or dependent, so state.

20. $x - y + 2z = 5$
$x + y + z = 6$
$2x - y - z = -3$

21. $7x - 2y + 9z = -3$
$4x - 6y + 8z = -5$
$8x + y + z = 1$

22. $2u + v + 3w = -2$
$5u + 2v = 5$
$2v - 3w = -7$

23. $p - r = -1$
$-q + r = 2$
$-2p + q = -4$

24. Find the values of a, b, and c such that the points $(0,2)$, $(1,-3)$, and $(2,-2)$ lie on the graph of $y = ax^2 + bx + c$.

25. Three forces on a beam are related by the system of equations
$0.2F_1 + 0.3F_2 = 2$
$0.2F_1 - 0.1F_3 = 1$
$0.4F_2 + 0.2F_3 = 3$.
Find forces F_1, F_2, and F_3.

26. If the middle-sized angle of a triangle measures 16° more than the smallest angle and the largest angle is twice the size of the smallest angle, find the measure of the three angles of the triangle.

27. Ron has a total of 285 coins—nickels, dimes, and quarters—in his piggy bank. If there is a total of $26.25 in the bank and there are four times as many dimes and quarters, how many of each coin does he have in the bank?

[8–3]
Directions Evaluate each determinant.

28. $\begin{vmatrix} 1 & 4 \\ -1 & 3 \end{vmatrix}$

29. $\begin{vmatrix} 0 & -6 \\ 4 & 3 \end{vmatrix}$

30. $\begin{vmatrix} 5 & 7 \\ -8 & -3 \end{vmatrix}$

Directions Evaluate each determinant using expansion about any row or column.

31.
$$\begin{vmatrix} 3 & 4 & 3 \\ -2 & 2 & 0 \\ 1 & -5 & 6 \end{vmatrix}$$

32.
$$\begin{vmatrix} -3 & 3 & 2 \\ -1 & 0 & -4 \\ -2 & 0 & 5 \end{vmatrix}$$

33.
$$\begin{vmatrix} 0 & 1 & -3 \\ -2 & 0 & 5 \\ 6 & 7 & 8 \end{vmatrix}$$

Directions Use Cramer's Rule to find the solution set of each system of linear equations.

34. $2x - y = 3$
$x - 3y = 4$

35. $4x + 5y = 0$
$2x - 4y = -1$

36. $-4x + 2y = 3$
$y = 9$

37. $6x - 3y = -2$
$3x = 8$

38. $x - 3y + 2z = 0$
$2x - y + z = 0$
$x + 4y - 3z = 0$

39. $-4x + y - 3z = -2$
$3x - 2y + z = 4$
$x + 3y - 2z = 1$

[8–4]
Directions Use an augmented matrix to solve the following systems.

40. $3x - y = 2$
$x + 2y = 0$

41. $x - y + 2z = 3$
$x + 3y - z = 1$
$2x - y + 2z = 0$

Chapter 8 cumulative test

[1–4]   **1.** Simplify $3[-4(12 - 3) - 14 + 6(3 - 9)]$.

Directions Evaluate the following formulas.

[1–5]   **2.** $V = k + gt$ when $k = 14$, $g = 32$, and $t = 3$

[1–5]   **3.** $C = \dfrac{C_1 C_2}{C_1 + C_2}$ when $C_1 = 8$ and $C_2 = 12$

Directions Perform the indicated operations and simplify.

[4–5]   **4.** $\dfrac{3a^3 - 11a^2 + 12a - 3}{a - 3}$

[4–3]   **5.** $\dfrac{30a^7 - 25a^5 + 15a^3}{5a^2}$

[3–3]   **6.** $(5x + 1)(x - 9) - (x + 6)^2$

[3–4]   **7.** $\dfrac{3^{-1}x^0 y^{-2}}{6^{-1}x^{-4}y^3}$

Directions Find the solution set of the following equations and inequalities.

[2–1]   **8.** $5(6y - 1) - 3(4y + 3) = 5y$

[6–3]   **9.** $3x^2 - 2x - 4 = 0$ (Use the quadratic formula.)

[6–2]   **10.** $x^2 + 6x - 5 = 0$ (Use completing the square.)

[6–2]   **11.** $2y^2 + y = 5$ (Use completing the square.)

[2–5]   **12.** $3(z + 1) - 1 \le 2(3z + 3)$

[6–7]   **13.** $x^2 + 2x - 3 \le 0$

[6–7] **14.** $2y^2 + 9y > 5$ [2–5] **15.** $|2x - 1| = 4$

[2–5] **16.** $|5 - 6x| > 2$ [4–1] **17.** Reduce the expression $\dfrac{8y + 12}{8y^3 + 27}$ to lowest terms.

Directions Perform the indicated operations and simplify.

[4–2] **18.** $\dfrac{x^2 - 16}{4x - 3} \cdot \dfrac{16x^2 - 9}{3x + 12}$ [4–3] **19.** $\dfrac{11}{x^2 - x - 20} - \dfrac{7}{x^2 + x - 12}$

[4–3] **20.** $\dfrac{3y}{y - 7} + \dfrac{y}{7 - y}$ [4–2] **21.** $\dfrac{p^3q}{16ab^2} \div \dfrac{7pq^2}{24a^2b^3}$

Directions Leave all answers with positive exponents. Assume all variables are positive.

[5–1] **22.** $(-27)^{\frac{2}{3}}$ [5–1] **23.** $(y^{\frac{3}{2}})^{-2}$ [5–1] **24.** $\dfrac{a^{\frac{3}{4}}}{a^{-\frac{1}{2}}}$ [5–5] **25.** $\sqrt{8} \cdot \sqrt{10}$

[5–5] **26.** $(\sqrt{6} + 3)(\sqrt{2} - 4)$ [5–6] **27.** $(4 + 5i)^2$ [5–2] **28.** $\sqrt[3]{8xy^3} - \sqrt[3]{64xy^3}$

[6–5] **29.** Find the solution set of the equation $p + 7\sqrt{p} + 6 = 0$. Identify any extraneous solutions that exist.

[7–1] **30.** Graph the equation $4x + 5y = -20$ using the intercepts.

[7–3] **31.** Graph the equation $3x - y = 6$ using the slope and y-intercept.

[7–3] **32.** Find the equation of the line through the points $(0,4)$ and $(-7,9)$.

[7–3] **33.** Find the equations of the line through $(1,3)$ and perpendicular to the line $4x - y = 5$.

[7–4] **34.** Sketch the graph of the inequality $2y + x < 2$.

[8–3] **35.** Evaluate $\begin{vmatrix} 1 & -2 & 3 \\ 0 & 2 & 1 \\ 5 & -1 & 4 \end{vmatrix}$.

[8–1] **36.** $\begin{aligned} 3x + y &= 6 \\ x - y &= 2 \end{aligned}$ [8–1] **37.** $\begin{aligned} 2y + 3x &= 1 \\ y - x &= -4 \end{aligned}$ [8–3] **38.** $\begin{aligned} 4x + 3y &= 0 \\ 3x - 2y &= 1 \end{aligned}$
(By determinants)

[8–4] **39.** $\begin{aligned} -3x - y &= 2 \\ 4x + 5y &= -5 \end{aligned}$
(By augmented matrix)   [8–2] **40.** $\begin{aligned} x + 4y - z &= -3 \\ -2x + y + 2z &= 0 \\ 3x - 2y + z &= 1 \end{aligned}$

9

Conic Sections

In chapter 7 we graphed first-degree equations in two variables, equations of the form $Ax + By = C$, which we called linear equations since the graph was a straight line. In this chapter we will consider the graphs of second-degree equations, that is, equations in which one or more terms are second-degree. These graphs result from slicing a cone with a plane as shown in figure 9.1 on page 354. For that reason, the graphs are called the **conic sections.**

Conic sections have many applications in the physical world. The arches of many bridges are parabolic in shape, and a projectile fired into the air follows an approximate parabolic path. An object on the face of the earth rotates along a circular path. Elliptic paths are followed by the orbits of the planets about the sun.

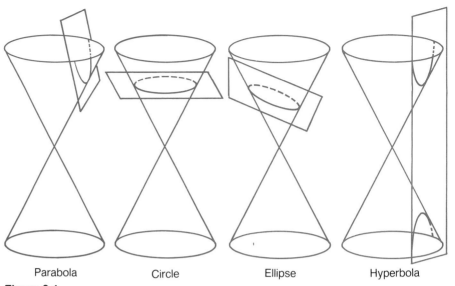

Parabola   Circle   Ellipse   Hyperbola

Figure 9.1

9–1 The parabola

If we assign coordinates to the cutting plane, each conic section can be described by a second-degree (quadratic) equation. The second-degree equation for a **parabola** has the following form:

> ■ **Equation for a parabola**
> $$y = ax^2 + bx + c,$$
> where a, b, and c are real numbers, $a \neq 0$.

Let us consider the quadratic equation in two variables

$$y = x^2 - 4.$$

If we choose the values for x in the set $\{-3, -2, -1, 0, 1, 2, 3\}$ and compute the corresponding values of y, the ordered pairs we obtain are

$$(-3,5), \ (-2,0), \ (-1,-3), \ (0,-4), \ (1,-3), \ (2,0), \text{ and } (3,5).$$

Plotting these ordered pairs and connecting the points with a smooth curve, we obtain the graph, a *parabola*, shown in figure 9.2. This parabola is the graph of the *equation* $y = x^2 - 4$.

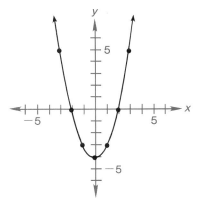

Figure 9.2

Similarly, consider the graph of the equation $y = -x^2 + 1$. Choosing values of x and then finding the corresponding values of y, we get the results shown in this table of related values.

x	-2	-1	0	1	2
y	-3	0	1	0	-3

Plot the points and connect them with a smooth curve to obtain the parabola shown in figure 9.3.

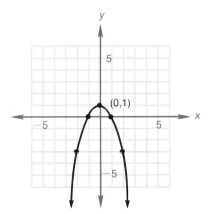

Figure 9.3

As we examine the parabolas in figures 9.2 and 9.3, we see some of the basic properties of the parabola.

1. For each x, there is a unique value of y.
2. There is a *lowest* point [$(0,-4)$ in figure 9.2] or a *highest* point [$(0,1)$ in figure 9.3] of the graph. We call this the **vertex** of the parabola.
3. A vertical line through the vertex, in this case the y-axis, divides the parabola into two identical (mirror-image) parts. This line is called the **line of symmetry.**
4. For each point on the parabola to the right of the line of symmetry, there is a corresponding point to the left of the line of symmetry.
5. When the coefficient of the squared term is
   a. positive, as in $y = x^2 - 4$, the parabola "opens upward" and the vertex is the lowest point of the graph; and
   b. negative, as in $y = -x^2 + 1$, the parabola "opens downard" and the vertex is the highest point of the graph.

Now let us consider the graph of the general quadratic equation $y = ax^2 + bx + c$. We could choose a number of values for x and obtain the corresponding values for y as we did before. However the parabola has certain special points (when they exist) that we can use to sketch the graph when we know the general shape of the curve. These points are (1) the vertex and (2) the x- and y-intercepts. These points, together with the previously observed characteristics of the parabola, enable us to obtain a reasonably accurate sketch of the curve.

The vertex

To obtain the coordinates of the vertex, we use the procedure of completing the square that we learned in chapter 6. Completing the square in the right member, we can write the equation $y = ax^2 + bx + c$ in the form

$$y = a(x - h)^2 + k \qquad (a \neq 0),$$

where the point (h,k) is the vertex of the parabola and the line $x = h$ is the line of symmetry.

Note
If $a > 0$, the parabola opens upward, and if $a < 0$, the parabola opens downard.

Example 9–1 A Determine the coordinates of the vertex of the parabola.

1. $y = (x - 2)^2 + 3$
Since $h = 2$ and $k = 3$, the vertex is the point $(2,3)$ and since $a = 1$, the parabola opens upward.

2. $y = -3(x + 4)^2 - 6$

The equation is not exactly in the form $y = a(x - h)^2 + k$, so we rewrite the equation as

$$y = -3[x - (-4)]^2 + (-6).$$

Now $h = -4$ and $k = -6$, so the vertex is the point $(-4, -6)$ and since $a = -3$, the parabola opens downward.

The intercepts

Recall that when we were graphing linear equations in two variables, we found the x-intercept by letting $y = 0$ and solving for x, and we found the y-intercept by letting $x = 0$ and solving for y. The same method is used to find the x- and y-intercepts when graphing a quadratic equation (or *any* equation) in two variables.

Example 9–1 B

Find the x- and y-intercepts of the equation $y = x^2 + 2x - 8$.
Let $y = 0$, then

$$0 = x^2 + 2x - 8$$
$$0 = (x + 4)(x - 2)$$
$$x + 4 = 0 \text{ or } x - 2 = 0$$
$$x = -4 \text{ or } x = 2.$$

The x-intercepts are -4 and 2 and the graph crosses the x-axis at $(-4,0)$ and $(2,0)$.

Let $x = 0$, then $y = (0)^2 + 2(0) - 8 = -8$.
The y-intercept is -8 and the graph crosses the y-axis at $(0,-8)$.

Observe that the y-intercept is the constant c in the equation $y = ax^2 + bx + c$.

Let us now graph some quadratic equations that yield parabolas, using the vertex and the x- and y-intercepts.

Example 9–1 C

Graph the following equations using the vertex, the intercepts, and the fact that a is either positive or negative. Give the equation of the line of symmetry.

1. $y = x^2 + 2x - 8$

We previously determined that the y-intercept is -8 and the x-intercepts are -4 and 2. To complete the square to find the vertex, we write the equation as

$$y = (x^2 + 2x + \quad) - 8 - (\quad).$$

To complete the square inside the parentheses, we use the method we learned in section 6–2.

$$y = (x^2 + 2x + 1) - 8 - 1$$

Notice that to compensate for the 1 we added inside the parentheses, we subtracted 1 from -8.

$$y = (x + 1)^2 - 9$$
$$y = 1[(x - (-1)]^2 + (-9)$$

Thus $a = 1$, $h = -1$, and $k = -9$, so the vertex is the point $(-1, -9)$ and the parabola opens upward since a is positive. The line of symmetry is $x = -1$.

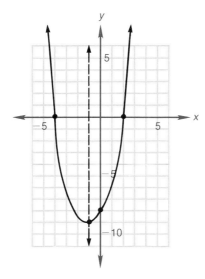

To graph the equation, plot the vertex $(-1, -9)$ together with the points $(-4, 0)$, $(2, 0)$, and $(0, -8)$. Then draw a smooth curve through these points.

2. $y = -3x^2 + x + 2$

The y-intercept is 2.

Let $y = 0$, then

$$0 = -3x^2 + x + 2$$
$$0 = 3x^2 - x - 2 \qquad \text{Multiply each term by } -1.$$
$$0 = (3x + 2)(x - 1)$$
$$3x + 2 = 0 \text{ or } x - 1 = 0$$
$$x = -\frac{2}{3} \text{ or } x = 1.$$

The x-intercepts are $-\dfrac{2}{3}$ and 1.

Since the coefficient of x^2 is -3, we must factor -3 from each of the first two terms to complete the square.

$$y = -3\left(x^2 - \frac{1}{3}x + \quad\right) + 2 - (\quad)$$

$$= -3\left(x^2 - \frac{1}{3}x + \frac{1}{36}\right) + 2 - \left(-\frac{1}{12}\right)$$

$$= -3\left(x - \frac{1}{6}\right)^2 + \frac{25}{12}$$

The number added inside the parentheses was $-3\left(\frac{1}{36}\right)$ or $-\frac{1}{12}$, so we must add $\frac{1}{12}$ outside the parentheses. The vertex is the point $\left(\frac{1}{6}, \frac{25}{12}\right)$ and since $a = -3$, the parabola opens downward. The line of symmetry is $x = \frac{1}{6}$.

To graph the equation, plot the vertex $\left(\frac{1}{6}, \frac{25}{12}\right)$ together with the points $\left(-\frac{2}{3}, 0\right)$, $(1,0)$, and $(0,2)$. Then draw a smooth curve through these points.

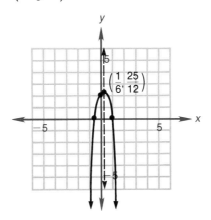

When considering the quadratic equation $y = ax^2 + bx + c$, we see that the graph has a *low point* (or *minimum value* of y) when $a > 0$ and a *high point* (or *maximum value* of y) when $a < 0$. We can use these values of y to answer questions in a physical situation.

Example 9-1 D

1. A projectile is fired upward from the ground so that its distance s in feet above the ground t seconds after firing is given by $s = -16t^2 + 96t$. Find the maximum height it will reach and the number of seconds it takes to reach that height.

Since $a = -16$, the graph will have a high point or maximum value of s that occurs at t. We must find the vertex by completing the square in the right member.

$$\begin{aligned} s &= -16t^2 + 96t \\ &= -16(t^2 - 6t + \quad) + (\quad) \\ &= -16(t^2 - 6t + 9) + (144) \\ &= -16(t - 3)^2 + 144 \end{aligned}$$

The vertex of the parabola is (3,144) and the projectile will reach the maximum height s of 144 feet in $t = 3$ seconds after firing.

2. Harry has 48 feet of fencing to use to fence off a rectangular area behind his house for his dog Nappy. What are the dimensions of the largest area that he can fence off with his 48 feet of fencing?

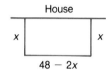

House

x x

$48 - 2x$

Let $x =$ the length of one of the equal sides. Then $48 - 2x =$ the length of the side opposite the house since Harry has 48 feet of fencing and the other two sides are x in length.

Using area $(A) =$ length $(\ell) \times$ width (w), where width $(w) = x$ and length $(\ell) = 48 - 2x$, we have the equation

$$A = x(48 - 2x) = -2x^2 + 48x.$$

Completing the square in the right member,

$$\begin{aligned} A &= -2x^2 + 48x \\ &= -2(x^2 - 24x + \quad) + (\quad) \\ &= -2(x^2 - 24x + 144) + (288) \\ &= -2(x - 12)^2 + 288. \end{aligned}$$

The vertex is the point (12,288) and the maximum area is 288 square feet when $x = 12$. The dimensions of the rectangle are 12 feet by $48 - 2(12) = 24$ feet.

Mastery points
Can you
- Find the x- and y-intercepts of a quadratic equation in two variables?
- Find the coordinates of the vertex of the parabola?
- Sketch the graph of a quadratic equation in two variables?
- Find the maximum and minimum values of a quadratic equation of the form $y = ax^2 + bx + c$?

Exercise 9–1

Directions Find the coordinates of the vertex of the parabola for the graph of each quadratic equation. See example 9–1 A.

1. $y = (x - 3)^2 + 4$
2. $y = 3(x + 1)^2 - 7$
3. $y = x^2 - 16$
4. $y = x^2 - 3$
5. $y = x^2 - 3x - 10$
6. $y = -x^2 + 3x + 6$
7. $y = x^2 + 22x + 21$
8. $y = 2x^2 - 7x + 3$
9. $y = -5x^2 - x + 1$

Directions Sketch the graph of each given quadratic equation using the intercepts and the coordinates of the vertex where possible. See example 9–1 C.

10. $y = x^2 - 5$
11. $y = x^2 - 1$
12. $y = x^2 - 2x - 24$
13. $y = x^2 + 5$
14. $y = x^2 - 10x + 9$
15. $y = x^2 + 9x + 8$
16. $y = -x^2 + 4x + 10$
17. $y = 4x^2 - 12x - 7$
18. $y = -5x^2 + 6x - 1$
19. $y = -3x^2 + 14x + 5$
20. $y = 2x^2 + 2x - 1$

See example 9–1 D–1.

21. The equation $s = 32t - 16t^2$ determines the distance s in feet that an object thrown vertically upward is above the ground in time t seconds. Find the time at which the object reaches its greatest height. Sketch the graph of the equation for $0 \le t \le 4$. (*Hint:* The greatest height is at the vertex.) When will the object hit the ground again? (*Hint:* When $s = 0$)

22. Given the equation $s = 64t - 8t^2$, do as instructed in exercise 21.

23. An arrow is shot vertically into the air with an initial velocity of 96 feet per second. If the height h in feet of the arrow at any time t seconds is given by
$h = 96t - 16t^2$,
find the maximum height that the arrow will attain. When will the arrow come back to the ground?

24. A company's profit P when producing x units of a commodity in a given week is given by
$P = -x^2 + 100x - 1,000$.
How many units must be produced to attain maximum profit?

25. Helen Nance owns a dress shop. She finds the profits from the shop are approximately given by
$P = -x^2 + 16x + 42$,
where P is the profit when x dresses are sold daily. How many dresses must Helen sell daily to produce the maximum profit?

26. Russell Sanderson owns a pizza shop. From past results, he determines that the cost C of running the shop is given by the equation
$C = 3n^2 - 36n + 140$,
where n is the number of pizzas sold daily. Find the number of pizzas Russell must sell to produce the lowest cost.

27. Tim Wesner sets up a Kool-Aid stand. He finds that the cost C of operating the stand is given by the equation
$C = 2x^2 - 20x + 100$,
where x denotes the number of glasses of Kool-Aid sold. How many glasses of Kool-Aid should he sell for his cost to be the lowest?

28. The velocity distribution of natural gas flowing smoothly in a pipeline is given by

$V = 6x - x^2$,

where V is the velocity in meters per second and x is the distance in meters from the inside wall of the pipe. What is the maximum velocity of the gas? (*Hint:* We want the second component V of the vertex of the graph of the equation.)

See example 9–1 D–2.

30. A farmer wishes to fence in a rectangular piece of ground along a river bank for grazing his cattle. If he has 400 feet of fencing, what should be the dimensions of the rectangle to fence in the greatest area, using the river as one side of the rectangle?

31. Find the two numbers whose sum is 56 and whose product is the greatest possible value. (*Hint:* Let x be one number and $56 - x$ be the other number. Maximize their product.)

32. The sum of the length and the width of a rectangle is 48 inches. Find the length of the rectangle that would yield the greatest area. (*Hint:* Let $x =$ the length and $48 - x =$ the width. Use the formula $A = \ell w$.)

29. The power output P of an automobile alternator that generates 14 volts and has an internal resistance of 0.20 ohms is given by

$P = 14I - 0.20I^2$.

At what current I does the generator generate maximum power and what is the maximum power?

33. What are the dimensions of the largest rectangular plot of ground that can be enclosed by 42 meters of fence? (*Hint:* $2\ell + 2w = 42$. Solve for either ℓ or w and substitute in the formula $A = \ell w$. Maximize A.)

34. Find the maximum area of a rectangular plot of ground that can be enclosed with 28 rods of fencing.

35. Given that the difference between two numbers is 8, what is the minimum product of the two numbers? (*Hint:* Let x be one number and $x - 8$ be the other number.)

36. Find the minimum product of two numbers whose difference is 10.

9–2 The circle

In section 9–1 we discussed second-degree equations having just one second-degree term. Now we will consider second-degree equations that contain two second-degree terms, x^2 and y^2. The second conic section, the circle, contains these terms in its equation. A **circle** is defined to be the set of all points that are a given distance from a fixed point. We call the fixed point the **center** of the circle, denoted by C, and the equal distance from the center to all points on the circle the **radius,** denoted by r.

Suppose we let C, the center of the circle, be denoted by the point (h,k) and choose an arbitrary point P on the circle that has coordinates (x,y). See figure 9.4.

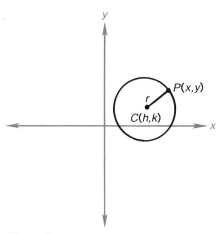

Figure 9.4

Using the distance formula and the points $P(x,y)$ and $C(h,k)$, we find the equation of the circle.

> ■ **Equation of the circle**
> $$(x - h)^2 + (y - k)^2 = r^2,$$
> where (h,k) is the center of the circle and r is the radius.

We call this the **standard form** of the equation of the circle.

Example 9–2 A

The equation $(x - 3)^2 + (y - 2)^2 = 16$, which can be written

$(x - 3)^2 + (y - 2)^2 = 4^2$, is the equation of a circle with center $C(3,2)$ and radius $r = 4$ units.

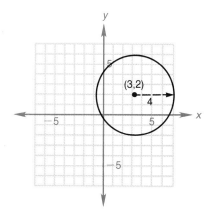

A special case exists when the center is at the origin. Then
$$(h,k) = (0,0)$$
and the equation of the circle is given by
$$(x - 0)^2 + (y - 0)^2 = r^2$$
$$x^2 + y^2 = r^2.$$

Example 9–2 B

$x^2 + y^2 = 25$ is the equation of a circle with center $C(0,0)$ and radius $r = 5$.

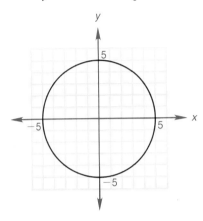

If $r = 1$, then $x^2 + y^2 = 1$, which is known as the *unit circle*.

If we are given the length of the radius of a circle and know the coordinates of the center, we can determine an equation of the circle using the standard form $(x - h)^2 + (y - k)^2 = r^2$.

Example 9–2 C

Find the equation of the circle with radius $r = 7$ units and center C at $(-3,4)$. We are given $r = 7$, $h = -3$, and $k = 4$. Substituting into the equation,

$$(x - h)^2 + (y - k)^2 = r^2$$
$$[x - (-3)]^2 + (y - 4)^2 = 7^2$$
$$(x + 3)^2 + (y - 4)^2 = 49.$$

If we expand $(x + 3)^2$ and $(y - 4)^2$ and get all nonzero terms in the left member of the equation, then $(x + 3)^2 + (y - 4)^2 = 49$ is equivalent to $x^2 + y^2 + 6x - 8y + 25 = 49$ or $x^2 + y^2 + 6x - 8y - 24 = 0$.
We call this the **general form** of the equation of a circle. That is, the equation of a circle is written in general form when it is written in the following form:

■ **General form of the equation of a circle**
$$Ax^2 + Ay^2 + Bx + Cy + D = 0.$$

Conic Sections

Example 9-2 D	Determine the coordinates of the center and the length of the radius of the given circle. Sketch the graph.

1. $x^2 + y^2 - 10x - 8y - 23 = 0$

We first rewrite this into the form $(x - h)^2 + (y - h)^2 = r^2$ by completing the square. Grouping the terms having the same variables, we obtain

$(x^2 - 10x) + (y^2 - 8y) - 23 = 0$ Add 23 to both members.
$(x^2 - 10x +\quad) + (y^2 - 8y +\quad) = 23 + (\quad) + (\quad)$.

Completing the square in both x and y (adding same quantities to both members), we have

$(x^2 - 10x + 25) + (y^2 - 8y + 16) = 23 + 25 + 16$
$$(x - 5)^2 + (y - 4)^2 = 64$$
$$(x - 5)^2 + (y - 4)^2 = 8^2.$$

The circle has center $C(5,4)$ and radius $r = 8$.
Plot the center $(5,4)$ and draw a circle that is 8 units in each direction from this point.

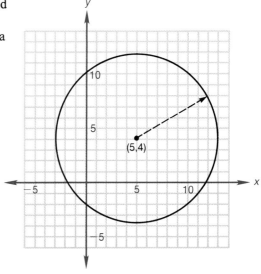

2. $3x^2 + 3y^2 - 18x + 12y + 27 = 0$

To complete the square, we must make the coefficients of the squared terms 1. To accomplish this, divide *all* terms by 3 to obtain

$$x^2 + y^2 - 6x + 4y + 9 = 0$$
$$(x^2 - 6x) + (y^2 + 4y) + 9 = 0$$
$$(x^2 - 6x) + (y^2 + 4y) = -9.$$

Completing the square and adding the same quantities to each member of the equation,

$(x^2 - 6x + 9) + (y^2 + 4y + 4) = -9 + 9 + 4$
$$(x - 3)^2 + (y + 2)^2 = 4$$
$$(x - 3)^2 + (y + 2)^2 = 2^2.$$

Therefore the circle has center $C(3,-2)$ and radius $r = 2$.
Plot the center $(3,-2)$ and draw a circle that is 2 units in each direction from this point.

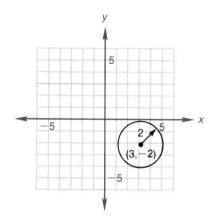

Mastery points

Can you
- Write the equation of a circle given the coordinates of the center and the radius?
- Find the coordinates of the center and the length of the radius of a circle given the equation?
- Given the general form of the equation of a circle, find the standard form of the equation of the circle?
- Graph the equation of a circle?

Exercise 9–2

Directions Write the equation for each of the following circles. See example 9–2 C.

1. Center $C(1,2)$ and radius 2
2. Center $C(-5,7)$ and radius 4
3. Center $C(4,-3)$ and radius $\sqrt{6}$
4. Center $C(-4,-5)$ and radius $2\sqrt{2}$
5. Center at the origin and radius 3
6. Center at the origin and radius $\sqrt{5}$

Directions Write the equation in the form $x^2 + y^2 + Ax + By + C = 0$, where A, B, and C are integers.

7. Center $C(-5,2)$ and radius 1
8. Center $C(1,-3)$ and radius $\sqrt{10}$

Directions Determine the coordinates of the center and the radius of each of the following circles. See example 9–2 D.

9. $(x - 3)^2 + (y - 2)^2 = 49$
10. $(x + 4)^2 + (y - 7)^2 = 11$
11. $x^2 + y^2 = 36$
12. $x^2 + y^2 = 18$
13. $x^2 + y^2 + 4x - 6y - 23 = 0$
14. $x^2 + y^2 + 4x - 4y - 8 = 0$
15. $2x^2 + 2y^2 + 8x - 12y = 74$
16. $3x^2 + 3y^2 - 24x + 12y - 87 = 0$

Directions Graph each of the following circles. See examples 9–2 A and B.

17. $x^2 + y^2 = 1$　　　　**18.** $x^2 + y^2 = 36$　　　　**19.** $3x^2 + 3y^2 = 12$

20. $4x^2 + 4y^2 = 24$　　　**21.** $(x - 4)^2 + (y + 3)^2 = 4$　　**22.** $(x + 2)^2 + (y - 5)^2 = 5$

23. $x^2 + y^2 - 2x = 15$　　**24.** $x^2 + y^2 - 4x = 0$　　**25.** $x^2 + y^2 - 2x + 4y - 20 = 0$

26. $x^2 + y^2 + 6x - 2y - 39 = 0$　**27.** $2x^2 + 2y^2 - 12x + 4y = -12$　**28.** $4x^2 + 4y^2 - 16x + 24y = 92$

29. Using the point (h,k), any point on the circle (x,y), r for the radius, and the distance formula, derive the equation of the circle $(x - h)^2 + (y - k)^2 = r^2$.

9–3 The ellipse and the hyperbola

The ellipse

Two other conics containing x^2 and y^2 terms are the ellipse and the hyperbola. In this book we shall consider only ellipses and hyperbolas with the center at the origin, $(0,0)$.

> ■ **Definition of an ellipse**
> An **ellipse** is the set of all points in the plane such that the sum of the distances from each point on the ellipse to two fixed points in the plane is a positive constant, k.

We call the fixed points the *foci* of the ellipse. Each fixed point is a **focus** of the ellipse, which we denote by F_1 and F_2. See figure 9.5.

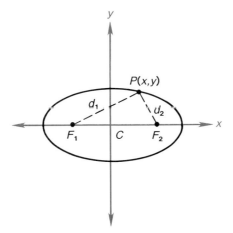

Figure 9.5

By definition, $d_1 + d_2 =$ a positive constant for any point $P(x,y)$ on the ellipse. Given the x-intercepts of the ellipse at $(a,0)$ and $(-a,0)$ and the y-intercepts at $(0,b)$ and $(0,-b)$ (a and b are both positive), we can show that the equation of the ellipse with its center at the origin $(0,0)$ is given as follows:

> ■ **Equation of the ellipse**
> The **standard form** of the equation of the ellipse centered at the origin is
> $$\frac{x^2}{a^2} + \frac{y^2}{b^2} = 1,$$
> where the intercepts are $(a,0)$, $(-a,0)$, $(0,b)$, and $(0,-b)$.

To graph an ellipse with its center at $(0,0)$, plot the four intercepts $a, -a, b, -b$ and sketch the ellipse through the intercepts. See figure 9.6.

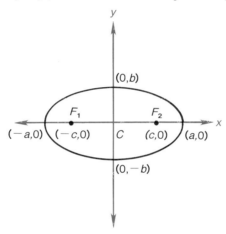

Figure 9.6

Example 9–3 A

Find the coordinates of the intercepts of the following ellipses and sketch the graph.

1. $\dfrac{x^2}{25} + \dfrac{y^2}{9} = 1$

(1) $a^2 = 25$, so $a = 5$ and
(2) $b^2 = 9$, so $b = 3$.
The x-intercepts are -5 and 5, the y-intercepts are -3 and 3.

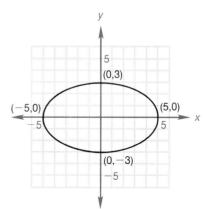

Conic Sections

2. $16x^2 + 9y^2 = 144$

Divide both members by 144 to obtain the standard form of the equation.

$$\frac{x^2}{9} + \frac{y^2}{16} = 1$$

(1) $a^2 = 9$, so $a = 3$ and (2) $b^2 = 16$, so $b = 4$. The x-intercepts are -3 and 3, the y-intercepts are -4 and 4.

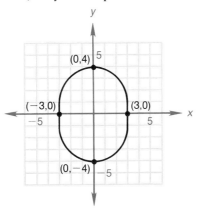

The hyperbola

The equation of the last conic section, the hyperbola, is similar to the equation of the ellipse.

> ### ■ Definition of a hyperbola
> A **hyperbola** is the set of all points in the plane such that the absolute value of the *difference* between the distances from each point on the hyperbola to two fixed points is a constant, k.

The fixed points are again called the **foci** of the hyperbola. See figure 9.7.

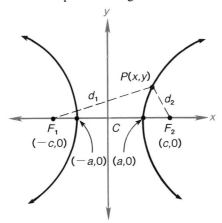

Figure 9.7

Thus $|d_1 - d_2|$ = a positive constant for any point (x,y) on the hyperbola. Using the above relationship, we can show that the standard form of the equation of a hyperbola is as follows:

■ **Equations of hyperbolas**
The **standard form** of the equation of a hyperbola with the center at the origin is given by

$$\frac{x^2}{a^2} - \frac{y^2}{b^2} = 1$$

with x-intercepts of a and −a and no y-intercepts (see figure 9.7) or

$$\frac{y^2}{b^2} - \frac{x^2}{a^2} = 1$$

with y-intercepts of b and −b and no x-intercepts (see figure 9.8).

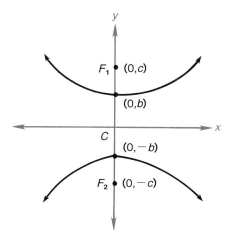

Figure 9.8

Solving the equations

$$\frac{x^2}{a^2} - \frac{y^2}{b^2} = 1 \quad \text{and} \quad \frac{y^2}{b^2} - \frac{x^2}{a^2} = 1 \quad \text{for } y \text{ in terms of } a, b, \text{ and } x,$$

we can show that the graph of this hyperbola will approach (get closer and closer to) but not cross the lines whose equations are

$$y = \frac{b}{a}x \quad \text{and} \quad y = -\frac{b}{a}x.$$

We call such lines **asymptotes.** They are indicated by dashed lines. These lines are useful for sketching the graph of the hyperbola and are found in the following way:

Given the equation $\dfrac{x^2}{4} - \dfrac{y^2}{25} = 1$, then $a^2 = 4$, $b^2 = 25$, and by using the distances $a = 2$ and $b = 5$, from the origin, along the positive and negative x- and y-axes, respectively, we can construct a rectangle with corners at $(2,5)$, $(2,-5)$, $(-2,5)$, and $(-2,-5)$. See figure 9.9.

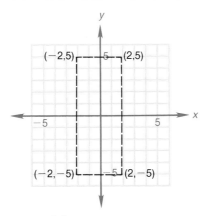

Figure 9.9

Drawing the two diagonals of the resulting rectangle, these lines *are* the asymptotes of the graph of the hyperbola. See figure 9.10. Using the intercepts and the asymptotes, we can now sketch the graph of a hyperbola with its center at the origin $(0,0)$.

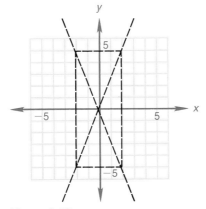

Figure 9.10

Example 9-3 B

Determine the intercepts and equations of the asymptotes. Sketch the graph.

1. $\dfrac{x^2}{16} - \dfrac{y^2}{4} = 1$

(1) $a^2 = 16$, so $a = 4$ and (2) $b^2 = 4$, so $b = 2$. The x-intercepts are -4 and 4. The equations of the asymptotes are $y = \dfrac{2}{4}x = \dfrac{1}{2}x$ and $y = -\dfrac{2}{4}x = -\dfrac{1}{2}x$. There are no y-intercepts. The rectangle is formed by the points $(4,2)$, $(4,-2)$, $(-4,2)$, and $(-4,-2)$.

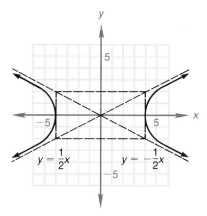

2. $9y^2 - 4x^2 = 36$

To get the equation into one of the given forms, divide each term by 36. Then

$$\dfrac{y^2}{4} - \dfrac{x^2}{9} = 1,$$

and (1) $b^2 = 4$, so $b = 2$ and (2) $a^2 = 9$, so $a = 3$. The y-intercepts are -2 and 2. The equations of the asymptotes are $y = \dfrac{2}{3}x$ and $y = -\dfrac{2}{3}x$. The rectangle is formed by the points $(3,2)$, $(3,-2)$, $(-3,2)$, and $(-3,-2)$.

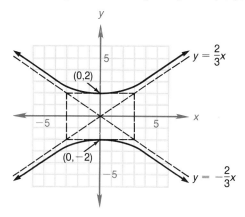

Conic Sections

In summary, to graph a hyperbola of either of the two standard forms, we follow this procedure:

Step 1 Locate the intercepts at $x = a$ and $x = -a$ if the coefficient of x^2 is positive and at $y = b$ and $y = -b$ if the coefficient of y^2 is positive.

Step 2 Draw the rectangle with corners at (a,b), $(a,-b)$, $(-a,b)$, and $(-a,-b)$.

Step 3 Sketch the asymptotes (in dashed lines) that are extensions of the diagonals of the rectangle.

Step 4 Sketch the hyperbola through the intercepts and approach the lines that are the asymptotes.

Let us now summarize the characteristics of the four conic sections—the parabola, the circle, the ellipse, and the hyperbola. Given the general equation $Ax^2 + By^2 + Cx + Dy + E = 0$, where a and b are not both 0, the graph will be

1. a *parabola* if either $a = 0$ or $b = 0$ but not both, and if one variable is linear;
2. a *circle* if $a = b$;
3. an *ellipse* if the signs of a and b are the same; and
4. a *hyperbola* if the signs of a and b are different.

Example 9–3 C

Identify each equation as a parabola, a circle, an ellipse, or a hyperbola.

1. $x^2 - 2y = 0$
Since the equation contains the second power of x and the first power of y, then y is linear and the graph is a **parabola.**

2. $4x^2 = 12 - 3y^2$
Adding $3y^2$ to each member, we obtain the equation

$4x^2 + 3y^2 = 12.$

Since we have $ax^2 + by^2$ and the coefficients of x^2 and y^2 are different, and have the same signs, $4x^2 = 12 - 3y^2$ is the equation of an **ellipse.**

3. $3y^2 = 9 + 3x^2$
Adding $-3x^2$ to each member, we obtain

$-3x^2 + 3y^2 = 9.$

Since we have $ax^2 + by^2$ and the signs of a and b are different, the equation $3y^2 = 9 + 3x^2$ is the equation of a **hyperbola.**

Mastery points
Can you
• Find the coordinates of the x- and y-intercepts for an ellipse and a hyperbola?
• Sketch the graphs of ellipses and hyperbolas?
• Find the equations of the asymptotes of a hyperbola and graph them?
• Identify the equation of a parabola, a circle, an ellipse, and a hyperbola and sketch their graphs?

Exercise 9–3

Directions Find the x- and y-intercepts of each given ellipse. Sketch the graph of each equation. See example 9–3 A.

1. $\dfrac{x^2}{9} + \dfrac{y^2}{25} = 1$
2. $\dfrac{x^2}{4} + \dfrac{y^2}{9} = 1$
3. $\dfrac{x^2}{49} + \dfrac{y^2}{25} = 1$
4. $\dfrac{x^2}{121} + \dfrac{y^2}{36} = 1$

5. $x^2 + \dfrac{y^2}{9} = 1$
6. $x^2 + \dfrac{y^2}{4} = 1$
7. $\dfrac{x^2}{16} + y^2 = 1$
8. $36x^2 + 9y^2 = 324$

9. $x^2 + 25y^2 = 100$
10. $16x^2 + y^2 = 64$
11. $3x^2 + 4y^2 = 12$
12. $9x^2 + 2y^2 = 18$

13. $8x^2 + y^2 = 16$
14. $x^2 + 3y^2 = 27$

Directions Find the x-intercepts, or y-intercepts, and the equations of the asymptotes of the hyperbola for the given equation. Sketch the graph of the equation. See example 9–3 B.

15. $\dfrac{x^2}{16} - \dfrac{y^2}{9} = 1$
16. $\dfrac{x^2}{4} - \dfrac{y^2}{25} = 1$
17. $\dfrac{y^2}{9} - \dfrac{x^2}{4} = 1$
18. $\dfrac{y^2}{16} - \dfrac{x^2}{25} = 1$

19. $y^2 - \dfrac{x^2}{9} = 1$
20. $y^2 - \dfrac{x^2}{4} = 1$
21. $x^2 - \dfrac{y^2}{16} = 1$
22. $9x^2 - y^2 = 36$

23. $y^2 - 25x^2 = -25$
24. $y^2 - 16x^2 = 64$
25. $25y^2 - 16x^2 = 400$
26. $16y^2 - 9x^2 = 144$

27. $25x^2 - 4y^2 = -100$
28. $2y^2 - 3x^2 = -18$
29. $2x^2 - 9y^2 = -36$
30. $6x^2 - 6y^2 = -1$

Directions Write each of the following equations in standard form. Identify as a parabola, a circle, an ellipse, or a hyperbola and sketch the graph. See example 9–3 C.

31. $x^2 = 9 - y^2$
32. $9x^2 = 36 + 4y^2$
33. $x^2 + 3 = y$
34. $2x^2 + y^2 = 8$

35. $4y^2 = 25 - 4x^2$
36. $y^2 - x^2 = 25$
37. $y^2 = 121 - x^2$
38. $8y^2 = 24 + 8x^2$

39. $x^2 + y = 8$
40. $18 - 2y^2 = 3x^2$

41. Sketch a cone and show how a plane can cut it in only *one point*. This point is sometimes called a *degenerate circle* (or *degenerate ellipse*). In the same way, a plane can cut a cone in *exactly* one *straight* line. This is sometimes called a *degenerate parabola*.

9-4 Systems of nonlinear equations

In sections 8–1 and 8–2 we discussed the simultaneous solution(s) of systems of linear equations. Now we will consider systems of two equations in which at least one of the equations is quadratic. These are called *systems of nonlinear equations*. The methods used to solve systems of linear equations can be used to solve systems involving quadratic equations. The following examples illustrate this kind of system.

Example 9–4 A

Find the solution set of the given system of equations.

$y = x^2 - 2x + 1$
$x + y = 3$

We first solve the linear equation for y to obtain the equivalent system

$y = x^2 - 2x + 1$
$y = 3 - x.$

Substitute $3 - x$ for y in the quadratic equation and solve for x.

$3 - x = x^2 - 2x + 1$

Adding $x - 3$ to each member of this equation, we obtain the quadratic equation

$0 = x^2 - x - 2$ or $x^2 - x - 2 = 0.$

Factoring the left member,

$(x - 2)(x + 1) = 0$ and so

$x = -1$ or $x = 2.$

Substituting for x in the equation $y = 3 - x$, we find

(1) when $x = 2$,
$\quad y = 3 - 2$
$\quad y = 1$

(2) when $x = -1$,
$\quad y = 3 - (-1)$
$\quad y = 4.$

Therefore when $x = 2$, $y = 1$, and when $x = -1$, $y = 4$. Thus $(x,y) = (2,1)$ and $(x,y) = (-1,4)$ are simultaneous solutions.
The solution set is $S = \{(2,1), (-1,4)\}$.

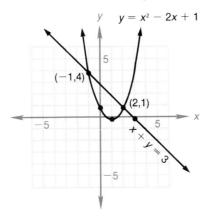

When both equations making up the system are quadratic equations, we use the method of solution by **elimination.** The result is then a quadratic equation in one variable or two variables and we proceed as in our preceding examples.

Example 9–4 B

Find the solution set of the system.

1. $x^2 + y^2 = 25$
$x^2 + 4y^2 = 52$

To eliminate the variable x^2, we multiply the first equation by -1 and add to the second equation. Thus we obtain

$$
\begin{aligned}
-x^2 - y^2 &= -25 \\
\underline{x^2 + 4y^2} &= \underline{52} \\
3y^2 &= 27 \\
y^2 &= 9 \\
y &= \pm 3.
\end{aligned}
$$

Substituting 3 and -3 for y in either original equation, say $x^2 + y^2 = 25$, we obtain corresponding values of x.

(1) When $y = 3$,
$x^2 + (3)^2 = 25$
$x^2 + 9 = 25$
$x^2 = 16$
$x = \pm 4.$

(2) When $y = -3$,
$x^2 + (-3)^2 = 25$
$x^2 + 9 = 25$
$x^2 = 16$
$x = \pm 4.$

Thus for (1) $y = 3$, $x = 4$; (2) $y = 3$, $x = -4$; (3) $y = -3$, $x = 4$; and (4) $y = -3$, $x = -4$. The solution set is

$S = \{(4,3), (-4,3), (4,-3), (-4,-3)\}.$

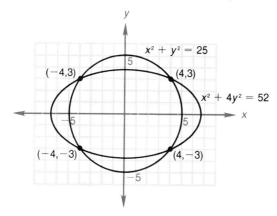

Mastery points
Can you
• Solve a system of equations that contain one linear and one quadratic equation?
• Solve a system of equations that contain two quadratic equations?

Exercise 9-4

Directions Solve each system of equations by substitution. Sketch the graphs of the systems in exercises 1–6. See example 9–4 A.

1. $x^2 + y^2 = 5$
$x + y = 1$

2. $x^2 + y^2 = 1$
$x + 2y = 2$

3. $x^2 - y^2 = 24$
$x + 2y = -3$

4. $x^2 - y^2 = 9$
$x + y = 5$

5. $y = x^2 - 2x + 1$
$y = 3 - x$

6. $y = x^2 - 4x + 4$
$x + y = 2$

7. $x^2 + 2y^2 = 12$
$2x = y + 2$

8. $x^2 + 3y^2 = 3$
$x - 3y = 0$

9. $x^2 - y^2 = 15$
$xy = 4$

10. $x^2 - y^2 = 35$
$xy = 6$

11. $x^2 + y^2 = 8$
$xy = 4$

12. $3x^2 - 4y^2 = 12$
$x = 4$

13. $2x^2 + y^2 = 4$
$y = -1$

14. $y = x^2 - 6x - 8$
$y = 10$

Directions Solve each system of equations by elimination or by a combination of elimination and substitution. See example 9–4 B.

15. $x^2 + y^2 = 9$
$x^2 - y^2 = 9$

16. $x^2 - y^2 = 24$
$x^2 + y^2 = 8$

17. $x^2 + 4y^2 = 64$
$x^2 + y^2 = 25$

18. $x^2 + 2y^2 = 22$
$2x^2 + y^2 = 17$

19. $x^2 - 2y^2 = -9$
$x^2 + y^2 = 18$

20. $3x^2 + 4y^2 = 16$
$x^2 - y^2 = 8$

21. $2x^2 - 9y^2 = 18$
$4x^2 + 9y^2 = 36$

22. $3x^2 + 4y^2 = 39$
$5x^2 - 2y^2 = -13$

23. $x^2 + y^2 = 4$
$2x^2 + 3y^2 = 5$

24. $x^2 + y^2 = 10$
$x^2 - 3xy + y^2 = 1$

25. $x^2 + y^2 = 6$
$2x^2 + 3xy + 2y^2 = 21$

26. $x^2 + xy - y^2 = 1$
$x^2 - y^2 = 3$

Directions For each of the following exercises, translate the verbal statements into a system of equations and solve.

Example

Find the dimensions of a rectangular plot of ground if the perimeter is 22 meters and the area is 24 square meters.

Solution

We use the formulas perimeter $P = 2\ell + 2w$ and area $A = \ell w$, where ℓ is the length and w is the width of the rectangle. Replacing P by 22 and A by 24, we have the system of equations

$2\ell + 2w = 22$
$\ell w = 24.$

w

ℓ

Solving the second equation for ℓ and substituting into the first equation, we obtain

$$\ell = \frac{24}{w}, \text{ so } 2\left(\frac{24}{w}\right) + 2w = 22$$

$$\frac{48}{w} + 2w = 22$$

$$48 + 2w^2 = 22w$$

$$2w^2 - 22w + 48 = 0$$

$$2(w^2 - 11w + 24) = 0$$

$$2(w - 3)(w - 8) = 0.$$

Then $w = 3$ or $w = 8$.

Substituting for w in the equation $\ell = \dfrac{24}{w}$, when

$w = 3$ or $w = 8$

$\ell = \dfrac{24}{3} = 8$ $\ell = \dfrac{24}{8} = 3.$

In either case, the dimensions of the rectangle are 8 meters by 3 meters.

Directions   Solve the following verbal problems.

27. Find the dimensions of a rectangle whose area is 120 square inches and whose perimeter is 46 inches.

28. The area of a rectangle is 80 square centimeters and the perimeter is 42 centimeters. What are the dimensions of the rectangle?

29. The sum of two numbers is 16 and the difference between their squares is 32. Find the numbers.

30. The sum of two numbers is 12 and the sum of their squares is 80. Find the numbers.

31. A manufacturer determines that the relationship between the demand (x) and the price (p) for their commodity is $xp = 6$ and the relationship between the supply (x) and price (p) of the same commodity is $4x = 23 - 5p$. The **equilibrium point** is where the supply equals the demand. Solve the system of equations to find the equilibrium point.

32. A piece of cardboard is in the form of a rectangle and has an area of 260 square inches. A square 2 inches on each side is cut out of each corner and an open box is formed by folding up the ends and sides. If the volume of the resulting box is 288 cubic inches, find the length and width of the cardboard.

Chapter summary

1. **Conic sections**—the parabola, the circle, the ellipse, and the hyperbola—are obtained by slicing a cone with a plane in four different ways.

2. The general form of the equations of conic sections is
$Ax^2 + By^2 + Cx + Dy + E = 0.$

3. The graph of a **quadratic equation** of the form $y = ax^2 + bx + c$ is a parabola opening upward if $a > 0$ and opening downward if $a < 0$.

4. The equation of the **circle** with center $C(h,k)$ and having radius r is given by
$(x - h)^2 + (y - k)^2 = r^2.$
When C is at $(0,0)$, the equation is given by
$x^2 + y^2 = r^2.$

5. The general form of the equation of a circle is given by
$Ax^2 + Ay^2 + Bx + Cy + D = 0.$

6. The equation of the ellipse with center $C(0,0)$ is given by
$$\frac{x^2}{a^2} + \frac{y^2}{b^2} = 1,$$
where the x-intercepts are a and $-a$ and the y-intercepts are b and $-b$.

7. The equation of the hyperbola with center $C(0,0)$ and having x-intercepts a and $-a$ is given by
$$\frac{x^2}{a^2} - \frac{y^2}{b^2} = 1$$
and having y-intercepts b and $-b$ is given by
$$\frac{y^2}{b^2} - \frac{x^2}{a^2} = 1.$$

8. The equations of the **asymptotes** for the graph of either hyperbola are given by

$$y = \frac{b}{a}x \text{ and } y = -\frac{b}{a}x.$$

They are the diagonals of a rectangle whose vertices are (a,b), $(a,-b)$, $(-a,b)$, and $(-a,-b)$.

Chapter review

[9–1]

Directions Determine the vertex, x-intercept(s), and y-intercept(s) of the following parabolas.

1. $y = -3(x + 1)^2$

2. $y = -x^2 + 4x + 45$

Directions Graph each of the following equations.

3. $y = 3x^2 - 2$

4. $y = -x^2 - 3x + 4$

5. $y = x^2 + 3x + 1$

[9–2]

Directions Write the equation for each of the following circles in the (1) form $(x - h)^2 + (y - k)^2 = r^2$ and (2) form $x^2 + y^2 + Ax + By + C = 0$, where A, B, and C are integers.

6. Center $C(-2,5)$ and radius 5

7. Center $C(4,0)$ and radius $\sqrt{11}$

8. Center at the origin and radius 9

Directions Determine the coordinates of the center and the length of the radius of the following circles.

9. $(x - 5)^2 = 36 - (y + 1)^2$

10. $y^2 = 7 - x^2$

11. $x^2 + y^2 + 6x - 8y + 1 = 0$

12. $x^2 + y^2 - 3x + 5y - 3 = 0$

13. $4x^2 + 4y^2 - 16y + 12x - 20 = 0$

14. $x^2 + y^2 - 4y - 15 = 0$

Directions Graph each of the following circles.

15. $x^2 + y^2 = 2$

16. $5x^2 = 20 - 5y^2$

17. $(x - 3)^2 + (y + 4)^2 = 1$

18. $x^2 + y^2 + 6x = 0$

19. $x^2 - 8y = 6x - y^2$

20. $x^2 + y^2 + 6x - 4y - 3 = 0$

[9–3]

Directions Determine the x- and y-intercepts for each given ellipse. Graph the equation.

21. $\dfrac{x^2}{16} + \dfrac{y^2}{100} = 1$

22. $x^2 + \dfrac{y^2}{25} = 1$

23. $16x^2 + 9y^2 = 144$

24. $4x^2 + 8y^2 = 16$

25. $6x^2 + y^2 = 24$

26. $x^2 + 8y^2 = 8$

Directions Determine the x-intercepts, or y-intercepts, and the equations of the asymptotes for the given hyperbola. Graph the equation.

27. $\dfrac{x^2}{36} - \dfrac{y^2}{49} = 1$

28. $\dfrac{y^2}{25} - \dfrac{x^2}{9} = 1$

29. $x^2 - \dfrac{y^2}{36} = 1$

30. $y^2 = 1 + \dfrac{x^2}{9}$

31. $16x^2 - 25y^2 = 400$

32. $x^2 - 9y^2 = -9$

Directions Identify each equation as a parabola, a circle, an ellipse, or a hyperbola. Graph each equation.

33. $x^2 - 16 = y^2$

34. $y + 4x^2 = 8x - 3$

35. $4x^2 = 100 - 25y^2$

36. $9y^2 = 225 - 9x^2$

37. $2y^2 = 8 - 2x^2$

38. $x^2 = y - 2x + 3$

[9–4]
Directions Solve the following systems of nonlinear equations.

39. $x^2 + y^2 = 2$
    $x + y = 2$

40. $x^2 + y^2 = 25$
    $x = 7 - y$

41. $x^2 + y^2 = 5$
    $x = 3$

42. $2x^2 - 3y^2 = 4$
    $y = -2$

43. $x^2 - y^2 = -1$
    $x + y = 3$

Directions Solve each system of equations by elimination or a combination of elimination and substitution.

44. $2x^2 + y^2 = 4$
    $x^2 - y^2 = 8$

45. $x^2 - y^2 = 16$
    $x^2 + 2y^2 = 25$

46. Find the length and the width of a rectangle whose perimeter is 52 inches and whose area is 153 square inches.

Chapter 9 cumulative test

Directions Given $x = 3$, $y = -5$, and $z = -2$, evaluate the following expressions.

[1–5]   1. $2x - 3y + 4z$

[1–5]   2. $4x^2 + 2y - 5$

[1–5]   3. $4x^2 + y^2 - z^2$

[1–5]   4. Given $C = \dfrac{5}{9}(F - 32)$, find C when $F = 212$.

[1–5]   5. What is the degree of the polynomial $4x^4 - 3x^2 + 7$?

[3–1]   6. From $4x^2 + 6x - 9$ subtract $-2x^2 - x + 5$.

Directions Given $P(x) = x^2 + 3x - 1$, $Q(x) = 4x - 1$, and $R(x) = x^2 + 3$ find

[1–5]   7. $P(x) - Q(x) + R(x)$

[1–5]   8. $P(x) + Q(x) - R(x)$

[1–5]   9. $P(3)$

[1–5]   10. $Q(-4)$

[1–5]   11. $R\left(\dfrac{1}{4}\right)$

[3–2]   12. Simplify the expression $x^{2 + n}x^{3n - 1}$ by performing the indicated operation.

Directions Perform the indicated operations.

[3–3] **13.** $(2a - 1)(3a^2 + 4a - 1)$

[4–5] **14.** $\dfrac{28a^2b^3c^5 - 35ab^3c^3}{7abc}$

[4–5] **15.** $(3x^3 - 4x^2 - 5x - 4) \div (x + 2)$

[4–3] **16.** $(2x + 9) - \dfrac{x + 3}{x - 5}$

[4–2] **17.** $\dfrac{16 - b^2}{2b + 1} \div (b + 4)$

[4–2] **18.** $\dfrac{x^2 - x - 6}{x^2 + x - 12} \cdot \dfrac{x^2 + 3x - 4}{x^2 + 2x - 3}$

[4–4] **19.** Simplify the complex rational expression

$$\dfrac{\dfrac{1}{y} - \dfrac{4}{x}}{\dfrac{2x - 8y}{xy}}.$$

Directions Solve the following equations.

[4–6] **20.** $\dfrac{y}{y - 3} + \dfrac{4}{5} = \dfrac{3}{y - 3}$

[6–2] **21.** $3x^2 - 2x + 2 = 0$

[2–5] **22.** $|4x - 3| = 7$

[6–6] **23.** $x^4 - 3x^2 - 10 = 0$

[6–5] **24.** $\sqrt[3]{2x + 1} - 3 = 0$

[7–4] **25.** Graph the inequality $3y - 2x \leq 12$.

Directions Find the slope of the given line.

[7–2] **26.** Passing through $(-5,6)$ and $(1,2)$

[7–2] **27.** Whose equation is $3x - 5y = 10$

[7–2] **28.** Determine if the lines $2x + 4y = 3$ and $4x - 8y = 6$ are parallel, perpendicular, or neither.

[7–3] **29.** Find an equation of the horizontal line passing through the point $(-1,7)$.

Directions Solve the following systems of equations.

[8–1] **30.** $3x - 4y = 2$
$2x + y = -1$

[8–2] **31.** $x + y - z = 1$
$x + y + z = 3$
$-x - y + z = -6$

[8 3] **32.** Evaluate $\begin{vmatrix} -3 & 0 & 1 \\ 2 & 4 & 1 \\ -1 & -1 & 0 \end{vmatrix}.$

[9–2] **33.** Given the circle $x^2 + y^2 - 6x + 4y - 3 = 0$, find the coordinates of the center and the length of the radius.

[9–1] **34.** Find the vertex and intercepts of the parabola $y = 2x^2 - 4x + 3$.

[9–3] **35.** Find the x- and y-intercepts of the ellipse $3x^2 = 12 - 2y^2$.

[9–3] **36.** Find the intercepts and equations of the asymptotes of the hyperbola
$$\dfrac{x^2}{9} - \dfrac{y^2}{4} = 1.$$

Directions Identify each equation as a circle, a parabola, an ellipse, or a hyperbola. Sketch the graph.

[9–3] **37.** $4x^2 - y = x + 3$ [9–3] **38.** $4y^2 = 2x^2 + 8$ [9–3] **39.** $3x^2 + 3y^2 = 12$

10

Functions

10–1 Relations and functions

Relations

In our study of mathematics, and out in the physical world, it is often useful to describe one quantity in terms of another quantity. The following examples illustrate this:

1. the demand for a product is related to the price of the product and
2. the cost to send mail is related to the weight of the piece of mail.

The relationships between these quantities are often stated as ordered pairs. Such relationships are called **relations.**

> ■ **Definition of a relation**
> A **relation** is any set of ordered pairs.

Example 10–1 A

1. The set of ordered pairs $\{(1,2), (-3,5), (0,6), (-7,9)\}$ is a relation consisting of four ordered pairs.

2. If it costs 15¢ per mile to operate an automobile, some ordered pairs that form this relation are

   $(10,1.50)$, $(100,15)$, $(200,30)$, and $(500,75)$,

   where the first component is the miles traveled and the second component is the cost in dollars.

3. $\{(x,y)\,|\,y = 2x + 1\}$ forms a relation that contains all ordered pairs such that x is a real number and y is found by adding 1 to twice the value of x. This relation contains the following ordered pairs.

   $(-4,-7)$, $(-3,-5)$, $(-1,-1)$, $(0,1)$, $(2,5)$

   There are infinitely many more ordered pairs in this relation.

We see in examples 2 and 3 that the value of the second component of each ordered pair is dependent on the value found in the first component. For this reason, in example 3, for instance, we call x the **independent variable** and y the **dependent variable,** since its value is dependent on what value of x we choose. Where two variables are involved in the definition of a relation, one of them will be independent and the other will be dependent.

Example 10–1 B

Given that it costs 15¢ per mile to operate an automobile, a set that defines this relation is

$\{(d,C)\,|\,C = 0.15d\}$,

where C is the cost in dollars to travel d miles. Then C is the dependent variable and d is the independent variable.

The set of all first components of the ordered pairs in a relation is called the **domain** of the relation, and the set of all second components of the ordered pairs in a relation is called the **range** of the relation. Thus in example 10–1 A number 1, the relation was defined by the set of ordered pairs $\{(1,2), (-3,5), (0,6), (-7,9)\}$. The *domain* of this relation is the set $\{1,-3,0,-7\}$ and the *range* is the set $\{2,5,6,9\}$. In example 10–1 A number 3, the set of all values of x form the *domain* and the set of all corresponding values of y form the *range* of the relation.

Functions

In all phases of mathematics, from the most elementary to the most sophisticated, the idea of a function is a cornerstone for each mathematical development.

Consider the distance d in miles that an automobile travels in time t hours. Suppose that the auto is traveling at an average rate of 50 miles per hour. The variables d and t are related by the equation

$$\{(t,d) \mid d = 50t\}.$$

We say that "d is a function of t" since a change in the value of t will cause a change in the value of d. For example,

$$\text{when } t = 2 \text{ hours, then } d = 50(2) = 100 \text{ miles,}$$
$$\text{when } t = 5 \text{ hours, then } d = 50(5) = 250 \text{ miles,}$$
$$\text{when } t = 10 \text{ hours, then } d = 50(10) = 500 \text{ miles.}$$

Note
t is the independent variable and d is the dependent variable, since the value of d is dependent on the chosen value of t.

Notice that for any chosen value of the independent variable t, we get a *unique* (one and only one) value of the dependent variable d. For this reason, the equation (or correspondence) $d = 50t$ defines d as a function of t.

A function

■ **Definition of a function**
A **function** is a relation that associates with each first component of the ordered pairs *exactly one* value of the second component.

Concept
A function is a set of ordered pairs (a relation) in which no two distinct ordered pairs have the same first components.

The variables x and y are generally used when defining a mathematical function. By the above definition, for y *to be a function of* x, the two variables must be related so that for each value of x, there is assigned a unique (only one) corresponding value of y. Then x is an element of the domain and y is an element of the range of the function. This is a very important concept because when we use an equation to determine the outcome in a given situation, we want the equation to be of the type that produces only one result.

Example 10–1 C

1. The following relations are functions.

$A = \{(1,2), (3,4), (-4,8), (0,-5)\}$ is a function with domain $\{-4,0,1,3\}$.
$B = \{(1,3), (-4,3), (9,3), (0,2)\}$ is a function with domain $\{-4,0,1,9\}$.

We see that three of the ordered pairs in the set B have the same second component, but this does not violate the definition of a function.

Functions

2. The relation $C = \{(3,0), (2,9), (3,-1), (-1,7)\}$ is *not* a function since the two ordered pairs $(3,0)$ and $(3,-1)$ have the same first components but different second components. The domain element 3 is associated with more than one range element.

Since a function defines a set of ordered pairs and is a special type of a relation, it has a *domain* and a *range*. The domain of a function must always be stated or implied by the nature of the equation defining the function. If the domain is not stated, we will assume it to be the set of all real numbers for which the function is defined or makes sense.

Example 10-1 D

Determine the domain of the function defined by the following.

1. $\{(x,y) \,|\, 3x + y = 4\}$, $x \in \{-3,-1,0,1,3\}$
Since replacement values of x make up the domain of a function, the domain in this case is restricted to $\{-3,-1,0,1,3\}$.

2. $\left\{ (x,y) \,|\, y = \dfrac{3}{2x - 1} \right\}$
We want the domain to be the set of all real numbers, unless for some reason we must restrict the values chosen for x. Thus, since division by 0 is undefined and the denominator of our equation $2x - 1 = 0$ when $x = \dfrac{1}{2}$, we

determine the domain to be all real numbers except $\dfrac{1}{2}$. In set-builder

notation the

$$\text{domain} = \left\{ x \,|\, x \in R, \, x \neq \dfrac{1}{2} \right\}.$$

3. $\{(x,y) \,|\, y = \sqrt{x - 1}\}$
Since $\sqrt{x - 1}$ is a real number only when the radicand $x - 1$ is nonnegative, we must restrict the domain of the function to values of x for which $x - 1 \geq 0$; that is, $x \geq 1$. Then the

$$\text{domain} = \{x \,|\, x \geq 1\}.$$

Recall that two or more points having the same abscissa lie on the same vertical line. Also we have learned that a function cannot have two or more ordered pairs having the same first component (the abscissa of the point). These facts lead us to a visual test for determining if a particular graph does or does not represent a function.

> ■ **Vertical line test for a function**
> If every vertical line drawn in the plane intersects the graph of a relation in *at most one point*, the relation is a function.

Example 10–1 E   Determine by the vertical line test if the following graphs represent functions.

1.

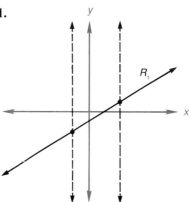

Since any vertical line drawn in the plane will intersect the graph of relation R_1 in *only one point*, R_1 is a function.

2.

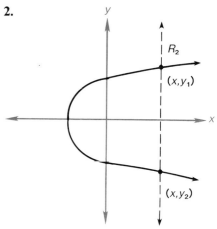

Many lines can be drawn in the plane that will intersect the graph of relation R_2 *in two points*, so R_2 is *not a function*, assuming x is the independent variable.

Mastery points

Can you
- Determine if a relation defines a function?
- Find the domain of a relation?
- Determine if a given graph of a relation represents a function by using the vertical line test?

Exercise 10–1

Directions   Determine whether or not the given relation defines a function. If not, show why not by an example. See example 10–1 C.

1. $\{(1,4),\ (3,6),\ (-1,5),\ (8,3)\}$

2. $\{(-4,0),\ (0,0),\ (6,7),\ (8,-6)\}$

3. $\{(1,2),\ (2,3),\ (-1,6),\ (2,-7)\}$

4. $\{(-6,2),\ (4,7),\ (7,4),\ (-6,0)\}$

5. $\{(-1,2),\ (-1,6),\ (-1,8),\ (-1,0)\}$

6. $\{(1,4),\ (2,4),\ (-3,4),\ (-6,4)\}$

7. $\{(-3,1),\ (1,1),\ (-7,1),\ (9,1)\}$

8. $\{(1,1),\ (2,2),\ (3,3),\ (4,4)\}$

9. $\{(x,y)\,|\,y = x + 7\}$ 10. $\{(x,y)\,|\,y = 3 - 4x\}$ 11. $\{(x,y)\,|\,y = x^2\}$ 12. $\{(x,y)\,|\,y = x^2 - x + 1\}$

13. $\{(x,y) \mid x = y^2\}$ **14.** $\{(x,y) \mid x = y^2 + 2\}$ **15.** $\{(x,y) \mid y = -3\}$ **16.** $\{(x,y) \mid y = 4\}$

17. $\{(x,y) \mid x = -10\}$ **18.** $\{(x,y) \mid x = 0\}$ **19.** $\{(x,y) \mid x - 3 = 0\}$ **20.** $\{(x,y) \mid x + 4 = 0\}$

Directions   Determine the domain of each relation. State the range in exercises 21–28. See examples 10–1 C and D.

21. $\{(1,2), (3,4), (5,6), (7,8)\}$

22. $\{(1,1), (2,2), (3,3), (4,4)\}$

23. $\{(-1,2), (-1,6), (-1,7), (-1,-8)\}$

24. $\{(4,-3), (4,0), (4,4), (4,8)\}$

25. $\{(-3,2), (2,2), (3,2), (-10,2)\}$

26. $\{(-6,0), (1,0), (2,0), (5,0)\}$

27. $\{(-6,7), (4,7), (-1,3), (9,8)\}$

28. $\{(-7,1), (10,-3), (4,-3), (9,1)\}$

29. $\{(x,y) \mid y = 2x - 3\}; x \in \{-3,-1,0,1,3\}$

30. $\{(x,y) \mid y = 4 - 3x\}; x \in \{-4,-2,0,2,4\}$

31. $\{(x,y) \mid y = x\}; x \in \{-5,-3,0,3,5\}$

32. $\{(x,y) \mid y = -x\}; x \in \{-2,-1,0,7,8\}$

33. $\left\{(x,y) \mid y = \dfrac{1}{x}\right\}; x \in \{-5,-1,1,2,4\}$

Directions   Determine the domain of each of the given relations. Write the answers in set-builder notation. See example 10–1 D.

34. $\{(x,y) \mid y = 4x - 3\}$ **35.** $\{(x,y) \mid x + y = 8\}$ **36.** $\{(x,y) \mid y = 3x^2\}$

37. $\{(x,y) \mid y = x^2 + 2x + 1\}$ **38.** $\{(x,y) \mid x = y^2\}$ **39.** $\{(x,y) \mid x = y^2 + 4\}$

40. $\{(x,y) \mid xy = 2\}$ **41.** $\left\{(x,y) \mid x = \dfrac{1}{y}\right\}$ **42.** $\left\{(x,y) \mid y = \dfrac{3}{x + 7}\right\}$

43. $\left\{(x,y) \mid y = \dfrac{5}{2x + 3}\right\}$ **44.** $\{(x,y) \mid y = \sqrt{x - 9}\}$ **45.** $\{(x,y) \mid y = \sqrt{3x + 4}\}$

Directions   Use the vertical line test to identify which of the following graphs represent functions where y is a function of x. Explain the answers. See example 10–1 E.

46.

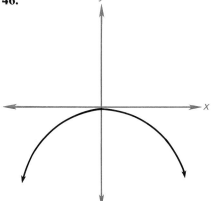

47.

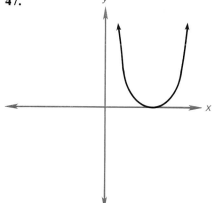

48.

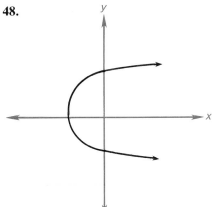

49.

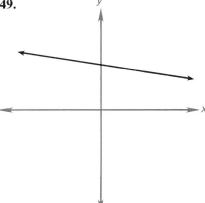

50.

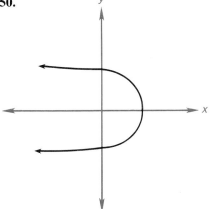

51.

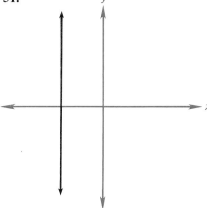

52.

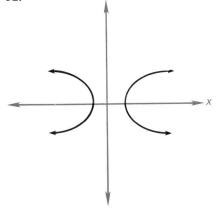

53.

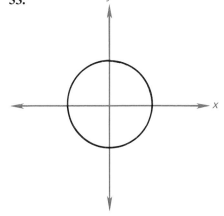

10-2 Functional notation

The lowercase letters f, g, and h are commonly used to denote a function. For example, the function defined by the equation $y = 2x - 5$, where x is the independent variable, is often written $f(x) = 2x - 5$. We read the symbol "$f(x)$" as "f at x" or "f of x," which means "the value of the function at x."

> **Note**
> We have named the dependent variable "$f(x)$." This new symbol represents a value of the function. Remember, $f(x)$ *does not* mean f times x.

The function has been given the name f and since $f(x)$ replaced y in the equation, $f(x)$ represents the element in the range of the function f that is associated with the chosen domain element, x.

To illustrate, consider the above-mentioned function f defined by

$$f(x) = 2x - 5.$$

If we let $x = 3$, then when x is replaced by 3 in the statement, $f(x) = 2x - 5$ becomes

$$f(3) = 2(3) - 5 = 6 - 5 = 1.$$

Thus we have determined that $f(3) = 1$, which we read "the value of the function f when $x = 3$ is 1." We usually just shorten this to read "f at 3 is 1" or "f of 3 is 1." In like fashion,

1. for $x = -2$, $f(-2) = 2(-2) - 5 = -4 - 5 = -9$. So $f(-2) = -9$, which we read "the value of the function f when $x = -2$ is -9" or "f at -2 is -9" or "f of -2 is -9."
2. for $x = 0$, $f(0) = 2(0) - 5 = 0 - 5 = -5$. So $f(0) = -5$, which we read "the value of the function f when $x = 0$ is -5" or "f at 0 is -5." Then for chosen domain element 0, the corresponding range element in f is -5.

> **Note**
> We have just determined that the function f defined by $f(x) = 2x - 5$ contains the three ordered pairs
>
> $$(3,1),\ (-2,-9),\ \text{and}\ (0,-5),$$
>
> as well as infinitely many other ordered pairs.

Example 10-2 A

Given $g(x) = 3x^2 - 4x + 1$, find

1. $g(3)$

We are given the domain element 3, and we substitute 3 for x to get

$$g(3) = 3(3)^2 - 4(3) + 1 = 27 - 12 + 1 = 16.$$

Therefore $g(3) = 16$ and $(3,16)$ is an element of function g.

2. $g(1)$

We are given the domain element 1, and we substitute 1 for x to get

$g(1) = 3(1)^2 - 4(1) + 1 = 3 - 4 + 1 = 0.$

Thus $g(1) = 0$ and $(1,0)$ is an element of function g.

3. $g(a)$

Replace a for the domain element x to get

$g(a) = 3(a)^2 - 4(a) + 1 = 3a^2 - 4a + 1.$

Thus $g(a) = 3a^2 - 4a + 1.$

4. $g(y + 1)$

Replace $y + 1$ for the domain element x to get

$$\begin{aligned} g(y + 1) &= 3(y + 1)^2 - 4(y + 1) + 1 \\ &= 3(y^2 + 2y + 1) - 4y - 4 + 1 \\ &= 3y^2 + 6y + 3 - 4y - 4 + 1 \\ &= 3y^2 + 2y. \qquad \text{Combine like terms.} \end{aligned}$$

Thus $g(y + 1) = 3y^2 + 2y.$

5. $g(x + h)$

Replace x with $x + h$ to obtain

$$\begin{aligned} g(x + h) &= 3(x + h)^2 - 4(x + h) + 1 \\ &= 3(x^2 + 2xh + h^2) - 4x - 4h + 1 \\ &= 3x^2 + 6xh + 3h^2 - 4x - 4h + 1. \end{aligned}$$

6. $g(x + h) - g(x)$

We have already determined that

$g(x + h) = 3x^2 + 6xh + 3h^2 - 4x - 4h + 1.$

Then

$$\begin{aligned} g(x + h) - g(x) &= (3x^2 + 6xh + 3h^2 - 4x - 4h + 1) \\ &\quad - (3x^2 - 4x + 1) \\ &= 3x^2 + 6xh + 3h^2 - 4x - 4h + 1 - 3x^2 + 4x - 1 \\ &= 6xh + 3h^2 - 4h. \qquad \text{Combine like terms.} \end{aligned}$$

7. $\dfrac{g(x + h) - g(x)}{h} \qquad (h \neq 0)$

We have determined that

$g(x + h) - g(x) = 6xh + 3h^2 - 4h.$

Then

$$\begin{aligned} \frac{g(x + h) - g(x)}{h} &= \frac{6xh + 3h^2 - 4h}{h} \\ &= \frac{h(6x + 3h - 4)}{h} \qquad \text{Factor out } h. \\ &= 6x + 3h - 4. \qquad \text{Reduce by } h. \end{aligned}$$

The expression $f[g(x)]$ is the *composition* of functions f and g, while the expression $g[f(x)]$ is the *composition* of functions g and f.

Example 10–2 B	Given $f(x) = 3x - 1$ and $g(x) = x^2 + 5$, find the following.

1. $f[g(2)]$

$f[g(2)]$ means to evaluate $f(x)$ when $x = g(2)$. We first evaluate

$g(2) = (2)^2 + 5 = 4 + 5 = 9.$

Thus we want

$f(9) = 3(9) - 1 = 27 - 1 = 26.$

We have found that

$f[g(2)] = f(9) = 26.$

2. $g[f(-3)]$

$g[f(-3)]$ means to evaluate $g(x)$ when $x = f(-3)$. We first find

$f(-3) = 3(-3) - 1 = -9 - 1 = -10.$

Then we evaluate

$g(-10) = (-10)^2 + 5 = 100 + 5 = 105.$

Thus we have found that

$g[f(-3)] = g(-10) = 105.$

Mastery points
Can you
- Evaluate $f(x)$ for any value of x given the function f?
- Find the composition $f[g(x)]$ and the composition $g[f(x)]$?

Exercise 10–2

Directions Find the value of each of the following if $f(x) = 3x - 2$ and $g(x) = x^2 + 2x - 5$. See example 10–2 A.

1. $f(0)$

2. $f(-6)$

3. $f\left(\dfrac{2}{3}\right)$

4. $g(0)$

5. $g(7)$

6. $g(-3)$

7. $g\left(\dfrac{1}{2}\right)$

8. $f(a)$

9. $f(a + 1)$

10. $f\left(\dfrac{1}{a}\right)$

11. $f(a^2)$

12. $f(5) - f(2)$

13. $f(6) - f(-3)$

14. $g(3) - g(1)$

15. $g(4) - g(-4)$

Directions For each of the following functions, find (a) $f(x + h)$; (b) $f(x + h) - f(x)$; (c) $\dfrac{f(x + h) - f(x)}{h}$, $h \neq 0$. See example 10–2 A–5, 6, and 7.

16. $f(x) = 4x - 1$ **17.** $f(x) = 4x^2$ **18.** $f(x) = 3x^2 - 2x$ **19.** $f(x) = 2x^2 + 3x + 2$

Directions Find the indicated values of each given function. Write the answers as ordered pairs. See example 10–2 A.

20. $f(x) = 5x - 6$; find $f(-2), f(0), f(2)$

21. $f(x) = 3x - 2$; find $f(-5), f(0), f\left(\dfrac{2}{3}\right)$

22. $h(x) = x^2 - 5$; find $h(-5), h(0), h(\sqrt{5})$

23. $h(x) = 3x^2 - 2x + 1$; find $h\left(-\dfrac{1}{2}\right), h(0), h(3)$

24. $g(x) = -7$; find $g\left(-\dfrac{7}{8}\right), g(0), g(25)$

25. $g(x) = 10$; find $g(-15), g(0), g\left(\dfrac{6}{5}\right)$

26. $h(x) = x^3 + 4$; find $h(-5), h(0), h\left(\dfrac{1}{2}\right)$

Directions Given $f(x) = 3x^2 - 5$ and $g(x) = 4x + 2$, find the following compositions and simplify the results. See example 10–2 B.

27. $f[g(-1)]$ **28.** $f[g(4)]$ **29.** $g[f(1)]$ **30.** $g[f(-2)]$

31. $f[g(x)]$ **32.** $g[f(x)]$ **33.** $f[f(x)]$ **34.** $g[g(x)]$

Directions Solve the following verbal problems.

35. The temperature in degrees Celsius C can be expressed as a function of degrees Fahrenheit F by $C = g(F)$. If $g(F) = \dfrac{5}{9}(F - 32)$, find $g(14)$, $g(32)$, and $g(212)$. Express the answers as ordered pairs $(F, g(F))$.

36. The temperature in degrees Fahrenheit F can be expressed as a function of degrees Celsius C by $F = f(C)$. If $f(C) = \dfrac{9}{5}C + 32$, find $f(0)$, $f(100)$, and $f(-10)$. Express the answers as ordered pairs $(C, f(C))$.

37. The volume V of a cube can be expressed as a function of its side s by $V = f(s) = s^3$. Find $f(1), f(3)$, and $f(5)$. Write the answers as ordered pairs $(s, f(s))$. What is the domain of f?

38. The area A of a circle can be expressed as a function of its radius r by $A = g(r) = \pi r^2$, where π is a constant. Find $g\left(\dfrac{1}{2}\right)$, $g(2.1)$, and $g(8)$. (Leave the answers in terms of π.) Write the answers as ordered pairs $(r, g(r))$. What is the domain of g?

39. The cost C of sending a first-class letter can be expressed as a function of its weight w in ounces by $C = h(w) = 22 + 17(w - 1)$. Find $h(2)$, $h(3)$, and $h(5)$. Express the answers as ordered pairs $(w, h(w))$.

40. The cost C in dollars of gasoline can be expressed as a function of the number of gallons n by $C = f(n) = 0.96n$. Find $f(10), f(12)$, and $f(25)$. Express the answers as ordered pairs $(n, f(n))$.

10-3 Special functions and their graphs

In previous chapters we discussed and graphed such equations as

$$(1)\ y = 3x + 2 \text{ and } (2)\ y = x^2 + 4x - 3.$$

We found that the graph of (1) was a straight line while the graph of (2) was a parabola. These were called linear equations and quadratic equations, respectively. In this section we will classify these special functions and also study two other types of functions.

The linear function

The linear equation that yields a straight line graph (except for vertical lines) will define **linear functions.**

> ■ **Linear function**
> A function that can be written in the form
> $$f(x) = mx + b,$$
> where m and b are real numbers, is called a **linear function.**

Recall in chapter 7 we found that in the equation $y = mx + b$, m was the slope of the line and b was the y-intercept. In like fashion, in the linear function $f(x) = mx + b$, m and b represent the same quantities and they can be used to graph a linear function.

Example 10-3 A

1. Write the equation $y = 3x - 2$ as a function and graph.
Replacing the y with $f(x)$, we have

$$f(x) = 3x - 2.$$

Then $m = 3 = \dfrac{3}{1}$ and b (y-intercept) $= -2$. From the y-intercept, $y = -2$, move 1 unit to the right and up 3 units to obtain a second point.

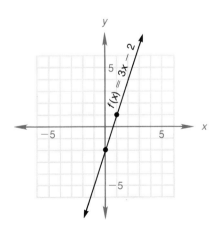

2. $y - 4 = 0$

If we solve this equation for y, we obtain $y = 4$. Then $f(x) = 4$.

The graph of this function is a horizontal straight line. This is a special linear function called the **constant function,** since the value of y is constant for any value of x that is chosen.

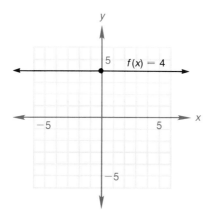

■ **Constant function**
Any function that can be written in the form
$$f(x) = k,$$
where k is a real number, is called a **constant function.**

The quadratic function

We found in section 9–1 that the graph of the quadratic equation $y = ax^2 + bx + c$ is a parabola that opens up when $a > 0$ and opens down when $a < 0$. So it is with the graph of the quadratic function.

■ **Quadratic function**
Any function that can be written in the form
$$f(x) = ax^2 + bx + c,$$
where a, b, and c are real numbers, $a \neq 0$, is called a **quadratic function.**

Example 10–3 B

Write the equation $y = x^2 + 3x$ as a function and graph.

Replace y with $f(x)$ and obtain $f(x) = x^2 + 3x$.

(1) Let $f(x) = 0$, then $0 = x^2 + 3x$
$$= x(x + 3).$$

Then $x = 0$ or $x = -3$ and the x-intercepts are the points $(0,0)$ and $(-3,0)$.

(2) Let $x = 0$, then
$$f(x) = 0^2 + 3(0) = 0$$

and the y-intercept is the point $(0,0)$.

(3) Find the vertex by completing the square.

$$f(x) = x^2 + 3x$$
$$= 1\left(x^2 + 3x + \frac{9}{4}\right) - \frac{9}{4}$$
$$= 1\left(x + \frac{3}{2}\right)^2 - \frac{9}{4}$$

The vertex is the point $\left(-\frac{3}{2}, -\frac{9}{4}\right)$.

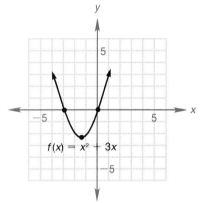

The linear and quadratic functions are special cases of a more general class of functions called the polynomial function.

The square root function

■ **Square root function**
Any function that can be stated in the form
$$f(x) = \sqrt{u},$$
where u is a polynomial in x, $u \geq 0$, is called a **square root function.**

Example 10–3 C

Graph the square root function $f(x) = \sqrt{x - 3}$.

If we plot a number of values of x and their corresponding values of $f(x)$, we obtain the graph below.

x	0	2	3	4	7	12
$f(x)$	$\sqrt{-3}$	$\sqrt{-1}$	0	1	2	3

Note

Since $\sqrt{-3}$ and $\sqrt{-1}$ are not real numbers, we cannot consider the first two ordered pairs. The domain must be $\{x \mid x - 3 \geq 0\}$ or $\{x \mid x \geq 3\}$, since x *cannot* be less than 3.

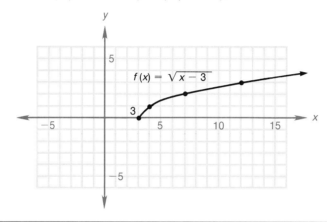

Mastery points

Can you
- Identify a linear function, a quadratic function, and a square root function?
- Graph each of the special functions stated above?

Exercise 10–3

Directions Identify each function and sketch its graph.

1. $f(x) = x + 5$

2. $f(x) = x - 2$

3. $f(x) = 4 - 3x$

4. $f(x) = 5 - x$

5. $f(x) = 6x - 1$

6. $f(x) = 2x + 3$

7. $f(x) = 2$

8. $f(x) = -3$

9. $f(x) = 0$

10. $f(x) = (x - 1)^2$

11. $f(x) = (x + 3)^2$

12. $f(x) = x^2 - 4x$

13. $f(x) = 5x^2 + 2x$ **14.** $f(x) = x^2 - x - 6$ **15.** $f(x) = -x^2 + 2x + 15$

16. $f(x) = -x^2 - 6x - 7$ **17.** $f(x) = x^2 + 8x + 12$ **18.** $f(x) = \sqrt{x - 5}$

19. $f(x) = \sqrt{4 - x}$ **20.** $f(x) = \sqrt{2 - x}$ **21.** $f(x) = \sqrt{4x - 1}$

22. $f(x) = \sqrt{2x + 3}$ **23.** $f(x) = \sqrt{x} - 1$ **24.** $f(x) = \sqrt{x} + 5$

25. $f(x) = -\sqrt{x + 7}$ **26.** $f(x) = -\sqrt{x - 6}$

Directions For the following functions, find $f(-2)$, $f(-1)$, $f(0)$, $f(1)$, and $f(2)$. Plot the points and connect them to graph each function. These are examples of the *absolute value function*.

27. $f(x) = |x| - 1$ **28.** $f(x) = |x| + 3$ **29.** $f(x) = |x|$

30. $f(x) = -|x|$ **31.** $f(x) = 5 - |x|$ **32.** $f(x) = 3 - |x|$

33. Given $f(x) = |x - 2|$, find values of $f(x)$ when x takes on values $-1, 0, 1, 2, 3$, and 4. Plot the points and connect them to get the graph of $f(x) = |x - 2|$.

34. Graph $f(x) = -|x - 2|$ using the same values of x as in exercise 33.

10–4 Inverse functions

Consider the function f defined by $f(x) = 3x - 5$. If we choose to let x take on the values $-4, -2, 0, 2$, and 4, we can determine that the ordered pairs

$$(-4, -17), (-2, -11), (0, -5), (2, 1), \text{ and } (4, 7)$$

belong to the function f. Now consider the function g defined by

$$g(x) = \frac{x + 5}{3}.$$

If we let x take on the values $-17, -11, -5, 1$, and 7 (range elements of f), we substitute to find that

$$(1) \quad g(-17) = \frac{-17 + 5}{3} = \frac{-12}{3} = -4$$

$$(2) \quad g(-11) = \frac{-11 + 5}{3} = \frac{-6}{3} = -2$$

$$(3) \quad g(-5) = \frac{-5 + 5}{3} = \frac{0}{3} = 0$$

$$(4) \quad g(1) = \frac{1 + 5}{3} = \frac{6}{3} = 2$$

$$(5) \quad g(7) = \frac{7 + 5}{3} = \frac{12}{3} = 4.$$

Thus the ordered pairs

$$(-17, -4), (-11, -2), (-5, 0), (1, 2), \text{ and } (7, 4)$$

are elements of the function g.

Observing the functions f and g, we see that when the components of the ordered pairs in f are interchanged, we have the ordered pairs in function g. If we continued to find other ordered pairs in the two functions, we would find that the same relationship would hold. When this relationship exists between two sets of ordered pairs, we say the functions are **inverses** of one another. Thus function g is the inverse of function f, and function f is the inverse of function g.

Given a function f, we denote the inverse of f by the symbol "f^{-1}" (read "the inverse of f" or "f inverse"). Thus $g = f^{-1}$ in the previous examples.

Note

The -1 in the symbol "f^{-1}" *should not* be interpreted as an exponent. Rather, it is a necessary part of the symbol as a whole and denotes the inverse of the function.

It is important to note that the inverse of a function is *not necessarily a function.*

Example 10–4 A

1. Consider function f defined by

$$f = \{(1,2), (4,5), (6,7), (-1,4)\}.$$

For every second component, there corresponds only one first component. Thus the inverse of f is a function and f^{-1} is defined by

$$f^{-1} = \{(2,1), (5,4), (7,6), (4,-1)\}.$$

2. Consider function g defined by

$$g = \{(1,2), (2,2), (3,-6), (4,-6)\}.$$

If we interchange range and domain, we obtain the set of ordered pairs

$$\{(2,1), (2,2), (-6,3), (-6,4)\},$$

which *does not* define a function since there are two distinct ordered pairs that have the same first component, 2 and also -6.

We then conclude that the inverse of a function is also a function if for each value of y there is *exactly one* value of x. We call this a **one-to-one function.**

> **■ Definition of a one-to-one function**
> A **one-to-one function** is any function that associates a unique value of x with each value of y.
>
> **Concept**
> A function is a one-to-one function if no two ordered pairs in the function have the *same second components.*

Thus the inverse of a function will be a function provided the function itself is one-to-one.

Recall that we have the vertical line test to determine if a graph represents a function. Since any two points having the same second component will lie on a horizontal line, it follows that any horizontal line drawn in the plane must intersect the graph in only one point if the function is one-to-one. We call this the **horizontal line test** for a one-to-one function, $y = f(x)$.

Example 10–4 B

Use the horizontal line test to determine if the given graphed function is one-to-one.

1.

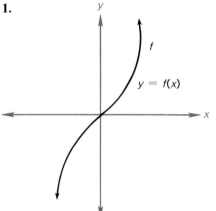

2.

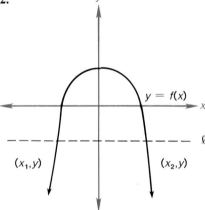

Since any horizontal line drawn in the plane will intersect the graph in only one point, the graph represents a one-to-one function.

Since there exists at least one horizontal line ℓ that can be drawn in the plane which will intersect the graph in two points, the graph does not represent a one-to-one function.

Now that we have a method for determining if a given function has an inverse function, let us consider a way of finding the inverse function, when it exists. Recall that the inverse of a function f can be obtained by interchanging x and y in the ordered pairs of f. Therefore we can find the inverse function f^{-1} by

1. replacing $f(x)$ by y,
2. interchanging x and y in the equation that defines the function f,
3. solving the resulting equation for y, and
4. replacing y by $f^{-1}(x)$ to define the function f^{-1}.

Example 10-4 C   Find the inverse of the given function.

1. $f(x) = 4x - 3$
We first rewrite the equation as
$$y = 4x - 3.$$
Interchange x and y to obtain the equation
$$x = 4y - 3.$$
Solving this equation for y, we have
$$x + 3 = 4y$$
$$y = \frac{x + 3}{4}.$$
Replacing y by $f^{-1}(x)$, we define the inverse function f^{-1} by
$$f^{-1}(x) = \frac{x + 3}{4}.$$

2. $g(x) = x^3 - 3$
Use the equation $y = x^3 - 3$ and interchange x and y to get the equation
$$x = y^3 - 3.$$
Solve this equation for y.
$$x = y^3 - 3$$
$$x + 3 = y^3$$
$$y = \sqrt[3]{x + 3} \qquad \text{Take the principal cube root of each member.}$$
Replacing y by $g^{-1}(x)$, we define the inverse function g^{-1} by
$$g^{-1}(x) = \sqrt[3]{x + 3}.$$

Consider another important fact that we should know about inverse functions f and f^{-1}. Suppose the ordered pair (a,b) belongs to the function f. Then the ordered pair (b,a) belongs to the function f^{-1}. Consider their positions in the plane relative to the graph of the equation $y = x$. To illustrate, suppose we let (2,4) be in f and then (4,2) is in f^{-1}. See figure 10.1.

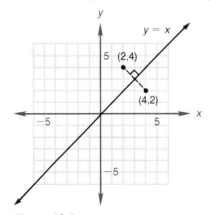

Figure 10.1

The line segment joining the two points is perpendicular to the line $y = x$, and the line $y = x$ divides the line segment in half (bisects the line segment). Thus the points (2,4) and (4,2) are "reflections" (or "mirror images") of each other with respect to the line $y = x$. This same relationship will exist with regard to all ordered pairs (a,b) and (b,a) in the functions f and f^{-1}, respectively. For this reason, we say the graphs of inverse functions f and f^{-1} are reflections of each other with respect to the line $y = x$.

We can now graph f^{-1} from the graph of f by locating a reflection point of each point of f with respect to the line $y = x$. In figure 10.2 we show the graphs of two one-to-one functions in solid lines and the graphs of their inverse functions in dashed lines.

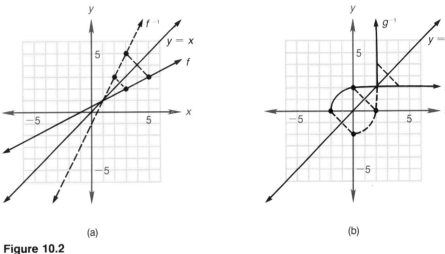

(a)　　　　　　　　　　　　(b)

Figure 10.2

Example 10–4 D

Sketch the graph of f^{-1} on the same coordinate plane as $f(x) = x^3 - 1$. We are not familiar with the graph of $y = x^3 - 1$, so we find a number of ordered pairs and plot their graph.

x		-2	-1	0	1	2
$y = x^3 - 1$		-9	-2	-1	0	7

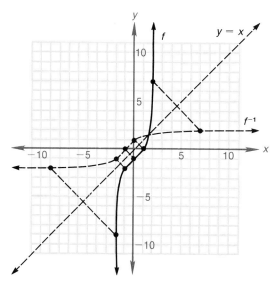

To determine points on the graph of f^{-1} that correspond to the points on f that we used, draw a line perpendicular to $y = x$ from the point on f. The corresponding point on f^{-1} must be at the same distance from $y = x$ on this perpendicular line as is the distance from the point on f to $y = x$.

Mastery points

Can you
- Determine if a given function is one-to-one and has an inverse function?
- Use the horizontal line test to determine if a graph represents a one-to-one function?
- Find the inverse function of a one-to-one function?
- Sketch the graph of f^{-1}, using the graph of f?

Exercise 10–4

Directions Determine if the given functions are one-to-one functions. Use the horizontal line test on the graph where necessary. Explain the answer. See example 10–4 A.

1. $f = \{(-4,3), (2,1), (-6,-4), (0,0)\}$

2. $g = \{(-8,0), (1,2), (7,8), (1,11)\}$

3. $h = \{(1,-6), (2,3), (4,-6), (5,-7)\}$

4. $F = \{(0,-7), (3,-4), (7,6), (1,-4)\}$

See example 10–4 B.

5. $f(x) = 2x - 7$

6. $g(x) = 8 - 3x$

7. $G(x) = x^2 - 3x + 1$

8. $H(x) = -x^2 + 2x + 4$

9. $f(x) = |x - 3|$

10. $g(x) = |2x + 4|$

11. $h(x) = \sqrt{x - 3}$

12. $F(x) = \sqrt{4 - 2x}$

See example 10–4 B.

13.

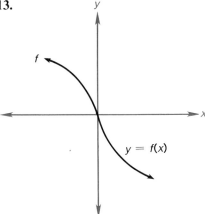

14.

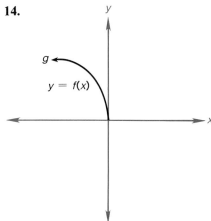

15.

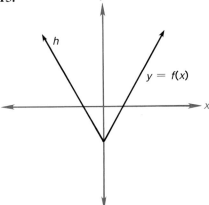

16.

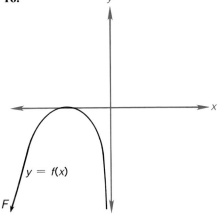

17.

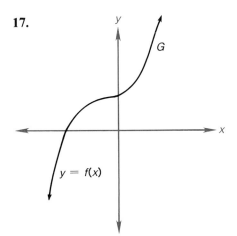

18.

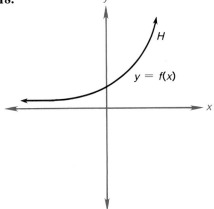

Directions Find the inverse function of each given one-to-one function. See example 10–4 C.

19. $f = \{(-3,4), (2,-3), (0,7)\}$ **20.** $g = \{(10,-4), (-6,3), (8,2)\}$ **21.** $h(x) = 5x$

22. $F(x) = -4x$ **23.** $G(x) = 3x + 7$ **24.** $H(x) = 5x - 1$

25. $f(x) = x^3 + 2$ **26.** $g(x) = 2x^3 - 3$ **27.** $h(x) = \sqrt{x + 5}, x \geq -5$

28. $F(x) = \sqrt{3x - 4}, x \geq \dfrac{4}{3}$ **29.** $G(x) = \sqrt{x + 2}, x \geq -2$ **30.** $H(x) = \sqrt{x} - 5, x \geq 0$

31. $f(x) = \sqrt[3]{x - 2}$ **32.** $g(x) = \sqrt[3]{5x + 2}$ **33.** $h(x) = \sqrt[3]{x} + 7$

34. $F(x) = \sqrt[3]{2x} - 1$

Directions Sketch the graphs of f and f^{-1} on the same coordinate plane. See example 10–4 D.

35. $f(x) = -4x$ **36.** $f(x) = 2x$ **37.** $f(x) = x + 4$ **38.** $f(x) = x - 2$

39. $f(x) = 5x - 6$ **40.** $f(x) = 2x - 3$ **41.** $f(x) = -\sqrt{3 - 2x}$ **42.** $f(x) = \sqrt{x + 2}$

43. $f(x) = x^3 + 1$ **44.** $f(x) = x^3$ **45.** $f(x) = x^2 - 2, x \geq 0$ **46.** $f(x) = x^3 - 4$

47. $f(x) = -x^2 + 4, x \leq 0$

Directions Show that $f[g(x)] = x$ and $g[f(x)] = x$ for the following functions f and g. This shows they are inverse functions.

48. $f(x) = 5x$ and $g(x) = \dfrac{x}{5}$ **49.** $f(x) = 4x + 7$ and $g(x) = \dfrac{x - 7}{4}$

50. $f(x) = x^3 - 3$ and $g(x) = \sqrt[3]{x + 3}$ **51.** $f(x) = \sqrt{x - 2}$ and $g(x) = x^2 + 2, x \geq 2$

10–5 Variation

If two variables are so related that as one variable changes, a change will also take place with the other variable (as with a function), then we have what is called a **variation**. This relationship is the basis for many formulas that are used in the scientific and physical world. Such formulas often determine functions. For example, consider the formula for the circumference C of a circle, $C = 2\pi r$, where 2π is a constant and r is the radius of the circle. As the radius r varies, the circumference C will also vary. That is, as

1. the radius r becomes longer, the circumference C becomes longer; and as
2. the radius r becomes shorter, the circumference C becomes shorter.

This variation is an example of a **direct variation** and we say "C varies directly as r."

> **■ Definition of direct variation**
> A variable y is said to *vary directly* as a variable x if
> $$y = kx \ (k > 0),$$
> where k is a positive constant.
>
> **Concept**
> Two variables vary directly if one variable, y, is some positive multiple of the other variable, x.

In our previous example, $C = 2\pi r$, $k = 2\pi$. Another, more general, example of direct variation is where y varies directly as the nth power of x if

$$y = kx^n, \ n > 0, \ k > 0.$$

The constant k is called the **constant of variation** (or the **constant of proportionality**).

Example 10–5 A

The formula for the area A of the surface of a sphere having radius r is given by $A = 4\pi r^2$. In this example $k = 4\pi$ and "A varies directly as the square of r." An alternate terminology is "y is *directly proportional* to the nth power of x."

Given the direct variation $y = kx$ and a pair of corresponding values of x and y, we can determine the value of the constant of variation k.

Example 10–5 B

Find the constant of variation k if the variable y varies directly as the variable x, and $y = 6$ when $x = 12$.
Because y varies directly as x, then $y = kx$, and if we substitute 6 for y and 12 for x, we have the equation $6 = k \cdot 12$.

Then $$k = \frac{1}{2} .$$

If we substitute the value for the constant of proportionality in the equation $y = kx$, we can then find the value of y given a specific value of x, and vice versa.

Example 10–5 C

Let y vary directly as x and suppose $y = 20$ when $x = 5$. Then find y when $x = 16$.

Since y varies directly as x, then $y = kx$. To find the constant of variation k, since $y = 20$ when $x = 5$, we substitute to obtain $20 = k \cdot 5$

$$4 = k.$$

Then we have the equation

$$y = 4x,$$

and replacing x by 16, we find

$$y = 4(16) = 64.$$

Therefore $y = 64$ when $x = 16$.

A second kind of variation between two variables occurs when an assignment of increasing positive values of one variable results in decreasing positive values of the other variable. Such a variation is called **inverse variation.**

> ■ **Definition of inverse variation**
> A variable y **varies inversely** as the variable x if
> $$y = \frac{k}{x}, k > 0.$$

Example 10–5 D

Boyle's law for gases states that the volume V of a gas varies inversely as the pressure P, provided the mass and temperature are constant. Then

$$V = \frac{k}{P}.$$

> **Note**
> Alternately, if we solve the formula for P and write the formula as $P = \dfrac{k}{V}$, then "P varies inversely with V."

Again we can determine the value of the constant of proportionality k given a pair of corresponding values of the variables. Using this, we can find the value of one of the variables, given a value of the other.

Example 10–5 E

Given $V = \dfrac{k}{P}$, find the constant of variation k if $V = 12$ cubic feet when $P = 100$ pounds per square foot.
Substituting we have

$12 = \dfrac{k}{100}$ and so $k = 1{,}200$. Then $V = \dfrac{1{,}200}{P}$.

Find V when $P = 75$ pounds per square foot.
Substituting 75 for P, we have

$V = \dfrac{1{,}200}{75} = 16$ cubic feet.

Note
Decreasing the pressure P from 100 pounds per square foot to 75 pounds per square foot caused the volume V to increase from 12 cubic feet to 16 cubic feet. This is what we expect with inverse variation.

Extending on this, if $n > 0$, then y *varies inversely* as the nth power of x, if

$$y = \dfrac{k}{x^n}, k > 0.$$

Again an alternate terminology for the above equation is "y *is inversely proportional* to the nth power of x."

A third variation relates one variable to two or more other variables. We call this a **joint variation.**

> ■ **Definition of joint variation**
> A variable z is said to **vary jointly** as variables x and y if
> $$z = kxy, k > 0.$$

Example 10–5 F

1. The formula for the area A of a triangle with base b and altitude h is given by

$$A = \dfrac{1}{2} bh.$$

The constant of proportionality is $\dfrac{1}{2}$ and "A varies jointly as b and h."

Alternately, we could say that "A varies directly as b and h" or "A varies directly as the product of b and h."

2. Coulomb's law says that the magnitude F of a force between two charges of electricity is given by

$$F = k \cdot \frac{q_a \cdot q_b}{r^2},$$

where q_a and q_b are two electrical charges that are at a distance r apart. Then the force "F varies directly as the product of q_a and q_b and inversely as the square of r." We have a joint variation.

In these examples, we can determine the constant k if we are given values of the variables. We can then evaluate any one of the variables by knowing the values of the others.

Example 10–5 G

1. The diametral pitch P of a gear varies directly as the number of teeth N and inversely as the pitch diameter D. Find k if $P = 10$ when $N = 12$ and $D = 4$.

If P varies directly as N and inversely as D, then

$$P = \frac{kN}{D}.$$

We substitute 10 for P, 12 for N, and 4 for D to obtain

$$10 = \frac{k \cdot 12}{4}$$

$40 = k \cdot 12$ and $k = \frac{40}{12} = \frac{10}{3}$. Then

$$P = \frac{\frac{10}{3} N}{D}.$$

2. We can now use the above to find N when $P = 9$ and $D = 20$. We substitute 9 for P and 20 for D in the equation to get

$$9 = \frac{\frac{10}{3} N}{20}$$

$$180 = \frac{10}{3} N$$

$$N = 180 \cdot \frac{3}{10} = 18 \cdot 3 = 54.$$

Therefore $N = 54$.

Exercise 10–5

Directions Express the given statements as equations using the constant of variation k. See examples 10–5 A, D, and F.

1. The speed S of a falling object varies directly as the time t.

2. The perimeter P of an equilateral triangle varies directly as the side s.

3. The time t required for an auto to travel a fixed distance is inversely proportional to the rate r.

4. The resistance R to the flow of electricity in a conductor varies inversely as the diameter d of the wire.

5. The momentum M of a body varies directly as the product of its mass m and its velocity v.

6. The current I varies directly as the product of the voltage E and the resistance R.

7. The maximum force F exerted on the vane of a wind generator varies jointly as the area A of the vane and the square of the wind velocity v.

8. The resistance R of an electrical conductor varies directly as the length ℓ and inversely as the cross-section area A.

9. The maximum torsinal stress S_{max} on a circular shaft varies directly as the torque T and inversely as the cube of the radius r.

10. The ideal gas law states that the pressure P of a gas varies directly with the absolute temperature T of the gas and inversely as the volume V of the gas.

11. The force of impact F varies directly as the product of the mass m and the velocity v and inversely as the product of the acceleration of gravity g and time t.

Directions Find the constant of variation for the stated conditions. See examples 10–5 B and C.

12. y varies directly as x, and $x = 20$ when $y = 5$.

13. p varies directly as h, and $p = 36$ when $h = 9$.

14. A varies directly as the square of r, and $A = 154$ when $r = 7$.

15. V is directly proportional to the cube of s, and $V = 81$ when $s = 3$.

16. P varies inversely as V, and $P = 120$ when $V = 5$.

17. s varies inversely as T, and $s = 48$ when $T = 2.5$.

18. n varies inversely as the square of p, and $n = 14$ when $p = 9$.

19. p varies directly as T and inversely as V, and $p = 24$ when $T = 6$ and $V = 5$.

20. A is directly proportional to the square of b and inversely proportional to c, and $A = 42$ when $b = 3$ and $c = 3$.

21. A varies directly as the product of h and $(a + b)$, and $A = 36$ when $h = 9$, $a = 2$, and $b = 6$.

Directions Find the value of the indicated variable. See examples 10–5 C, E, and G.

22. x varies directly as y. If $x = 36$ when $y = 4$, find x when $y = 13$.

23. w varies directly as ℓ. If $w = 9$ when $\ell = 6$, find w when $\ell = 15$.

24. R_1 varies inversely as R_2. If $R_1 = 48$ when $R_2 = 4$, find R_1 when $R_2 = 12$.

25. If y varies inversely as the square of z, find y when $z = 3$ if $y = 12$ when $z = 2$.

26. Find A when $r = 3$ if A is directly proportional to the square of r, and $A = 48$ when $r = 4$.

27. If v varies directly as s and inversely as t, find v when $s = 8$ and $t = 6$ if $v = 20$ when $s = 4$ and $t = 2$.

28. If T varies directly with the square of s and inversely as the cube of m, find T when $s = 2$ and $m = 3$, if $T = 14$ when $s = 3$ and $m = 6$.

Directions Solve the following verbal problems.

29. The volume of sales V varies inversely as the unit price P of a commodity. If $V = 4,000$ units when $P = \$1.50$, find P when $V = 3,000$ units.

30. The velocity v of a soundwave varies directly as the product of frequency n and the wavelength ℓ. If $v = 12$ feet per second when $n = 2$ and $\ell = 4$, find v when $n = 6$ and $\ell = 8$.

31. The intensity of illumination on a surface E in foot-candles varies inversely as the square of the distance d of the light source from the surface. If $E = 6.4$ foot-candles when $d = 8$ feet, find E when $d = 6$ feet (round to decimal point).

32. The field intensity H of a magnetic field varies directly as the force F acting on it and inversely as the strength of the pole m. If $H = 3$ oersteds when $F = 750$ dynes and $m = 200$, find H when $F = 500$ dynes and $m = 175$.

33. The electrical resistance R of a wire is directly proportional to the length L of the wire and inversely proportional to the square of the diameter d of the wire. If $R = 8$ ohms when $L = 8$ feet and $d = 1$ inch, find R when $L = 6$ feet and $d = 2$ inches.

34. The simple interest I earned by a savings deposit in a given time t varies jointly as the principal P and the interest rate r. If $I = \$180$ when $P = \$1,000$ and $r = 0.06$, find I when $P = \$750$ and $r = 0.06$ in the same amount of time.

35. The maximum force F that a rectangular cantilever beam (a beam supported at one end) can withstand at its free end varies directly as the beam's width w and the square of its height h and inversely as its length ℓ. If $F = 5,000$ pounds when $w = 0.5$ feet, $h = 0.7$ feet, and $\ell = 12$ feet, find F when $w = 0.45$ feet, $h = 0.6$ feet, and $\ell = 18$ feet (round to decimal point).

Chapter summary

1. A **relation** is any set of ordered pairs.

2. A **function** is a relation in which no two distinct ordered pairs have the same first components. For each x there is a unique y.

3. The **domain** of a function is the set of the first components of the ordered pairs in the function.

4. The **range** of a function is the set of the second components of the ordered pairs in the function.

5. A graph represents a function if any vertical line drawn in the plane intersects the graph in only one point, $y = f(x)$.

6. A **one-to-one function** is a function in which no two ordered pairs have the same second components. For each y there is a unique x.

7. The inverse of a function f, denoted by f^{-1}, is obtained by interchanging the components in each ordered pair of the function f.

8. Given function f, f^{-1} is a function if and only if f is one-to-one.

9. The graphs of f and f^{-1} are mirror images (reflections) of each other with respect to the line $y = x$.

10. The composition of f and g is given by $f[g(x)]$ and the composition of g and f is given by $g[f(x)]$.

11. The following are special functions.
    a. A linear function is any function of the form $f(x) = mx + b$.
    b. A quadratic function is any function of the form $f(x) = ax^2 + bx + c, a \neq 0$.
    c. A constant function is any function of the form $f(x) = k$, where k is a real number.
    d. A square root function is any function of the form $f(x) = \sqrt{x}$.

12. Two variables x and y **vary directly** if $y = kx$ or $x = ky$ for some positive constant k.

13. Two variables x and y **vary inversely** if $y = \dfrac{k}{x}$ or $x = \dfrac{k}{y}$ for some positive constant k.

14. The variable z **varies jointly** as variables x and y if $z = kxy$ for some $k > 0$.

Chapter review

[10–1]

Directions Determine whether or not each relation defines a function. If not, explain why not.

1. $\{(1,3), (2,-4), (1,0)\}$
2. $\{(4,3), (-2,3), (1,3)\}$
3. $\{(0,7), (-7,1), (6,-3)\}$
4. $2x + y = 3$
5. $\{(x,y) \mid y = 5\}$
6. $\{(x,y) \mid x = -7\}$

Directions Determine the domain and range of each relation.

7. $\{(-3,4), (4,2), (0,0), (-2,7)\}$
8. $\{(x,y) \mid y = -4x + 1\}; x \in \{-3,-2,-1,0,1,2,3\}$
9. $\{(x,y) \mid y = x^2 + 2x - 1\}; x \in \{-3,0,1,4\}$
10. $\{(x,y) \mid y = -7\}; x \in \{-8,-2,0,3,7\}$

Directions Determine the domain of each relation.

11. $\{(x,y) \mid y = 2x + 7\}$
12. $\{(x,y) \mid y = 2x^2 + 9\}$
13. $\{(x,y) \mid xy = -3\}$
14. $\left\{(x,y) \mid y = \dfrac{3}{3x - 4}\right\}$
15. $\{(x,y) \mid y = \sqrt{x + 2}\}$

[10–2]

Directions Given $f(x) = x + 2$ and $g(x) = x^2 - 3$, find each of the following.

16. $f(-2)$
17. $f(-4)$
18. $g(0)$
19. $f(3) - g(2)$
20. $f[g(-3)]$
21. $g[f(0)]$
22. $\dfrac{f(x + h) - f(x)}{h}, h \neq 0$.

[10–3]

Directions Identify each function by name and sketch the graph.

23. $f(x) = 3x + 5$

24. $f(x) = -7$

25. $f(x) = x^2 - 5x - 14$

26. $f(x) = \sqrt{x + 5}$

27. $f(x) = |x| + 7$

[10–4]

Directions In each of the following, determine if the given function is one-to-one. Use the horizontal line test if necessary.

28. $f = \{(-3,1), (4,3), (2,1)\}$

29. $g = \{(0,2), (4,-3), (6,3)\}$

30. $f(x) = 5 - 4x$

31. $g(x) = x^2 + 1$

32. $h(x) = |2x - 3|$

33. $F(x) = \sqrt{x + 4}$

Directions Find the inverse function, f^{-1}, of each given function.

34. $f(x) = 4x + 3$

35. $f(x) = x^2 - 2, x \geq 0$

36. $f(x) = \sqrt{2x - 3}$

Directions Show that functions f and g are inverse functions of each other by showing that $f[g(x)] = x$ and $g[f(x)] = x$.

37. $f(x) = 5 - 2x$ and $g(x) = \dfrac{5 - x}{2}$

38. $f(x) = 3x^3 - 4$ and $g(x) = \sqrt[3]{\dfrac{x + 4}{3}}$

Directions Sketch the graph of each function f and its inverse f^{-1} on the same set of coordinate axes.

39. $f(x) = 3x - 2$

40. $f(x) = \sqrt{4 - x}$

41. $f(x) = x^2 - 7, x \geq 0$

[10–5]

Directions Express each statement as an equation using a constant of variation k.

42. x varies directly as the square of y.

43. The current I is inversely proportional to the resistance R.

44. The angular momentum H of a particle varies jointly as the linear velocity v and the radius of rotation r.

45. The resistance R of a conductor varies directly as the length ℓ and inversely as the area A.

46. If s varies directly as w^3, find the constant of variation k if $s = 16$ when $w = 2$.

47. If y varies inversely as the square of x, find k if $y = 24$ when $x = 3$.

48. If R varies directly as ℓ and inversely as A, find k if $R = 0.45$ when $\ell = 5.0$ and $A = 0.055$.

49. P varies jointly as R and I^2. If $P = 1,500$ when $R = 20$ and $I = 5$, find P when $R = 20$ and $I = 4$.

50. d varies jointly as a and b and inversely as c. If $d = 15$ when $a = 3$, $b = 7$, and $c = 14$, find d when $a = 5$, $b = 8$, and $c = 20$.

Chapter 10 cumulative test

[1–4]   **1.** Perform the indicated operations and simplify.
$$\frac{4(2 + 5) - 4^2 + 9}{(-2)(-7)}$$

[1–5]   **2.** Given $P(x) = x^3 - 4x + 1$, find
(a) $P(1)$, (b) $P(-3)$, (c) $P(0)$.

[1–5]   **3.** Given $S = V_o t + \dfrac{1}{2}at^2$, find S when $V_o = 20$, $t = 3$, $a = 32$.

Directions Perform the indicated operations. Assume all variable exponents are positive integers.

[3–2] **4.** $(x^4y^5)^2 (-x^5y^8)$ [3–2] **5.** $\dfrac{x^{n-1}}{x^{2-3n}}$ [3–2] **6.** $(2x-1)^0$ [3–4] **7.** $\dfrac{x^{-4}y^{-5}}{x^2y^2}$

[3–4] **8.** 4^{-2} [3–4] **9.** $\left(\dfrac{1}{2}\right)^{-3}$ [3–4] **10.** $\dfrac{4^{-6}}{4^{-4}}$

Directions Completely factor the following expressions.

[3–5] **11.** $3a^2 - 27b^2$ [3–8] **12.** $y^4 - x^4$ [3–5] **13.** $5x(2x+3) - 10(2x+3)$

[3–6] **14.** $6a^2 + 7a - 3$ [3–6] **15.** $2a^2b^2 - ab - 1$ [3–8] **16.** $2x^3 - 54y^3$

Directions Find the solution set of the following equations and inequalities.

[2–2] **17.** $ax - by = 2x + 1$ for x [2–1] **18.** $-3(a+5) + 4(2a-1) = 4$

[2–4] **19.** $-3 \le 2x + 1 < 5$ [6–7] **20.** $x^2 - 6x - 16 \ge 0$

[4–6] **21.** $\dfrac{3}{y} + \dfrac{2}{y} = \dfrac{7}{3}$ [6–7] **22.** $y - \dfrac{3}{y} = 2$

Directions Perform the indicated operations.

[5–1] **23.** $(a^{-\frac{3}{4}})^{\frac{1}{2}}$ [5–1] **24.** $\dfrac{b^{\frac{5}{4}}}{b}$ [5–1] **25.** $(x^4y^6)^{\frac{1}{2}}$

Directions Find the solution set of the following systems of equations.

[8–1] **26.** $x - 2y = 6$
$3x + y = -1$

[8–4] **27.** $3y - 2x = 0$
$2y + 4x = 3$ Use an augmented matrix.

Directions Given $f(x) = 5 - 6x$ and $g(x) = x^2 + x - 1$, find

[10–2] **28.** $f(-3)$ [10–2] **29.** $f[g(2)]$ [10–2] **30.** $g[f(-1)]$

[10–4] **31.** Given $f(x) = 4x + 3$, find f^{-1}.   [10–4] **32.** Given $f(x) = x^3 + 1$ and $g(x) = x - 1$, show that f and g are inverse functions by showing that (a) $f[g(x)] - x$, (b) $g[f(x)] - x$.

Directions Find the x- and y-intercepts of the graphs of the following equations, if they exist. What type of figure is each?

[9–1] **33.** $y = x^2 + 5x - 40$ [9–3] **34.** $x^2 + 2y^2 = 8$

[9–3] **35.** $\dfrac{x^2}{16} - \dfrac{y^2}{25} = 1$ [9–1] **36.** $y = -x^2 - 8x + 20$

Directions Give a name to each function. Sketch its graph.

[10–3] **37.** $f(x) = 7$ [10–3] **38.** $f(x) = 4x - 6$ [10–3] **39.** $f(x) = 3x^2 - 11x - 4$

[10–3] **40.** $f(x) = \sqrt{x + 9}$ [10–3] **41.** $f(x) = |x| - 3$

11

Exponential and Logarithmic Functions

11–1 The exponential function

The exponential function

Expressions such as 2^x and $(1.05)^x$, where the exponent is a variable and the base is a constant, are important mathematical tools. These expressions can be used to

1. solve problems that involve the growth of bacteria in a culture, or the decay of a substance, and
2. to compute the amount to which a savings account increases when interest is compounded periodically.

We now wish to consider a function that is defined by an expression in which the exponent is a variable and the base is a constant. Such a function is called an **exponential function.**

> ■ **Definition of exponential function**
> Given $b > 0$ and $b \neq 1$, **the exponential function with base b**
> is the function f defined by
> $$f(x) = b^x.$$

In our work with exponents thus far, we have defined 2^x when x is a rational number. That is, we know

$$2^0 = 1;\ 2^2 = 4;\ 2^{-2} = \frac{1}{2^2} = \frac{1}{4};\ 2^{\frac{1}{3}} = \sqrt[3]{2};\ \text{and } 2^{-\frac{1}{2}} = \frac{1}{\sqrt{2}}.$$

What we have not previously defined is

$$2^x \text{ when } x \text{ is irrational.}$$

For example, consider the value of

$$2^{\sqrt{2}}.$$

The definition of this is beyond the scope of our text, but we can intuitively interpret the meaning of this number.

We know the irrational number $\sqrt{2}$ is approximately equal to 1.41421, correct to five decimal places. Then

$$2^1 < 2^{\sqrt{2}} < 2^2,$$

and since

$$1.414 < \sqrt{2} < 1.415,$$

then

$$2^{1.414} < 2^{\sqrt{2}} < 2^{1.415}$$
$$2.665 < 2^{\sqrt{2}} < 2.667. \qquad (2^{\sqrt{2}} \approx 2.67)$$

This same type of argument can be made for any irrational exponent x. We then assume that, for $b > 0$, b^x is defined for all real numbers. We also assume that all properties of exponents previously learned extend to real numbers.

Graph of the exponential function

Exponential functions can be graphed by finding several ordered pairs that belong to the function, plotting the points, and connecting the points with a smooth curve.

Example 11-1 A

Sketch the graphs of the following exponential functions.

1. $f(x) = 3^x$

We make a table of related values.

x	-3	-2	-1	0	1	2
$y = 3^x$	$\dfrac{1}{27}$	$\dfrac{1}{9}$	$\dfrac{1}{3}$	1	3	9

Plot the points and connect them with a smooth curve.

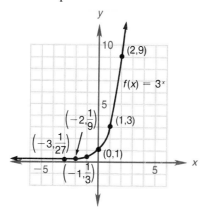

2. $g(x) = \left(\dfrac{1}{3}\right)^x$

We make a table of related values.

x	-2	-1	0	1	2	3
$y = \left(\dfrac{1}{3}\right)^x$	9	3	1	$\dfrac{1}{3}$	$\dfrac{1}{9}$	$\dfrac{1}{27}$

Plot the graphs of the ordered pairs and draw a smooth curve.

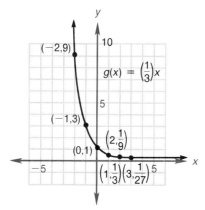

We observe the following characteristics about the exponential function from our two graphs.

1. In example 1, when $f(x) = 3^x$, as x increases in value, y also increases in value. This is called **exponential growth** and occurs when base $b > 1$.

2. In example 2, when $f(x) = \left(\dfrac{1}{3}\right)^x$, as x increases in value, y decreases in value. This is called **exponential decay** and occurs when $0 < b < 1$.

In each case we see that the domain of the function is the set of all real numbers, and since the graph of each function is entirely above the x-axis, the range is the set of positive real numbers.

Note
The vertical line test for functions and the horizontal line test for one-to-one functions are both met. Thus the exponential function is a one-to-one function and so has an inverse that is a function. We will study this inverse function in section 11–2.

The following property of exponents is used to solve some equations involving exponential expressions, called **exponential equations.**

■ **Property of exponents**
For $b > 0$, $b \neq 1$, and x and y are real numbers,
$$\text{if } b^x = b^y, \text{ then } x = y.$$

The following examples illustrate how this property is used to solve exponential equations.

Example 11–1 B

Find the solution set S for the following exponential equations.

1. $3^x = 81$
   Since $81 = 3^4$, then we have $3^x = 3^4$, which is true when $x = 4$. The solution set is $S = \{4\}$.

2. $5^{4x+1} = (25)^{3x}$
   Since $25 = 5^2$, we replace $(25)^{3x}$ by $(5^2)^{3x} = 5^{2 \cdot 3x} = 5^{6x}$,

   then $\qquad 5^{4x+1} = 5^{6x}$
   and so $\quad 4x + 1 = 6x$
   $$1 = 2x$$
   $$x = \frac{1}{2}.$$

   The solution set is $S = \left\{\dfrac{1}{2}\right\}$.

   This procedure will not work for all exponential equations. Others are covered later.

We mentioned earlier that exponential equations occur in bacteria growth. The following example illustrates that growth.

Example 11–1 C

The number of bacteria in a culture that initially contains 3,000 bacteria triples every 15 hours. How many bacteria are there in the culture after (a) 15 hours, (b) 30 hours, (c) 45 hours? Write an equation for the number of bacteria in the culture, N, after t hours.

Since the bacteria *triples* (3 times) every 15 hours, after
(a) 15 hours, there are $3,000(3) = 9,000$ bacteria;
(b) 30 hours, the number of bacteria triples *twice,* so there are

$$[3,000(3)](3) = 3,000(3)^2 = 3,000(9)$$
$$= 27,000 \text{ bacteria};$$

(c) 45 hours, the number of bacteria triples *three* times, so there are

$$[3,000(3)](3)(3) = 3,000(3)^3$$
$$= 3,000(27) = 81,000 \text{ bacteria.}$$

In t hours, the number of times the bacteria triples is found by $\dfrac{t}{15}$, since it triples every 15 hours. Thus the number of bacteria is given by

$$N = 3,000(3)^{\frac{t}{15}}.$$

Mastery points
Can you
- Sketch the graph of an exponential function?
- Find the solution set of exponential equations?

Exercise 11–1

Directions Sketch the graph of each given exponential function. See example 11–1 A.

1. $f(x) = 2^x$

2. $g(x) = 4^x$

3. $h(x) = \left(\dfrac{1}{2}\right)^x$

4. $f(x) = \left(\dfrac{1}{4}\right)^x$

5. $g(x) = 3^{-x}$

6. $h(x) = 2^{-x}$

7. $f(x) = 2^{2x}$

8. $g(x) = 3^{2x}$

9. $h(x) = \left(\dfrac{1}{3}\right)^{-x}$

10. $f(x) = 2^{x-1}$

11. $f(x) = 3^{x^2}$

12. $h(x) = 2^{x^2}$

Directions Find the solution set of each exponential equation. See example 11–1 B.

13. $2^x = 32$

14. $3^x = 27$

15. $4^x = 32$

16. $9^x = 243$

17. $16^x = 64$

18. $2^x = \dfrac{1}{16}$

19. $3^x = \dfrac{1}{9}$

20. $4^x = \dfrac{1}{64}$

Exponential and Logarithmic Functions

21. $5^{-x} = 25$ **22.** $3^{-x} = 81$ **23.** $16^{-x} = 32$ **24.** $9^{2x} = 27$

25. $8^{3x} = 4$ **26.** $\left(\dfrac{1}{2}\right)^x = 16$ **27.** $\left(\dfrac{1}{3}\right)^x = 27$ **28.** $\left(\dfrac{2}{3}\right)^x = \dfrac{4}{9}$

29. $\left(\dfrac{4}{3}\right)^x = \dfrac{64}{27}$ **30.** $(32)^{x-1} = 8$ **31.** $(27)^{2x+1} = 9$ **32.** $(25)^{4x+1} = 125$

See example 11–1 C.

33. The number of bacteria in a culture is initially 1,500 bacteria. If the number of bacteria triples every 12 hours, how many bacteria, N, are there in the culture after (a) 12 hours, (b) 36 hours, (c) 30 hours (use a calculator), (d) t hours?

34. The population of University City doubles every 10 years due to industries moving in. If the population was 40,000 in 1980, what is the population going to be in the year (a) 2000, (b) 2005, (c) t years from 1980?

35. Money deposited in a savings account will double every 72 years, where r is the rate of interest per year. If $5,000 is deposited at 8% interest, how much money is in the account after (a) 18 years, (b) 24 years, (c) t years?

36. A radioactive substance decays according to the equation
$$A = 200(3)^{-0.5t},$$
where A is the amount in grams and t is the time in months. How many grams of the substance are present after (a) 2 months, (b) 1 year?

37. Due to the introduction of an antibacterial substance into a culture that contains 8,000 bacteria, the bacteria are being destroyed according to the equation
$$N = 8,000(2)^{-t},$$
where N is the number of bacteria present and t is the time in hours. How many bacteria are present after (a) 3 hours, (b) 5 hours?

38. The production of an oil well is decreasing at a rate according to the equation
$$A = 1,000,000(2)^{-0.3t},$$
where A is the amount in barrels and t is the time in years. What is the production when (a) $t = 1$ year, (b) $t = 4$ years, (c) $t = 5$ years?

11–2 The logarithm

In section 11–1 we found that the exponential function f is defined by $f(x) = b^x$, $b > 0$ and $b \neq 1$, is a one-to-one function and thus has an inverse that is a function. This inverse function is called the **logarithmic function.** Recall that to find the inverse function of a one-to-one function, we interchange the variables x and y. Then given the exponential function with base b defined by $f(x) = b^x$, we replace $f(x)$ by y to get $y = b^x$ and interchange x and y to obtain the equation $x = b^y$. We introduce a new notation to define the inverse function.

> ■ **Definition of a logarithm**
> For every $b > 0$, $b \neq 1$,
> $$x = b^y \text{ is equivalent to } y = \log_b x.$$

The word "**log**" is the abbreviation of the word **logarithm.** The notation $\log_b x$ is read "the logarithm of x to the base b" or, in shortened form, "log base b of x."

Graph of the logarithmic function

We saw in chapter 10 that the graph of an inverse function f^{-1} is the reflection of the graph of function f with respect to the line $y = x$. We can then sketch the graph of the general logarithmic function $f^{-1}(x) = \log_b x$, using the graph of the general exponential function, $f(x) = b^x$ ($b > 0, b \neq 1$). (See figure 11.1.)

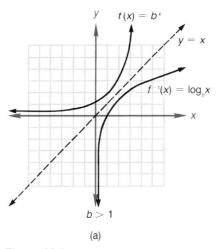

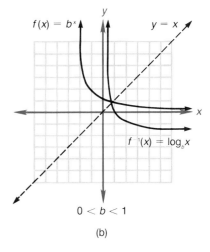

Figure 11.1

Since by definition $x = b^y$ is equivalent to $y = \log_b x$, we use this relationship to write the statements in example 11–2 A.

Example 11–2   A

1. $2^3 = 8$ is equivalent to $\log_2 8 = 3$.

2. $\left(\dfrac{1}{3}\right)^3 = \dfrac{1}{27}$ is equivalent to $\log_{\frac{1}{3}}\left(\dfrac{1}{27}\right) = 3$.

3. $4^{-2} = \dfrac{1}{16}$ is equivalent to $\log_4 \dfrac{1}{16} = -2$.

4. $b^u = v$ is equivalent to $\log_b v = u$.

It is important to remember that the **logarithm of a number to base b is** the *exponent* **to which the base b must be raised to get that number.** Thus the **logarithm of a number is an exponent.**

Using the previous equivalencies, we can state the following properties of the logarithmic function:

1. Since $b^1 = b$, then $\log_b b = 1$.
2. Since $b^0 = 1$, then $\log_b 1 = 0$.
3. Since $x = b^y$ is equivalent to $y = \log_b x$, then $b^{\log_b x} = x$.
4. Since $\log_b y = x$ means $y = b^x$, then $\log_b b^x = x$.

Example 11–2 B

By the above properties,

1. $\log_3 3 = 1$ **2.** $\log_{\frac{1}{4}}\left(\dfrac{1}{4}\right) = 1$ **3.** $\log_6 1 = 0$

4. $\log_{\frac{1}{2}} 1 = 0$ **5.** $5^{\log_5 9} = 9$ **6.** $4^{\log_4 7} = 7$

7. $\left(\dfrac{1}{2}\right)^{\log_{\frac{1}{2}} 2} = 2$ **8.** $\left(\dfrac{1}{5}\right)^{\log_{\frac{1}{5}} 100} = 100$

Note

In examples 5–8 we use property 3. In example 5, since $b^{\log_b x} = x$, then $5^{\log_5 9} = 9$.

9. $\log_2 2^5 = 5$ **10.** $\log_6 6^7 = 7$.

Logarithmic equations

We can use the definition $x = b^y$ is equivalent to $y = \log_b x$ to find the solution set S of equations involving logarithms, called **logarithmic equations.** The following examples illustrate how this is done.

Example 11–2 C

Find the solution set S for each logarithmic equation.

1. $\log_4 64 = x$
By definition, $\log_4 64 = x$ is equivalent to $4^x = 64$.
Since $64 = 4^3$, then we have the equation $4^x = 4^3$, which is true when $x = 3$.
Thus the solution set is $S = \{3\}$.

2. $\log_{10} x = 5$
By definition, $\log_{10} x = 5$ is equivalent to $10^5 = x$.
Since $10^5 = 100,000$, then $x = 100,000$ and the solution set is
$S = \{100,000\}$.

3. $\log_x 512 = 3$
By definition, $\log_x 512 = 3$ is equivalent to $x^3 = 512$.
Since $512 = 8^3$, then we have the equation $x^3 = 8^3$ and taking the cube root of each member, $x = 8$. Thus the solution set is $S = \{8\}$.

Exercise 11–2

Directions Sketch the graph of $f(x) = \log_b x$ using the graph of $y = b^x$.

Example
$f(x) = \log_3 x$

Solution
We sketch the graph of $y = 3^x$. Because $g(x) = 3^x$ and $f(x) = \log_3 x$ are inverse functions, we may reflect this graph with respect to the line $y = x$.

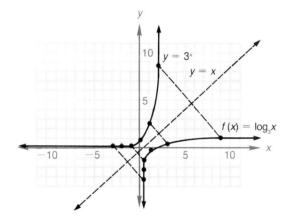

1. $f(x) = \log_2 x$ **2.** $g(x) = \log_4 x$ **3.** $h(x) = \log_5 x$ **4.** $F(x) = \log_{10} x$

5. $G(x) = \log_{\frac{1}{2}} x$ **6.** $H(x) = \log_{\frac{1}{3}} x$

Directions Write the following exponential equations in logarithmic form. See example 11–2 A.

7. $3^4 = 81$ **8.** $4^2 = 16$ **9.** $2^6 = 64$ **10.** $8^3 = 512$

11. $\left(\dfrac{1}{2}\right)^3 = \dfrac{1}{8}$ **12.** $\left(\dfrac{1}{3}\right)^4 = \dfrac{1}{81}$ **13.** $\left(\dfrac{3}{2}\right)^{-3} = \dfrac{8}{27}$ **14.** $\left(\dfrac{5}{6}\right)^{-2} = \dfrac{36}{25}$

15. $\log_2 16 = 4$ **16.** $\log_3 81 = 4$ **17.** $\log_{10} 1{,}000 = 3$ **18.** $\log_5 625 = 4$

19. $\log_4\left(\dfrac{1}{16}\right) = -2$ **20.** $\log_2\left(\dfrac{1}{32}\right) = -5$ **21.** $\log_{10} 0.0001 = -4$ **22.** $\log_{10} 0.00001 = -5$

23. $\log_{\frac{1}{2}} 8 = -3$ **24.** $\log_{\frac{1}{3}} 81 = -4$

Directions Evaluate each logarithmic expression.

Example
$\log_{10} 1{,}000$

Solution
Since $\log_{10} 1{,}000 = y$ means $(10)^y = 1{,}000$, we write both members with the same base. Now $1{,}000 = (10)^3$, so we have the equation
$$(10)^y = (10)^3$$
and $y = 3$.
Thus $\log_{10} 1{,}000 = 3$.

25. $\log_2 32$ **26.** $\log_3 81$ **27.** $\log_5 25$ **28.** $\log_7 1$ **29.** $\log_2\left(\dfrac{1}{4}\right)$

30. $\log_4\left(\dfrac{1}{64}\right)$ **31.** $\log_6\left(\dfrac{1}{216}\right)$ **32.** $\log_5\dfrac{1}{25}$ **33.** $\log_8 8$ **34.** $\log_{\frac{1}{2}}\left(\dfrac{1}{8}\right)$

35. $\log_{\frac{1}{3}}\left(\dfrac{1}{81}\right)$ **36.** $\log_{\frac{3}{2}}\left(\dfrac{27}{8}\right)$ **37.** $\log_{\frac{5}{4}}\left(\dfrac{125}{64}\right)$ **38.** $\log_5 \sqrt{5}$ **39.** $\log_7 \sqrt[3]{7}$

Directions Find the solution set of each logarithmic equation. See example 11–2 C.

40. $\log_x 25 = 2$ **41.** $\log_x 32 = 5$ **42.** $\log_x 512 = 3$ **43.** $\log_x\left(\dfrac{1}{27}\right) = -3$

44. $\log_x\left(\dfrac{1}{64}\right) = -6$ **45.** $\log_x 7 = 1$ **46.** $\log_x 5 = 1$ **47.** $\log_b 1 = 0$

48. $\log_{10} x = -2$ **49.** $\log_3 x = 4$ **50.** $\log_6 x = 3$ **51.** $\log_6 x = -2$

52. $\log_7 x = -3$ **53.** $\log_{\frac{1}{4}} x = 4$

54. For what values of x is $\log_6(x - 2)$ defined? (*Hint:* The domain of the logarithmic function is the set of positive real numbers.)

55. For what values of p is $\log_7(p + 1)$ defined?

56. For what values of x is $\log_8(x^2 - 1)$ defined?

57. For what values of y is $\log_9(y^2 - y - 12)$ defined?

11–3 Properties of logarithms

Logarithms have been used to simplify complex numerical computations. The advent of the hand-held calculator has made logarithms obsolete for these calculations. However in this section we shall see how logarithms can be used to perform these calculations.

> ■ **Product property of logarithms**
> If $b > 0$, $b \neq 1$, and x and y are positive real numbers, then
> $$\log_b(xy) = \log_b x + \log_b y.$$
>
> **Concept**
> The logarithm of the product of two positive real numbers is equal to the sum of the logarithms of the numbers.

To prove this property, let $\log_b x = m$ and $\log_b y = n$ and then, by definition, $b^m = x$ and $b^n = y$. Therefore $x \cdot y = b^m \cdot b^n = b^{m+n}$ by a property of exponents. Writing this as a logarithmic equation, we have

$$\log_b(xy) = m + n \text{ since } b^{m+n} = xy.$$

Substituting $\log_b x$ for m and $\log_b y$ for n, we obtain the property

$$\log_b(xy) = \log_b x + \log_b y.$$

Example 11–3 A

By the preceding property, the following relationships are true.

1. $\log_3(7 \cdot 4) = \log_3 7 + \log_3 4$

2. $\log_5 6 + \log_5 9 = \log_5(6 \cdot 9) = \log_5 54$

3. $\log_4 20 = \log_4(4 \cdot 5) = \log_2(2 \cdot 2 \cdot 5) = \log_2 2 + \log_2 2 + \log_2 5$
$$= 1 + 1 + \log_2 5$$
$$= 2 + \log_2 5$$

The following two properties state the remaining properties of the logarithms. Their proofs are similar to the one just completed and they are left as exercises.

> ■ **Quotient property of logarithms**
> If $b > 0$, $b \neq 1$, and x and y are positive real numbers, then
> $$\log_b\left(\frac{x}{y}\right) = \log_b x - \log_b y.$$
>
> **Concept**
> The logarithm of the quotient of two positive real numbers is equal to the logarithm of the numerator minus the logarithm of the denominator.

Note
A common error is to assume $\log_b(x \pm y) = \log_b x \pm \log_b y$.
This is not true.

Example 11–3 B

1. $\log_6\left(\dfrac{8}{9}\right) = \log_6 8 - \log_6 9$ **2.** $\log_3 5 - \log_3 7 = \log_3\left(\dfrac{5}{7}\right)$

3. $\log_5\left(\dfrac{5}{8}\right) = \log_5 5 - \log_5 8$ but $\log_5 5 = 1$, so we have

$\log_5\left(\dfrac{5}{8}\right) = \log_5 5 - \log_5 8 = 1 - \log_5 8.$

> ■ **Power property of logarithms**
> If $b > 0$, $b \ne 1$, r is any real number and x is a positive real number, then
> $$\log_b(x^r) = r \log_b x.$$
>
> **Concept**
> The logarithm of any real power of a positive real number is equal to the exponent times the logarithm of the number.

Example 11–3 C

1. $\log_3 2^3 = 3 \log_3 2$ **2.** $6 \log_5 2 = \log_5(2^6)$
$\qquad\qquad\qquad\qquad\qquad\qquad\qquad\qquad = \log_5 64$

3. $\log_7(7^2) = 2 \log_7 7$
Since $\log_7 7 = 1$, then $2 \log_7 7 = 2 \cdot 1 = 2$ and $\log_7(7^2) = 2$.

Using these properties, we can find the solution set of other logarithmic equations.

Example 11–3 D

Find the solution set of the following logarithmic equations.

1. $\log_5(x - 2) = 3$
Writing this equation in its equivalent exponential form, we have

$x - 2 = 5^3$

and so $x - 2 = 125$.

Adding 2 to each member, we obtain

$x = 127.$

The solution set is $S = \{127\}$.

2. $\log_2 x + \log_2(x + 2) = 3$

Since the bases are the same, we can use the property $\log_b(xy) = \log_b x + \log_b y$ in reverse order.

$\log_2 x + \log_2(x + 2) = \log_2 x(x + 2) = \log_2(x^2 + 2x)$

Now $\log_2(x^2 + 2x) = 3$ means $x^2 + 2x = 2^3$.

Then
$$x^2 + 2x = 8$$
$$x^2 + 2x - 8 = 0$$
$$(x + 4)(x - 2) = 0.$$
Thus $x - -4$ or $x = 2$.

However for $x = -4$, $\log_2(-4 + 2) = \log_2(-2)$, which is undefined since the domain must be positive. The solution set is $S = \{2\}$.

Note

When solving logarithmic equations, we must *always* check our solutions for values that may not be in the domain of the function defined by the expression in the original equation.

Mastery points

Can you
- Write a logarithmic expression in alternate forms using the properties?
- Solve logarithmic equations using the properties
  1. $\log_b(xy) = \log_b x + \log_b y$,
  2. $\log_b\left(\dfrac{x}{y}\right) = \log_b x - \log_b y$, and
  3. $\log_b x^r = r \log_b x$?

Exercise 11–3

Directions Use the properties of logarithms to write the following logarithms as a sum or difference of two, or more, logarithms, or as a product of a real number times a logarithm. Write all numbers as a product of prime factors. All variables represent positive real numbers. See examples 11–3 A and B.

1. $\log_b(3 \cdot 5)$

2. $\log_2(7 \cdot 5)$

3. $\log_3\left(\dfrac{7}{13}\right)$

4. $\log_7\left(\dfrac{23}{17}\right)$

5. $\log_b(5^3)$

6. $\log_6(11^2)$

7. $\log_{10}(5^4)$

8. $\log_b(24)$

9. $\log_{\frac{1}{2}}(36)$

10. $\log_{10}\left(\dfrac{15}{16}\right)$

11. $\log_5\left(\dfrac{24}{25}\right)$

12. $\log_b\left(\dfrac{x^3}{y^2}\right)$

13. $\log_b(xy)^3$

14. $\log_b \sqrt[3]{x^2}$

15. $\log_b(\sqrt{x} \cdot \sqrt[3]{y})$

16. $\log_b\left(\dfrac{\sqrt[3]{x} \cdot \sqrt[4]{y}}{z}\right)$

17. $\log_b(2x + 3y)$

Directions Write the following logarithms as the logarithm of a single number or expression. Assume that all variables are positive real numbers. See examples 11–3 A, B, and C.

18. $\log_4 5 + \log_4 (11)$

19. $\log_3 (17) - \log_3 3$

20. $\log_{10}(25) - \log_{10}(12)$

21. $\log_e (11) - \log_e (16)$

22. $2 \log_5 (11)$

23. $3 \log_6 4$

24. $2 \log_2 7 + \log_2 5$

25. $3 \log_{10} 3 + 2 \log_{10} 4$

26. $4 \log_e 2 - \log_e (10)$

27. $\dfrac{1}{5} \log_4 (32) + \dfrac{1}{4} \log_4 (81)$

28. $\log_e 4 + 2 \log_e 3 - 3 \log_e 2$

29. $2 \log_{10} 5 - \dfrac{1}{2} \log_{10} 9 + 3 \log_{10} 2$

30. $4 \log_2 3 - \dfrac{1}{3} \log_2 8 + \log_2 7 - 2 \log_2 5$

31. $\dfrac{1}{3} \log_b (x^2)$

32. $\dfrac{1}{2} \log_b x + \dfrac{1}{2} \log_b y$

33. $\dfrac{1}{4} \log_b (x^3) - \dfrac{1}{4} \log_b y$

34. $\log_{10}(x + 3) + \log_{10}(x - 4)$

35. $\log_5 (x + 6) - 2 \log_5 (3x) + \log_5 (x - 4)$

36. $2 \log_b x - \log_b (3x - 1) + \log_b (x + 1)$

Directions Given $\log_b 2 = 0.3010$, $\log_b 3 = 0.4771$, $\log_b 7 = 0.8451$, and $\log_b 10 = 1$, compute the following logarithms and then write the statement in equivalent exponential form.

Example
Given $\log_b 2 = 0.3010$ and $\log_b 3 = 0.4771$, find $\log_b 6$.

Solution
Since $\log_b 6 = \log_b (2 \cdot 3) = \log_b 2 + \log_b 3$, then we replace $\log_b 2$ by 0.3010 and $\log_b 3$ by 0.4771 to obtain
$\log_b 6 = \log_b (2 \cdot 3)$
$= \log_b 2 + \log_b 3 = 0.3010 + 0.4771 = 0.7781$.
Thus $\log_b 6 = 0.7781$.

37. $\log_b (14)$

38. $\log_b (20)$

39. $\log_b (49)$

40. $\log_b 8$

41. $\log_b (42)$

42. $\log_b \left(\dfrac{27}{4} \right)$

43. $\log_b \left(\dfrac{7}{10} \right)$

44. $\log_b \sqrt[3]{3}$

45. $\log_b \sqrt{7}$

46. $\log_b \sqrt{\dfrac{3}{4}}$

47. $\log_b \sqrt[4]{9}$

48. In exercises 37–47, what is b?

Directions Find the solution set of the following logarithmic equations. See example 11–3 D.

49. $\log_4 x + \log_4 5 = 3$

50. $\log_3 x + \log_3 8 = 2$

51. $\log_5 x - 2 \log_5 2 = 1$

52. $\log_2 x - 3 \log_2 3 = 5$

53. $\log_3 (2x + 1) - \log_3 5 = 3$

54. $\log_4 (x - 5) - \log_4 3 = 2$

55. $\log_{10}(x + 21) + \log_{10} x = 2$

56. $\log_{10}(3x + 2) + \log_{10} 2 = 2$

57. $\log_3 (x + 6) - \log_3 (x - 2) = 2$

58. $\log_{10}(x) + \log_{10}(x + 3) = 1$

59. Prove that for $b > 0$, $b \neq 1$, and positive real numbers u and v,
$$\log_b \left(\frac{u}{v} \right) = \log_b u - \log_b v.$$

60. Prove that for $b > 0$, $b \neq 1$, positive real number u and any real number r,
$\log_b (u^r) = r \log_b u$.

61. Show by an example that for positive real numbers u and v,
$\log_b (u + v) \neq \log_b u + \log_b v$.

11–4 The common logarithms

We have worked thus far with logarithms to different bases. There are two logarithms that are most commonly used and are of great importance in mathematics. They are logarithms to base 10 and to base e, defined, respectively, by (1) $y = \log_{10}x$ and (2) $y = \log_e x$. Because 10 is the base of our number system, we will begin our discussion with base 10 logarithms. In section 11–5 we will discuss base e logarithms. The two bases have similar properties.

The function values of the logarithmic functions with base 10 are called the **common logarithms.** When we use 10 as a base, it is customary to omit the base when writing a logarithm. That is, we will use

$$\log x \quad \text{instead of} \quad \log_{10}x, \ x > 0,$$

and the base will be understood to be 10.

From our previous work with logarithms, we realize that a common logarithm of a number is the power to which 10 must be raised to obtain the number. For example,

$$\log 100 = 2 \text{ since } 10^2 = 100.$$

In like fashion,

$\log 1{,}000 = 3$ since $10^3 = 1{,}000$
$\log 100 = 2$ since $10^2 = 100$
$\log 10 = 1$ since $10^1 = 10$
$\log 1 = 0$ since $10^0 = 1$

$\log 0.1 = -1$ since $10^{-1} = \dfrac{1}{10} = 0.1$

$\log 0.01 = -2$ since $10^{-2} = \dfrac{1}{100} = 0.01$

$\log 0.001 = -3$ since $10^{-3} = \dfrac{1}{1{,}000} = 0.001$

Table 11.1

Use of a calculator

To find the common logarithm of a number other than a power of 10, we use a calculator with a log key. To illustrate, to find log 724 by using a calculator, we press $\boxed{7}\ \boxed{2}\ \boxed{4}\ \boxed{\log}$ and read 2.8597386 on the display. Thus

$$\log 724 = 2.8597 \text{ (correct to four decimal places)}.$$

That is, $10^{.8597} = 7.24$ and so $10^{2.8597} = 10^{2+.8597} = 10^2 \cdot 10^{.8597} = (100)(7.24) = 724.$

Example 11–4 A

Find the following common logarithms using a calculator.

1. log 715,000

Using the calculator, press $\boxed{7}\ \boxed{1}\ \boxed{5}\ \boxed{0}\ \boxed{0}\ \boxed{0}\ \boxed{\log}$ and read 5.854306 on the display. Rounding off to four decimal places, log 715,000 = 5.8543, and we have found $10^{5.8543} \approx 715{,}000$.

Exponential and Logarithmic Functions

2. log 0.0749

Press $\boxed{.}\,\boxed{0}\,\boxed{7}\,\boxed{4}\,\boxed{9}\,\boxed{\text{log}}$ and read -1.125518 on the display. Thus log $0.0749 = -1.1255$ and $10^{-1.1255} \approx 0.0749$.

The antilogarithm

Now suppose we know the logarithm of a number N and wish to determine N. That is, we know log N and wish to find N. The number N is called the **antilogarithm** (or abbreviated **antilog**) of log N. Thus

$$\text{antilog } x = N \text{ is equivalent to log } N = x.$$

In example 11–4 A, 715,000 is the antilog of 5.8543, and 0.0749 is the antilog of -1.1255.

We can use the calculator to find the antilogarithm of x.

Example 11–4   B   Find the antilogarithm of the following.

1. Given log $N = 3.8603$
This means $10^{3.8603} = N$. Press $\boxed{1}\,\boxed{0}\,\boxed{y^x}\,\boxed{3}\,\boxed{.}\,\boxed{8}\,\boxed{6}\,\boxed{0}\,\boxed{3}\,\boxed{=}$ and read 7249.37 on the display; $N \approx 7{,}249.37$.

2. Given log $N = 0.8745 - 2 = -1.1255$
This means $10^{-1.1255} = N$. Press $\boxed{1}\,\boxed{0}\,\boxed{y^x}\,\boxed{1}\,\boxed{.}\,\boxed{1}\,\boxed{2}\,\boxed{5}\,\boxed{5}\,\boxed{+/-}\,\boxed{=}$ and read 0.0749 on the display; $N = 0.0749$.

On some calculators antilogarithms can be found by using the INV and log keys.

Mastery points

Can you
- Find the common logarithm of a number using a calculator?
- Find the antilogarithm of a number using a calculator?

Exercise 11–4

Directions   Find each of the following logarithms using a calculator. Then write the equivalent exponential statement. See example 11–4 A.

1. log 8

2. log 5.7

3. log 53

4. log 547

5. log 80,200

6. log 523,000

7. log 794,000,000

8. log 0.157

9. log 0.00863

10. log 0.0000941

11. log 0.000000107

12. log 0.0431

13. An artificial earth satellite is orbiting above the earth's surface at a height of 504,000 miles. Find log 504,000.

14. The resistive force F exerted by water on a ship traveling at $1\frac{1}{2}$ statute miles per hour is 244,000 pounds. Find log 244,000.

15. The angular velocity of the earth is 0.000073 radians per second and the earth's moment of inertia is 9.8×10^{37} kilograms per cubic meter. Find (a) log 0.000073 and (b) log (9.8×10^{37}).

16. The current rate of interest on money market certificates is 6.75%.. Find log (6.75%). (*Hint:* Write the percent in decimal form first.)

17. Given the formula $H = w \log T$, find (a) H when $w = 110$ and $T = 360$ and (b) w when $H = 16.4$ and $T = 31.1$. (Round to two decimal places.)

18. Given $\log R = \dfrac{D}{10}$, find D when $R = 1.47$.

19. Given $P_H = -\log A_H$, find P_H when $A_H = 0.00167$.

20. The number of decibels of sound, N_{db}, in comparing the "loudness" of two sounds, is given by
$$N_{db} = 10 \log\left(\frac{P_1}{P_2}\right),$$
where P_1 and P_2 are the signal power levels to be compared. Find N_{db} when $P_1 = 950$ and $P_2 = 25$.

21. Using exercise 20, find N_{db} when $P_1 = 1,070$ and $P_2 = 35$.

22. A decibel power gain, G_{db}, of an electronic amplifier is given by
$$G_{db} = 10 \log\left(\frac{P_o}{P_i}\right),$$
where P_o = the power output and P_i = the power input. Find G_{db} when $P_o = 20$ and $P_i = 0.005$.

23. Using exercise 22, find G_{db} when $P_o = 18$ and $P_i = 0.006$.

Directions Find the antilogarithm of each number, using a calculator. See example 11-4 B.

24. 0.5416

25. 3.8445

26. 4.4997

27. 7.9138

28. $8.7528 - 10$

29. $5.2175 - 10$

30. 8.8340

31. 2.5283

32. $3.9896 - 10$

33. $7.7699 - 10$

34. $6.7609 - 10$

35. $3.5273 - 10$

Directions Solve the following statements as indicated.

36. Given $P_H = -\log A_H$, find A_H when $P_H = 7.78$.

37. Given $H = w \log T$, find T when $w = 1.87$ and $H = 3.15$.

38. Given $\log R = \dfrac{D}{10}$, find R when $D = 44.5$.

39. Given $G_{db} = 10 \log\left(\dfrac{P_o}{P_i}\right)$, find P_o when $G_{db} = 15$ and $P_i = 0.60$. (*Hint:* Use the property $\log\left(\dfrac{P_o}{P_i}\right) = \log P_o - \log P_i$.)

40. Given $z_o = 276 \log\left(\dfrac{b}{a}\right)$, find (a) b when $z_o = 487$ and $a = 0.112$ and (b) a when $z_o = 552.1$ and $b = 5.7$.

41. A chemist defines the pH (hydrogen potential) of a solution by
$$pH = -\log[H^+],$$
where $[H^+]$ denotes the numerical value for the concentration of hydrogen ions in aqueous solution in moles per liter. Find the pH of the solution with hydrogen concentration, $[H^+]$, as 5.0×10^{-4}.

42. Using exercise 41, find pH with the hydrogen concentration 6.1×10^{-7}.

43. Using exercise 41, find the hydrogen ion concentration with a pH 5.8.

44. Using exercise 41, find the hydrogen ion concentration with a pH 6.8.

11–5 Logarithms to the base e

Natural logarithms

In section 11–4 we mentioned two important logarithmic bases: base 10 and base e. Also in section 11–4 we studied the base 10 logarithms, called the **common logarithms.** In applications, logarithms to base e are more frequently used than logarithms to base 10. We call the base e logarithms the **natural logarithms.**

The number e is an irrational number, as is the irrational number π. Computed $e \approx 2.7182818284 \ldots$, but we generally round off this value and use the value $e \approx 2.718$.

Just as a special symbol was used to represent the logarithm to base 10, that is, "log," we also have a special symbol to represent the natural logarithm of positive number x. We denote

$$\log_e x \text{ by } \ln x.$$

That is, by definition

$$\ln x = y \text{ is equivalent to } e^y = x.$$

Changing the base

To develop methods for finding decimal approximations of $\ln x$, we first develop a special formula for converting logarithms from one base to another. Since we can find the logarithm to the base 10 for any positive number, we will then be able to find the logarithm for any positive number to any other base, such as base e.

■ **Change of base property**
Let $a > 0$, $b > 0$, $x > 0$, $a \neq 1$, and $b \neq 1$. Then

$$\log_a x = \frac{\log_b x}{\log_b a} = \frac{\log x}{\log a}.$$

Example 11–5 A

Find the value of each logarithm, rounding to four decimal places.

1. $\log_5 13$

Using $\log_5 13 = \dfrac{\log 13}{\log 5}$, since

$\log 13 = 1.1139$ and $\log 5 = 0.6990$,

$\log_5 13 = \dfrac{1.1139}{0.6990} = 1.5936$.

2. $\log_{14} 33.5$

Using $\log_{14} 33.5 = \dfrac{\log 33.5}{\log 14}$, we find

$\log 33.5 = 1.5250$ and $\log 14 = 1.1461$.

Then $\log_{14} 33.5 = \dfrac{1.5250}{1.1461} = 1.3306$.

We can determine that $\log e = \log 2.718 = 0.4343$. Then by the relationship for changing bases,

$$\log_e x = \frac{\log x}{\log e} \text{ or } \ln x = \frac{\log x}{\log e} = \frac{\log x}{0.4343} = \log x \cdot \frac{1}{0.4343}.$$

Since $\dfrac{1}{0.4343} = 2.3026$ we can find $\ln x$ using the following:

■ **Finding ln x**

$$\ln x = \frac{\log x}{\log e} \approx 2.3026 \log x.$$

Example 11–5 B

Find the value of each natural logarithm.

1. $\log_e 1,000$
   Using $\log_e x = 2.3026 \log x$, since $x = 1,000$ and $\log 1,000 = 3$, then we have

   $\log_e 1,000 = 2.3026 \log 1,000 = (2.3026)\, 3 = 6.9078.$

   Thus $\log_e 1,000 = 6.9078$. This means that $e^{6.9078} = 1,000$.

2. $\ln 0.576$
   Using $\ln x = 2.3026 \log x$, we have

   $\ln 0.576 = 2.3026 \log 0.576.$

   Since $\log 0.576 = -0.2396$
   then $\ln 0.576 = 2.3026(-0.2396) = -0.5517$, correct to four decimal places. This means that $e^{-0.5517} = 0.576$.

Natural logarithms can be found directly on a calculator that has a key labeled ln. For example, to find $\ln 57.6$, enter $\boxed{5}\,\boxed{7}\,\boxed{.}\,\boxed{6}\,\boxed{\ln}$ and the result showing on the display is 4.0535225. For numbers between 0 and 1, the natural logarithm will be negative as with common logarithms. For example, to find $\ln 0.5731$, enter $\boxed{.}\,\boxed{5}\,\boxed{7}\,\boxed{3}\,\boxed{1}\,\boxed{\ln}$ and read $5.5669505 - 01$ or -0.5566951 on the display.

The properties of logarithms that we used with the common logarithms also apply to the natural logarithms.

Growth and decay

Problems involving natural growth and natural decay require the use of the exponential and logarithmic functions. The general equations for natural growth and natural decay are given by

$$\text{growth, } q = q_0 e^{rt}$$
$$\text{decay, } q = q_0 e^{-rt},$$

where q is the quantity (or number) present at time $t \neq 0$, q_0 is the quantity (or number) present when time $t = 0$, and r is the rate of growth or decay.

Note
1. e is the base of the natural logarithm.
2. We assume $r > 0$ at all times. Then $-r < 0$ and this causes the decrease in quantity.

We will use these equations to find the time it takes for a quantity q_0 at time $t = 0$ to *grow* or *decay* to a quantity q. Solving for t, we obtain the following equations:

■ **Growth and decay formulas**
$$t = \frac{1}{r} \ln\left(\frac{q}{q_0}\right), \text{ for growth; } t = -\frac{1}{r} \ln\left(\frac{q}{q_0}\right), \text{ for decay.}$$

Example 11–5 C

1. How long will it take the earth's population to double if it continues to grow at the rate of 2 percent per year compounded continuously? Find the answer to the nearest tenth.
   Since the earth's population is to double, we are given that $q = 2q_0$. Also we are given $r = 0.02$ and we want time t. Using the equation

   $$t = \frac{1}{r} \ln\left(\frac{q}{q_0}\right),$$
   $$t = \frac{1}{0.02} \ln\left(\frac{2q_0}{q_0}\right) = 50 \ln 2.$$

   Now $\ln 2 = 0.6901$. Therefore

   $$t = 50(0.6931) = 34.7.$$

   The population will double in approximately 34.7 years.

2. Radioactive strontium 90 is used in nuclear reactors and decays according to the equation

   $$q = q_0 e^{-0.0248t}.$$

   Find the half-life of strontium 90.

   ### Note
   Half-life is the time necessary for a substance to decay to one-half of its original quantity.

We are given $q = \frac{1}{2}q_0$, $r = 0.0248$ and we want t in years. Using the equation

$$t = -\frac{1}{r}\ln\left(\frac{q}{q_0}\right),$$

$$t = -\frac{1}{0.0248}\ln\left(\frac{\frac{1}{2}q_0}{q_0}\right) = -40.3\ln\left(\frac{1}{2}\right).$$

Now $\ln\left(\frac{1}{2}\right) = \ln 1 - \ln 2 = 0 - 0.6931 = -0.6931.$ $\left(\begin{array}{l}\ln 1 = 0 \\ \ln 2 = 0.6931\end{array}\right)$

Substituting, we obtain

$$t = -40.3(-0.6931) = 27.9.$$

Therefore the half-life of strontium 90 is approximately 27.9 years.

Mastery points
Can you
- Find the logarithm of any positive number to any given base?
- Find the natural logarithm of a positive number using the common logarithms?
- Use natural logarithms in working applications of natural growth and natural decay?

Exercise 11–5

Directions Find each of the following logarithms to the nearest hundredth. Round to four decimal places. See example 11–5 A.

1. $\log_5 8$ 2. $\log_7 9$ 3. $\log_4 25$ 4. $\log_7 47$

5. $\log_2 76.3$ 6. $\log_3 95.7$ 7. $\log_{23} 45.6$ 8. $\log_{18} 157$

See example 11–5 B.

9. $\log_e 2.73$ 10. $\log_e 107$ 11. $\ln 6.73$ 12. $\ln 5.13$ 13. $\ln 56.4$

14. $\ln 97.1$ 15. $\ln 347$ 16. $\ln 197$ 17. $\log_e 5$ 18. $\log_e 3$

19. $\log 5e$ 20. $\log 3e$ 21. $\ln 7e$ 22. $\log 6e^3$ 23. $\ln 3e^3$

24. $\ln 4e$

25. When money is invested at interest rate r percent per year compounded continuously, after t years it is worth

$q_0 e^{\frac{rt}{100}}$dollars,

where q_0 is the amount invested. How long will it take for an investment of $200,000 to triple at 6% per year compounded continuously?

(*Hint:* Use $q = q_0 e^{\frac{rt}{100}}$, where $r = 6$.)

26. Using exercise 25, how long will it take for $15,000 to become one and one-half times as great?

27. Using the equation for the decay of strontium 90,

$q = q_0 e^{-0.0248t}$,

how many years will it take until only one-fourth of the original amount of the substance is left?

28. Using the formula of exercise 27, how long will it take for 90% of the strontium 90 to decay?

29. Radioactive carbon 14 diminishes by radioactive decay according to the equation

$q = q_0 e^{-0.000124t}$,

where t is in years. Carbon 14 enters all living tissue through carbon dioxide and is maintained to be constant as long as the plant or animal is alive. The decay takes place when the tissues are dead. Estimate the age of the skull uncovered if 20% of the original amount of carbon 14 is still present. (*Hint:* $q = 0.2q_0$.)

30. Using the information of exercise 29, estimate the age of the skull if only 5% of the original amount of carbon 14 is still present.

31. For relatively clear bodies of water, light intensity is reduced according to the equation $I = I_0 e^{-kd}$,

where I is the intensity at d feet below the surface and I_0 is the intensity at the surface. In a particular body of water, $k = 0.00853$. Find the depth to the nearest tenth of a foot at which the light is reduced to 10% of that at the surface.

32. Using exercise 31, when $k = 0.0585$, find the depth at which the light is reduced to 2% of that at the surface.

33. Suppose a certain species of bees increases in number according to the exponential equation $q = 25e^{0.2t}$,

where t is measured in days. In how many days, correct to the nearest tenth, will there be 375 bees?

11–6 Exponential equations

In section 11–1 we found the solution set of some exponential equations, that is, equations in which the exponent is a variable. Now we consider some exponential equations whose solution set can be found by using logarithms. To do this we use a property of logarithms.

> ■ **Property of logarithms**
> Given positive real numbers x, y, and b, $b \neq 1$, $b > 0$,
> $$\text{if } x = y, \text{ then } \log_b x = \log_b y. \quad (1)$$

Example 11–6 A

Find the solution set of each exponential equation, correct to the nearest hundredth.

1. $3^{x+7} = 11$

Using property (1) and common logarithms, if $3^{x+7} = 11$, then

$$\log(3^{x+7}) = \log (11)$$
$$(x + 7)\log 3 = \log (11). \qquad \text{(Power property of logarithms)}$$

Dividing each member by log 3, we obtain the equation

$$x + 7 = \frac{\log (11)}{\log 3}$$
$$x = \frac{\log (11)}{\log 3} - 7$$
$$= \frac{1.0414}{0.4771} - 7$$
$$= 2.1828 - 7 = -4.8172 \approx -4.82.$$

Therefore the solution set is $S = \{-4.82\}$.

2. An amount of money P (called the principal) is invested at r percent compounded annually. The amount of money A after t years is given by the equation $A = P(1 + r)^t$. How long will it take $P = \$1,000$ to be worth $A = \$2,500$ at 6% compounded annually?

We want t in years when $r = 0.06$, $A = \$2,500$, and $P = \$1,000$. Substituting, we have the equation

$$2,500 = 1,000(1 + 0.06)^t.$$

Dividing both members by 1,000 and simplifying, we have

$$2.5 = (1.06)^t.$$

Applying property (1) and using common logarithms,

$$\log (2.5) = \log (1.06)^t$$

and so

$$\log (2.5) = t \log (1.06).$$

Solving for t, we have

$$t = \frac{\log (2.5)}{\log (1.06)}.$$

Evaluating the logarithms and substituting,

$$t = \frac{0.3979}{0.0253} = 15.7.$$

Therefore an investment of $1,000 would take approximately 15.7 years to grow to $2,500 when compounded annually at 6%.

Note

$t = \dfrac{\ln 2.5}{\ln 1.06}$ will work just as well.

Exercise 11–6

Directions Find the solution set of the given exponential equation. Find solutions correct to the nearest hundredth.

1. $2^x = 9$

2. $7^x = 8$

3. $4^{x+1} = 6$

4. $5^{x-2} = 11$

5. $3^{-x} = 10$

6. $9^{-x} = 19$

7. $6^{2x-1} = 12$

8. $8^{3x+1} = 9$

9. $3^{x+1} = 4$

10. $5^{x+2} = 3$

11. $16^{1-x} = 13$

12. $19^{2-x} = 17$

13. $4^{5-2x} = 9$

14. $2^{x^2} = 7$

15. $\left(\dfrac{1}{3}\right)^{1-x} = 5$

16. $\left(\dfrac{3}{5}\right)^{2x} = 2$

17. Solve the equation $y = x^{3n}$ for n using the common logarithms, log.

18. Solve the equation $y = e^{kt}$ for t using the natural logarithms, ℓn.

19. What rate of interest, to the nearest tenth of a percent, is required for $50 to yield $61.50 after 5 years, compounded annually? [*Hint:* Use $A = P(1 + r)^t$.]

20. If $2, compounded annually, yields $2.98 after 8 years, what is the rate of interest, to the nearest hundredth of a percent?

21. If $65 yields $82.50 when the rate of interest is 6.25% compounded annually, how many years was the money invested? (to the nearest tenth)

22. For how many years, to the nearest hundredth, must $1,050 be invested at 8.5% interest compounded annually to have a return of $1,760?

23. A certain bacteria divides every 15 minutes to produce two new bacteria. If the number increases according to the equation
$A = A_0 2^{4t}$,
where t is in hours, how long would it take for 2,000 bacteria to increase to 750,000 bacteria?

24. Work exercise 23 for 5,000 bacteria to increase to 1,000,000 bacteria.

25. Using the formula of exercise 23, how long would it take for the bacteria to triple?

26. Radium disintegrates in such a way that of q_0 milligrams present at a given time,
$q = q_0(0.96)^t$
milligrams will remain after t centuries. What is the half-life of radium, to the nearest tenth of a century?

27. Using exercise 26, in how many centuries (to the nearest tenth) will 200 milligrams of radium disintegrate to 125 milligrams?

28. A machine depreciates according to the formula
$V = V_0(1 - r)^t$,
where V is the value after t years, V_0 is the original value, and r percent is the constant rate of depreciation. In how many years (to the nearest tenth) will the machine depreciate to one-half its original value at a depreciation rate of 15%?

29. Using exercise 28, in how many years (to the nearest hundredth) will the value depreciate from a value of $6,500 to a value of $5,000 at a depreciation rate of 12%?

30. Using exercise 28, what is the rate of depreciation (to the nearest tenth) if a machine depreciates to one-fourth its original value in 9 years?

31. Using exercise 28, what is the rate of depreciation (to the nearest tenth) if the machine reduces in value from $1,600 to $1,280 in 6 years?

Chapter summary

1. An **exponential function** with base b is a function f defined by $f(x) = b^x$, $b > 0$, $b \neq 1$.

2. The **logarithmic function** with base b, $b > 0$, $b \neq 1$, is the function defined by $f(x) = \log_b x$ such that $f(x) = y = \log_b x$ if and only if $x = b^y$, $x > 0$.

3. The exponential and logarithmic functions are inverse functions of each other.

4. The logarithm of a number to base b is the exponent to which the base b must be raised to get that number.

5. If x and y are positive real numbers, $b > 0$, $b \neq 1$, then
   a. $\log_b(xy) = \log_b x + \log_b y$
   b. $\log_b\left(\dfrac{x}{y}\right) = \log_b x - \log_b y$
   c. $\log_b(x^r) = r \log_b x$ (r is a real number)
   d. $\log_b 1 = 0$
   e. $\log_b b = 1$
   f. $b^{\log_b x} = x$
   g. $\log_b b^x = x$

6. The **common logarithms** are logarithms with base 10, denoted by log.

7. The **natural logarithms** are logarithms with base e, denoted by ℓn.

8. $\log_a x = \dfrac{\log_b x}{\log_b a}$, $a > 0$, $b > 0$, $x > 0$, $a \neq 1$, $b \neq 1$.

Chapter review

[11–1]

1. Sketch the graph of the exponential function
   (a) $f(x) = 3^x$, (b) $g(x) = \left(\dfrac{1}{3}\right)^x$, (c) $h(x) = 4^{-x}$.

2. Find the solution set of each exponential equation (a) $5^x = 625$, (b) $16^{2x-1} = 8$.

[11–2]

3. Sketch the graph of $f(x) = \log_6 x$ by using the graph of $y = 6^x$.

5. Write the logarithmic equation $\log_{\frac{1}{3}} 27 = -3$ as an exponential equation.

4. Write the exponential equation $4^{-3} = \dfrac{1}{64}$ as a logarithmic equation.

Directions Evaluate each logarithmic expression.

6. $\log_6 216$

7. $\log_{\frac{1}{4}}(256)$

8. $\log_9 \sqrt[3]{9}$

Directions Find the solution set of each logarithmic equation.

9. $\log_x 125 = 3$

10. $\log_3 x = -5$

11. $\log_2 x = \dfrac{5}{2}$

12. For what values of x is $\log_{10}(x^2 + 2x - 15)$ defined?

[11-3]

Directions Use the properties of logarithms to write each logarithmic expression as a sum or difference of logarithms, or as a product of a real number times a logarithm. Use the prime factor form of each number. All variables are positive real numbers.

13. $\log_b(56)$

14. $\log_4\left(\dfrac{5}{6}\right)$

15. $\log_5\left(\dfrac{9}{4}\right)$

Directions Write each of the following logarithms as the logarithm of a single number. All variables are positive real numbers.

16. $\log_5 7 + \log_5 4$

17. $\log_6 15 - \log_6 3$

18. $4 \log_b x + 2 \log_b y - \log_b z$

19. $\log_4(x + 3) - \log_4(x - 4)$

20. $3 \log_b x + \log_b(2x - 1) - 3 \log_b(x + 1)$

Directions Given $\log_b 2 = 0.3010$, $\log_b 3 = 0.4771$, and $\log_b 11 = 1.0413$, compute the following logarithms and then write the statement in equivalent exponential form.

21. $\log_b 22$

22. $\log_b 121$

23. $\log_b 96$

24. $\log_b \sqrt[3]{11}$

25. $\log_b\left(\dfrac{3}{22}\right)$

26. $\log_b\left(\dfrac{27}{\sqrt[4]{11}}\right)$

Directions Find the solution set of each logarithmic equation.

27. $\log_4 x - \log_4 3 = 2$

28. $2 \log_2 3 + \log_2 x = 1$

29. $\log_5(2x + 1) + \log_5 2 = 2$

30. $\log_3(x - 3) - \log_3(x + 2) = -3$

[11-4]

Directions Find the following common logarithms.

31. $\log 342$

32. $\log 507,000$

33. $\log 0.00736$

34. Given the formula pH $= -\log[\text{H}^+]$, find pH when $[\text{H}^+] = 6.00312$.

35. The formula for compound interest is given by $A = P(1 + i)^n$, where A is the amount after n years that the principal P will grow to at interest rate i percent (per year). Use common logarithms to find how many years it will take for the principal to triple ($A = 3P$) at 7.5% interest compounded annually.

Directions Evaluate the following logarithms (correct to two decimal places) using $\log_a x = \dfrac{\log x}{\log a}$.

36. $\log_6 7$ **37.** $\log_{32} 4.73$ **38.** $\ln 47.3$

39. The number of a radioactive substance, present at time t, is given by $q = q_0 e^{-0.6t}$, where t is measured in days. What is the half-life of the substance? (Round to the nearest tenth.)

40. Using the formula in exercise 39, how long will it take for 2.3 grams of a substance to decay to 0.7 grams? (Round to the nearest tenth.)

[11–6]
Directions Find the solution set of each exponential equation (correct to two decimal places).

41. $5^x = 17$ **42.** $4^{-x} = 3$ **43.** $15^{x-2} = 9$ **44.** $(4)^{x^2} = 3$ **45.** $\left(\dfrac{4}{5}\right)^{3x} = 2$

46. A certain bacteria increases in number according to the equation $A = A_0 3^{2t}$, where t is in hours, A_0 is the initial number, and A is the number of bacteria at time t. How long will it take for 1,000 bacteria to increase to 125,000 bacteria?

Chapter 11 cumulative test

[1–1]   **1.** Given $\{x \mid -4 < x \le 5\}$, list the integers in the set.

[1–2]   **2.** Simplify the expression $4 - 12 \div 3 + 2^3 - \sqrt{16} \cdot 4$.

Directions Perform the indicated operations and simplify.

[1–2]   **3.** $3[-6(9 - 5) - 3 \cdot 4 + 9]$

[3–3]   **4.** $(2a - b)^2$

[3–3]   **5.** $(5x - 3y)(5x + 3y)$

[3–3]   **6.** $(4y + 2)(3y - 5)$

Directions Perform the indicated operations. Give the answer with positive exponents.

[3–2]   **7.** $(3a^2 b)(-6a^3 b^2)$ [3–4]   **8.** $(-3a^{-1} b^2 c^{-3})^3$ [3–4]   **9.** $\left(\dfrac{4a^2 b^{-3}}{12a^{-4} b^2}\right)^2$

Directions Find the solution set of the following equations or inequalities.

[2–1]   **10.** $3(x + 2) - 4x = 5(2x - 1)$

[2–4]   **11.** $4y - 3 \le 2y + 1$

[2–4]   **12.** $-5 \le 2x - 1 < 4$

[2–5]   **13.** $|4y - 3| = 2$

[2–5]   **14.** $|2 - 3x| \le 4$

[2–5]   **15.** $|x + 5| > 6$

[6–1]   **16.** $2y^2 - y = 3$

[4–6]   **17.** $\dfrac{3}{x} - \dfrac{2}{x} = \dfrac{4}{5}$

[4–3]   **18.** Add $\dfrac{5}{4ab^2} + \dfrac{6}{8a^2 b}$.

[4–3]   **19.** Subtract $\dfrac{6y}{y - 7} - \dfrac{9y}{7 - y}$.

[4–2] **20.** Divide $\dfrac{x^2 - 5x + 4}{x^2 - 25} \div \dfrac{x^2 - 4x + 3}{x^2 - 10x + 25}$. **[4–4]** **21.** Simplify the complex fraction $\dfrac{\dfrac{4}{x} - \dfrac{3}{y}}{\dfrac{4y - 3x}{xy}}$.

Directions Perform the indicated operations and simplify.

[5–5] **22.** $(2\sqrt{3} - 1)^2$ **[5–6]** **23.** $(4 + 3i)(4 - 3i)$ **[5–6]** **24.** i^{23}

[5–4] **25.** $\sqrt{50} - 3\sqrt{8} + 2\sqrt{32}$ **[5–5]** **26.** Rationalize the denominator of the expression $\dfrac{3 + \sqrt{2}}{\sqrt{3} - 2}$.

Directions Find the solution set of the following equations.

[6–2] **27.** $y^2 - 3y = 7$ **[6–5]** **28.** $\sqrt{5x + 9} = x - 1$

[6–7] **29.** Find the solution set of the inequality $z^2 - 3z \le 40$. Graph the solution set on the number line.

[7–3] **30.** Find the equation of the line that satisfies the given conditions. Write the answer in standard form.
(a) Passes through the points $(-1,4)$ and $(0,6)$
(b) Passes through the point $(-9,8)$ and parallel to $2x - 3y = 1$
(c) Passes through $(5,-3)$ and having slope $\dfrac{4}{3}$

[10–2] **31.** Given $f(x) = 5x + 1$ and $g(x) = x^3 + 2$, find
(a) $f(5)$, (b) $g(-3)$, (c) $f[f(-2)]$,
(d) $\dfrac{f(x + h) - f(x)}{h}$.

[7–3] **32.** Sketch the graph of the equation $2x - 3y = -12$ using the slope and y-intercept.

[9–1] **33.** Sketch the graph of the equation $x = y^2 - 4y + 3$.

[9–3] **34.** Determine if the graph of the given equation is a parabola, a circle, an ellipse, or a hyperbola:
(a) $3y^2 + 3x^2 = 14$, (b) $y^2 + 2y = x$,
(c) $4y^2 = 3x^2 + 9$, (d) $x^2 + 4y^2 = 4$.

[8–1] **35.** Solve the system of equations
$4x + 3y = 2$
$2x - 5y = -1$.

[11–6] **36.** Solve the exponential equation $4^x = 16$.

[11–3] **37.** Find x, given (a) $\log_x 32 = 5$,
(b) $\log_6 x = -2$.

[11–5] **38.** Evaluate $\log_7 14$ using natural logarithms and a calculator.

[11–3] **39.** Solve the equation
$\log_3(x - 1) - \log_3(x - 2) = 2$.

[11–3] **40.** Write the expression
$\log_b 4 - 3 \log_b 5 + \log_b 2$ as the logarithm of a single expression.

12

Sequences and Series

12–1 Sequences

A sequence

In chapter 10 we discussed the concept of a function. Now we wish to consider a very special function—a function whose domain is the set of positive integers. Such a function is called a **sequence.**

> ■ **Definition of a sequence**
> An infinite **sequence** is a function whose domain is the set of positive integers {1, 2, 3, · · ·}.

There are two kinds of sequences, finite and infinite. A **finite sequence** is a sequence whose domain is the first n positive integers.

Sequences play an important role in the fields of science, finance, and mathematics. For example, the periodic amount of money in a savings account that is compounded at regular intervals is a special kind of sequence.

For any positive integer n, the function value (value of y) of the sequence is written as a_n (read "a sub-n"). The function values a_1, a_2, a_3, $\cdots$, written in order, are called the **terms** of the sequence, with a_1 the first term, a_2 the second term, a_3 the third term, and so on. The expression a_n that defines the sequence is called the *general term* of the sequence.

Example 12–1 A

1. Write the first five terms of the sequence.

$a_n = 5n + 2$

Now $a_1 = 5(1) + 2 = 7$

$\qquad a_2 = 5(2) + 2 = 12$

$\qquad a_3 = 5(3) + 2 = 17$

$\qquad a_4 = 5(4) + 2 = 22$

$\qquad a_5 = 5(5) + 2 = 27.$

Thus the first five items of the sequence are

$7,12,17,22,27, \cdots .$

2. $a_n = (-1)^n(n + 7)$

Now $a_1 = (-1)^1(1 + 7) = (-1)(8) = -8$

$\qquad a_2 = (-1)^2(2 + 7) = (1)(9) = 9$

$\qquad a_3 = (-1)^3(3 + 7) = (-1)(10) = -10$

$\qquad a_4 = (-1)^4(4 + 7) = (1)(11) = 11$

$\qquad a_5 = (-1)^5(5 + 7) = (-1)(12) = -12.$

The first five terms of the sequence are

$-8,9,-10,11,-12.$

Finding the general term

On occasion, we may be given the first few terms of a sequence and asked to find an expression for the general term, a_n. There are no rules for finding this from the first few terms of the sequence. We do this mostly by inspection or trial and error. However we should always be aware of one clue. Consider again the sequences

1. $a_n = 5n + 2$, whose terms are $7,12,17,22,27, \cdots$; each term of the sequence differs by 5, and the coefficient of n is 5; and
2. $a_n = (-1)^n(n + 7)$, whose terms are $-8,9,-10,11,-12, \cdots$; each term alternates in sign, caused by the factor -1, to some positive integer power; the numerical value of each term differs by 1 and so the coefficient of n is 1.

Example 12–1 B   Find an expression for the general term of the given sequence.

1. $5,7,9,11, \cdot \; \cdot \; \cdot$

The difference between each term is 2, so we conclude that $2n$ must be part of the general term. Now if we consider the first term a_1, let $n = 1$ and ask ourselves, $2(1) + $ (what number?) $= 5$. Since $2(1) + 3 = 5$, we think $a_n = 2n + 3$. Check this with the succeeding terms.

$$a_2 = 2(2) + 3 = 4 + 3 = 7 \text{ (true)}$$
$$a_3 = 2(3) + 3 = 6 + 3 = 9 \text{ (true)}$$

Therefore $a_n = 2n + 3$.

2. $\dfrac{3}{4}, -\dfrac{9}{8}, \dfrac{27}{16}, -\dfrac{81}{32}, \; \cdot \; \cdot \; \cdot$

First, the signs alternate and the first term a_1 is positive. We must then have a factor of -1 to an even power when $n = 1$. Therefore one factor of the nth term could be $(-1)^{n+1}$ since $(-1)^{1+1} = (-1)^2 = 1$. Inspection shows us that the numerator of each term is a power of 3, starting with 3^1. Inspecting the denominator, we see that each denominator is a power of 2. When $n = 1$, the denominator of the first term could be

$$2^{n+1} = 2^{1+1} = 2^2 = 4.$$

The denominator of the second term, when $n = 2$, is then

$$2^{n+1} = 2^{2+1} = 2^3 = 8,$$

and so on. Therefore we conclude that the numerator is 3^n and the denominator is 2^{n+1}. Thus

$$a_n = (-1)^{n+1} \, \frac{3^n}{2^{n+1}}.$$

Mastery points
Can you
- Find the terms of a sequence, given the general term?
- Find any given term of the sequence?
- Define an expression for the general term, given a sequence?

Exercise 12–1

Directions Write the first five terms of the sequence whose general term a_n is given. See example 12–1 A.

1. $a_n = 4n + 3$

2. $a_n = 5n - 4$

3. $a_n = \dfrac{2}{3n}$

4. $a_n = \dfrac{5}{2n + 1}$

5. $a_n = \dfrac{n + 5}{3n - 4}$

6. $a_n = \dfrac{4n}{5n + 2}$

7. $a_n = \dfrac{2^n}{5n}$

8. $a_n = \dfrac{3^{n+1}}{n + 2}$

9. $a_n = (-1)^n(6n - 5)$ **10.** $a_n = (-1)^{n+1}(n + 1)^2$ **11.** $a_n = (-1)^{n-1}\dfrac{3^n}{2^n + 1}$

12. $a_n = (-1)^n \dfrac{3}{4^n + 2}$ **13.** $a_n = (-1)^{2n}$ **14.** $a_n = (-1)^{3n-1}$

Directions Find the indicated term of the given sequence. See example 12–1 A.

15. $a_n = 3n + 5$, find a_9.   **16.** $a_n = 7 - 2n$, find a_{11}.   **17.** $a_n = \dfrac{1}{3n}$, find a_{37}.

18. $a_n = \dfrac{3}{5n}$, find a_{51}.   **19.** $a_n = \dfrac{4n - 3}{2n + 7}$, find a_{17}.   **20.** $a_n = \dfrac{9 - 5n}{-6 - 3n}$, find a_{12}.

21. $a_n = (-1)^n(6n + 5)$, find a_{14}.   **22.** $a_n = (-1)^{n-1}(5n - 6)$, find a_{15}.

23. $a_n = (-1)^{n+2}n(n + 6)$, find a_{13}.   **24.** $a_n = 2n^2(3n - 1)$, find a_8.

Directions Given the terms of the following sequences, find an expression for the general term a_n. See example 12–1 B.

25. 6,8,10,12,14, $\cdot\ \cdot\ \cdot$ **26.** 1,4,7,10,13, $\cdot\ \cdot\ \cdot$ **27.** 2,7,12,17,22, $\cdot\ \cdot\ \cdot$

28. 1,4,9,16, $\cdot\ \cdot\ \cdot$ **29.** 3,6,11,18,27, $\cdot\ \cdot\ \cdot$ **30.** $\dfrac{1}{3}, \dfrac{1}{9}, \dfrac{1}{27}, \dfrac{1}{81}, \dfrac{1}{243}, \cdot\ \cdot\ \cdot$

31. $1, \dfrac{1}{2}, \dfrac{1}{4}, \dfrac{1}{8}, \dfrac{1}{16}, \cdot\ \cdot\ \cdot$ **32.** $\dfrac{2}{3}, \dfrac{3}{4}, \dfrac{4}{5}, \dfrac{5}{6}, \dfrac{6}{7}, \cdot\ \cdot\ \cdot$ **33.** $\dfrac{3}{5}, \dfrac{4}{7}, \dfrac{5}{9}, \dfrac{6}{11}, \dfrac{7}{13}, \cdot\ \cdot\ \cdot$

34. $3, \dfrac{8}{3}, \dfrac{7}{3}, 2, \dfrac{5}{3}, \cdot\ \cdot\ \cdot$ **35.** $-6,10,-14,18,-22, \cdot\ \cdot\ \cdot$ **36.** $-\dfrac{3}{2}, 3, -\dfrac{9}{2}, 6, \dfrac{15}{2}, \cdot\ \cdot\ \cdot$

Directions Solve the following verbal problems.

37. A culture of bacteria triples every hour if the original culture has 1,000 bacteria. How many bacteria were there after (a) 3 hours, (b) 5 hours, (c) n hours?

38. A pendulum swings a distance of 20 inches on its first swing. If each subsequent swing back is three-fourths of the previous swing, how far does it swing on the (a) third swing, (b) seventh swing?

39. A ball is dropped from a height of 10 feet. If the ball rebounds one-half the height of its previous fall, how high does it rebound on the (a) second bounce, (b) sixth bounce, (c) nth bounce?

40. Jim Jarrett gives his son an allowance each month of 10¢ on the first day, 15¢ on the second day, 20¢ on the third day, and so on. Write a sequence for the first ten days of the month. Write an expression for the amount received on the nth day of the month. How much does he receive on the thirtieth day?

41. Steve Navarro begins a new job at a starting yearly salary of $16,000, with a promise of a salary increase of $1,500 per year for the first 6 years. (a) Write a sequence showing his salary during the first 6 years. (b) Write a general term for the sequence. (c) If this yearly pay increase were to continue beyond 6 years, what would his salary be after 20 years?

12–2 Series

A series

Associated with any sequence is the sum of the first n terms of the sequence, called a **series.** That is, the series associated with the infinite sequence

$$a_1, a_2, a_3, \cdots, a_n, \cdots$$

is given by

$$a_1, a_1 + a_2, a_1 + a_2 + a_3, a_1 + a_2 + a_3 + a_4, \cdots.$$

Note
$a_1 + a_2 + a_3 + a_4 + \cdots + a_n$ is called the nth partial sum.

Example 12–2 A

Given the sequence whose general term is $a_n = 2n + 1$, the *sequence* is

$3, 5, 7, 9, \cdots,$

the *series* of this sequence is

$3, 3 + 5, 3 + 5 + 7, 3 + 5 + 7 + 9, \cdots$ or $3, 8, 15, 24, \cdots.$

Summation notation

A compact way of representing a sum when the general term is known is by **sigma** (or **summation**) **notation.** To do this, we use the Greek letter sigma, Σ, in conjunction with the general term of the related sequence. To illustrate, consider the sequence

$$3, 7, 11, 15, 19, 23, 27, \cdots.$$

We can determine that a general term of this sequence is $4n - 1$. Suppose we want the sum of the first 7 terms of the sequence, called the **partial sum.** We want

$$\sum_{i=1}^{7} (4i - 1) = 3 + 7 + 11 + 15 + 19 + 23 + 27 = 105,$$

where an expression for the general term, $4n - 1$, becomes $4i - 1$ when n is replaced by i. We call the letter i, as used in this situation, the **index of summation.** Other letters often used for this purpose are j and k.

Note
This use of i has no connection with its use in our work with complex numbers.

We read the expression $\sum_{i=1}^{7} (4i - 1)$ "the summation as i goes from 1 to 7 of $4i - 1$." The first and last integers used to replace the index of summation, in this case 1 and 7, are called the **lower and upper limits of summation,** respectively.

Example 12–2 B	Expand the following indicated partial sums. Find the indicated sum.

1. $\displaystyle\sum_{j=1}^{4} (5j + 2)$

Now the index of summation is j, and we successively replace j by the integers 1, 2, 3, and 4. Thus

$$\sum_{j=1}^{4} (5j + 2) = [5(1) + 2] + [5(2) + 2] + [5(3) + 2] + [5(4) + 2]$$
$$= 7 + 12 + 17 + 22 = 58.$$

2. $\displaystyle\sum_{k=1}^{3} (-1)^k(2k + 5)$

We successively replace the index of summation k by the integers 1, 2, and 3. Therefore

$$\sum_{k=1}^{3} (-1)^k(2k + 5) = (-1)^1 [2(1) + 5] + (-1)^2 [2(2) + 5]$$
$$+ (-1)^3 [2(3) + 5]$$
$$= (-1)(7) + 1(9) + (-1)(11)$$
$$= -7 + 9 - 11 = -9.$$

We should note that there is nothing unique about the way the above-indicated sums have been stated using sigma notation.

Reversing the procedure, it is sometimes desirable to express a sum in the compact summation notation form.

Example 12–2 C	Write the following partial sums in sigma notation.

1. $-5 + 25 - 125 + 625$

There are four terms, so we can use the limits of summation 1 and 4. Since the operations alternate with a negative first term, a general term will contain -1 to an odd power. Use $(-1)^j$ for the first factor of a general term. By inspection, the numerical value of the terms of the series are powers of 5. Thus

$$-5 + 25 - 125 + 625 = \sum_{j=1}^{4} (-1)^j 5^j.$$

2. $\dfrac{2}{3} + \dfrac{4}{7} + \dfrac{6}{11} + \dfrac{8}{15} + \dfrac{10}{19}$

The limits of summation can be 1 and 5, since there are five terms. Inspection tells us a general term of the numerator is $2n$ and of the denominator is $4n - 1$. Then

$$a_n = \frac{2n}{4n - 1} \text{ and}$$

$$\frac{2}{3} + \frac{4}{7} + \frac{6}{11} + \frac{8}{15} + \frac{10}{19} = \sum_{k=1}^{5} \frac{2k}{4k - 1}.$$

Exercise 12–2

Directions Expand the following indicated partial sums. Find the indicated sum. See example 12–2 B.

1. $\sum\limits_{j=1}^{5} j^2$

2. $\sum\limits_{k=1}^{4} k^3$

3. $\sum\limits_{i=1}^{6} (2i + 3)$

4. $\sum\limits_{i=1}^{7} (i - 2)$

5. $\sum\limits_{k=1}^{5} k(k - 3)$

6. $\sum\limits_{j=1}^{6} (j^2 + 2)$

7. $\sum\limits_{i=1}^{4} i(2i - 1)$

8. $\sum\limits_{i=1}^{4} (i + 1)(i - 2)$

9. $\sum\limits_{j=1}^{5} (j - 3)(j + 2)$

10. $\sum\limits_{k=1}^{4} k^2(k + 2)$

11. $\sum\limits_{k=1}^{5} \dfrac{1}{k + 3}$

12. $\sum\limits_{i=1}^{4} \dfrac{3}{3i - 1}$

13. $\sum\limits_{k=1}^{5} \dfrac{2k + 1}{k + 3}$

14. $\sum\limits_{j=1}^{4} \dfrac{j^2}{3j - 2}$

15. $\sum\limits_{i=1}^{5} \dfrac{4}{i^2}$

16. $\sum\limits_{j=1}^{5} (-1)^j \cdot \dfrac{3}{2j}$

17. $\sum\limits_{k=1}^{4} (-1)^{k + 1} \cdot \dfrac{1}{3k}$

18. $\sum\limits_{i=1}^{5} (-1)^i i$

19. $\sum\limits_{j=1}^{3} (-1)^{j - 1}(j)^j$

20. $\sum\limits_{k=1}^{5} (-1)^{k + 1}(k + 1)^k$

Directions Expand and find the following indicated partial sums.

Example

$\sum\limits_{i=2}^{5} (3i + 1)$

Solution

$\sum\limits_{i=2}^{5} (3i + 1) = [3(2) + 1] + [3(3) + 1] + [3(4) + 1] + [3(5) + 1] = 7 + 10 + 13 + 16 = 46$

21. $\sum\limits_{i=3}^{8} (i + 2)$

22. $\sum\limits_{j=0}^{4} (2j + 1)$

23. $\sum\limits_{k=2}^{5} \dfrac{1}{k}$

24. $\sum\limits_{i=4}^{9} (i^2 + 3)$

25. $\sum\limits_{j=0}^{5} \dfrac{2j + 1}{j + 1}$

26. $\sum\limits_{k=2}^{6} (-1)^k$

27. $\sum\limits_{i=3}^{7} (-1)^i(3i - 2)$

28. $\sum\limits_{i=4}^{7} (-1)^i \dfrac{2}{2i + 3}$

Directions Write the following partial sums in sigma notation. See example 12–2 C.

29. $1 + 2 + 3 + 4 + 5$

30. $3 + 6 + 9 + 12$

31. $1 + 8 + 27 + 64 + 125 + 216$

32. $5 + 9 + 13 + 17 + 21$

33. $\dfrac{2}{3} + \dfrac{3}{4} + \dfrac{4}{5} + \dfrac{5}{6}$

34. $\dfrac{1}{3} + \dfrac{3}{5} + \dfrac{5}{7} + \dfrac{7}{9} + \dfrac{9}{11}$

35. $2 + \dfrac{4}{3} + \dfrac{6}{9} + \dfrac{8}{27}$

36. $\dfrac{1}{1} + \dfrac{2}{2} + \dfrac{3}{4} + \dfrac{4}{8} + \dfrac{5}{16}$

37. $\dfrac{5}{4} + \dfrac{7}{7} + \dfrac{9}{10} + \dfrac{11}{13}$

38. $2 - 5 + 8 - 11 + 14$

39. $-2 + 4 - 8 + 16 - 32 + 64$

40. $\dfrac{1}{2} - \dfrac{2}{3} + \dfrac{3}{4} - \dfrac{4}{5} + \dfrac{5}{6}$

12–3 Arithmetic sequences

An arithmetic sequence

Consider the sequence defined by

$$3,7,11,15,19, \cdot \cdot \cdot , 4n - 1, \cdot \cdot \cdot ,$$

whose major characteristic is that the difference between any two successive terms is 4. Recall that this determines the coefficient of n in the general term. Such a sequence is called an **arithmetic sequence** (or **arithmetic progression**).

> ■ **Definition of an arithmetic sequence**
> An **arithmetic sequence** is a sequence in which each term after the first differs from the preceding term by the same constant number.

We call this constant number the **common difference,** to be denoted by d. Thus in the above sequence, the common difference $d = 4$. That is, for any term of an arithmetic sequence,

$$d = a_{n+1} - a_n,$$

where d is the common difference and a_n and a_{n+1} are successive terms of the sequence.

Example 12–3 A

1. The sequence $7,11,15,19$ is arithmetic since

$$11 - 7 = 4, 15 - 11 = 4, 19 - 15 = 4,$$

and the common difference $d = 4$.

2. $-10,-4,2,8,14$ is an arithmetic sequence since

$$-4 - (-10) = -4 + 10 = 6$$
$$2 - (-4) = 2 + 4 = 6$$
$$8 - 2 = 6$$

and $14 - 8 = 6$. The common difference $d = 6$.

3. $1,3,9,12,36$ is not an arithmetic sequence since $3 - 1 = 2$ whereas $9 - 3 = 6$.

The general term of an arithmetic sequence

To find an expression for the nth term, a_n, consider the expressions

$$a_1, \quad a_2 = a_1 + d, \quad a_3 = a_1 + 2d, \quad a_4 = a_1 + 3d, \quad a_5 = a_1 + 4d.$$

This suggests the following:

> ■ **General term of an arithmetic sequence**
> A general term of an arithmetic sequence with first term a_1
> and common difference d is given by
> $$a_n = a_1 + (n - 1)d.$$

Example 12–3 B

1. Find the 21st term of the arithmetic sequence whose first term $a_1 = 3$ and common difference $d = 4$.
   Using $a_n = a_1 + (n - 1)d$ and substituting 3 for a_1, 4 for d, and 21 for n, we have

   $$\begin{aligned} a_{21} &= 3 + (21 - 1)4 \\ &= 3 + (20)4 = 3 + 80 = 83. \end{aligned}$$

   Thus the 21st term $a_{21} = 83$.

2. Given the arithmetic sequence $5, -1, -7, -13, \cdots$, find a_{20}.
   Since $-1 - 5 = -6$, then $d = -6$. Now $a_1 = 5$ and $n = 20$, so we use $a_n = a_1 + (n - 1)d$ and substitute to obtain

   $$\begin{aligned} a_{20} &= 5 + (20 - 1)(-6) \\ &= 5 + (19)(-6) = 5 + (-114) = -109. \end{aligned}$$

   The 20th term $a_{20} = -109$.

Given a finite arithmetic sequence, it is possible to determine the number of terms in the sequence, n, if we can determine d and a_1.

Example 12–3 C

Find the number of terms in the arithmetic sequence $-9, -4, 1, 6, \cdots, 111$.
Now from the first 4 terms of the sequence, we can determine $a_1 = -9$, $d = -4 - (-9) = -4 + 9 = 5$, and $a_n = 111$. We want the value of n.
Replacing a_1 by -9, d by 5, and a_n by 111 in the formula $a_n = a_1 + (n - 1)d$, we get

$$\begin{aligned} 111 &= -9 + (n - 1)5 \\ 111 &= -9 + 5n - 5 \\ 111 &= 5n - 14 \\ 125 &= 5n \end{aligned}$$

and so

$$25 = n.$$

Thus the sequence has $n = 25$ terms.

Sum of the terms of an arithmetic sequence

Now consider the sum of the terms in an arithmetic sequence having n terms. We denote this by S_n (called the nth partial sum). This sum can be written

$$S_n = a_1 + (a_1 + d) + (a_1 + 2d) + (a_1 + 3d) + \cdots + [a_1 + (n-1)d].$$

Another way to write this sum would be to add in reverse order, starting with the nth term, a_n, and subtracting multiples of the common difference d. Then we obtain

$$S_n = a_n + (a_n - d) + (a_n - 2d) + (a_n - 3d) + \cdots + [a_n - (n-1)d].$$

Now if we add the corresponding terms of both members of the two equations, we obtain

$$S_n = a_1 + (a_1 + d) + (a_1 + 2d) + \cdots + [a_1 + (n-1)d]$$
$$+ \ S_n = a_n + (a_n - d) + (a_n - 2d) + \cdots + [a_n - (n-1)d]$$
$$\overline{2S_n = (a_1 + a_n) + (a_1 + a_n) + (a_1 + a_n) + \cdots + (a_1 + a_n).}$$

$$n \text{ terms of } (a_1 + a_n)$$

We can write the right member by the product $n(a_1 + a_n)$ and the sum of these two equations is given by

$$2S_n = n(a_1 + a_n).$$

Dividing each member of the equation by 2, we obtain the sum of the first n term.

■ **Sum of the first n terms of an arithmetic sequence**
The sum of the first n terms of an arithmetic sequence is given by

$$S_n = \frac{n}{2}(a_1 + a_n).$$

Example 12–3 D

1. Find the sum of the first 25 terms of the arithmetic sequence whose general term is $a_n = 3n + 5$.
   Given $a_n = 3n + 5$, we can determine $a_1 = 3(1) + 5 = 8$, $d = 3$ (the coefficient of n) and, using $a_n = a_1 + (n-1)d$, we substitute to obtain

   $$a_{25} = 8 + (25 - 1)3$$
   $$= 8 + (24)(3) = 8 + 72 = 80.$$

   Since $n = 25$, using the formula

   $$S_n = \frac{n}{2}(a_1 + a_n),$$

   replace n by 25, a_{25} by 80, and a_1 by 8 to obtain

   $$S_{25} = \frac{25}{2}(8 + 80)$$

   $$= \frac{25}{2}(88) = 25(44) = 1,100.$$

2. Find $\sum\limits_{i=1}^{13} (3i - 1)$.

We want S_{13} (the thirteenth partial sum). Now $n = 13$, $a_1 = 3(1) - 1 = 2$, and $d = 3$ (the coefficient of i). We substitute into the formula $a_n = a_1 + (n - 1)d$ to determine a_{13}. Then

$$a_{13} = 2 + (13 - 1)3$$
$$= 2 + 12(3) = 2 + 36 = 38.$$

Substituting 13 for n, 2 for a_1, and 38 for a_n in $S_n = \dfrac{n}{2}(a_1 + a_n)$, we obtain

$$S_{13} = \frac{13}{2}(2 + 38)$$

$$= \frac{13}{2}(40) = 13(20) = 260.$$

Therefore $\sum\limits_{i=1}^{13} (3i - 1) = 260$.

Another form of the formula

$$S_n = \frac{n}{2}(a_1 + a_n)$$

can be obtained by replacing a_n by $a_1 + (n - 1)d$.

Replacing a_n by $a_1 + (n - 1)d$ in the formula $S_n = \dfrac{n}{2}(a_1 + a_n)$, we have

$$S_n = \frac{n}{2}[a_1 + a_1 + (n - 1)d]$$

$$S_n = \frac{n}{2}[2a_1 + (n - 1)d].$$

Example 12–3 E

Find the sum of the first 12 terms of the arithmetic sequence whose first term $a_1 = -11$ and common difference $d = 4$.

Using the formula $S_n = \dfrac{n}{2}[2a_1 + (n - 1)d]$, we want S_{12}. Substituting 12 for n, -11 for a_1, and 4 for d, we have

$$S_{12} = \frac{12}{2}[2(-11) + (12 - 1)4]$$
$$= 6[-22 + (11)4] = 6[-22 + 44] = 6(22) = 132.$$

Thus the sum of the first 12 terms is $S_{12} = 132$.

Exercise 12–3

Directions Determine whether or not the given sequence is arithmetic. If it is arithmetic, find the common difference d. See example 12–3 A.

1. $2,3,4,5,6, \cdots$

2. $1,6,11,16,21, \cdots$

3. $-10,-8,-6,-4, \cdots$

4. $4,6,8,10,12, \cdots$

5. $3,5,8,10,12, \cdots$

6. $-16,-19,-22,-25, \cdots$

7. $\frac{3}{2},2,\frac{5}{2},3,\frac{7}{2}, \cdots$

8. $-\frac{1}{2},\frac{1}{2},\frac{3}{2},\frac{5}{2}, \cdots$

9. $-\frac{7}{3},-\frac{2}{3},1,\frac{8}{3}, \cdots$

10. $1,\frac{1}{2},\frac{1}{3},\frac{1}{4},\frac{1}{5}, \cdots$

Directions Find the indicated term of the arithmetic sequence having the following characteristics. See example 12–3 B.

11. $a_1 = 4, d = 5$; find a_{16}.

12. $a_1 = -4, d = 2$; find a_{21}.

13. $a_1 = -10, d = -3$; find a_{17}.

14. $a_1 = 6, d = -7$; find a_{23}.

15. $a_1 = 2, d = \frac{1}{3}$; find a_{12}.

16. $a_1 = 4, d = \frac{1}{2}$; find a_{10}.

17. $a_1 = \frac{1}{3}, d = \frac{2}{3}$; find a_{14}.

18. $a_1 = \frac{3}{4}, d = -\frac{1}{2}$; find a_{13}.

19. $5,9,13,17, \cdots$; find a_{16}.

20. $4,13,22,31, \cdots$; find a_{19}.

21. $-15,-10,-5,0, \cdots$; find a_{25}.

22. $-27,-25,-23,-21, \cdots$; find a_{20}.

Directions Find the number of terms in the given finite arithmetic sequence. See example 12–3 C.

23. $14,29,44,59, \cdots, 89$

24. $2,5,8, \cdots, 83$

25. $-3,-11,-19,-27, \cdots, -115$

26. $7,1,-5,-11, \cdots, -101$

27. $\frac{1}{2},0,-\frac{1}{2},-1, \cdots, -\frac{27}{2}$

28. $\frac{5}{3},\frac{4}{3},1,\frac{2}{3}, \cdots, -2$

Directions Find the indicated partial sum for each of the given arithmetic sequences. The last given term is the nth term, a_n. (That is, $a_{16} = 33$ in exercise 29.) See example 12–3 D–1.

29. $3,5,7, \cdots , 33$; find S_{16}.

30. $0,3,6, \cdots , 63$; find S_{22}.

31. $7,4,1, \cdots , -32$; find S_{14}.

32. $1,-7,-15, \cdots , -111$; find S_{15}.

33. $\dfrac{1}{2} ,1,\dfrac{3}{2} , \cdots , 9$; find S_{18}.

34. $\dfrac{1}{4} ,1,\dfrac{7}{4} , \cdots , \dfrac{37}{4}$; find S_{13}.

35. $a_n = 2n + 5$; find S_{14}.

36. $a_n = 6n - 3$; find S_{15}.

37. $a_n = 5 - n$; find S_{19}.

38. $a_n = 7 - 3n$; find S_{17}.

Directions Find the indicated sum. See example 12–3 D–2.

39. $\sum\limits_{k=1}^{15} (2k - 5)$

40. $\sum\limits_{j=1}^{13} (3j + 9)$

41. $\sum\limits_{i=1}^{22} (3 - 2i)$

42. $\sum\limits_{i=1}^{19} (4 - i)$

43. $\sum\limits_{j=1}^{17} \dfrac{1}{3} j$

44. $\sum\limits_{k=1}^{14} \dfrac{1}{2} k$

45. $\sum\limits_{j=1}^{10} \left(\dfrac{3}{5} j - 2 \right)$

46. $\sum\limits_{i=1}^{11} \left(\dfrac{2}{3} i + 4 \right)$

Directions Find the indicated partial sum of the terms in the given arithmetic sequence. See example 12–3 E.

47. $2,8,14,22, \cdots$; find S_{13}.

48. $1,8,15,22, \cdots$; find S_{12}.

49. $24,19,14, \cdots$; find S_{15}.

50. $-6,-4,-2, \cdots$; find S_{14}.

51. $\dfrac{1}{6} ,-\dfrac{5}{6} ,-\dfrac{11}{6} , \cdots$; find S_{11}.

52. $-1,-4,-7, \cdots$; find S_{16}.

Directions Solve the following verbal problems.

53. A display of cans has 21 cans in the bottom row, 19 cans in the row above, 17 cans in the next row, and so on. How many cans are there if the top row contains 1 can?

54. A stock boy in a grocery store stacks a number of boxes of cereal so that there are 30 boxes in the first row, 27 boxes in the second row, 24 boxes in the third row, and so on. How many boxes of cereal does he have if there are 3 boxes in the top row?

55. In exercise 53, if there are 8 rows of cans, how many cans are in the display?

56. In exercise 54, if there are 9 rows of cereal, how many boxes of cereal are there?

57. Find the sum of the even integers from 2 to 116.

58. How many times will a clock strike in 12 hours if it strikes only on the hour?

59. A parachutist in free fall falls vertically 16 feet during the first second, 48 feet during the second second, 80 feet during the third second, and so on. How far will she fall during the eighth second? How far will she fall during the first 10 seconds?

60. Find the sum of the odd integers from 1 to 101.

61. Ron Line is offered a job as a mechanic starting at \$700 per month. If he is guaranteed a pay increase of \$10 per month every 3 months, what will his salary be after 8 years?

62. Neglecting air resistance, how long would it take before the parachutist pulls the rip cord to break her fall after falling 3,600 feet in exercise 59? (*Hint:* $S_n = 3,600$ and find n.)

63. In exercise 61, what total salary would Ron have earned in 8 years?

64. In exercise 61, how many years would it take for his salary to reach $1,000 per month?

65. Kenny Kranz opened a savings account for his daughter by depositing $50 on the day that she was born. On each subsequent birthday, he deposited $30 more than the previous year. How much money was deposited on his daughter's eighteenth birthday?

66. In exercise 65, how much money (disregarding interest) had Kenny deposited for his daughter after her eighteenth birthday?

12–4 Geometric sequences

A geometric sequence

Suppose a man offers to rent you his house under the conditions that the rent will be figured daily as follows:

first day	1¢
second day	2¢
third day	4¢
fourth day	8¢
fifth day	16¢
sixth day	32¢

The daily rent on the house forms the sequence

$$1,2,4,8,16,32, \cdot \cdot \cdot$$

in which case each term, after the first, is obtained by multiplying the preceding term by the constant 2. Such a sequence is called a **geometric sequence.**

> ■ **Definition of a geometric sequence**
> A **geometric sequence** is a sequence having the property that each term after the first term can be obtained by multiplying the preceding term by the same nonzero constant multiplier.

A geometric sequence is also called a **geometric progression.** We call the constant multiplier the **common ratio** since successive terms of the sequence form a "common ratio." We denote the common ratio by r and can obtain it by dividing any term after the first by the preceding term. This common ratio is the characteristic that distinguishes a geometric sequence from any other sequence.

Example 12–4   A

Determine whether the given sequence is geometric. If it is geometric, find the common ratio r.

1. $4,12,36,108, \cdot \cdot \cdot$
   Since $12 \div 4 = 3$, $36 \div 12 = 3$, and $108 \div 36 = 3$, the sequence is geometric and the common ratio $r = 3$.

2. $6,12,36,72, \cdot \cdot \cdot$
   Since $12 \div 6 = 2$ and $36 \div 12 = 3$, the sequence is not geometric.

General term of a geometric sequence

The definition of a geometric sequence shows that the first few terms of a general geometric sequence take the form

$$a_1, a_2 = a_1r, a_3 = a_1r^2, a_4 = a_1r^3, \cdot \cdot \cdot ,$$

from which we can determine the following about the general term of a geometric sequence.

> ■ **General term of a geometric sequence**
> The general term of a geometric sequence with first term a_1 and common ratio r is given by
> $$a_n = a_1r^{n-1}.$$

Example 12–4   B

1. Given the sequence
   $$2,10,50,250, \cdot \cdot \cdot ,$$
   we can determine that this is a geometric sequence with the common ratio $r = 5$ because $\dfrac{10}{2} = 5$, $\dfrac{50}{10} = 5$, and $\dfrac{250}{50} = 5$. Since $a_1 = 2$, the general term of the geometric sequence is given by
   $$a_n = 2(5)^{n-1}.$$

2. State the terms of the geometric sequence whose nth general term is
   $$a_n = 3\left(\frac{1}{2}\right)^{n-1}.$$

The terms of the sequence are

$$a_1 = 3\left(\frac{1}{2}\right)^0 = 3(1) = 3, \; a_2 = 3\left(\frac{1}{2}\right)^1 = \frac{3}{2}, \; a_3 = 3\left(\frac{1}{2}\right)^2 = 3\left(\frac{1}{4}\right) = \frac{3}{4},$$

$$a_4 = 3\left(\frac{1}{2}\right)^3 = 3\left(\frac{1}{8}\right) = \frac{3}{8},$$

and so on. The sequence then is given by

$$3, \frac{3}{2}, \frac{3}{4}, \frac{3}{8}, \; \cdots \; , 3\left(\frac{1}{2}\right)^{n-1}, \; \cdots \; ,$$

where $a_1 = 3$ and $r = \frac{1}{2}$.

Sum of the terms of a geometric sequence

Now consider the sum of the first n terms of a geometric sequence, denoted by S_n, and called the nth partial sum of the geometric sequence. With the geometric sequence,

$$a_1, \; a_1 r, \; a_1 r^2, \; a_1 r^3, \; \cdots \; , a_1 r^{n-1},$$

is associated the geometric series

$$S_n = a_1 + a_1 r + a_1 r^2 + a_1 r^3 + \; \cdots \; + a_1 r^{n-1}. \qquad (1)$$

If we multiply each member of this equation by the common ratio r, we obtain

$$r S_n = a_1 r + a_1 r^2 + a_1 r^3 + a_1 r^4 + \; \cdots \; + a_1 r^n. \qquad (2)$$

Subtracting term by term equation (2) from equation (1), we obtain

$$S_n = a_1 + a_1 r + a_1 r^2 + a_1 r^3 + \; \cdots \; + a_1 r^{n-1}$$
$$\underline{r S_n = \qquad a_1 r + a_1 r^2 + a_1 r^3 + \; \cdots \; + a_1 r^{n-1} + a_1 r^n}$$
$$S_n - r S_n = a_1 - a_1 r^n.$$

Thus

$$S_n - r S_n = a_1 - a_1 r^n$$
$$(1 - r)S_n = a_1 - a_1 r^n \qquad \text{(Factor } S_n.\text{)}$$
$$S_n = \frac{a_1 - a_1 r^n}{1 - r}. \qquad (r \neq 1)$$

■ **Sum of the first n terms of a geometric sequence**
The sum of the first n terms of a geometric sequence is given by

$$S_n = \frac{a_1(1 - r^n)}{1 - r} \; (r \neq 1),$$

where a_1 is the first term and r is the common ratio.

Example 12–4 C

1. Find the sum of the first seven terms of the geometric sequence whose first term $a_1 = 3$ and whose common ratio $r = 2$.

   Now we want S_7 where $n = 7$, $a_1 = 3$, and $r = 2$. Using the formula

   $$S_n = \frac{a_1(1 - r^n)}{1 - r},$$

   we substitute 7 for n, 3 for a_1, and 2 for r to obtain the equation

   $$S_7 = \frac{3(1 - 2^7)}{1 - 2}$$

   $$= \frac{3(1 - 28)}{-1}$$

   $$= \frac{3(-127)}{-1} = \frac{-381}{-1} = 381.$$

2. Find $\sum_{k=1}^{4} 2(3)^k$.

   The series is the fourth partial sum of a geometric sequence, where $a_1 = 2 \cdot 3 = 6$ and $r = 3$. We want S_4. Using the formula

   $$S_n = \frac{a_1(1 - r^n)}{1 - r},$$

   we substitute 6 for a_1, 3 for r, and 4 for n. Then

   $$S_4 = \frac{6(1 - 3^4)}{1 - 3}$$

   $$= \frac{6(1 - 81)}{-2} = \frac{6(-80)}{-2} = \frac{-480}{-2} = 240.$$

   Therefore $S_4 = \sum_{k=1}^{4} 2(3)^k = 240$.

Mastery points

Can you
- Identify a geometric sequence?
- Find the common ratio of a geometric sequence?
- Write the terms of a geometric sequence?
- Find the general term of a geometric sequence?
- Find the indicated term of a given geometric sequence?
- Find the nth partial sum of a geometric sequence?

Exercise 12-4

Directions Determine if the given terms form a geometric sequence. If they do, find the common ratio and write the next three terms of the sequence. See example 12–4 A.

1. 1, 3, 9, · · ·

2. 6, 12, 24, · · ·

3. $\dfrac{1}{2}, \dfrac{1}{6}, \dfrac{1}{18}, \cdots$

4. $\dfrac{1}{3}, \dfrac{1}{2}, \dfrac{3}{5}, \cdots$

5. $4, -2, 1, \cdots$

6. $-1, \dfrac{1}{2}, -\dfrac{1}{4}, \cdots$

7. $6, -2, \dfrac{2}{3}, \cdots$

8. $12, 4, \dfrac{4}{3}, \cdots$

9. $\dfrac{1}{2}, \dfrac{2}{3}, \dfrac{3}{4}, \cdots$

10. 16, 48, 80, · · ·

Directions Find a general term, a_n, of the given geometric sequence. See example 12–4 B–1.

11. 3, 6, 12, · · ·

12. 8, 12, 18, · · ·

13. 27, −18, 12, · · ·

14. −81, 27, −9, · · ·

15. $1, \sqrt{3}, 3, \cdots$

16. $32, 16\sqrt{2}, 16, \cdots$

17. $-\dfrac{1}{15}, \dfrac{1}{5}, -\dfrac{3}{5}, \cdots$

Directions Find the indicated term of the geometric sequence having the following characteristics. See example 12–4 B–2.

18. $a_1 = 3, r = 2$; find a_6.

19. $a_1 = 2, r = 3$; find a_5.

20. $a_1 = 16, r = \dfrac{1}{2}$; find a_7.

21. $a_1 = 81, r = \dfrac{1}{3}$; find a_4.

22. $a_1 = 1, r = -4$; find a_5.

23. $a_1 = 5, r = -2$; find a_6.

24. $a_1 = 25, r = -\dfrac{1}{5}$; find a_4.

25. $a_1 = -32, r = -\dfrac{1}{4}$; find a_5.

26. 3, 18, 108, · · · ; find a_6.

27. 9, 18, 36, · · · ; find a_7.

28. 10, −20, 40, · · · ; find a_8.

29. 7, −14, 28, · · · ; find a_9.

Directions Find the indicated partial sum of the geometric sequence having the following characteristics. See example 12–4 C.

30. $a_1 = 8, r = 2$; sum of the first 6 terms

31. $a_1 = 14, r = 3$; sum of the first 5 terms

32. $a_1 = -64, r = \dfrac{1}{4}$; sum of the first 4 terms

33. $a_1 = -10, r = \dfrac{1}{5}$; sum of the first 4 terms

34. 9, 18, 36, · · · ; find S_7.

35. 3, 18, 108, · · · ; find S_6.

36. $\dfrac{1}{3}, \dfrac{1}{9}, \dfrac{1}{27}, \cdots$; find S_5.

37. $\dfrac{1}{2}, \dfrac{1}{4}, \dfrac{1}{8}, \cdots$; find S_9.

38. $\dfrac{4}{3}, \dfrac{8}{3}, \dfrac{16}{3}, \cdots$; find S_7.

39. −5, 15, −45, · · · ; find S_5.

See example 12–4 C–2.

40. $\displaystyle\sum_{i=1}^{9} 3^i$

41. $\displaystyle\sum_{j=1}^{8} 4^j$

42. $\displaystyle\sum_{k=1}^{7} (-2)^k$

43. $\displaystyle\sum_{i=1}^{5} (-3)^i$

44. $\displaystyle\sum_{j=1}^{6} \left(\frac{1}{4}\right)^{j}$

45. $\displaystyle\sum_{k=1}^{7} \left(\frac{2}{3}\right)^{k}$

46. $\displaystyle\sum_{i=1}^{8} \left(-\frac{1}{3}\right)^{i}$

47. $\displaystyle\sum_{k=1}^{6} 3\left(\frac{2}{5}\right)^{k}$

48. $\displaystyle\sum_{j=1}^{8} 4\left(\frac{2}{3}\right)^{j}$

49. $\displaystyle\sum_{i=1}^{6} -5\left(\frac{3}{5}\right)^{i}$

50. $\displaystyle\sum_{j=1}^{6} -3\left(-\frac{1}{3}\right)^{j}$

Directions Solve the following verbal problems.

51. A ball is dropped from a height of 9 feet. If on each rebound it rises two-thirds of the height from which it fell, what distance has it traveled when it strikes the ground for the sixth time?

52. A basketball rebounds to a height that is three-fourths of the height from which it fell. If the basketball is dropped initially from a height of 4 meters, what distance has it traveled when it strikes the floor for the fifth time?

53. On a visit to Las Vegas, Emmett Broughton doubled his bet each time that he lost. If his first bet was $2 and he lost 8 consecutive bets, how much did he bet on the ninth bet?

54. In exercise 53, if Emmett loses his ninth bet also, how much will he have lost after his ninth loss?

55. In the first paragraph of this section, a man is paid monthly rent for his house at the rate of 1¢ the first day, 2¢ the second day, 4¢ the third day, and so on. At this rate, what would the rent for the house be for a month of 30 days?

56. A certain bacteria culture under a given condition triples in number each hour. If there were originally 1,000 bacteria, how many hours would it take the number of bacteria present in the culture to surpass 1 million?

57. A pump used to expel air from a tank removes one-fifth of what remains in the tank with each stroke. What part of the air in the tank has been removed after the fifth stroke?

58. An automobile depreciates in value each year by one-fifth of its value at the beginning of the year. If an auto is purchased for $8,000, what is its value at the end of the fourth year?

12–5 Infinite geometric series

Consider the formula for the sum of the first n terms of a geometric sequence given by

$$S_n = \frac{a_1 - a_1 r^n}{1 - r} = \frac{a_1(1 - r^n)}{1 - r},$$

which can be written in the form

$$S_n = \frac{a_1}{1 - r}(1 - r^n).$$

Now let n get greater and greater given an infinite geometric series. Let $|r| < 1$ and recall that if $|r| < 1$, then $-1 < r < 1$. We can show that as n becomes increasingly large, r^n becomes closer to zero. To illustrate, if $r = \dfrac{1}{3}$,

then

$$r^2 = \left(\frac{1}{3}\right)^2 = \frac{1}{9}, r^3 = \left(\frac{1}{3}\right)^3 = \frac{1}{27}, r^4 = \left(\frac{1}{3}\right)^4 = \frac{1}{81}, \text{ and so on.}$$

It becomes obvious that r^n can be made as close to zero as we wish by choosing a sufficiently large numerical value of n. As n increases, we can see $r^n = \left(\dfrac{1}{3}\right)^n$ approaches (gets closer and closer to) the value zero.

Now since $|r| < 1$, r^n approaches the value zero as n increases, then

$$\frac{a_1}{1-r}(1-r^n)$$

approaches the value

$$\frac{a_1}{1-r}(1-0) = \frac{a_1}{1-r}.$$

■ **Definition of the sum of an infinite geometric series**
If $|r| < 1$, the sum of the terms of an infinite geometric series, denoted by S_∞, is given by

$$S_\infty = \frac{a_1}{1-r}.$$

If $|r| \geq 1$, the sum does not exist.

In sigma notation we write the preceding definition by

$$S_\infty = \sum_{i=1}^{\infty} a_1 r^{i-1} = \frac{a_1}{1-r},$$

where the upper limit n is replaced by ∞ in the statement $S_n = \sum_{i=1}^{n} a_1 r^{i-1}$.

Example 12–5 A

1. Find the sum of the terms of an infinite geometric series such that $a_1 = 2$ and $r = \dfrac{1}{2}$.

We want S_∞ and use the formula

$$S_\infty = \frac{a_1}{1-r} = \frac{2}{1-\dfrac{1}{2}} = \frac{2}{\dfrac{1}{2}} = 4.$$

2. Find $\displaystyle\sum_{i=1}^{\infty} 2\left(\frac{1}{4}\right)^i$.

We want $S_\infty = \displaystyle\sum_{i=1}^{\infty} 2\left(\frac{1}{4}\right)^i$, in which $a_1 = 2\left(\dfrac{1}{4}\right) = \dfrac{2}{4} = \dfrac{1}{2}$ and

$r = \dfrac{1}{4}$. Using $S_\infty = \dfrac{a_1}{1-r}$, substitute $\dfrac{1}{2}$ for a_1 and $\dfrac{1}{4}$ for r to obtain

$$S_\infty = \frac{\dfrac{1}{2}}{1-\dfrac{1}{4}} = \frac{\dfrac{1}{2}}{\dfrac{3}{4}} = \frac{1}{2} \cdot \frac{4}{3} = \frac{2}{3}.$$

The sum of the terms of an infinite geometric series, $|r| < 1$, has some practical uses. We now consider two of the most common applications.

Example 12–5 B

1. Write the repeating decimal $0.27\overline{27}$ as a rational number.

   Now $0.27\overline{27} = 0.27 + 0.0027 + 0.000027 + \cdots$.

   We have an infinite geometric series with $a_1 = 0.27$ and

   $$r = \frac{0.0027}{0.27} = 0.01.$$

   Using the formula $S_\infty = \frac{a_1}{1 - r}$, we substitute 0.27 for a_1 and 0.01 for r to obtain

   $$S_\infty = \frac{0.27}{1 - 0.01} = \frac{0.27}{0.99} = \frac{27}{99} = \frac{3}{11}.$$

   Thus the repeating decimal $0.27\overline{27} = \frac{3}{11}$.

2. A ball is dropped from a height of 12 meters. If each time it strikes the floor the ball rebounds to a height that is three-fourths of the height from which it fell, find the total distance that the ball travels before it comes to rest on the floor.

   Let d be the total distance the ball travels. After the initial 12-meter drop, the ball will travel the same distance up and back down after each striking on the floor. Therefore

   $$a_1 = 12\left(\frac{3}{4}\right) = 9 \text{ and } r = \frac{3}{4} \text{ and using the formula } S_\infty = \frac{a_1}{1 - r},$$

   $$S_\infty = \frac{9}{1 - \frac{3}{4}} = \frac{9}{\frac{1}{4}} = 36.$$

   Then $d = 12 + 2(36) = 12 + 72 = 84$, and the ball would travel a distance of 84 meters before coming to rest on the floor.

Mastery points
Can you
- Find the sum of the terms of an infinite geometric series with $|r| < 1$?
- Express a repeating decimal as a rational number using
  $$S_\infty = \frac{a_1}{1 - r}, \ |r| < 1?$$

Exercise 12-5

Directions Find the sum of the terms of the given infinite geometric series. If the series has no sum, indicate that condition. See example 12-5 A-1.

1. $a_1 = 1, r = \dfrac{2}{3}$

2. $a_1 = 2, r = \dfrac{1}{3}$

3. $a_1 = -3, r = \dfrac{1}{2}$

4. $a_1 = \dfrac{1}{5}, r = \dfrac{1}{10}$

5. $a_1 = \dfrac{3}{5}, r = \dfrac{1}{3}$

6. $a_1 = 4, r = -\dfrac{1}{2}$

7. $a_1 = -\dfrac{5}{6}, r = -\dfrac{2}{3}$

8. $14 + 7 + \dfrac{7}{2} + \cdots$

9. $12 + 4 + \dfrac{4}{3} + \cdots$

10. $3 + \dfrac{3}{4} + \dfrac{3}{16} + \cdots$

11. $4 + \dfrac{4}{5} + \dfrac{4}{25} + \cdots$

12. $1 + \dfrac{2}{3} + \dfrac{4}{9} + \cdots$

13. $6 - 8 + \dfrac{32}{3} - \cdots$

14. $\displaystyle\sum_{i=1}^{\infty} \left(\dfrac{1}{2}\right)^{i-1}$

15. $\displaystyle\sum_{j=1}^{\infty} \left(\dfrac{4}{5}\right)^{j}$

16. $\displaystyle\sum_{k=1}^{\infty} \left(\dfrac{7}{8}\right)^{k+1}$

17. $\displaystyle\sum_{i=1}^{\infty} \left(-\dfrac{2}{3}\right)^{i}$

18. $\displaystyle\sum_{k=1}^{\infty} \left(-\dfrac{5}{3}\right)^{k}$

19. $\displaystyle\sum_{i=1}^{\infty} 3\left(\dfrac{1}{5}\right)^{i}$

20. $\displaystyle\sum_{j=1}^{\infty} 4\left(\dfrac{1}{6}\right)^{j-1}$

Directions Express the given repeating decimal as a rational number. See example 12-5 B-1.

21. $0.333\overline{3}$ 22. $0.4242\overline{42}$ 23. $0.281818\overline{1}$ 24. $0.4727\overline{272}$ 25. $0.036363\overline{36}$

Directions Solve the following verbal problems. See example 12-5 B-2.

26. A ball returns to two-thirds of its previous height with each bounce. If the ball is dropped from a height of 6 feet, what is the total distance the ball will travel before coming to rest?

27. When a weight on an attached spring is dropped, it falls a distance of 30 inches before the spring stretches to its limit and the weight springs back up. If the weight rebounds to nine-tenths of the preceding distance it fell, through what total distance does the weight travel before coming to rest?

28. A bob in a pendulum travels an arc length that is seven-eighths of its preceding arc length. If the first arc length is 16 centimeters, how far will the bob move before coming to rest?

29. If the first swing of a pendulum bob is 14 inches and each succeeding swing is five-sixths as long as the preceding one, what is the total distance the bob will travel before coming to rest?

30. A grant from an alumnus of Henry Ford Community College was such that the college was to receive $30,000 the first year and two-thirds of the preceding year's donation each year thereafter. What was the total amount of money the college would receive from the alumnus?

31. Muriel Lakey's cat, Epu, receives 5 milligrams of a medicine at 2 P.M. If Epu is to receive four-fifths of the preceding dose of the medicine every hour thereafter, how many milligrams of medicine does Epu receive altogether?

12–6 The binomial expansion

Consider the indicated product $(x + y)^n$, where n is a positive integer. By performing the indicated multiplication, we can obtain polynomial expressions for the positive integral powers of the binomial expression $x + y$. That is, we can multiply to show that

$$(x + y)^1 = x + y$$
$$(x + y)^2 = x^2 + 2xy + y^2$$
$$(x + y)^3 = x^3 + 3x^2y + 3xy^2 + y^3$$
$$(x + y)^4 = x^4 + 4x^3y + 6x^2y^2 + 4xy^3 + y^4$$
$$(x + y)^5 = x^5 + 5x^4y + 10x^3y^2 + 10x^2y^3 + 5xy^4 + y^5,$$

and so on. Each of the polynomials thus obtained is called the **binomial expansion** of the related power of the binomial $x + y$. In this section we shall develop a formula that will enable us to express any positive integral power of a binomial as a polynomial.

Before doing this, let us investigate the expansions given above to determine the properties that will hold for the expansion of the general binomial $(x + y)^n$.

1. The first term of each expansion is x raised to the power of the binomial itself, x^n.
2. The second term of each expansion is of the form $nx^{n-1}y$.
3. As we proceed term by term from this point, the exponent of x decreases by 1 and the exponent of y increases by 1 with each succeeding term.
4. The next to last term is of the form nxy^{n-1}.
5. The last term of each expansion is y raised to the power of the binomial, y^n.
6. In each term the sum of the exponents of x and y is always n.
7. There are $n + 1$ terms in each expansion.

Pascal's triangle

If we consider the coefficients of the five expansions stated previously, we can write them in a triangular pattern. (See figure 12.1.)

$(x + y)^1$					1		1			
$(x + y)^2$				1		2		1		
$(x + y)^3$			1		3		3		1	
$(x + y)^4$		1		4		6		4		1
$(x + y)^5$	1		5		10		10		5	1

Figure 12.1

This pattern was used by a seventeenth-century French mathematician named Blaise Pascal (1623–62) and is called **Pascal's Triangle.** When the coefficients are thus arranged, it is possible to determine the coefficients of the next expansion. The coefficients of the first and last terms are always 1. Each of the other coefficients is obtained by adding the two numbers above it, one to the left and one to the right. We can see that the coefficients of $(x + y)^6$ in the sixth row will then be 1, 6, 15, 20, 15, 6, and 1, as shown in figure 12.2.

$(x + y)^6$

1 + 5 + 10 + 10 + 5 + 1

6   15   20   15   6   1

Figure 12.2

Therefore

$$(x + y)^6 = x^6 + 6x^5y + 15x^4y^2 + 20x^3y^3 + 15x^2y^4 + 6xy^5 + y^6.$$

Factorial notation

Although it is possible to use Pascal's Triangle to determine the coefficients in any expansion $(x + y)^n$, where n is a positive integer, we need a more efficient way to do this for greater powers of the binomial. To do this, we must use **factorial notation.**

To write a product of n consecutive positive integers (starting with 1), we use the shorthand notation "$n!$," which is read "**n factorial**" or "**factorial n.**"

■ **Definiton of n factorial**
$$n! = n(n - 1)(n - 2)(n - 3) \cdot \cdot \cdot (3)(2)(1).$$

Equivalently,

$$n! = (1)(2)(3) \cdot \cdot \cdot (n - 2)(n - 1)n.$$

To illustrate,

$$5! = 5 \cdot 4 \cdot 3 \cdot 2 \cdot 1 = 120 \quad \text{or } 5! = 1 \cdot 2 \cdot 3 \cdot 4 \cdot 5 = 120.$$

Note
We agree that 0! = 1. Then 0! and 1! both equal 1.

We now state the general binomial expansion of $(x + y)^n$ for any positive integer n.

$$(x + y)^n = x^n + \frac{n}{1!}x^{n-1}y + \frac{n(n-1)}{2!}x^{n-2}y^2 + \frac{n(n-1)(n-2)}{3!}x^{n-3}y^3$$
$$+ \frac{n(n-1)(n-2)(n-3)}{4!}x^{n-4}y^4 + \cdots + y^n$$

This statement is called the **binomial expansion** (or **binomial theorem**).

Example 12–6 A

1. Expand and simplify $(a + 3b)^4$.
   Applying the binomial expansion, we replace n by 4, x by a, and y by $3b$ to obtain the statement

$$(a + 3b)^4 = a^4 + \frac{4}{1!}a^3 (3b) + \frac{4 \cdot 3}{2!}a^2(3b)^2 + \frac{4 \cdot 3 \cdot 2}{3!} a(3b)^3 + (3b)^4$$
$$= a^4 + 4a^3(3b) + 6a^2(9b^2) + 4a(27b^3) + 81b^4$$
$$= a^4 + 12a^3b + 54a^2b^2 + 108ab^3 + 81b^4.$$

2. Expand and simplify $(3c - 2d^2)^5 = [3c + (-2d^2)]^5$.
   From our binomial expansion $n = 5$, $x = 3c$, and $y = -2d^2$, so we substitute these values to obtain the statement

$$(3c - 2d^2)^5 = (3c)^5 + \frac{5}{1!}(3c)^4(-2d^2) + \frac{5 \cdot 4}{2!}(3c)^3(-2d^2)^2$$
$$+ \frac{5 \cdot 4 \cdot 3}{3!}(3c)^2(-2d^2)^3$$
$$+ \frac{5 \cdot 4 \cdot 3 \cdot 2}{4!}(3c)(-2d^2)^4 + (-2d^2)^5$$
$$= 243c^5 + 5(81c^4)(-2d^2) + 10(27c^3)(4d^4)$$
$$+ 10(9c^2)(-8d^6) + 5(3c)(16d^8) + (-32d^{10})$$
$$= 243c^5 - 810c^4d^2 + 1{,}080c^3d^4 - 720c^2d^6 + 240cd^8 - 32d^{10}.$$

Finding the rth term of an expansion

Sometimes we wish to find one of the terms in the expansion and we would like to do this without expanding the binomial fully. We determine the rth term of the expansion using the following expression:

■ **rth term of the binomial expansion $(x + y)^n$ is**
$$\frac{n!}{[n - (r - 1)]!(r - 1)!}x^{n-(r-1)}y^{r-1}.$$

In general, in the expansion of $(x + y)^n$, the term containing the variables $x^{n-k}y^k$ has coefficient $\dfrac{n!}{(n-k)!k!}$, which is sometimes written in the form $\dbinom{n}{k}$. That is, for the rth term in the expansion

$$\binom{n}{k} = \frac{n!}{(n-k)!k!},$$

where $k = r - 1$. To illustrate, by definition,

$$\binom{9}{5} = \frac{9!}{(9-5)!5!} = \frac{9!}{4!\,5!} = \frac{9 \cdot 8 \cdot 7 \cdot 6 \cdot 5!}{4 \cdot 3 \cdot 2 \cdot 1 \cdot 5!}$$
$$= \frac{9 \cdot 8 \cdot 7 \cdot 6}{4 \cdot 3 \cdot 2 \cdot 1} = 9 \cdot 2 \cdot 7 = 126.$$

Example 12-6 B

Find the sixth term in the expansion of $(2a + b)^9$.

Here $n = 9$, $r = 6$, $x = 2a$, and $y = b$. Since $r = 6$, then $r - 1 = 5$. We substitute 9 for n, 6 for r, 5 for $r - 1$, $2a$ for x, and b for y to obtain the expression for the sixth term as

$$\frac{9!}{(9-5)!5!}(2a)^{9-5}(b)^5 = \frac{9!}{4!5!}(2a)^4(b)^5$$
$$= 126(16a^4)(b^5) = 2{,}016a^4b^5.$$

We can use the binomial expansion to approximate a power of a decimal number.

Example 12-6 C

Use the binomial expansion to evaluate $(2.01)^6$ correct to four decimal places. Expand to the first four terms.

Now $(2.01)^6 = (2 + 0.01)^6$ and applying the binomial expansion to this expression,

$$(2 + 0.01)^6 = 2^6 + \frac{6}{1!}(2)^5(0.01) + \frac{6 \cdot 5}{2!}(2)^4(0.01)^2 + \frac{6 \cdot 5 \cdot 4}{3!}(2)^3(0.01)^3$$
$$| \cdots$$

$$= 64 + 6(32)(0.01) + 15(16)(0.0001) + 20(8)(0.000001)$$
$$+ \cdots$$
$$= 64 + 192(0.01) + 240(0.0001) + 160(0.000001) + \cdots$$
$$= 64 + 1.92 + 0.024 + 0.00016 = 65.94416.$$

Rounding to four decimal places, $(2.01)^6 \approx 65.9442$.

Exercise 12–6

Directions Expand and simplify each expression.

1. $6!$

2. $7!$

3. $\dfrac{12!}{8!}$

4. $\dfrac{9!}{7!}$

5. $\dfrac{10!}{9!}$

6. $\dfrac{15!}{14!}$

7. $\dfrac{8!}{2!6!}$

8. $\dfrac{7!}{4!3!}$

9. $\dfrac{10!}{3!7!}$

10. $\dfrac{13!}{5!8!}$

Directions Expand and simplify the following binomials using the binomial expansion. See example 12–6 A.

11. $(a - 3)^4$

12. $(b + 2)^5$

13. $(p + q)^6$

14. $(a - b)^7$

15. $(2a + 3)^4$

16. $(3b + 2)^5$

17. $\left(\dfrac{p}{2} - q\right)^6$

18. $\left(2r - \dfrac{q}{3}\right)^4$

19. $(a^2 + b^2)^5$

Directions Find the indicated term of each binomial expansion. See example 12–6 B.

20. $(a + b)^{13}$, seventh term

21. $(a - b)^{14}$, sixth term

22. $(p + 3)^{11}$, eighth term

23. $(q - 2)^{12}$, fifth term

24. $(r - 2s)^{10}$, fifth term

25. $(6 - k)^9$, seventh term

Directions Use the binomial expansion to calculate the following expressions correct to four decimal places. Expand to the first four terms. See example 12–6 C.

26. $(1.01)^8$

27. $(1.002)^{13}$

28. $(2.02)^7$

29. $(0.97)^5$

30. In the expansion of $\left(p^2 - \dfrac{1}{4}\right)^{12}$, find the term involving p^{10}.

31. Find the middle term in the expansion of $(a + \sqrt{a})^{12}$.

Directions Evaluate the following. See example 12–6 B.

32. $\dbinom{5}{2}$

33. $\dbinom{8}{4}$

34. $\dbinom{12}{7}$

Chapter summary

1. An infinite **sequence** is a function whose domain is the set of the positive integers.

2. A sequence is **finite** when its domain is the set $\{1,2,3, \cdot \cdot \cdot ,n\}$ for some fixed n and **infinite** when the domain is the set of positive integers.

3. A **series** is the sum of the first r terms of a sequence.

4. The sum of the first n terms of a sequence whose general term is a_n, in **sigma notation**, is given by

$$\sum_{i=1}^{n} a_i,$$

where i is the **index of summation**, 1 is the **lower limit**, and n is the **upper limit** of summation.

5. An **arithmetic sequence** is a sequence in which each term after the first differs from the preceding term by the same common difference d.

6. The nth term a_n of an arithmetic sequence is given by
$$a_n = a_1 + (n - 1)d,$$
where a_1 is the first term and d is the common difference.

7. The nth partial sum S_n of the terms of an arithmetic sequence is given by
$$S_n = \frac{n}{2}(a_1 + a_n) = \frac{n}{2}[2a_1 + (n - 1)d].$$

8. A **geometric sequence** is a sequence in which each term after the first term can be obtained by multiplying the preceding term by the same nonzero constant multiplier, called the **common ratio** and denoted by r.

9. The general term of a geometric sequence with first term a_1 and common ratio r is given by $a_n = a_1 r^{n-1}$.

10. The nth partial sum S_n of the terms of a geometric sequence is given by
$$S_n = \frac{a_1 - a_1 r^n}{1 - r} = \frac{a_1(1 - r^n)}{1 - r} \ (r \neq 1).$$

11. The sum of the terms in an infinite geometric series is given by
$$S_n = \frac{a_1}{1 - r}, \ |r| < 1.$$

12. The product "n factorial," denoted by $n!$, is defined by
$$n! = n(n - 1)(n - 2) \cdot \cdot \cdot (3)(2)(1).$$

13. The **binomial expansion** of the binomial $(x + y)^n$ is given by $(x + y)^n$
$$= x^n + \frac{n}{1!}x^{n-1}y + \frac{n(n - 1)}{2!}x^{n-2}y^2$$
$$+ \frac{n(n - 1)(n - 2)}{3!}x^{n-3}y^3$$
$$+ \cdot \cdot \cdot nxy^{n-1} + y^n.$$

14. $\displaystyle \binom{n}{k} = \frac{n!}{(n - k)!k!}$

Chapter review

[12–1]
Directions Write the first five terms of each sequence whose general term a_n is given.

1. $a_n = 4n + 3$

2. $a_n = \dfrac{5n}{2n - 1}$

3. $a_n = (-1)^n \cdot \dfrac{4}{2n + 5}$

4. $a_n = (-1)^{n+1} \cdot 2^n$

Directions Find the indicated term of the sequence whose general term a_n is given.

5. $a_n = 4 - 3n$, find a_6.

6. $a_n = (-1)^n(3n-4)$, find a_7.

7. $a_n = \dfrac{3^{n+1}}{2n}$, find a_9.

8. $a_n = (-1)^{n-1} \cdot \dfrac{2^n + 1}{3^n}$, find a_{11}.

Directions Given the following sequences, find an expression for the general term a_n.

9. $5,7,9,11, \cdots$

10. $3,8,13,18, \cdots$

11. $\dfrac{2}{3}, \dfrac{3}{7}, \dfrac{4}{11}, \dfrac{5}{15}, \cdots$

12. $-4,9,-14,19, \cdots$

13. Dockage fees for a boat at a marina are $3.00 for the first night, $3.25 for the second night, $3.50 for the third night, and so on. Write an expression for the general term of the sequence. How much did it cost to dock the boat on the eighth night?

[12–2]

Directions Expand each indicated sum and find the sum.

14. $\displaystyle\sum_{i=1}^{4} (4i - 1)$

15. $\displaystyle\sum_{i=1}^{6} i(i + 5)$

16. $\displaystyle\sum_{k=1}^{5} \dfrac{k^2}{k + 1}$

17. $\displaystyle\sum_{j=1}^{6} (-1)^j \cdot \dfrac{4}{5j}$

Directions Write each sum in sigma notation.

18. $5 + 8 + 11 + 14$

19. $\dfrac{4}{5} - \dfrac{5}{6} + \dfrac{6}{7} - \dfrac{7}{8} + \dfrac{8}{9}$

[12–3]

Directions Find the indicated term of each arithmetic sequence having the following characteristics.

20. $a_1 = 5$, $d = 4$; find a_{15}.

21. $a_1 = -3$, $d = 5$; find a_{17}.

22. $3,7,11,15, \cdots$; find a_{21}.

23. $-6,-8,-10,-12$; find a_{19}.

Directions Find the number of terms in each given finite arithmetic sequence.

24. $-2,3,8, \cdots, 68$

25. $4,0,-4,-8, \cdots, -76$

Directions Find the indicated partial sum of each given arithmetic sequence.

26. $-3,3,9, \cdots, 81$; find S_{15}.

27. $10,7,4, \cdots, -50$; find S_{21}.

28. $\displaystyle\sum_{j=1}^{29} \dfrac{2}{3} j$

29. $\displaystyle\sum_{k=1}^{25} \left(\dfrac{1}{2} k + 1 \right)$

30. Company B starts offers of a beginning wage of $12,000 with a raise of $450 each year thereafter. What would the wage be after 11 years?

[12-4]

Directions Find the general term a_n of each given geometric sequence.

31. 3,6,12, $\cdots$

32. $-\dfrac{3}{4}, \dfrac{9}{16}, -\dfrac{27}{64}, \cdots$

Directions Find the indicated term of each geometric sequence having the following characteristics.

33. $a_1 = 5, r = 3$; find a_5.

34. $a_1 = -24, r = \dfrac{1}{3}$; find a_4.

35. $a_1 = 36, r = -\dfrac{2}{3}$; find a_3.

36. $-9, 18, -36, \cdots$; find a_7.

Directions Find the indicated partial sum of each geometric sequence having the following characteristics.

37. $a_1 = 3, r = 3$; find S_5.

38. $a_1 = -24, r = \dfrac{1}{2}$; find S_6.

39. $\displaystyle\sum_{j=1}^{5} \left(-\dfrac{3}{4}\right)^j$

40. $\displaystyle\sum_{k=1}^{7} 4\left(\dfrac{1}{3}\right)^k$

[12-5]

Directions Find the sum of the terms of each given infinite geometric series.

41. $a_1 = 3, r = \dfrac{3}{4}$

42. $a_1 = -2, r = \dfrac{1}{5}$

43. $\displaystyle\sum_{j=1}^{\infty} 3\left(-\dfrac{1}{4}\right)^j$

44. $\displaystyle\sum_{k=1}^{\infty} \left(-\dfrac{2}{3}\right)\left(-\dfrac{1}{5}\right)^{k+1}$

Directions Use the infinite geometric series to write each repeating decimal as a rational number.

45. $0.353\overline{535}$

46. $0.4323\overline{232}$

47. A boat at anchor experiences a series of waves, each wave having 25% less amplitude (height) than the previous one. If the first wave has amplitude 3 meters, how much vertical distance does the boat travel before coming to rest? (*Hint:* The boat travels *up* and *down* the same distance with each wave.)

[12-6]

Directions Expand and simplify each binomial.

48. $(x + 5)^7$

49. $(2a - 3b)^5$

50. $\left(\dfrac{1}{2}a - 3b\right)^4$

Directions Find the indicated term in the expansion of each given binomial.

51. $(a - 4)^{11}$, fifth term

52. $(3a + b)^{14}$, seventh term

53. $(2x + 3y)^{12}$, the term in which y^2 appears

54. The pendulum on a Seth Thomas antique regulator clock swings such that, after the first swing, each swing the pendulum travels is three-fourths of the previous swing. If it travels 2 feet on the first swing, how far does it travel on the fourth swing?

55. Evaluate $\dbinom{13}{8}$.

Final examination

[1-1] **1.** Given $\{y \mid y$ is an integer between -8 and $4\}$, list the elements in the set.

[1-1] **2.** Given $A = \{-1,2,4,7\}$, $B = \{0,1,2,7,9\}$, and $C = \{-4,-1,0,9\}$, find (a) $A \cap B$, (b) $A \cup C$, (c) $(B \cup C) \cap \emptyset$.

[1-1] **3.** Given the set $\{x \mid -4 \leq x < 9\}$, list the integers in the set.

[1-4] **4.** Perform the indicated operations and simplify.
$$5 - \{6 - [3 + 18 \div 2 - 3^2] - (4 + 7)\}$$

[3-4] **5.** Simplify the following expressions. Leave all answers with positive exponents.
a. $(-a^3b^{-3})(2a^4b^{-2})$
b. $(-2a^{-3}b^2)^{-2}$
c. $\dfrac{a^{-4}b^3}{a^2b^{-1}}$

[3-3] **6.** Multiply as indicated.
a. $(2y + 7)^2$
b. $(4x + 2y)(4x - 2y)$
c. $(x - 2)(x^2 + 2x + 4)$

Directions Completely factor the following expressions.

[3-5] **7.** $12xy - 4x^2y^3 + 8x^3y^2$

[3-6] **8.** $7y^2 - 34y - 5$

[3-7] **9.** $4a^2 - 20ab + 25b^2$

[3-8] **10.** $8x^2 - 50y^2$

[3-8] **11.** $16a^3 - 2b^3$

[3-8] **12.** $6ax - 3ay - 2bx + by$

Directions Find the solution set of the following equations and inequalities.

[2-1] **13.** $5(3y - 1) + 2y = 8(2 - y)$

[2-4] **14.** $3(2x + 3) < 4(x - 5)$

[2-4] **15.** $-9 < 4x + 1 \leq 5$

[2-5] **16.** $|2x - 5| = 3$

[2-5] **17.** $|5 - 3x| < 4$

[2-5] **18.** $|5c - 4| \geq 6$

[6-1] **19.** $x^2 - 5x = 24$

[4-6] **20.** $\dfrac{3}{x - 1} - \dfrac{4}{3} = \dfrac{6}{x - 1}$

[4-3] **21.** Add $\dfrac{5}{2a - 1} + \dfrac{6}{4a + 3}$.

[4-3] **22.** Subtract $\dfrac{2y + 1}{y^2 - y - 42} - \dfrac{y + 1}{y^2 - 36}$.

[4-2] **23.** Divide $\dfrac{3a^2 - 13a + 4}{a^2 + 2a + 1} \div \dfrac{a^2 - 8a + 16}{a^2 - 1}$.

[4-4] **24.** Simplify the complex fraction $\dfrac{\dfrac{4}{a} - \dfrac{2}{b}}{\dfrac{1}{b} - \dfrac{2}{a}}$.

[5-4] **25.** Combine the expression
$3\sqrt{75} + 2\sqrt{27} - \sqrt{48}$.

[5-5] **27.** Multiply $(3 - 2\sqrt{5})(4 + 3\sqrt{2})$.

[6-5] **29.** Find the solution set of the radical equation $\sqrt{x - 1} = x - 3$. Indicate extraneous solutions.

[7-3] **31.** Find the equation of the line having the following conditions. Write the equation in standard form.
   a. Through points $(1,2)$ and $(-3,-4)$
   b. Through $(4,-3)$ and perpendicular to the line $y - 4x = 3$
   c. Through $(3,-2)$ and having slope $-\dfrac{2}{3}$

[10-2] **33.** Given $f(x) = 5x - 3$ and $g(x) = x^2 - x + 1$, find
   a. $f(-3)$
   b. $g(4)$
   c. $\dfrac{f(x + h) - f(x)}{h}$.

Directions Sketch the graph of the following equations and inequalities.

[7-3] **34.** $2y - 5x = 10$

[7-4] **36.** $4y - 3x < -24$

[9-3] **38.** Determine if the given equation represents a circle, a parabola, an ellipse, or a hyperbola.
   a. $x^2 + y^2 - 2x + 4y - 3 = 0$
   b. $2y^2 + 6 = x^2$
   c. $4x - x^2 = y$
   d. $8y^2 = 6 - 5x^2$

[8-3] **40.** Evaluate $\begin{vmatrix} 4 & -1 & 3 \\ 5 & 0 & -2 \\ 3 & 6 & 1 \end{vmatrix}$.

[11-3] **42.** Write the expression $\log_b 5 + \log_b 6 - 3 \log_b 2$ as a logarithm of a single number.

[11-6] **44.** Find the solution set of the equation $3^{2-x} = 4$. Round to the nearest tenth.

[5-6] **26.** Rationalize the denominator $\dfrac{2 + i}{3 - 2i}$.

[6-3] **28.** Find the solution set of the quadratic equation $3p^2 - 2 = 7p$.

[6-7] **30.** Find the solution set of the quadratic inequality $y^2 \geq 4y + 3$.

[7-2] **32.** Find the slope and y-intercept of the line $5y - 3x = 9$.

[9-1] **35.** $y = x^2 - 3x - 10$

[9-3] **37.** $5x^2 + y^2 = 20$

[8-1] **39.** Solve the system of equations
$3y + 5x = 1$
$2y - 3x = -3$.

[8-3] **41.** Solve the system of equations
$x - 3y = 4$
$2x + 5y = 3$ by determinants.

[11-5] **43.** Find $\log_3 7$ using the common logarithms.

[11-5] **45.** Find ℓn 36.

[12–1] 46. Write the sum $3 + 9 + 15 + 21 + 27$ in summation notation.

[12–2] 47. Find the indicated sum $\sum\limits_{i=1}^{7} (2i - 1)$.

[12–3] 48. Given $a_1 = -2$ and $d = -3$, find a_{15} of the arithmetic sequence.

[12–3] 49. Find S_{26} of the sequence
$$\frac{1}{2}, 1, \frac{3}{2}, 2, \cdot \ \cdot \ \cdot \ \cdot$$

[12–4] 50. Find a_4 of the geometric sequence given
$$a_1 = \frac{1}{2} \text{ and } r = -\frac{1}{3}.$$

[12–5] 51. Find $\sum\limits_{i=1}^{\infty} 2\left(-\frac{1}{2}\right)^{i}$.

[12–6] 52. Expand the binomial $(3x - 2y)^4$.

[12–6] 53. Find the seventh term of the expansion of $(a - 5)^{12}$.

[12–6] 54. Find the rational equivalent of $0.234234\overline{234}$.

[12–6] 55. Evaluate $\begin{pmatrix} 11 \\ 5 \end{pmatrix}$.

Appendix A
Synthetic Division

In section 4–5 we studied the procedure for dividing a multinomial by another multinomial. Many times the divisor is a binomial of the form $x - k$, k is a constant. We can shorten the process considerably by using **synthetic division.**

Consider the following example: $(2x^3 + 5x^2 + x - 1) \div (x + 2)$. Performing the indicated division, we obtain

$$
\begin{array}{r}
2x^2 + x - 1 \quad \text{(quotient)} \\
(\text{divisor}) \quad x + 2 \ \overline{)\,2x^3 + 5x^2 + x - 1\,} \quad \text{(dividend)} \\
\underline{2x^3 + 4x^2} \\
x^2 + x \\
\underline{x^2 + 2x} \\
-x - 1 \\
\underline{-x - 2} \\
1 \ \text{(remainder).}
\end{array}
$$

We can write the exact same problem using only the coefficients of the terms.

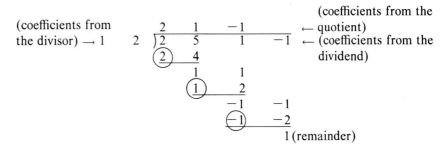

(coefficients from the divisor) → 1

(coefficients from the quotient)

← (coefficients from the dividend)

Observe that the circled numbers are repetitions of the numbers directly above them. We can rewrite the problem without them.

$$
\begin{array}{cccc}
 & 2 & 1 & -1 \\
1 \quad 2 \ \overline{)\,2} & 5 & 1 & -1 \\
 & \underline{4} \\
 & 1 & \textcircled{1} \\
 & & 2 \\
 & & \underline{-1} & \textcircled{-1} \\
 & & & \underline{-2} \\
 & & & 1
\end{array}
$$

The circled numbers are again the same as the numbers directly above them. Therefore we can omit them.

$$
\begin{array}{cccc}
 & 2 & 1 & -1 \\
1 \quad 2 \ \overline{)\,2} & 5 & 1 & -1 \\
 & \underline{4} \\
 & 1 \\
 & & 2 \\
 & & \underline{-1} \\
 & & & -2 \\
 & & & \underline{} \\
 & & & 1
\end{array}
$$

All of the numbers can be condensed. We omit the top row since it duplicates the bottom row. We shall also omit the 1 at the upper left.

$$
\begin{array}{r|rrrr}
2 & 2 & 5 & 1 & -1 \\
  &   & 4 & 2 & -2 \\
\hline
  & 2 & 1 & -1 & 1
\end{array}
$$

By changing the 2 at the upper left to its additive inverse, -2, and adding the additive inverse in each step, instead of subtracting, we obtain the same result. This is the final form to be used when performing synthetic division.

$$
\begin{array}{r|rrrr}
-2 & 2 & 5 & 1 & -1 \\
   &   & -4 & -2 & 2 \\
\hline
   & 2 & 1 & -1 & 1 \\
   & x^2 & x & \text{(constant)} & \text{(remainder)}
\end{array}
$$

The bottom line of our synthetic division problem represents the coefficients of the terms of the quotient along with the remainder. That is, 2   1 -1 1 represents $2x^2 + x - 1$ with a remainder of 1.

Note
The degree of the quotient, 2, is one less than the degree of the dividend, 3. This will *always* be the case.

Example A–1

Divide the following using synthetic division.

1. $(3x^2 + 7x + 6) \div (x + 1)$

To use synthetic division our divisor *must* be of the form $x - k$. Therefore we write $x + 1$ as $x - (-1)$ and $k = -1$. Now write the coefficients of the terms in $3x^2 + 7x + 6$ with -1 to the left.

$$
\begin{array}{r|rrr}
-1 & 3 & 7 & 6 \\
\hline
\end{array}
$$

First, we bring down the 3 and multiply that with the -1. This product, -3, is added to the 7.

$$
\begin{array}{r|rrr}
-1 & 3 & 7 & 6 \\
   & \downarrow & -3 & \\
\hline
   & 3 & 4 &
\end{array}
$$

The 4 is now multiplied with -1 and the product, -4, is added to 6.

$$
\begin{array}{r|rrr}
-1 & 3 & 7 & 6 \\
   &   & -3 & -4 \\
\hline
   & 3 & 4 & 2
\end{array}
$$

We can now read the coefficients of the quotient and the remainder from the last row. The quotient is $3x + 4 + \dfrac{2}{x + 1}$. (2 is the remainder.)

2. $(2x^3 - 7x^2 - x + 12) \div (x - 3)$

Our divisor, $x - 3$, is already in the form $x - k$, where $k = 3$. We set up the problem as follows:

$$3 \,\,\underline{)\,2 \quad\quad -7 \quad\quad -1 \quad\quad 12}$$

We bring down the 2 and begin the repetitive process of multiplying the last number in the bottom row times k, 3, and adding this to the value in the next column.

$$
\begin{array}{r|rrrr}
3 & 2 & -7 & -1 & 12 \\
  & \downarrow & 6 & -3 & -12 \\
\hline
  & 2 & -1 & -4 & 0
\end{array}
$$

Since we have a zero in the last position of the bottom row, there is no remainder and the quotient is $2x^2 - x - 4$.

3. $(x^3 - 11x + 8) \div (x - 3)$

Since the x^2 term is missing in the dividend, we think of the expression as $x^3 + 0x^2 - 11x + 8$ when writing down the coefficients.

$$
\begin{array}{r|rrrr}
3 & 1 & 0 & -11 & 8 \\
  & \downarrow & 3 & 9 & -6 \\
\hline
  & 1 & 3 & -2 & 2
\end{array}
$$

Our quotient is then $x^2 + 3x - 2 + \dfrac{2}{x - 3}$.

4. $(2x^4 - 10x^3 + 11x^2 - 11x + 8) \div (x - 2)$

$$
\begin{array}{r|rrrrr}
2 & 2 & -10 & 11 & -11 & 8 \\
  & \downarrow & 4 & -12 & -2 & -26 \\
\hline
  & 2 & -6 & -1 & -13 & -18
\end{array}
$$

The quotient is $2x^3 - 6x^2 - x - 13 - \dfrac{18}{x - 2}$.

Exercise A-1

Directions Perform the indicated divisions using synthetic division. Assume that no divisor is equal to zero.

1. $\dfrac{a^2 + 7a + 10}{a + 5}$

2. $\dfrac{x^2 + 7x + 6}{x + 1}$

3. $\dfrac{3a - 10 + a^2}{a - 2}$

4. $\dfrac{y^2 - 6 + y}{y - 2}$

5. $\dfrac{x^2 + 5x + 9}{x - 2}$

6. $\dfrac{y^2 - 6y - 4}{y - 3}$

7. $\dfrac{2a + 3a^2 - 1}{a + 1}$

8. $\dfrac{5 - 3x + 2x^2}{x + 2}$

9. $\dfrac{a^3 + 2a^2 - a - 4}{a - 1}$

10. $\dfrac{y^3 - 3y^2 + y + 2}{y + 1}$

11. $\dfrac{x^3 - 1}{x - 1}$

12. $\dfrac{a^3 + 1}{a + 1}$

13. $\dfrac{y^3 + 3y^2 - 4}{y - 2}$

14. $\dfrac{x^3 + 5x^2 - 1}{x - 1}$

15. $\dfrac{2a^3 - 9a^2 - 8a + 17}{a - 5}$

16. $\dfrac{3x^3 + 11x^2 + 12}{x + 4}$

17. $\dfrac{x^4 - 5x^3 + 5x^2 + 6x - 10}{x - 3}$

18. $\dfrac{2a^4 + 6a^3 + a^2 + 2a - 5}{a + 3}$

19. $\dfrac{3y^4 - 7y^3 - 10y^2 + 11y + 6}{y - 3}$

20. $\dfrac{a^5 + 2a^4 + 3a^3 + 6a^2 + a + 2}{a + 2}$

Answers to exercise A–1

1. $a + 2$

2. $x + 6$

3. $a + 5$

4. $y + 3$

5. $x + 7 + \dfrac{23}{x - 2}$

6. $y - 3 - \dfrac{13}{y - 3}$

7. $3a - 1$

8. $2x - 7 + \dfrac{19}{x + 2}$

9. $a^2 + 3a + 2 - \dfrac{2}{a - 1}$

10. $y^2 - 4y + 5 - \dfrac{3}{y + 1}$

11. $x^2 + x + 1$

12. $a^2 - a + 1$

13. $y^2 + 5y + 10 + \dfrac{16}{y - 2}$

14. $x^2 + 6x + 6 + \dfrac{5}{x - 1}$

15. $2a^2 + a - 3 + \dfrac{2}{a - 5}$

16. $3x^2 - x + 4 - \dfrac{4}{x + 4}$

17. $x^3 - 2x^2 - x + 3 - \dfrac{1}{x - 3}$

18. $2a^3 + a - 1 - \dfrac{2}{a + 3}$

19. $3y^3 + 2y^2 - 4y - 1 + \dfrac{3}{y - 3}$

20. $a^4 + 3a^2 + 1$

Appendix B
Answers and Solutions

Chapter 1

Exercise 1-1

Answers to odd-numbered problems

1. {Sunday,Saturday}   **3.** {10,12,14}   **5.** ∅   **7.** $A \subseteq D$ **9.** ∅ ⊆ B   **11.** {6,7,8}   **13.** true   **15.** true   **17.** false
19. true   **21.** true   **23.** {1,2,3,4}   **25.** {1,2,3,4,7,8,9}   **27.** ∅   **29.** {2,3,4}   **31.** {1,2,3,4}

39. 3   **41.** 0   **43.** −2   **45.** −4   **47.** <   **49.** >   **51.** <   **53.** <   **55.** $H \cap Q = \emptyset$ **57.** {8} ⊆ N   **59.** {8} ∉ N
61. yes   **63.** no   **65.** yes   **67.** yes   **69.** no   **71.** yes   **73.** yes   **75.** yes

Solutions to trial exercise problems

4. ∅. There are no months with less than 28 days.   **11.** {6,7,8}. The elements are inside braces and separated by commas.

31. {1,2,3,4} ∪ ({2,3,4} ∩ {7,8,9}) = {1,2,3,4} ∪ ∅ = {1,2,3,4}   **36.**

we approximate $-\sqrt{3}$ as −1.732 and $\sqrt{2}$ as 1.414.   **43.** $-|-2| = -2$. The negative sign in front of the absolute value bar remains.
52. $0 > -5$. 0 is greater than −5 since 0 is to the right of −5 on the number line.   **58.** 8 ⊄ N. We place a slash mark, /,
through ⊆ to negate it.

Exercise 1-2

Answers to odd-numbered problems

1. 2   **3.** −14   **5.** 1   **7.** 18   **9.** −7   **11.** −4   **13.** −20   **15.** −12   **17.** 13   **19.** +9   **21.** −17   **23.** 10
25. −28   **27.** −27   **29.** −24   **31.** 48   **33.** −240   **35.** 60   **37.** 0   **39.** −64   **41.** 36   **43.** −64   **45.** −8
47. 2   **49.** −4   **51.** 3   **53.** −9   **55.** indeterminate   **57.** 10   **59.** 0   **61.** 4   **63.** undefined   **65.** −4   **67.** −$88
69. 3   **71.** 5 yd   **73.** $7(-10) = -70$ **75.** −24° F   **77.** $68   **79.** 32   **81.** 240¢ = $2.40   **83.** 290 ml   **85.** 8,708 feet
87. $1.80   **89.** 34 miles   **91.** 133 over 78   **93.** 27   **95.** 2 mb gain

Solutions to trial exercise problems

9. $(-7) - 0 = -7$. We can subtract zero from any number and the number will be our answer.   **14.** $6 - 9 + 11 - 8$
$= 6 + (-9) + 11 + (-8) = (-3) + 11 + (-8) = 8 + (-8) = 0$ **21.** $[(-6) - 5] + 4 - (16 - 6)$
$= [(-6) + (-5)] + 4 - [16 + (-6)] = (-11) + 4 - 10 = (-11) + 4 + (-10) = (-7) + (-10) = -17$
36. $(-9)(+2)(0)(-4) = 0$. When zero is one of the factors, the product will be zero.   **38.** $-5^2 = -(5 \cdot 5) = -(25) = -25$
59. $\dfrac{(-6)(0)}{(-2)} = \dfrac{0}{(-2)} = 0$ **62.** $\dfrac{(-8)(-6)}{(-2)(0)} = \dfrac{48}{0} = $ undefined

Exercise 1-3

Answers to odd-numbers problems

1. $3y + 4x$ **3.** $4 + (2 + 6)$ **5.** $a \cdot 5$ **7.** $2(xy)$ **9.** 0   **11.** $x + 5$ **13.** $2a - 3$ **15.** 7   **17.** $a \geq 5$ **19.** $a < b$
21. $x > 5$ or $x < 5$ **23.** $a > 6$ **25.** closure property for addition   **27.** identity property of multiplication   **29.** multiplicative
inverse property   **31.** identity property of addition   **33.** associative property of addition   **35.** associative property of multiplication
37. identity property of addition   **39.** commutative property of addition   **41.** associative property of addition   **43.** identity property
of addition   **45.** $-6, \dfrac{1}{6}$ **47.** $-5, \dfrac{1}{-5}$ **49.** $-x, \dfrac{1}{x}$ **51.** $7, -7$ **53.** $3, \dfrac{1}{3}$ **55.** $16x$ **57.** $8a$ **59.** z **61.** $8x + 13$

63. $9x - 4$ **65.** a. given, b. closure property of multiplication, c. reflexive property of equality, d. given, e. substitution property of equality **67.** a. given, b. additive inverse property, c. additive inverse property, d. uniqueness of additive inverse

Solutions to trial exercise problems

22. $4 \geq y$. From trichotomy, if 4 is not less than y, it must be greater than or equal to y. **36.** commutative property of addition because the order in which we added the numbers was changed

Exercise 1–4

Answers to odd-numbered problems

1. 2 **3.** 37 **5.** -8 **7.** 41 **9.** 0 **11.** 10 **13.** 1 **15.** -2 **17.** 25 **19.** 41 **21.** 19 **23.** 9.38 **25.** $-\dfrac{7}{24}$

27. -10.74 **29.** 108.05 **31.** $\dfrac{2}{17}$ **33.** 155 **35.** 45 **37.** 17 **39.** 5 **41.** 80 **43.** 1 **45.** 100 square meters

47. 5,313.6 pounds per square inch **49.** 262 square feet **51.** 77 **53.** $93 **55.** $176,000 **57.** $2,547 **59.** 4,900 words

Solutions to trial exercise problems

10. $8 + 0(5 - 7) = 8 + 0(-2) = 8 + 0 = 8$ **20.** $12 + 3 \cdot 16 \div 4^2 - 5 = 12 + 3 \cdot 16 \div 16 - 5 = 12 + 48 \div 16 - 5$
$= 12 + 3 - 5 = 15 - 5 = 10$ **33.** $5[20 - 3(4 - 6) + 5] = 5[20 - 3(-2) + 5] = 5[20 - (-6) + 5] = 5[26 + 5]$

$= 5[31] = 155$ **41.** $\left[\dfrac{10 + (-2)}{2(-1)} \right]\left[\dfrac{(-10)(-4)}{(-2)} \right] = \left[\dfrac{8}{(-2)} \right]\left[\dfrac{40}{(-2)} \right] = (-4)(-20) = 80$

Exercise 1–5

Answers to odd-numbered problems

1. 3 terms, polynomial **3.** 3 terms, polynomial **5.** 1 term, not a polynomial since a variable appears under the radical symbol
7. $4, -7, 1$ **9.** $5, 1, -7$ **11.** trinomial, 3rd degree **13.** binomial, 4th degree **15.** monomial, 5th degree **17.** -12 **19.** 60
21. 31 **23.** -7 **25.** -22 **27.** $Q(2) = 2, Q(-3) = 22, Q(0) = 4$ **29.** $P(1) = 0, P(-2) = -9, P(0) = -1$ **31.** 6

33. 31 **35.** -42 **37.** 4 **39.** -9 **41.** 60 **43.** 55 **45.** 83 **47.** 144 **49.** $\dfrac{79}{19}$ or $4\dfrac{3}{19}$ **51.** 12 **53.** 21 **55.** 38 m

57. $426\dfrac{2}{3}$ rpm **59.** 2.5 amperes **61.** $3x$ **63.** $x + 8$ **65.** $4(x + 7)$ **67.** $x - 4$ **69.** Let $n =$ the number; $n - 15$.

71. Let $n =$ the number; $8n + 14$. **73.** Let $n =$ the number; $4(n + 3)$.

Solutions to trial exercise problems

4. $\dfrac{4a^2 - b^2}{10}$ has one term since the fraction bar is a grouping symbol, and it is a polynomial because division by a constant is allowed.

8. 3 is the numerical coefficient of x^3, -2 is the numerical coefficient of x^2, and 1 is understood to be the numerical coefficient of x.
19. $[4(\ ) + 3(\ )][(\ ) - 2(\ )] = [4(-3) + 3(2)][(-2) - 2(4)] = [(-12) + 6][(-2) - 8] = [(-12) + 6][(-2) + (-8)]$
$= (-6)(-10) = 60$ **24.** $[3(\ ) - 2(\ )] - [2(\ ) + (\ )][(\ ) + (\ )] = [3(-3) - 2(2)] - [2(-3) + (2)][(-2) + (4)]$
$= [(-9) - (4)] - [(-6) + (2)][2] = [(-9) + (-4)] - [-4][2] = [-13] - [-8] = [-13] + [8] = -5$ **36.** $P[Q(2)]$
$= P[2(2) - 1] = P(4 - 1) = P(3) = (3)^2 + 2(3) + 1 = 9 + 2(3) + 1 = 9 + 6 + 1 = 15 + 1 = 16$ **42.** $I = (\ )(\ )(\ )$
$= (2,000)(0.09)(3) = 540$

49. $A = \dfrac{(\ )(\ ) + (\ )(\ )}{(\ ) + (\ )} = \dfrac{(80)(3) + (110)(5)}{(80) + (110)} = \dfrac{240 + 550}{190} = \dfrac{790}{190} = \dfrac{79}{19}$ or $4\dfrac{3}{19}$

Chapter 1 review exercises

1. $\{c,o,l,e,g\}$ **2.** $\{January, February, March\}$ **3.** $\emptyset$ **4.** $B \subseteq D$ **5.** $\emptyset \subseteq A$ **6.** $5 \in C$ **7.** $B \not\subseteq C$ **8.** true **9.** true
10. false **11.** $\{1,2,3,4,5,6\}$ **12.** $\{2,3,4\}$ **13.** $\emptyset$ **14.** $<$ **15.** $>$ **16.** $<$ **17.** -13 **18.** 8 **19.** -3 **20.** -9
21. 6 **22.** -10 **23.** 5 **24.** -6 **25.** -3 **26.** 0 **27.** -16 **28.** 6 **29.** -5 **30.** additive inverse property
31. distributive property of multiplication over addition **32.** multiplicative inverse property **33.** associative property of multiplication
34. $11a$ **35.** $14x$ **36.** $9b$ **37.** $8y$ **38.** $8x + 17$ **39.** $4a - 4$ **40.** $5x - 2$ **41.** $11c - 6$ **42.** 2 **43.** 4 **44.** -4

45. 29 **46.** -71 **47.** -72 **48.** $-\dfrac{16}{3}$ **49.** 12 **50.** $\dfrac{31}{24}$ or $1\dfrac{7}{24}$ **51.** 13.5 **52.** 9.79 **53.** 4 terms, polynomial **54.** 1

term, polynomial **55.** -8 **56.** 1 **57.** -60 **58.** -364 **59.** (a) 5, (b) -1, (c) 5, (d) 11 **60.** 70 **61.** 75 **62.** $\dfrac{12}{5}$ or $2\dfrac{2}{5}$

63. $y + 4$ **64.** $a - 12$ **65.** $7(a + 6)$ **66.** $5x - 3$ **67.** Let $n =$ the number; $5(n + 12)$. **68.** Let $n =$ the number; $\dfrac{n + 4}{8}$.

Chapter 2

Exercise 2–1

Answers to odd-numbered problems

1. $\{3\}$ **3.** $\left\{\dfrac{5}{2}\right\}$ **5.** $\{6\}$ **7.** $\{3\}$ **9.** $\{12\}$ **11.** $\left\{\dfrac{32}{3}\right\}$ **13.** $\{12\}$ **15.** $\left\{\dfrac{27}{2}\right\}$ **17.** $\{3\}$ **19.** $\{-1\}$ **21.** $\left\{\dfrac{5}{3}\right\}$

23. $\left\{\dfrac{1}{3}\right\}$ **25.** $\left\{\dfrac{15}{14}\right\}$ **27.** $\left\{\dfrac{5}{3}\right\}$ **29.** $\emptyset$ **31.** $\left\{-\dfrac{7}{2}\right\}$ **33.** $\{8\}$ **35.** $\{24\}$ **37.** $\left\{\dfrac{9}{2}\right\}$ **39.** $\{2\}$ **41.** $\left\{\dfrac{31}{3}\right\}$ **43.** $\left\{\dfrac{-6}{7}\right\}$

45. $\{6\}$ **47.** $\{-8.4\}$ **49.** $\{8\}$ **51.** $\emptyset$ **53.** $\ell = 8$ **55.** $t = 4$ years **57.** $115c$ **59.** $\dfrac{18}{h}$ **61.** $50h + 25q + 10d + 5n$

63. $3n, 3n - 8$ **65.** $d + 464 - 5m$ **67.** $w + 1$ **69.** $j + 2$ **71.** $2d + 500$ **73.** $59c + 115b$ **75.** $20(w + 11)$

Solutions to trial exercise problems

18. $4x + 5 = 5$
$4x = 0$
$x = 0$
$S = \{0\}$

44. $5.6z - 22.15 = 24.33$
$5.6z = 46.48$
$z = 8.3$
$S = \{8.3\}$

50. $3(2x - 1) = 6x + 7$
$6x - 3 = 6x + 7$
$-3 = 7$ (false)
$S = \emptyset$

Exercise 2–2

Answers to odd-numbered problems

1. $R = 9$ **3.** $P = 3,000$ **5.** $n = 20$ **7.** $V_1 = 51$ **9.** $t = \dfrac{I}{pr}$ **11.** $m = \dfrac{E}{c^2}$ **13.** $m = \dfrac{F}{a}$ **15.** $b = \dfrac{A}{h}$ **17.** $R = \dfrac{W}{I^2}$

19. $w = \dfrac{P - 2\ell}{2}$ **21.** $g = \dfrac{V - k}{t}$ **23.** $q = \dfrac{D - R}{d}$ **25.** $\ell = \dfrac{px - m}{p}$ **27.** $W = R + 2bc + b^2$ **29.** $a = \dfrac{V + br^2}{r^2}$

31. $d = \dfrac{2S - 2an}{n^2 - n}$ **33.** $g = \dfrac{2Vt - 2S}{t^2}$ **35.** $d = \dfrac{\ell - a}{n - 1}$ **37.** $x = \dfrac{12 - 3y}{2}$ **39.** $y = \dfrac{-3x}{7}$ **41.** $x = \dfrac{5y + 6}{10}$

43. $y = \dfrac{4x - 3}{a}$ **45.** $y = \dfrac{ax + 3a + 4b}{b}$ **47.** $v = \dfrac{2s - gt^2}{2t}$ **49.** $P_1 = \dfrac{nP_2 - P - c}{n}$

Solution to trial exercise problem

26. $R = W - b(2c + b); c$
$R = W - 2bc - b^2$
$2bc + R = W - b^2$
$2bc = W - b^2 - R$
$c = \dfrac{W - b^2 - R}{2b}$

Exercise 2–3

Answers to odd-numbered problems

1. 17 **3.** 58,23 **5.** 13 **7.** 38,26 **9.** 79,47 **11.** 19,5 **13.** 22,23,24 **15.** 27,29,31 **17.** 8,10,12 **19.** 36,9 **21.** 24
23. 11,33,5 **25.** $10,000 at 8%; $5,000 at 6% **27.** $5,000 at 10%; $7,000 at 12% **29.** $12,285.71 at 5%; $5,714.29 at 9%
31. $18,000 at 10% **33.** $10,000 at 14%; $8,000 at 9% **35.** $\ell = 28$ feet, $w = 23$ feet **37.** 6 inches **39.** 7 cm, 14 cm, 17 cm

41. 40 cm³ of 10% solution, 80 cm³ of 4% solution   **43.** 3 ounces of 60% gold, 9 ounces of 80% gold   **45.** 100 3-grain capsules, 100 2-grain capsules   **47.** 10 liters of 60% solution, 20 liters of 30% solution

Solutions to trial exercise problems

30. x = the amount invested at 10%.

$5,000 at 8\%$	more invested at 10%	will be	total at 9%
$5,000(0.08)$	$+ x(0.10)$	$=$	$(5,000 + x)(0.09)$

Solving for x: $5,000(0.08) + 0.10x = (5,000 + x)(0.09)$

$$400 + 0.10x = 450 + 0.09x$$
$$400 + 0.01x = 450$$
$$0.01x = 50$$
$$x = 5,000$$

Therefore $5,000

37. x = the original length of a side. Increasing the original side by 4, the new side would be $x + 4$. The area of a square is found by $A = s^2$. Therefore the equation would be

area of new square	is	original area	increased by 64
$(x + 4)^2$	$=$	x^2	$+ 64$

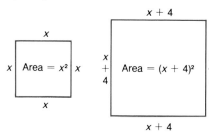

Solving for x: $(x + 4)^2 = x^2 + 64$

$$(x + 4)(x + 4) = x^2 + 64$$
$$x^2 + 4x + 4x + 16 = x^2 + 64$$
$$x^2 + 8x + 16 = x^2 + 64$$
$$8x + 16 = 64$$
$$8x = 48$$
$$x = 6$$

Therefore 6 inches

Exercise 2–4

Answers to odd-numbered problems

1. $[-3,1]$ **3.** $[-5,-1)$

5. $[4,+\infty)$ **7.** $(-\infty,-1]$

9. $\{x \mid x > 9\}$ or $(9,+\infty)$ **11.** $\{x \mid x \geq 12\}$ or $[12,+\infty)$ **13.** $\{x \mid x \geq -5\}$ or $[-5,\infty)$ **15.** $\left\{x \mid x < \dfrac{3}{2}\right\}$ or $\left(-\infty,\dfrac{3}{2}\right)$

17. $\{x \mid x > 1\}$ or $(1,+\infty)$ **19.** $\left\{x \mid x \geq -\dfrac{27}{2}\right\}$ or $\left[\dfrac{-27}{2},+\infty\right)$ **21.** $\left\{x \mid x \geq \dfrac{7}{6}\right\}$ or $\left[\dfrac{7}{6},+\infty\right)$ **23.** $\left\{x \mid x \geq \dfrac{1}{4}\right\}$ or $\left[\dfrac{1}{4},+\infty\right)$ **25.** $\{x \mid x < 3\}$ or $(-\infty,3)$ **27.** $\{x \mid x \geq -1\}$ or $[-1,+\infty)$ **29.** $\{x \mid x \geq 6\}$ or $[6,+\infty)$

31. $\{x \mid x \geq -14\}$ or $[-14,+\infty)$ **33.** $\{x \mid x \leq -1\}$ or $(-\infty,-1]$ **35.** $\left\{x \mid \dfrac{2}{3} < x < 3\right\}$ or $\left(\dfrac{2}{3},3\right)$ **37.** $\left\{x \mid \dfrac{2}{7} \leq x \leq \dfrac{8}{7}\right\}$ or $\left[\dfrac{2}{7},\dfrac{8}{7}\right]$ **39.** $\left\{x \mid \dfrac{-10}{3} \leq x \leq 0\right\}$ or $\left[\dfrac{-10}{3},0\right]$ **41.** $\{x \mid -1 < x < 1\}$ or $(-1,1)$ **43.** $\left\{x \mid 4 \leq x \leq \dfrac{13}{2}\right\}$ or $\left[4,\dfrac{13}{2}\right]$ **45.** $\left\{x \mid 2 \leq x \leq \dfrac{7}{2}\right\}$ or $\left[2,\dfrac{7}{2}\right]$ **47.** $\left\{x \mid \dfrac{-2}{3} \leq x \leq \dfrac{2}{3}\right\}$ or $\left[\dfrac{-2}{3},\dfrac{2}{3}\right]$ **49.** $(x$ = student's score), $x \geq 90$

51. $(t$ = temperature), $t \leq 42$ **53.** $(c$ = cars), $c \geq 10$ **55.** $(m$ = minutes), $96 \leq m \leq 384$ **57.** $\dfrac{3}{2}c \leq P \leq 3c$

59. $5x - 6 < 17$, $\left\{x \mid x < \dfrac{23}{5}\right\}$ or $\left(-\infty, \dfrac{23}{5}\right)$ **61.** $\dfrac{1}{2}x + 16 > 24$, $\{x \mid x > 16\}$ or $(16, +\infty)$ **63.** $19 - 2x \le 8$, $\left\{x \mid x \ge \dfrac{11}{2}\right\}$

or $\left[\dfrac{11}{2}, +\infty\right)$ **65.** $\dfrac{66 + 71 + 84 + x}{4} \ge 75$, minimum score is 79, $\{x \mid x \ge 79\}$ or $[79, +\infty)$ **67.** $12 < 3x + 2 < 23$,

$\left\{x \mid \dfrac{10}{3} < x < 7\right\}$ or $\left(\dfrac{10}{3}, 7\right)$ **69.** $16 < 4s < 84$, $\{s \mid 4 < s < 21\}$ or $(4, 21)$

Solutions to trial exercise problems

6. $(-\infty, 2)$ or

We use a hollow circle or a parenthesis at 2 to denote that we do not contain the endpoint (strict inequality).

20.
$$4 - 2(3x + 1) > 8x - 12$$
$$4 - 6x - 2 > 8x - 12$$
$$2 - 6x > 8x - 12$$
$$2 > 14x - 12$$
$$14 > 14x$$
$$1 > x$$
$$\{x \mid x < 1\} \text{ or } (-\infty, 1)$$

54. If $x = $ the temperature, then $18 \le x \le 41$.

13.
$$-4x \le 20$$
$$\dfrac{-4x}{-4} \ge \dfrac{20}{-4}$$
$$x \ge -5$$
$$\{x \mid x \ge -5\} \text{ or } [-5, +\infty)$$

40.
$$2 \le 1 - x \le 6$$
$$1 \le -x \le 5$$
$$\dfrac{1}{-1} \ge \dfrac{-x}{-1} \ge \dfrac{5}{-1}$$
$$-1 \ge x \ge -5$$
$$\{x \mid -5 \le x \le -1\} \text{ or } [-5, -1]$$

64. If $x = $ the score on the fourth quiz, then
$$\dfrac{7 + 10 + 8 + x}{4} \ge 8$$
$$\dfrac{25 + x}{4} \ge 8$$
$$25 + x \ge 32$$
$$x \ge 7.$$
Therefore she must score 7 or more on the fourth quiz to have an average of 8 or more.

Exercise 2–5

Answers to odd-numbered problems

1. $\{9, -9\}$ **3.** $\{4, -4\}$ **5.** $\{2, -10\}$ **7.** $\left\{\dfrac{-4}{3}, 4\right\}$ **9.** $\emptyset$ **11.** $\left\{-\dfrac{3}{4}, \dfrac{9}{4}\right\}$ **13.** $\{-1, 7\}$ **15.** $\{-24, 4\}$ **17.** $\{3, 15\}$

19. $\left\{0, \dfrac{3}{2}\right\}$ **21.** $\left\{-\dfrac{3}{2}, -\dfrac{1}{2}\right\}$ **23.** $\left\{-\dfrac{2}{3}\right\}$ **25.** $\{x \mid -4 < x < 4\}$, $(-4, 4)$

27. $\{x \mid x \le -2 \text{ or } x \ge 2\}$, $(-\infty, -2] \cup [2, +\infty)$

29. $\{x \mid -3 \le x \le 3\}$, $[-3, 3]$

31. $\{x \mid x < -5 \text{ or } x > 5\}$, $(-\infty, -5) \cup (5, +\infty)$

33. $\left\{x \mid -\dfrac{2}{3} < x < \dfrac{10}{3}\right\}$, $\left(-\dfrac{2}{3}, \dfrac{10}{3}\right)$

35. $\left\{x \mid x \ge 2 \text{ or } x \le -\dfrac{4}{5}\right\}$, $\left(-\infty, -\dfrac{4}{5}\right] \cup [2, +\infty)$ **37.** $\left\{x \mid -3 \le x \le \dfrac{17}{3}\right\}$, $\left[-3, \dfrac{17}{3}\right]$ **39.** $\left\{x \mid \dfrac{-19}{5} < x < 1\right\}$, $\left(\dfrac{-19}{5}, 1\right)$

41. $\left\{x \mid -3 \le x \le \dfrac{17}{3}\right\}$, $\left[-3, \dfrac{17}{3}\right]$ **43.** $\{x \mid -5 < x < 10\}$, $(-5, 10)$ **45.** $\emptyset$ **47.** all real numbers, $\{x \mid x \in R\}$, $(-\infty, +\infty)$

49. $\left\{x \mid \frac{2}{7} < x < \frac{6}{7}\right\}, \left(\frac{2}{7}, \frac{6}{7}\right)$ **51.** $\left\{x \mid x \le -2 \text{ or } x \ge \frac{8}{3}\right\}, (-\infty, -2] \cup \left[\frac{8}{3}, +\infty\right)$ **53.** $\emptyset$ **55.** all real numbers,

$\{x \mid x \in R\}, (-\infty, +\infty)$ **57.** $\left\{x \mid x > 1 \text{ or } x < \frac{1}{5}\right\}, \left(-\infty, \frac{1}{5}\right) \cup (1, +\infty)$ **59.** $\{x \mid x > 6.2 \text{ or } x < 1.475\}, (-\infty, 1.475) \cup (6.2, +\infty)$

61. $\{x \mid -0.5 \le x \le 2\}, [-0.5, 2]$ **63.** $\{x \mid -1 < x < 7\}, (-1, 7)$ **65.** Let x = the number; $|x| \le 6, \{x \mid -6 \le x \le 6\}, [-6, 6]$.
67. Let x = the number; $|2x| < 14, \{x \mid -7 < x < 7\}, (-7, 7)$. **69.** Let x = the number; $|2x + 5| > 15, \{x \mid x > 5 \text{ or } x < -10\}$,
$(-\infty, -10) \cup (5, +\infty)$.

Solutions to trial exercise problems

9. $|5x - 3| = -4$
$\emptyset$ since the absolute value can only be greater than or equal to zero.

18. $|5 - 3x| - 4 = 8$
$|5 - 3x| = 12$

$\begin{array}{lcl} 5 - 3x = 12 & \text{or} & 5 - 3x = -12 \\ -3x = 7 & & -3x = -17 \\ x = -\dfrac{7}{3} & & x = \dfrac{17}{3} \end{array}$

$S = \left\{-\dfrac{7}{3}, \dfrac{17}{3}\right\}$

20. $|3x - 7| = |5x + 3|$

$\begin{array}{lcl} 3x - 7 = 5x + 3 & \text{or} & 3x - 7 = -(5x + 3) \\ -7 = 2x + 3 & & 3x - 7 = -5x - 3 \\ -10 = 2x & & 8x - 7 = -3 \\ -5 = x & & 8x = 4 \end{array}$

$S = \left\{-5, \dfrac{1}{2}\right\}$ $\qquad x = \dfrac{4}{8} = \dfrac{1}{2}$

40. $|1 - 2x| \le 5$
$-5 \le 1 - 2x \le 5$
$-6 \le -2x \le 4$
$\dfrac{-6}{-2} \ge \dfrac{-2x}{-2} \ge \dfrac{4}{-2}$
$3 \ge x \ge -2$
$S = \{x \mid -2 \le x \le 3\}, [-2, 3]$

44. $|4x - 9| < 0$
$\emptyset$. The absolute value cannot be less than zero.

68. Let x = the number, then $|3x - 4| > 11$.
$\begin{array}{lcl} 3x - 4 \ge 11 & \text{or} & 3x - 4 \le -11 \\ 3x \ge 15 & & 3x \le -7 \\ x \ge 5 & & x \le -\dfrac{7}{3} \end{array}$

$S = \left\{x \mid x \le -\dfrac{7}{3}, x \ge 5\right\}, \left(-\infty, -\dfrac{7}{3}\right] \cup [5, +\infty)$

Chapter 2 review exercises

1. $\{8\}$ **2.** $\{6\}$ **3.** $\{28\}$ **4.** $\{4\}$ **5.** $\left\{\dfrac{15}{7}\right\}$ **6.** $\left\{\dfrac{10}{3}\right\}$ **7.** $\left\{\dfrac{5}{3}\right\}$ **8.** $\left\{-\dfrac{13}{5}\right\}$ **9.** $\{9\}$ **10.** $w = \dfrac{v}{\ell h}$ **11.** $t = \dfrac{v - k}{g}$

12. $d = \dfrac{D - R}{q}$ **13.** $b = \dfrac{ar^2 - v}{r^2}$ **14.** $v = \dfrac{2s + gt^2}{2t}$ **15.** $n = \dfrac{\ell - a + d}{d}$ **16.** $x = \dfrac{3y}{2}$ **17.** 12 **18.** 36 **19.** $\dfrac{33}{5}, \dfrac{99}{5}, \dfrac{3}{5}$

20. 11 feet, 30 feet **21.** $15,000 at 10%; $9,000 at 8% **22.** 40 cl of 42% solution, 60 cl of 12% solution **23.** $\{x \mid x \le 6\}, (-\infty, 6]$

24. $\{x \mid x > 16\}, (16, +\infty)$ **25.** $\left\{x \mid x > -\dfrac{9}{2}\right\}, \left(-\dfrac{9}{2}, +\infty\right)$ **26.** $\left\{x \mid x \le \dfrac{9}{8}\right\}, \left(-\infty, \dfrac{9}{8}\right]$ **27.** $\left\{x \mid x > -\dfrac{3}{2}\right\}, \left(-\dfrac{3}{2}, +\infty\right)$

28. $\left\{x \mid x \le \dfrac{25}{3}\right\}, \left(-\infty, \dfrac{25}{3}\right]$ **29.** $\left\{x \mid -\dfrac{7}{2} < x < 1\right\}, \left(-\dfrac{7}{2}, 1\right)$ **30.** $\left\{x \mid -\dfrac{4}{5} \le x \le 0\right\}, \left[-\dfrac{4}{5}, 0\right]$

31. $\{x \mid -5 < x < -2\}, (-5, -2)$ **32.** $\left\{x \mid \dfrac{2}{3} < x \le \dfrac{10}{3}\right\}, \left(\dfrac{2}{3}, \dfrac{10}{3}\right]$ **33.** $\{x \mid -1 \le x < 0\}, [-1, 0)$ **34.** Let x = the number;

$4x - 5 \ge 19, \{x \mid x \ge 6\}, [6, +\infty)$. **35.** Let x = the number; $22 < 3x + 7 < 34, \{x \mid 5 < x < 9\}, (5, 9)$. **36.** $\{-15, 15\}$

37. $\left\{-\dfrac{17}{3}, \dfrac{7}{3}\right\}$ **38.** $\left\{-\dfrac{3}{2}, \dfrac{17}{2}\right\}$ **39.** $\{0, 3\}$ **40.** $\left\{-7, -\dfrac{1}{5}\right\}$ **41.** $\left\{-\dfrac{11}{2}, -\dfrac{1}{6}\right\}$ **42.** $\{x \mid x \ge 10 \text{ or } x \le -10\}$,

$(-\infty, -10] \cup [10, +\infty)$ **43.** $\left\{x \mid -\dfrac{11}{2} < x < \dfrac{1}{2}\right\}, \left(-\dfrac{11}{2}, \dfrac{1}{2}\right)$ **44.** $\left\{x \mid -\dfrac{6}{5} \le x \le \dfrac{8}{5}\right\}, \left[-\dfrac{6}{5}, \dfrac{8}{5}\right]$

45. $\left\{x \mid x > \dfrac{1}{2} \text{ or } x < -4\right\}, (-\infty, -4) \cup \left(\dfrac{1}{2}, +\infty\right)$ **46.** $\left\{x \mid x \le -\dfrac{4}{3} \text{ or } x \ge 2\right\}, \left(-\infty, -\dfrac{4}{3}\right] \cup [2, +\infty)$

47. $\left\{x \mid -\dfrac{9}{4} < x < \dfrac{15}{4}\right\}, \left(-\dfrac{9}{4}, \dfrac{15}{4}\right)$ **48.** all real numbers, $\{x \mid x \in R\}, (-\infty, +\infty)$ **49.** $\left\{x \mid -\dfrac{3}{2} < x < \dfrac{5}{2}\right\}, \left(-\dfrac{3}{2}, \dfrac{5}{2}\right)$

50. $\left\{x \mid x \ge 2 \text{ or } x \le -\dfrac{9}{2}\right\}, \left(-\infty, -\dfrac{9}{2}\right] \cup [2, +\infty)$ **51.** $\emptyset$

Chapter 2 cumulative test

1. false   **2.** false   **3.** true   **4.** true   **5.** true   **6.** true   **7.** true   **8.** 24   **9.** -1 **10.** 12   **11.** 0   **12.** 37

13. -6 **14.** -49 **15.** commutative property of multiplication   **16.** commutative property of multiplication   **17.** $\{1,2,3,4,5\}$

18. $\{4,6,8,9,10,11\}$ **19.** $\{10,12\}$ **20.** $\left\{\dfrac{7}{4}\right\}$ **21.** $P = \dfrac{-M}{\ell - x}$ or $\dfrac{M}{x - \ell}$ **22.** $P_2 = \dfrac{P + nP_1 + c}{n}$ **23.** $\{x \mid x \ge -5\}$

24. $\left\{x \mid x \ge -\dfrac{1}{4} \text{ or } x \le -\dfrac{11}{4}\right\}$ **25.** $\left\{\dfrac{21}{17}\right\}$ **26.** $\left\{x \mid x \ge \dfrac{15}{4}\right\}$ **27.** $\{-1,1\}$ **28.** $\left\{\dfrac{16}{7}\right\}$ **29.** $\emptyset$ **30.** $\left\{x \mid -1 \le x \le \dfrac{11}{3}\right\}$

31. 16,8,48   **32.** $26,000 at 11%; $14,000 at 8%

Chapter 3

Exercise 3–1

Answers to odd-numbered problems

1. $-x^2 + 8x$ **3.** $-5a^3 + a^2 + 5a$ **5.** $4x^2y + 3xy$ **7.** $-x^2y + 12xy$ **9.** $3b^2$ **11.** $-4x + 22yz - 26z$

13. $-9a^2b + 13ab - 16c$ **15.** $-28x + 37y - 3z$ **17.** $10x^2 - 7x - 7$ **19.** $2y^2 - xy - 3yz$ **21.** $4x^2 - 3xy - 2y^2$

23. $x^2y - 2x^2y^2 - xy + 8xy^2$ **25.** $-2a^2b^2 + ab + 3$ **27.** $2x^2 + 5x - 8$ **29.** $8a - 10$ **31.** $9x^2 + 6x + 3$

33. $-9y^2 + 16y + 8$ **35.** $12a^2 - 7$ **37.** $-3a^2 + 13a - 15$ **39.** $7x^2 - 10xy$ **41.** $-x - 16$ **43.** $3t - 2$ **45.** $2a - 5$

47. $2x - y$ **49.** $-9x + 6y$ **51.** $-3x - 8y$ **53.** $2a + b$ **55.** $11x$ **57.** $2a - 2b - 3c$ **59.** $8x^2 - 7x$

61. $-8x^2 + 10x + 7$ **63.** $-6x^2y^2 + 3x^2y - 2xy^2$ **65.** $5x^2 - 2x$ **67.** $-5x^2 + 2x$ **69.** $-5x^2 + 2x$ **71.** $-5x^2 + 2x$

Solutions to trial exercise problems

13. $-(5a^2b - 6ab + 16c) + (7ab - 4a^2b) = -5a^2b + 6ab - 16c + 7ab - 4a^2b = (-5a^2b - 4a^2b) + (6ab + 7ab) - 16c$
$= -9a^2b + 13ab - 16c$ **23.** $\begin{matrix}(3x^2y - 2x^2y^2 + 3xy^2) \\ - (2x^2y + \quad xy - 5xy^2)\end{matrix} = \begin{matrix}3x^2y - 2x^2y^2 \quad + 3xy^2 \\ -2x^2y \qquad\quad - xy + 5xy^2 \\ \hline x^2y \quad 2x^2y^2 - xy + 8xy^2\end{matrix}$

27. $(5x^2 + 3x - 7) - (3x^2 - 2x + 1) = 5x^2 + 3x - 7 - 3x^2 + 2x - 1 = 2x^2 + 5x - 8$ **32.** $(6t^2 - 7t + 14) - (8t^2 - 11t + 6)$
$= 6t^2 - 7t + 14 - 8t^2 + 11t - 6 = -2t^2 + 4t + 8$ **36.** $[(6x^2 + 3x) + (5x^2 - 4x + 2)] - (2x^2 - 9x + 4)$
$= [6x^2 + 3x + 5x^2 - 4x + 2] - (2x^2 - 9x + 4) = (11x^2 - x + 2) - (2x^2 - 9x + 4) = 11x^2 - x + 2 - 2x^2 + 9x - 4$
$= 9x^2 + 8x - 2$ **40.** $[(8a^2 + 11) + (2a^2 - 7a + 6)] - (4a^2 - a - 1) = [8a^2 + 11 + 2a^2 - 7a + 6] - (4a^2 - a - 1)$
$- [10u^3 - 7u + 17]$ $(4u^3 \quad u \quad 1) \quad 10u^3 \quad 7u + 17 \quad 4a^2 + a + 1 - 6a^2 = 6a + 18$
49. $-(3x - 2y) - (x + y) - [2x - y + (3x - 4y)] = -3x + 2y - x - y - [2x - y + 3x - 4y] = -4x + y - [5x - 5y]$
$= -4x + y - 5x + 5y = -9x + 6y$ **54.** $-\{5x - 3y + [2x - (5x - 7y)] + 4y\} = -\{5x - 3y + [2x - 5x + 7y] + 4y\}$
$= -\{5x - 3y + [-3x + 7y] + 4y\} = -\{5x - 3y - 3x + 7y + 4y\} = -\{2x + 8y\} = -2x - 8y$
60. $-[-3a^2 + (4a - 3) + 5a] - \{7a^2 + [4a - (a^2 - 3) + 5a^2]\} = -[-3a^2 + 4a - 3 + 5a] - \{7a^2 + [4a - a^2 + 3 + 5a^2]\}$
$= -[-3a^2 + 9a - 3] - \{7a^2 + [4a^2 + 4a + 3]\} = 3a^2 - 9a + 3 - \{7a^2 + 4a^2 + 4a + 3\} = 3a^2 - 9a + 3 - \{11a^2 + 4a + 3\}$
$= 3a^2 - 9a + 3 - 11a^2 - 4a - 3 = -8a^2 - 13a$ **66.** $(2x^2 - x + 3) - [(5x - 4) + (3x^2 + 4x - 7)]$
$= 2x^2 - x + 3 - [5x - 4 + 3x^2 + 4x - 7] = 2x^2 - x + 3 - [3x^2 + 9x - 11] = 2x^2 - x + 3 - 3x^2 - 9x + 11 = -x^2 - 10x + 14$

Exercise 3–2

Answers to odd-numbered problems

1. $(-2)^4$, -2 base, 4 exponent   **3.** x^5, x base, 5 exponent   **5.** $(2x)^4$, $2x$ base, 4 exponent   **7.** $(x^2 + 3y)^3$, $x^2 + 3y$ base, 3 exponent
9. a^9 **11.** y^3 **13.** $(-2)^6 = 64$ **15.** $(-2)^4 = 16$ **17.** -36 **19.** x^8 **21.** x^7y^5 **23.** $6a^3b^2$ **25.** $6x^5$ **27.** $24x^3y^9$
29. $2^6 = 64$ **31.** -729 **33.** 64 **35.** a^{12} **37.** $x^{10}y^5z^{15}$ **39.** $49s^8t^4$ **41.** $x^{36}y^{48}z^{32}$ **43.** a^9b^{17} **45.** $x^{13}y^{24}$ **47.** $-x^{10}y^{18}$
49. $-75x^{10}y^{11}$ **51.** $17x^9$ **53.** $-76a^{13}$ **55.** $9x^{10} - 8x^8$ **57.** $24a^{13} + 18a^{11}$ **59.** $3x^{17} + 96x^{14}$ **61.** a^{9b} **63.** a^{9b}
65. $a^{5b + 3}$ **67.** $x^{3y + 5}$ **69.** x^{15y^2} **71.** $x^{3n^2 + 3n}$ **73.** $x^{n^2 + 3n + 2}$ **75.** \$5,634.13   **77.** 2.5 grams

Solutions to trial exercise problems

16. $(-2)(-2^2) = (-2)(-4) = 8$ **30.** $(-2^2)^3 = (-4)^3 = -64$ **42.** $(3x^2y)^2(2xy^3) = 9x^4y^2 \cdot 2xy^3 = 18x^5y^5$
50. $(2a^2)^2a^3 + (3a)^3a^4 = 4a^4a^3 + 27a^3a^4 = 4a^7 + 27a^7 = 31a^7$ **55.** $(3x^5)^2 - (2x^2)^3x^2 = 9x^{10} - 8x^6x^2 = 9x^{10} - 8x^8$.
The subtraction cannot be performed because we do not have similar terms.   **60.** $x^{5n} \cdot x^{4n} = x^{5n + 4n} = x^{9n}$ **64.** $x^{2n + 1} \cdot x^{n + 4}$
$= x^{(2n + 1) + (n + 4)} = x^{2n + 1 + n + 4} = x^{3n + 5}$ **68.** $(a^{3n})^{4n} = a^{3n \cdot 4n} = a^{12n^2}$ **71.** $(x^{3n})^{n + 1} = x^{3n(n + 1)} = x^{3n^2 + 3n}$

Exercise 3–3

Answers to odd-numbered problems

1. $2a^3 - 3a^2 + 4a$ **3.** $-6y^3 + 10y^2 - 8y$ **5.** $12a^4 - 6a^3b + 9a^2b^2$ **7.** $30x^3y^2 - 24x^2y^4 + 12x^2y^2$
9. $-10x^5y^5 - 35x^4y^{10} + 15x^5y^9$ **11.** $a^2 + 8a + 15$ **13.** $b^2 - b - 20$ **15.** $2x^2 + xy - y^2$ **17.** $10x^2 + 3xy - y^2$
19. $42x^2 - 2xy - 20y^2$ **21.** $x^2 + 6x + 9$ **23.** $9x^2 + 6xy + y^2$ **25.** $16x^2 + 24xy + 9y^2$ **27.** $4a^2 - 20a + 25$
29. $16x^2 - 24xy + 9y^2$ **31.** $x^2 - 9y^2$ **33.** $25a^2 - 4b^2c^2$ **35.** $x^3 - 2x^2y - xy^2 + 2y^3$ **37.** $15a^3 - 31a^2b + 23ab^2 - 6b^3$
39. $x^4 + x^3 - 5x^2 - 17x - 12$ **41.** $5b^4 + 2b^3 + 19b^2 - 2b + 21$ **43.** $x^3 + 6x^2y + 12xy^2 + 8y^3$
45. $64a^3 - 48a^2b + 12ab^2 - b^3$ **47.** $2a^3 - a^2b - 8ab^2 + 4b^3$ **49.** $2a^2 + 6a + 29$ **51.** $2x^2 + 21x - 26$ **53.** $y^2 + y + 8$
55. $-6b + 9$ **57.** $-14x^2 - 4x$ **59.** $26x - 28$ **61.** $a^{n + 2} + 3a^n$ **63.** $3x^{2n} + x^n$ **65.** $x^{2n + 3} - x^n$ **67.** $b^{2n} - 5b^n + 6$
69. $9x^{2n} + 6x^ny^n + y^{2n}$ **71.** $a^{2n} - b^{6n}$ **73.** $\pi R^2 - \pi r^2$ **75.** $2\pi rh + 2\pi r^2$
77. $\dfrac{Wx^4}{24EI} - \dfrac{\ell Wx^3}{16EI} - \dfrac{\ell^3 Wx}{48EI}$

Solutions to trial exercise problems

8. $-a^3b(3a^2b^5 - ab^4 - 7a^2b) = -3a^5b^6 + a^4b^5 + 7a^5b^2$ **30.** $(a - 3b)(a + 2b) = a^2 + 2ab - 3ab - 6b^2 = a^2 - ab - 6b^2$
38. $(x^2 - 2x + 1)(x^2 + 3x + 2) = x^4 + 3x^3 + 2x^2 - 2x^3 - 6x^2 - 4x + x^2 + 3x + 2 = x^4 + x^3 - 3x^2 - x + 2$
42. $(a - 3b)^3 = (a - 3b)(a - 3b)(a - 3b) = [(a - 3b)(a - 3b)](a - 3b) = [a^2 - 3ab - 3ab + 9b^2](a - 3b)$
$= [a^2 - 6ab + 9b^2](a - 3b) = a^3 - 3a^2b - 6a^2b + 18ab^2 + 9ab^2 - 27b^3 = a^3 - 9a^2b + 27ab^2 - 27b^3$
56. $-2[5a - (2a + 3) - 3(3a + 7)] = -2[5a - 2a - 3 - 9a - 21] = -2[-6a - 24] = 12a + 48$
64. $x^{n + 2}(x^{n + 1} + x) = x^{n + 2 + n + 1} + x^{n + 2 + 1} = x^{2n + 3} + x^{n + 3}$ **68.** $(2a^n - b^n)^2 = (2a^n - b^n)(2a^n - b^n)$
$= 4a^{n + n} - 2a^nb^n - 2a^nb^n + b^{n + n} = 4a^{2n} - 4a^nb^n + b^{2n}$

Exercise 3–4

Answers to odd-numbered problems

1. 1 **3.** 1 **5.** 7 **7.** $\dfrac{1}{a^4}$ **9.** $\dfrac{3}{a^2}$ **11.** $\dfrac{a}{b^4c^3}$ **13.** $\dfrac{x^5}{2}$ **15.** $\dfrac{1}{a^8}$ **17.** $4x^3y^6$ **19.** $\dfrac{1}{27x^3}$ **21.** $\dfrac{1}{4}$ **23.** $-\dfrac{1}{81}$ **25.** $\dfrac{1}{4}$
27. $\dfrac{1}{x^3}$ **29.** $\dfrac{20}{x^5}$ **31.** $\dfrac{10b}{a}$ **33.** $\dfrac{1}{9x^8}$ **35.** $\dfrac{a^9}{64b^{12}}$ **37.** $\dfrac{8}{a^6}$ **39.** $\dfrac{b^9}{27a^6}$ **41.** $\dfrac{27a^6}{b^3}$ **43.** $\dfrac{27x^{12}y^{15}}{z^{27}}$ **45.** $\dfrac{a^{12}}{8b^{21}}$ **47.** $\dfrac{a^6}{b^7}$
49. $\dfrac{x^2}{y^3}$ **51.** $\dfrac{4}{x^4}$ **53.** $\dfrac{18}{a^7b^3}$ **55.** $\dfrac{1}{x^4z^8}$ **57.** $\dfrac{27x^9z^6}{y^{15}}$ **59.** $8a^3b^3$ **61.** $\dfrac{b^{13}c^8}{a^{12}}$ **63.** $\dfrac{y^6}{x^8z^8}$ **65.** a^{3b-5} **67.** $x^{6n - 8}$ **69.** $a^{n + 4}$
71. $a^{n + 1}b^{2n + 3}$ **73.** $x^{-3n + 3}$ **75.** 1.55×10^5 **77.** 8.63×10^{-2} **79.** 8.06×10^{21} **81.** 5.787×10^{-4} **83.** 2.2046×10^{-3}
85. 1.102×10^{-3} **87.** 5.872×10^9 **89.** -0.0437 **91.** $4,990,000$ **93.** $48,300$ **95.** 3.17×10^{11} **97.** 1.25×10^{-9}
99. 1.83×10^1 **101.** 1.19×10^{14} **103.** 2.52×10^{-6} **105.** 2.74×10^{-2} **107.** 2.73×10^{-6} **109.** 5.93×10^5
111. 3.18×10^9

Solutions to trial exercise problems

9. $3a^{-2} = 3 \cdot \dfrac{1}{a^2} = \dfrac{3}{a^2}$ **12.** $\dfrac{1}{4a^{-3}} = \dfrac{1}{4 \cdot \dfrac{1}{a^3}} = \dfrac{1}{\dfrac{4}{a^3}} = 1 \cdot \dfrac{a^3}{4} = \dfrac{a^3}{4}$ **22.** $-2^{-2} = -(2^{-2}) = -\left(\dfrac{1}{2^2}\right) = -\dfrac{1}{4}$

28. $(2a^{-3})(3a^{-5}) = 2 \cdot 3 \cdot a^{-3 + (-5)} = 6a^{-8} = 6 \cdot \dfrac{1}{a^8} = \dfrac{6}{a^8}$ **44.** $\left(\dfrac{2x^5y^2}{4x^3y^7}\right)^4 = \left(\dfrac{x^2}{2y^5}\right)^4 = \dfrac{(x^2)^4}{(2y^5)^4} = \dfrac{x^8}{2^4(y^5)^4} = \dfrac{x^8}{16y^{20}}$

51. $\left(\dfrac{3x^{-2}}{12x^{-4}}\right)\left(\dfrac{16x^{-5}}{x}\right) = \left(\dfrac{1}{4} \cdot x^{(-2)-(-4)}\right)(16x^{(-5)-1}) = \left(\dfrac{1}{4}x^2\right)(16x^{-6}) = \dfrac{x^2}{4} \cdot \dfrac{16}{x^6} = \dfrac{4}{x^4}$

58. $(3x^2y^{-2})^2 \, (x^{-4}y^3)^3 = \left(\dfrac{3x^2}{y^2}\right)^2\left(\dfrac{y^3}{x^4}\right)^3 = \dfrac{3^2(x^2)^2}{(y^2)^2} \cdot \dfrac{(y^3)^3}{(x^4)^3} = \dfrac{9x^4}{y^4} \cdot \dfrac{y^9}{x^{12}} = \dfrac{9y^5}{x^8}$ **89.** $-4.37 \times 10^{-2} = -0.0437$

Exercise 3–5

Answers to odd-numbered problems

1. $2^2 \cdot 3$ **3.** $2^3 \cdot 3$ **5.** $2^3 \cdot 7$ **7.** prime   **9.** $3 \cdot 13$ **11.** $-(-2a + 3b)$ **13.** $-3(-a^3 + 3b^2)$
15. $-4(x^3 + 9xy - 4xy^2)$ **17.** $-3a^2b^2(-a - 4b - 5a^2)$ **19.** $-3x(-x + 3y)$ **21.** $-8R(3S + 2 - 4R)$
23. $5(x^2 + 2xy - 4y)$ **25.** $3a(6b - 9 + c)$ **27.** $3(5R^2 - 7S^2 + 12T)$ **29.** $3RS(R - 2S + 4)$ **31.** $4x^3y(3xy^2 - 2xy + 4)$
33. $3xy(x - 2y^3 + 5x^2y)$ **35.** $(3a + b)(x - y)$ **37.** $7(2b - 1)(a + c)$ **39.** $4a(b - 2c)(2x + y)$ **41.** $(5a - 1)(b + 3)$
43. $x^n(y^2 + z)$ **45.** $x^ny^n(x^ny^n - 1)$ **47.** $y^3(y^n - 1)$ **49.** $(a + 3b)(2x - y)$ **51.** $(2a + 3b)(x + 4y)$ **53.** $(2a - b)(x^2 + 3)$
55. $(c - 2d)(3a + b)$ **57.** $(2x^2 + 3)(x + 5)$ **59.** $(2x - 1)(4x^2 + 3)$ **61.** $\pi r(s + r)$ **63.** $P(1 + r)$

Solutions to trial exercise problems

4. $28 = 2 \cdot 2 \cdot 7 = 2^2 \cdot 7$ **18.** $-ab^2 - ac^2 = (-a)b^2 + (-a) \cdot c^2 = -a(b^2 + c^2)$ **36.** $5a(3x - 1) + 10(3x - 1)$
$= \underline{5(3x - 1)} \cdot a + \underline{5(3x - 1)} \cdot 2 = 5(3x - 1)(a + 2)$ **46.** $x^{n+4} + x^4 = x^4 \cdot x^n + x^4 = x^4(x^n + 1)$ **53.** $2ax^2 - 3b - bx^2 + 6a$
$= 2ax^2 - bx^2 + 6a - 3b = (2ax^2 - bx^2) + (6a - 3b) = x^2(2a - b) + 3(2a - b) = (2a - b)(x^2 + 3)$

Exercise 3–6

Answers to odd-numbered problems

1. $-4, -1$ **3.** $-5, -4$ **5.** $8, 8$ **7.** $-9, 9$ **9.** $-8, 3$ **11.** $6, 10$ **13.** $(3a - 2)(a + 3)$ **15.** $(4x - 1)(x - 1)$
17. $(x + 2)(x + 16)$ **19.** $(2y + 1)(y - 1)$ **21.** $(8x - 1)(x - 2)$ **23.** $(2a - 3)(a - 4)$ **25.** $(2x + 9)(x + 2)$
27. $(7x - 1)(x + 3)$ **29.** $(2x + 1)(2x - 3)$ **31.** $(6x + 1)(x - 4)$ **33.** $(2z + 1)(5z + 2)$ **35.** will not factor
37. $(5x + 1)(x - 2)$ **39.** $(3x - 4)(2x - 3)$ **41.** $3(x + 2)(x + 2) = 3(x + 2)^2$ **43.** will not factor   **45.** $2(3x - 4)(x - 5)$
47. $(2x + 3)(-2x - 3) = -(2x + 3)^2$ **49.** $(-7x + 1)(x - 5)$ or $(7x - 1)(-x + 5)$ or $-(7x - 1)(x - 5)$
51. $(3x + 4)(4x - 1)$ **53.** $2x(x + 2)(2x + 1)$ **55.** $2(3x - 8)(3x + 1)$ **57.** $(4x - 3)(2x - 3)$ **59.** $(x - 10)(x + 8)$
61. $4x(x - 3)(x - 3) = 4x(x - 3)^2$

Solutions to trial exercise problems

2. $x^2 + 10x + 25$; $a = 1, b = 10, c = 25$.
  Then $m \cdot n = a \cdot c = 1 \cdot 25 = 25$ and $m + n$
  $= b = 10$. m and n are 5 and 5.

7 $a^2 - 81$; $a = 1, b = 0, c = -81$. Then
  $m \cdot n = a \cdot c = 1 \cdot -81 = -81$ and $m + n = b = 0$.
  m and n are 9 and -9.

25	+
$25 \cdot 1$	$25 + 1 = 26$
$5 \cdot 5$	$5 + 5 = 10$

81	−
$81 \cdot 1$	$81 - 1 = 80$
$27 \cdot 3$	$27 - 3 = 24$
$9 \cdot 9$	$9 - 9 = 0$

38. $6x^2 + 21x + 18 = 3(2x^2 + 7x + 6)$; m and n are 3 and 4. $3[2x^2 + 3x + 4x + 6] = 3[(2x^2 + 3x) + (4x + 6)]$
$= 3[x(2x + 3) + 2(2x + 3)] = 3(2x + 3)(x + 2)$ **47.** $-4x^2 - 12x - 9$; m and n are -6 and -6. $-4x^2 - 6x - 6x - 9$
$= (-4x^2 - 6x) + (-6x - 9) = -2x(2x + 3) - 3(2x + 3) = (2x + 3)(-2x - 3)$ **53.** $4x^3 + 10x^2 + 4x = 2x(2x^2 + 5x + 2)$
m and n are 4 and 1. $2x(2x^2 + 4x + x + 2) = 2x[(2x^2 + 4x) + (x + 2)] = 2x[2x(x + 2) + 1(x + 2)] = 2x(x + 2)(2x + 1)$

Exercise 3–7

Answers to odd-numbered problems

1. $(x + 15)(x - 2)$ **3.** $(y + 8)(y - 3)$ **5.** $(x - 6)(x + 4)$ **7.** $3(x + 12)(x - 1)$ **9.** $4(x - 3)(x + 2)$

11. $3(y^2 - 9y + 4)$ **13.** $(xy + 3)(xy - 7)$ **15.** $(xy + 1)(xy + 12)$ **17.** $(xy - 2)(xy - 12)$ **19.** $4(ab + 4)(ab + 2)$

21. $-2(xy - 5)(xy + 2)$ **23.** $-4(ab + 3)(ab + 2)$ **25.** $(3xy - 2)(xy + 3)$ **27.** $(2ab - 3)(ab - 2)$

29. $(2xy + 1)(xy - 1)$ **31.** $(3xy + 2)(2xy + 3)$ **33.** $(x - 2y + 9)(x - 2y + 1)$ **35.** $(y + 3z - 7)(y + 3z + 2)$

37. $(R + 5S - 4)(R + 5S - 2)$ **39.** $(x + 5)^2$ **41.** $(b - 6)^2$ **43.** $(c + 9)^2$ **45.** $(3y + 2)^2$ **47.** $(x - 7y)^2$

49. $(2x - 3y)^2$ **51.** $(3x + 5y)^2$ **53.** $(x + 3y)(x + 8y)$ **55.** $(R + 8S)(R - 3S)$ **57.** $(a + 7b)(a - 5b)$

59. $(x + 8y)(x + 2y)$ **61.** $(4y - z)(y + 3z)$ **63.** $(3R + 2S)(R - 2S)$ **65.** $(a - 2b)(x + 6)(x + 2)$

67. $(y + 2z)(x - 10)(x - 3)$ **69.** $(3y - z)(x + 12)(x + 3)$ **71.** $(3x - y)(a - 12)(a + 5)$ **73.** $(x + 5y)(a - 7)(a - 2)$

75. $(x^n + 7)(x^n + 2)$ **77.** $(x^n - 5)(x^n - 3)$ **79.** $(2x^n + 1)(x^n - 4)$ **81.** $(2x^n - 3)(x^n - 2)$ **83.** $(W - 17)(W - 9)$

85. $(C_2 + 7)(C_2 - 6)$

Solutions to trial exercise problems

10. $5y^2 - 15y - 55 = 5(y^2 - 3y - 11)$; m and n do not exist, therefore $5(y^2 - 3y - 11)$ is the completely factored form.
21. $-2x^2y^2 + 6xy + 20 = -2(x^2y^2 - 3xy - 10)$; m and n are -5 and 2. Then $-2(x^2y^2 - 3xy - 10) = -2(xy - 5)(xy + 2)$.
24. $2a^2b^2 + ab - 6$; m and n are 4 and -3. $2a^2b^2 + 4ab - 3ab - 6 = (2a^2b^2 + 4ab) + (-3ab - 6) = 2ab(ab + 2) - 3(ab + 2)$
$= (ab + 2)(2ab - 3)$ **32.** $(a + b)^2 - 4(a + b) - 12$; m and n are -6 and 2. $[(a + b) - 6][(a + b) + 2] = (a + b - 6)(a + b + 2)$
52. $a^2 + 7ab + 10b^2$; m and n are 5 and 2. $(a + 5b)(a + 2b)$ **65.** $x^2(a - 2b) + 8x(a - 2b) + 12(a - 2b) = (a - 2b)(x^2 + 8x + 12)$.
In the second parentheses m and n are 6 and 2, and we have $= (a - 2b)(x + 6)(x + 2)$.   **74.** $x^{2n} + 5x^n + 6$; m and n are 2 and 3.
$x^{2n} + 2x^n + 3x^n + 6 = (x^{2n} + 2x^n) + (3x^n + 6) = x^n(x^n + 2) + 3(x^n + 2) = (x^n + 2)(x^n + 3)$ *Note:* We could have used our
alternate procedure, but when in doubt, always use the four-step procedure.

Exercise 3–8

Answers to odd-numbered problems

1. $(a + 7)(a - 7)$ **3.** $(8 + S)(8 - S)$ **5.** $(6x + y^2)(6x - y^2)$ **7.** $(4a + 7b)(4a - 7b)$ **9.** $8(a + 2b)(a - 2b)$

11. $2(5 + x)(5 - x)$ **13.** $2(7ab + 5xy)(7ab - 5xy)$ **15.** $(x^2 + 9)(x + 3)(x - 3)$ **17.** $(x^2 + 4y^2)(x + 2y)(x - 2y)$

19. $(a^n + 2)(a^n - 2)$ **21.** $(x^n + y^n)(x^n - y^n)$ **23.** $(x^{2n} + 9)(x^n + 3)(x^n - 3)$ **25.** $(a + 5b)(2x + y)(2x - y)$

27. $3(3x - y)(a + 3)(a - 3)$ **29.** $3(3x - y)(2a + b)(2a - b)$ **31.** $(2a + b + x - 2y)(2a + b - x + 2y)$

33. $(3a - b + 2a - b)(3a - b - 2a + b) = (5a - 2b)(a) = a(5a - 2b)$ **35.** $(x + 2y + x - 3y)(x + 2y - x + 3y)$

$= (2x - y)(5y) = 5y(2x - y)$ **37.** $(3a - b)(9a^2 + 3ab + b^2)$ **39.** $(x + y)(x^2 - xy + y^2)$ **41.** $(a + 2b)(a^2 - 2ab + 4b^2)$

43. $(2y - 3x)(4y^2 + 6xy + 9x^2)$ **45.** $3(3a - b)(9a^2 + 3ab + b^2)$ **47.** $3(x + 2)(x^2 - 2x + 4)$ **49.** $(4z + 5)(16z^2 - 20z + 25)$

51. $2(a - 3b)(a^2 + 3ab + 9b^2)$ **53.** $R^2(R + 4S)(R^2 - 4RS + 16S^2)$ **55.** $3x^2(x - 3y)(x^2 + 3xy + 9y^2)$

57. $(xy^4 + z^3)(x^2y^8 - xy^4z^3 + z^6)$ **59.** $(a^6b^3 - 3c)(a^{12}b^6 + 3a^6b^3c + 9c^2)$ **61.** $3(xy^2 + 3z)(x^2y^4 - 3xy^2z + 9z^2)$

63. $(x + 3y + z)(x^2 + 6xy + 9y^2 - xz - 3yz + z^2)$ **65.** $(4a + b - 3c)(16a^2 + 8ab + b^2 + 12ac + 3bc + 9c^2)$

67. $(a - 2b - 2x - y)(a^2 - 4ab + 4b^2 + 2ax - 4bx + ay - 2by + 4x^2 + 4xy + y^2)$ **69.** $(5x + 2y)(7x^2 - xy + 13y^2)$

71. $9x(x^2 - xy + y^2)$ **73.** (a) $\dfrac{V}{8I}(h + 2v_1)(h - 2v_1)$, (b) $\dfrac{3V}{2A}\left(1 + \dfrac{2V_1}{H}\right)\left(1 - \dfrac{2V_1}{H}\right)$

Solutions to trial exercise problems

9. $8a^2 - 32b^2 = 8(a^2 - 4b^2) = 8[(a)^2 - (2b)^2] = 8(a + 2b)(a - 2b)$ **14.** $x^4 - y^4 = (x^2)^2 - (y^2)^2 = (x^2 + y^2)(x^2 - y^2)$
$= (x^2 + y^2)(x + y)(x - y)$ **18.** $x^{2n} - 1 = (x^n)^2 - (1)^2 = (x^n + 1)(x^n - 1)$ **24.** $a^2(x + 2y) - b^2(x + 2y)$
$= (x + 2y)(a^2 - b^2) = (x + 2y)(a + b)(a - b)$ **31.** $(2a + b)^2 - (x - 2y)^2 = [(2a + b) + (x - 2y)][(2a + b) - (x - 2y)]$
$= (2a + b + x - 2y)(2a + b - x + 2y)$ **44.** $64a^3 - 8 = 8(8a^3 - 1) = 8[(2a)^3 - (1)^3]$. Then $8(\quad - \quad)[(\quad)^2 + (\quad)(\quad) + (\quad)^2]$
and $8(2a - 1)[(2a)^2 + (2a)(1) + (1)^2] = 8(2a - 1)(4a^2 + 2a + 1)$.   **55.** $3x^5 - 81x^2y^3 = 3x^2(x^3 - 27y^3) = 3x^2[(x)^3 - (3y)^3]$.
Then $3x^2(\quad - \quad)[(\quad)^2 + (\quad)(\quad) + (\quad)^2]$ and $3x^2(x - 3y)[(x)^2 + (x)(3y) + (3y)^2] = 3x^2(x - 3y)(x^2 + 3xy + 9y^2)$.
62. $(a + 2b)^3 - c^3$. Then $(\quad - \quad)[(\quad)^2 + (\quad)(\quad) + (\quad)^2]$ and $[(a + 2b) - c][(a + 2b)^2 + (a + 2b)(c) + (c)^2]$
$= (a + 2b - c)(a^2 + 4ab + 4b^2 + ac + 2bc + c^2)$.

Exercise 3–9

Answers to odd-numbered problems

1. $(m - 7)(m + 7)$ **3.** $(x + 5)(x + 1)$ **5.** $(7a + 1)(a + 5)$ **7.** $(2a + 3)(a + 6)$ **9.** $(ab + 4)(ab - 2)$
11. $(3a + b)(9a^2 - 3ab + b^2)$ **13.** $5(3x + y)(5x^2 + a)$ **15.** $10(x - y)^2$ **17.** $4(m - 2n)(m + 2n)$
19. $(a - b - 2x - y)(a - b + 2x + y)$ **21.** $3(x^2 - 3y)(x^4 + 3x^2y + 9y^2)$ **23.** $2xy^2(6x^2 - 9x + 8y^2)$
25. $(2x + 3y)(2x - 3y)$ **27.** $(x + 2y)(3a - b)$ **29.** $(3a^3 - bc)(9a^6 + 3a^3bc + b^2c^2)$ **31.** $(5a + 3)(a - 7)$
33. $(a - 1)(a + 1)(a - 2)(a + 2)$ **35.** $(2a + 3b)(2a - 5b)$ **37.** $(y - 2)(y + 2)(y^2 + 4)$ **39.** $2(a + 2)(2a + 1)$
41. $(x + y - 9)(x + y + 1)$ **43.** $(2a - 1)(3a + 5)$ **45.** $4ab(x + 3y)(1 - 2ab)$ **47.** $(2a - 5b)^2$
49. $5x(4x^2 + 1)(2x + 1)(2x - 1)$ **51.** $3ab(a^2 - 3b^2)^2$ **53.** $(3a - x - 5y)(3a + x + 5y)$ **55.** $(3x - 13)(x + 7)$
57. $3x^2(xy^3 + 3z^2)(x^2y^6 - 3xy^3z^2 + 9z^4)$ **59.** $(3 - x)(3 + x)(a - 3)^2$

Chapter 3 review exercises

1. $13y^2 - 9y$ **2.** $14x^2y + 5x^2y^2 - xy^2$ **3.** $2x^2 - 5x + 11$ **4.** $8a + 4$ **5.** $-10x$ **6.** $-15x^5$ **7.** $6a^3b^5$ **8.** $8a^9b^{12}c^3$
9. $x^8y^9z^2$ **10.** $17a^7$ **11.** $4b^{10} - 27b^8$ **12.** $10a^4b - 15a^3b^2 + 20a^2b^3$ **13.** $2a^2 + ab - b^2$ **14.** $y^2 - 49$

15. $4a^2 + 4ab + b^2$ **16.** $a^3 - 2a^2b - ab^2 + 2b^3$ **17.** $2x^2 + 16x - 17$ **18.** $\dfrac{b^3}{4a^3}$ **19.** $\dfrac{6}{x^9}$ **20.** $\dfrac{9y^6}{x^4}$ **21.** $\dfrac{64x^9}{y^{15}}$ **22.** $\dfrac{a^9}{b^3}$

23. $\dfrac{a^3c^7}{2b}$ **24.** $\dfrac{4}{y^5}$ **25.** $6x^2(2x - 3)$ **26.** $4a^2b(3a^2 - b^2 + 6ab)$ **27.** $(x - 2y)(a + b)$ **28.** $5x(2x + 1)(x - 3z)$
29. $(2x - y)(3a - b)$ **30.** $(2y + x)(4a + 3b)$ **31.** $(2x + y)(2a + 3b)$ **32.** $(a + 3b)(x - 2y)$ **33.** $(2x - y)(3a - b)$
34. $(a + 3b)(2x + y)$ **35.** $(2a - 3)(a - 2)$ **36.** $(4x - 5)(2x - 1)$ **37.** $(2y + 1)(3y - 4)$ **38.** $(3a - 1)(a + 1)$
39. $(4x - 1)(x + 3)$ **40.** $(2x + 1)(2x + 1) = (2x + 1)^2$ **41.** $(6a + 1)(4a + 3)$ **42.** $(4x - 3)(2x - 3)$
43. $(2x + 3)(x + 6)$ **44.** $3(2x + 1)(x + 8)$ **45.** $(-2x - 1)(x - 6)$ or $(2x + 1)(-x + 6)$ or $-(2x + 1)(x - 6)$
46. $(x + 2)(x + 12)$ **47.** $(x - 2)(x - 7)$ **48.** $2a(a - 5)(a + 1)$ **49.** $x(x - 3)(x + 2)$ **50.** $(xy + 6)(xy + 4)$
51. $(ab - 10)(ab + 2)$ **52.** $(4ab + 3)(ab + 2)$ **53.** $(2xy - 1)(2xy - 9)$ **54.** $(x + 3y - 1)(x + 3y + 2)$ **55.** $(x + 2y - 7)^2$
56. $(x + 4)(x - 4)(x + 1)(x - 1)$ **57.** $(x^2 - 5)(x + 2)(x - 2)$ **58.** $(2x + 3)(2x - 3)(x^2 + 4)$ **59.** $(3x - 2y)(x + 3y)$
60. $(a - 2b)(a + b)$ **61.** $(x + 2y)(x + 3y)$ **62.** $(x + 9)(x - 9)$ **63.** $4(x + 3y)(x - 3y)$ **64.** $3(a + 3b)(a - 3b)$
65. $(a^2 + 9)(a + 3)(a - 3)$ **66.** $(a + 2b)(x + y)(x - y)$ **67.** $(x - 3z)(2a + 3b)(2a - 3b)$ **68.** $(a + 2b)(a^2 - 2ab + 4b^2)$
69. $(3x - y)(9x^2 + 3xy + y^2)$ **70.** $8(2a + b)(4a^2 - 2ab + b^2)$ **71.** $x^2(3x + y)(9x^2 - 3xy + y^2)$
72. $2(x - 2b)(x^2 + 2bx + 4b^2)$ **73.** $(x^4y^5 - z^3)(x^8y^{10} + x^4y^5z^3 + z^6)$ **74.** $(3x + 4)^2$ **75.** $(a^2 - b^3)(a^4 + a^2b^3 + b^6)$
76. $3a(a + 5)(a - 5)$ **77.** $(a + 6)^2(3 + x)(3 - x)$ **78.** $(6a - 1)(a + 3)$ **79.** $(x - 2y)(4m + 3n)$
80. $5x(2x + y)(4x^2 - 2xy + y^2)$ **81.** $ab^2(a - 2b)^2$ **82.** $(x + 4y)(x - 2y)$ **83.** $(3a + 2)(5a - 2)$

Chapter 3 cumulative test

1. false **2.** false **3.** true **4.** 2 **5.** undefined **6.** 4 **7.** 0 **8.** 81 **9.** 37 **10.** -34 **11.** $\{x \mid x \geq 7\}$ **12.** $\{5\}$
13. $\left\{-3, \dfrac{1}{3}\right\}$ **14.** $\left\{x \mid -\dfrac{1}{3} < x < \dfrac{17}{3}\right\}$ **15.** $\left\{x \mid x > \dfrac{1}{6} \text{ or } x < \dfrac{-11}{6}\right\}$ **16.** $\left\{x \mid -\dfrac{3}{2} < x < \dfrac{7}{2}\right\}$ **17.** $\left\{x \mid -5 \leq x \leq \dfrac{-3}{2}\right\}$
18. Let $x =$ the number; $\{x \mid x > 9\}$. **19.** Let $r =$ the number; $\{r \mid 3 < r < 7\}$. **20.** $6a^5b^5$ **21.** $16x^{12}v^{16}$ **22.** $\dfrac{3a^3b^5}{2c^6}$

23. $\dfrac{y^4}{x^6}$ **24.** $\dfrac{y^9}{27x^9}$ **25.** $4x^2 - 2x - 3$ **26.** $\dfrac{a^9c^3}{27b^6}$ **27.** $7ab - a - 5b$ **28.** $9a^2 - 6ab + b^2$ **29.** $8a^{18}b^{11}$
30. $12x^6y^3 - 18x^6y^4 + 24x^4y^5$ **31.** $(x + 6)(x - 6)$ **32.** $(x + 3y)(x^2 - 3xy + 9y^2)$ **33.** $(2a - 3)(a - 6)$
34. $(y + 6)(y + 1)$ **35.** $(xy + 2)(xy - 4)$ **36.** $(7x + 1)(x - 5)$ **37.** $(2a + 3b)(x - 3y)$ **38.** $5x(2a - b)(3x + 1)$
39. $(2x - 5y)^2$ **40.** $3(a + 2b)(a - 2b)$

Chapter 4

Exercise 4-1

Answers to odd-numbered problems

1. (a) The domain is the set of all real numbers except 4. (b) domain $= \{x \mid x \in R, x \neq 4\}$ **3.** (a) The domain is the set of all real numbers except -1. (b) domain $= \{x \mid x \in R, x \neq -1\}$ **5.** (a) The domain is the set of all real numbers except $\dfrac{3}{2}$. (b) domain $= \left\{z \mid z \in R, z \neq \dfrac{3}{2}\right\}$ **7.** (a) The domain is the set of all real numbers except $-\dfrac{4}{7}$. (b) domain $= \left\{x \mid x \in R, x \neq -\dfrac{4}{7}\right\}$ **9.** (a) The domain is the set of all real numbers except 0 and 4. (b) domain $= \{x \mid x \in R, x \neq 0,4\}$ **11.** (a) The domain is the set of all real numbers except -4. (b) domain $= \{x \mid x \in R, x \neq -4\}$ **13.** (a) The domain is the set of all real numbers except $-\dfrac{5}{2}$ and $\dfrac{5}{2}$. (b) domain $= \left\{y \mid y \in R, y \neq -\dfrac{5}{2}, \dfrac{5}{2}\right\}$ **15.** (a) The domain is the set of all real numbers except $-\dfrac{1}{2}$ and 3. (b) domain $= \left\{x \mid x \in R, x \neq -\dfrac{1}{2}, 3\right\}$.   **17.** (a) The domain is the set of all real numbers. (b) domain $= \{x \mid x \in R\}$.

19. $\dfrac{5}{3x}$ **21.** $\dfrac{p^2}{q^2}$ **23.** $-\dfrac{ab^3}{c^4}$ **25.** $\dfrac{3n^2p^2}{4m^3}$ **27.** $\dfrac{3}{7}$ **29.** $\dfrac{3}{a-2}$ **31.** $2(x-1)$ **33.** $\dfrac{4(y+1)}{3}$ **35.** -5 **37.** $\dfrac{a-3}{4}$

39. $2y-1$ **41.** x^2-2x+4 **43.** $\dfrac{-3}{2(y^2+xy+x^2)}$ **45.** $\dfrac{y-7}{y+7}$ **47.** $\dfrac{m-6}{m-3}$ **49.** $\dfrac{y-8}{y-5}$ **51.** $\dfrac{a-1}{a+1}$ **53.** $\dfrac{2x-3}{4x+1}$

55. $\dfrac{3m+2}{2m+1}$ **57.** $\dfrac{2(a+2)}{3a+1}$ **59.** $\dfrac{5(y+1)}{4(y+3)}$ **61.** $\dfrac{-(a+6)}{3+a}$ **63.** $\dfrac{p+4q}{p+2q}$

Solutions to trial exercise problems

5. Since $2z-3=0$ when $z=\dfrac{3}{2}$, then the domain is (a) all real numbers except $\dfrac{3}{2}$, (b) $\left\{z \mid z \in R, z \neq \dfrac{3}{2}\right\}$.   **10.** Since $3m^2+6m=3m(m+2)$ and $3m=0$ when $m=0$, $m+2=0$ when $m=-2$, the domain is (a) all real numbers except 0 and -2, (b) $\{m \mid m \in R, m \neq 0, -2\}$.   **13.** Since $4y^2-25=(2y+5)(2y-5)$ and $2y+5=0$ when $y=-\dfrac{5}{2}$, $2y-5=0$ when $y=\dfrac{5}{2}$, the domain is (a) all real numbers except $-\dfrac{5}{2}$ and $\dfrac{5}{2}$, (b) $\left\{y \mid y \in R, y \neq -\dfrac{5}{2}, \dfrac{5}{2}\right\}$.   **17.** Since $x^2+4 \neq 0$ for any value of x, the domain is (a) all real numbers, (b) $\{x \mid x \in R\}$.

30. $\dfrac{-36x}{42x^3+24x} = \dfrac{-36x}{6x(7x^2+4)} = \dfrac{6x \cdot -6}{6x(7x^2+4)} = \dfrac{-6}{7x^2+4}$

35. $\dfrac{5x-5y}{y-x} = \dfrac{5(x-y)}{-(x-y)} = \dfrac{5}{-1} = -5$ **41.** $\dfrac{x^3+8}{x+2} = \dfrac{(x+2)(x^2-2x+4)}{x+2} = x^2-2x+4$

53. $\dfrac{2x^2+3x-9}{4x^2+13x+3} = \dfrac{(2x-3)(x+3)}{(x+3)(4x+1)} = \dfrac{2x-3}{4x+1}$ **63.** $\dfrac{p^2+7pq+12q^2}{p^2+5pq+6q^2} = \dfrac{(p+3q)(p+4q)}{(p+2q)(p+3q)} = \dfrac{p+4q}{p+2q}$

Exercise 4-2

Answers to odd-numbered problems

1. $\dfrac{8}{3x}$ **3.** $\dfrac{1}{2x^2y^2}$ **5.** $\dfrac{a^2}{7bc^3}$ **7.** $2ab^2$ **9.** $\dfrac{4a-8}{15a-60}$ **11.** $\dfrac{a-6}{6}$ **13.** $\dfrac{-4p^2-5p-1}{p-1}$ **15.** $\dfrac{x-5}{x-3}$ **17.** $\dfrac{3a^2-5a-2}{3a^2-4a+1}$

19. $\dfrac{x^2-2x+4}{2x+7}$ **21.** $\dfrac{-1}{x+2y}$ **23.** $\dfrac{a^2}{b^3}$ **25.** $\dfrac{1}{r^2-5r+4}$ **27.** $\dfrac{1}{3x+12}$ **29.** $\dfrac{x^3+x^2-x-1}{4}$ **31.** $\dfrac{x^3-125}{x+5}$

33. $\dfrac{-x^2-3x+40}{x^2+4x-21}$ **35.** $\dfrac{x+4}{x-4}$ **37.** $\dfrac{a-b}{2}$ **39.** $-(2m^2+5mn+3n^2)$ **41.** $\dfrac{a-6}{a+3}$ **43.** $\dfrac{x^2-9}{x^2-5x+4}$

45. $-(9y^2+6xy+4x^2)(x^2-xy+y^2)$ **47.** $\dfrac{b^2+b-2}{b^2-b-2}$ **49.** $\dfrac{z^2-2wz+w^2}{z^2+2wz+w^2}$

Solutions to trial exercise problems

9. $\dfrac{4a + 8}{3a - 12} \cdot \dfrac{a - 2}{5a + 10} = \dfrac{4(a + 2) \cdot (a - 2)}{3(a - 4) \cdot 5(a + 2)} = \dfrac{4(a - 2)}{15(a - 4)} = \dfrac{4a - 8}{15a - 60}$

17. $\dfrac{2a^2 - a - 6}{3a^2 - 4a + 1} \cdot \dfrac{3a^2 + 7a + 2}{2a^2 + 7a + 6} = \dfrac{(2a + 3)(a - 2)}{(3a - 1)(a - 1)} \cdot \dfrac{(3a + 1)(a + 2)}{(2a + 3)(a + 2)} = \dfrac{(a - 2)(3a + 1)}{(3a - 1)(a - 1)} = \dfrac{3a^2 - 5a - 2}{3a^2 - 4a + 1}$

25. $\dfrac{r + 4}{r^2 - 1} \div \dfrac{r^2 - 16}{r + 1} = \dfrac{(r + 4) \cdot (r + 1)}{(r^2 - 1) \cdot (r^2 - 16)} = \dfrac{(r + 4)(r + 1)}{(r + 1)(r - 1)(r + 4)(r - 4)} = \dfrac{1}{(r - 4)(r - 1)} = \dfrac{1}{r^2 - 5r + 4}$

33. $\dfrac{56 - x - x^2}{x^2 - 6x - 7} \div \dfrac{x^2 + 4x - 21}{x^2 - 4x - 5} = \dfrac{-(x^2 + x - 56) \cdot (x^2 - 4x - 5)}{(x^2 - 6x - 7) \cdot (x^2 + 4x - 21)} = \dfrac{-(x - 7)(x + 8) \cdot (x - 5)(x + 1)}{(x - 7)(x + 1) \cdot (x + 7)(x - 3)}$

$= \dfrac{-(x + 8)(x - 5)}{(x + 7)(x - 3)} = \dfrac{-(x^2 + 3x - 40)}{x^2 + 4x - 21} = \dfrac{-x^2 - 3x + 40}{x^2 + 4x - 21}$

39. $\dfrac{n^2 - m^2}{2m - 3n} \div \dfrac{m - n}{4m^2 - 9n^2} = \dfrac{(n^2 - m^2) \cdot (4m^2 - 9n^2)}{(2m - 3n) \cdot (m - n)}$

$= \dfrac{-(m - n)(m + n) \cdot (2m + 3n)(2m - 3n)}{(2m - 3n) \cdot (m - n)} = -(m + n)(2m + 3n) = -(2m^2 + 5mn + 3n^2)$

41. $\dfrac{a^2 - 8a + 15}{a^2 + 7a + 6} \cdot \dfrac{a^2 - 6a - 7}{a^2 - 9} \div \dfrac{a^2 - 12a + 35}{a^2 - 36} = \dfrac{(a^2 - 8a + 15) \cdot (a^2 - 6a - 7) \cdot (a^2 - 36)}{(a^2 + 7a + 6) \cdot (a^2 - 9) \cdot (a^2 - 12a + 35)}$

$= \dfrac{(a - 3)(a - 5) \cdot (a - 7)(a + 1) \cdot (a + 6)(a - 6)}{(a + 6)(a + 1) \cdot (a + 3)(a - 3) \cdot (a - 7)(a - 5)} = \dfrac{a - 6}{a + 3}$

47. $\dfrac{ab - a + 3b - 3}{ab + a - 4b - 4} \div \dfrac{ab - 2a + 3b - 6}{ab + 2a - 4b - 8}$

$= \dfrac{(ab - a + 3b - 3) \cdot (ab + 2a - 4b - 8)}{(ab + a - 4b - 4) \cdot (ab - 2a + 3b - 6)} = \dfrac{(a + 3)(b - 1) \cdot (a - 4)(b + 2)}{(a - 4)(b + 1)(a + 3)(b - 2)} = \dfrac{(b - 1)(b + 2)}{(b + 1)(b - 2)} = \dfrac{b^2 + b - 2}{b^2 - b - 2}$

Exercise 4–3

Answers to odd-numbered problems

1. $70x$ **3.** $80k^2$ **5.** $288a^3b^2$ **7.** $30x^2y^4$ **9.** $4a(a - 2)$ **11.** $6x(x - 7)$ **13.** $3(p + 3)^2(p - 4)$
15. $10(a + 5)(a - 5)$ **17.** $(q + 7)(q - 7)(q - 2)$ **19.** $5(a + b)(a - b)$ **21.** $(2a + 1)(3a - 1)(a - 7)(a + 7)$

23. $\dfrac{12x^2}{21x^3}$ **25.** $\dfrac{15x^3y}{24x^2y^2}$ **27.** $\dfrac{4p^2 - 12p}{p - 3}$ **29.** $\dfrac{p^2 - 5p + 6}{p^2 - 4}$ **31.** $\dfrac{16x^3 + 24x^2 + 36x}{8x^3 - 27}$ **33.** $\dfrac{2n^2 - 11n + 5}{n^2 + 2n - 35}$

35. $\dfrac{2y^2 - y - 10}{4y^2 + 7y - 2}$ **37.** $\dfrac{15m^2 + 12m}{25m^2 - 16}$ **39.** $\dfrac{9}{a - 9}$ **41.** $\dfrac{13}{q}\,(q \neq 0)$ **43.** $\dfrac{-3y}{y + 4}\,(y \neq -4)$ **45.** $\dfrac{5x - 8}{3x + 4}\left(x \neq \dfrac{-4}{3}\right)$

47. $\dfrac{2y - 7}{4y + 3}\left(y \neq -\dfrac{3}{4}\right)$ **49.** $\dfrac{47}{12x}$ **51.** $\dfrac{14a^2 - 2a}{(3a + 5)(2a - 3)}$ **53.** $\dfrac{2x - 4}{x - 9}$ **55.** $\dfrac{129y - 122}{10(y - 2)(y + 2)}$ **57.** $\dfrac{16x - 61}{(x - 6)(x + 1)(x - 1)}$

59. $\dfrac{5x^2 + 3xy + 6y^2}{(x - 3y)(x + y)^2}$ **61.** $\dfrac{2a^2 - 16a - 25}{(2a - 1)(4a + 3)(a + 5)}$ **63.** $\dfrac{20a^2 - 25a + 1}{5a - 2}$ **65.** $\dfrac{17p^2 - 20p - 75}{8p(p - 5)(p + 3)}$

67. $\dfrac{3y^2 - y + 15}{(y + 3)(y - 2)(y + 2)}$ **69.** $\dfrac{5b^2 + 3b - 30}{5(b + 2)(b - 2)}$ **71.** $\dfrac{3x^2 - 15xy + 14y^2}{(x - 6y)^2(x + 2y)}$ **73.** $\dfrac{15x^2 - 9x - 9}{(4x - 3)(2x - 5)(3x + 1)}$

75. $\dfrac{49m^2 - 20mn + 3n^2}{(8m - n)(m + 2n)(5m - 4n)}$ **77.** $\dfrac{qrs + prs + pqs + pqr}{pqrs}$ **79.** $\dfrac{f_1 + f_2 - d}{f_1 f_2}$ **81.** $\dfrac{(n - 1)(R_2 + R_1)}{R_1 R_2}$

Solutions to trial exercise problems

13. $p^2 - p - 12 = (p - 4)(p + 3)$; $p^2 + 6p + 9 = (p + 3)^2$; $3p - 12 = 3(p - 4)$. The LCM is $3(p + 3)^2(p - 4)$. **18.** $2y + 10$
$= 2(y + 5)$; $25 - y^2 = -(y + 5)(y - 5)$; $y^2 - 10y + 25 = (y - 5)^2$. The LCM is $2(y + 5)(y - 5)^2$. **21.** $2a^2 - 13a - 7$
$= (2a + 1)(a - 7)$; $6a^2 + a - 1 = (3a - 1)(2a + 1)$; $a^2 - 49 = (a + 7)(a - 7)$. The LCM is $(2a + 1)(3a - 1)(a - 7)(a + 7)$.

31. Since $8x^3 - 27 = (2x - 3)(4x^2 + 6x + 9)$, then $\dfrac{4x}{2x - 3} = \dfrac{4x \cdot (4x^2 + 6x + 9)}{(2x - 3) \cdot (4x^2 + 6x + 9)} = \dfrac{16x^3 + 24x^2 + 36x}{8x^3 - 27}$.

37. Since $25m^2 - 16 = (5m + 4)(5m - 4)$, then $-\dfrac{3m}{4 - 5m} = -\dfrac{3m}{-(5m - 4)} = \dfrac{3m}{5m - 4} = \dfrac{3m \cdot (5m + 4)}{(5m - 4)(5m + 4)} = \dfrac{15m^2 + 12m}{25m^2 - 16}$.

49. $\dfrac{8}{3x} + \dfrac{5}{4x} = \dfrac{8(4)}{3x(4)} + \dfrac{5(3)}{4x(3)} = \dfrac{32}{12x} + \dfrac{15}{12x} = \dfrac{32+15}{12x} = \dfrac{47}{12x}$

53. $\dfrac{x+2}{x-9} - \dfrac{x-6}{9-x} = \dfrac{x+2}{x-9} - \dfrac{x-6}{-(x-9)} = \dfrac{x+2}{x-9} + \dfrac{x-6}{x-9} = \dfrac{(x+2)+(x-6)}{x-9} = \dfrac{x+2+x-6}{x-9} = \dfrac{2x-4}{x-9}$

60. $\dfrac{8q}{4q^2 - 9p^2} + \dfrac{5p}{4q^2 - 12pq + 9p^2} = \dfrac{8q}{(2q+3p)(2q-3p)} + \dfrac{5p}{(2q-3p)^2} = \dfrac{8q(2q-3p)}{(2q+3p)(2q-3p)^2} + \dfrac{5p(2q+3p)}{(2q-3p)^2(2q+3p)}$

$= \dfrac{(16q^2 - 24pq) + (10pq + 15p^2)}{(2q+3p)(2q-3p)^2} = \dfrac{16q^2 - 14pq + 15p^2}{(2q+3p)(2q-3p)^2} \left(q \neq -\dfrac{3}{2}p, \dfrac{3}{2}p\right)$ **63.** $(4a-3) + \dfrac{2a+5}{5a-2} = \dfrac{4a-3}{1} + \dfrac{2a+5}{5a-2}$

$= \dfrac{(4a-3)\cdot(5a-2) + 1\cdot(2a+5)}{1\cdot(5a-2)} = \dfrac{20a^2 - 23a + 6 + 2a + 5}{5a-2} = \dfrac{20a^2 - 21a + 11}{5a-2}$

72. $\dfrac{a-b}{a^2 - 3ab - 4b^2} + \dfrac{2b-5a}{a^2 - 16b^2} = \dfrac{a-b}{(a-4b)(a+b)} + \dfrac{2b-5a}{(a+4b)(a-4b)} = \dfrac{(a-b)\cdot(a+4b)}{(a-4b)(a+b)(a+4b)} + \dfrac{(2b-5a)\cdot(a+b)}{(a+4b)(a-4b)(a+b)}$

$= \dfrac{(a^2 + 3ab - 4b^2) + (2b^2 - 3ab - 5a^2)}{(a+4b)(a-4b)(a+b)} = \dfrac{a^2 + 3ab - 4b^2 + 2b^2 - 3ab - 5a^2}{(a+4b)(a-4b)(a+b)} = \dfrac{-4a^2 - 2b^2}{(a+4b)(a-4b)(a+b)}$

Exercise 4–4

Answers to odd-numbered problems

1. $\dfrac{21}{20}$ **3.** $\dfrac{12}{11}$ **5.** $\dfrac{7}{8m}$ **7.** $\dfrac{-4y}{3y+3}$ **9.** $\dfrac{x-3}{x+7}$ **11.** $\dfrac{15x-20}{8x+4}$ **13.** $\dfrac{20}{27}$ **15.** $\dfrac{5}{7y}$ **17.** $\dfrac{35-7a}{4a+36}$ **19.** $\dfrac{8x-4y}{3}$

21. $\dfrac{4x^2 + 5x - 9}{5x^2 + 16x + 3}$ **23.** $\dfrac{xy^2 + 2x^2}{3y^2 - 7xy^2}$ **25.** $\dfrac{mn(m+n)}{n-m}$ **27.** $\dfrac{5q + 4p^2}{p^2 q (p-q)}$ **29.** $\dfrac{a^2 + 9a + 23}{a^2 + 7a + 7}$

31. $\dfrac{4y^3 + 17y^2 - 18y - 15}{4y^3 + 25y^2 + 31y - 39}$ **33.** $\dfrac{t^2 - 3t - 4}{t^2 + 3t - 18}$ **35.** $\dfrac{-3}{2x + 14}$ **37.** $\dfrac{5b^2 + 24b - 5}{4b - 22}$ **39.** $\dfrac{1}{2}$ **41.** $\dfrac{3c + 4a - 2b}{5}$

43. $\dfrac{V_s C - it}{RC}$ **45.** $CP = \dfrac{T_1}{T_2 - T_1}$ **47.** $r = \dfrac{2dr_1 r_2}{dr_2 + dr_1} = \dfrac{2r_1 r_2}{r_2 + r_1}$

Solutions to trial exercise problems

21. $\dfrac{4 - \dfrac{3}{x+3}}{5 + \dfrac{6}{x-1}} = \dfrac{4(x+3)(x-1) - \dfrac{3}{x+3}\cdot(x+3)(x-1)}{5(x+3)(x-1) + \dfrac{6}{x-1}\cdot(x+3)(x-1)}$

$= \dfrac{4(x^2 + 2x - 3) - 3(x-1)}{5(x^2 + 2x - 3) + 6(x+3)}$

$= \dfrac{4x^2 + 8x - 12 - 3x + 3}{5x^2 + 10x - 15 + 6x + 18}$

$= \dfrac{4x^2 + 5x - 9}{5x^2 + 16x + 3}$

24. $\dfrac{\dfrac{6}{a} - \dfrac{5}{b}}{\dfrac{1}{a^2} + \dfrac{1}{b^2}} = \dfrac{\dfrac{6}{a}\cdot a^2 b^2 - \dfrac{5}{b}\cdot a^2 b^2}{\dfrac{1}{a^2}\cdot a^2 b^2 + \dfrac{1}{b^2}\cdot a^2 b^2} = \dfrac{6ab^2 - 5a^2 b}{b^2 + a^2}$

29. $\dfrac{(a+5) + \dfrac{3}{a+4}}{(a+3) - \dfrac{5}{a+4}} = \dfrac{(a+5)(a+4) + \dfrac{3}{a+4}(a+4)}{(a+3)(a+4) - \dfrac{5}{a+4}(a+4)}$

$= \dfrac{a^2 + 9a + 20 + 3}{a^2 + 7a + 12 - 5} = \dfrac{a^2 + 9a + 23}{a^2 + 7a + 7}$

35.
$$\dfrac{\dfrac{3}{x^2 - x - 6}}{\dfrac{2}{x + 2} - \dfrac{4}{x - 3}} = \dfrac{\dfrac{3}{(x - 3)(x + 2)}}{\dfrac{2}{x + 2} - \dfrac{4}{x - 3}}$$

$$= \dfrac{\dfrac{3}{(x - 3)(x + 2)} \cdot (x - 3)(x + 2)}{\dfrac{2}{x + 2}(x - 3)(x + 2) - \dfrac{4}{x - 3}(x - 3)(x + 2)}$$

$$= \dfrac{3}{2(x - 3) - 4(x + 2)} = \dfrac{3}{2x - 6 - 4x - 8}$$

$$= \dfrac{3}{-2x - 14} = \dfrac{3}{-(2x + 14)} = \dfrac{-3}{2x + 14}$$

Exercise 4–5

Answers to odd-numbered problems

1. $5x^2 - 3x + 2$ **3.** $-x^3 + 2x^2 - 3$ **5.** $c - 1$ **7.** $10xy^2 + 7 - 6y^2$ **9.** $3a^6b - 5a^4b^4 + 7a^3b$ **11.** $2a + 3c$ **13.** $a - 4$

15. $y + 2 + \dfrac{1}{y + 5}$ **17.** $x - 4 - \dfrac{2}{x - 3}$ **19.** $a^2 - 2a + 4$ **21.** $2y^2 - 3y + 1 + \dfrac{2}{y + 1}$ **23.** $x^2 + 3x - 4$

25. $2a^2 + 3a - 4 + \dfrac{5}{2a + 1}$ **27.** $3y^2 - y + 4 + \dfrac{2}{3y + 1}$ **29.** $a^2 - 3a + 9$ **31.** $3x^3 - 5x^2 + 5x - 4 + \dfrac{3}{x + 1}$ **33.** $3a - 1$

35. $y^2 + y - 1$ **37.** $x^2 + 3x + 2 - \dfrac{4}{x^2 + x - 4}$ **39.** $3a^2 - 2a - 4$ **41.** $y^2 + 3 + \dfrac{2}{2y^2 - 3y + 2}$ **43.** 2, it is the same.

45. $2x - 3$ **47.** $3x - 2$

Solutions to trial exercise problems

10. $\dfrac{x(y - 2) - z(y - 2)}{y - 2} = \dfrac{x(y - 2)}{y - 2} - \dfrac{z(y - 2)}{y - 2} = x - z$

16. $(2x - 5 + x^2) \div (x + 4) = (x^2 + 2x - 5) \div (x + 4)$

$$
\begin{array}{r}
x - 2 \\
x + 4 \overline{)\, x^2 + 2x - 5} \\
\underline{x^2 + 4x} \\
-2x - 5 \\
\underline{-2x - 8} \\
3
\end{array}
$$

Answer: $x - 2 + \dfrac{3}{x + 4}$

27. $\dfrac{9y^3 + 11y + 6}{3y + 1}$

$$
\begin{array}{r}
3y^2 - y + 4 \\
3y + 1 \overline{)\, 9y^3 + 0y^2 + 11y + 6} \\
\underline{9y^3 + 3y^2} \\
-3y^2 + 11y \\
\underline{-3y^2 - y} \\
12y + 6 \\
\underline{12y + 4} \\
2
\end{array}
$$

Answer: $3y^2 - y + 4 + \dfrac{2}{3y + 1}$

32. $\dfrac{2x^3 + 5x^2 + 5x + 3}{x^2 + x + 1}$

$$
\begin{array}{r}
2x + 3 \\
x^2 + x + 1 \overline{)\, 2x^3 + 5x^2 + 5x + 3} \\
\underline{2x^3 + 2x^2 + 2x} \\
3x^2 + 3x + 3 \\
\underline{3x^2 + 3x + 3} \\
0
\end{array}
$$

Answer: $2x + 3$

38. $\dfrac{2x^4 - x^3 + 5x^2 - x + 3}{x^2 + 1}$

Answer: $2x^2 - x + 3$

$$
\begin{array}{r}
2x^2 - x + 3 \\
x^2 + 0x + 1 \overline{)\, 2x^4 - x^3 + 5x^2 - x + 3} \\
\underline{2x^4 + 0x^3 + 2x^2} \\
-x^3 + 3x^2 - x \\
\underline{-x^3 - 0x^2 - x} \\
3x^2 + 0x + 3 \\
\underline{3x^2 + 0x + 3} \\
0
\end{array}
$$

Exercise 4-6

Answers to odd-numbered problems

1. $\{-21\}$ **3.** $\left\{\dfrac{-15}{11}\right\}$ **5.** $\left\{\dfrac{10}{3}\right\}$ **7.** $\left\{\dfrac{67}{7}\right\}$ **9.** $\{12\}$ **11.** $\left\{\dfrac{-13}{6}\right\}$ **13.** $\emptyset$ **15.** $\left\{\dfrac{-3}{5}\right\}$ **17.** $\left\{\dfrac{-13}{5}\right\}$ **19.** $\emptyset$

21. $\left\{\dfrac{13}{3}\right\}$ **23.** $\left\{\dfrac{-16}{5}\right\}$ **25.** $\left\{\dfrac{13}{2}\right\}$ **27.** $\emptyset$ **29.** $\{-15\}$ **31.** $\left\{\dfrac{-17}{3}\right\}$ **33.** $\{26\}$ **35.** $\left\{\dfrac{-31}{5}\right\}$ **37.** $\{-1\}$

39. $\left\{-\dfrac{1}{12}\right\}$ **41.** $b = \dfrac{3a}{4 - 7a}$ **43.** $p = \dfrac{3q - 3}{3a + 4}$ **45.** $y = \dfrac{5x + 19}{3}$ **47.** $r = \dfrac{S - a}{S}$ **49.** $b_1 = \dfrac{2A - b_2 h}{h}$

51. $t = \dfrac{L_t - L_0}{kL_0}$; $L_0 = \dfrac{L_t}{kt + 1}$ **53.** $t = \dfrac{I}{Pr}$ **55.** $R_3 = \dfrac{RR_1 R_2}{R_1 R_2 - RR_2 - RR_1}$ **57.** $A = \dfrac{aF}{f}$

59. $T_2 = \dfrac{R_2 M + R_2 T_1 - R_1 M}{R_1}$; $R_1 = \dfrac{R_2(M + T_1)}{M + T_2}$

Solutions to trial exercise problems

5. $\dfrac{4}{a} - \dfrac{6}{3a} = \dfrac{3}{5}$

Multiply by the LCM, $15a$.

$15a \cdot \dfrac{4}{a} - 15a \cdot \dfrac{6}{3a} = 15a \cdot \dfrac{3}{5}$

$15 \cdot 4 - 5 \cdot 6 = 3a \cdot 3$

$60 - 30 = 9a$

$30 = 9a$

$\dfrac{10}{3} = a$

$S = \left\{\dfrac{10}{3}\right\}$

16. $1 + \dfrac{5}{3m - 9} = \dfrac{10}{m - 3}$

Multiply by the LCM, $3(m - 3)$.

$3(m - 3) \cdot 1 + 3(m - 3) \cdot \dfrac{5}{3(m - 3)} = 3(m - 3) \cdot \dfrac{10}{m - 3}$

$3m - 9 + 5 = 3 \cdot 10$

$3m - 4 = 30$

$3m = 34$

$m = \dfrac{34}{3}$

$S = \left\{\dfrac{34}{3}\right\}$

21. $4 - \dfrac{2x}{5 - x} = \dfrac{6}{x - 5}$

Since $5 - x = -(x - 5)$, then we have

$4 - \dfrac{2x}{-(x - 5)} = \dfrac{6}{x - 5}$

$4 + \dfrac{2x}{x - 5} = \dfrac{6}{x - 5}$. Multiply by $x - 5$.

$4(x - 5) + (x - 5) \cdot \dfrac{2x}{x - 5} = (x - 5) \cdot \dfrac{6}{x - 5}$

$4x - 20 + 2x = 6$

$6x - 20 = 6$

$6x = 26$

$x = \dfrac{26}{6} = \dfrac{13}{3}$

$S = \left\{\dfrac{13}{3}\right\}$

29. $\dfrac{6}{q^2 + q - 6} = \dfrac{5}{q^2 + 3q - 10}$

Since $q^2 + q - 6 = (q + 3)(q - 2)$ and $q^2 + 3q - 10 = (q + 5)(q - 2)$, the LCM is $(q + 3)(q - 2)(q + 5)$. Multiply each member by the LCM.

$(q + 3)(q - 2)(q + 5) \cdot \dfrac{6}{(q + 3)(q - 2)} = (q + 3)(q - 2)(q + 5) \cdot \dfrac{5}{(q + 5)(q - 2)}$

$(q + 5) \cdot 6 = (q + 3) \cdot 5$

$6q + 30 = 5q + 15$

$q = -15$

$S = \{-15\}$

41. $\dfrac{y-3}{x+2} = \dfrac{5}{3}$

Multiply by the LCM, $3(x+2)$.

$$3(x+2) \cdot \dfrac{y-3}{x+2} = 3(x+2) \cdot \dfrac{5}{3}$$
$$3(y-3) = (x+2) \cdot 5$$
$$3y - 9 = 5x + 10$$
$$3y = 5x + 19$$
$$y = \dfrac{5x + 19}{3}$$

48. $R_x = R_m\left(\dfrac{E_1}{E_2} - 1\right)$

Now $R_x = R_m \cdot \dfrac{E_1}{E_2} - R_m$

$$R_x + R_m = \dfrac{R_m E_1}{E_2}$$

Multiply each member by E_2.

$$E_2(R_x + R_m) = R_m E_1$$

Divide each member by R_m.

$$\dfrac{E_2(R_x + R_m)}{R_m} = E_1$$

Exercise 4–7

Answers to odd-numbered problems

1. 1 hour 12 minutes   **3.** 40 minutes   **5.** 45 hours   **7.** $13\dfrac{11}{13}$ hours ≈ 13.8 hours   **9.** 18 hours for pump B; 36 hours for pump A

11. 7 hour   **13.** $\dfrac{48}{17}$ mph   **15.** 30 km/hr   **17.** 1,600 miles   **19.** $1\dfrac{2}{3}$ miles   **21.** $\dfrac{2}{3}$ and 2   **23.** 13   **25.** 3

Solutions to trial exercise problems

4. Let $x =$ minutes taken to mix the 12 loaves working together.

$$\dfrac{1}{36} + \dfrac{1}{40} + \dfrac{1}{30} = \dfrac{1}{x} \qquad \text{Multiply by the LCM, } 360x.$$

$$360x \cdot \dfrac{1}{36} + 360x \cdot \dfrac{1}{40} + 360x \cdot \dfrac{1}{30} = 360x \cdot \dfrac{1}{x}$$
$$10x + 9x + 12x = 360$$
$$31x = 360$$
$$x = \dfrac{360}{31} = 11\dfrac{19}{31}$$

Working together, they could mix the 12 loaves in $11\dfrac{19}{31}$ minutes ≈ 11.6 minutes.

7. Let $x =$ number of hours necessary to fill the basin with all pipes open.

$$\dfrac{1}{10} + \dfrac{1}{12} - \dfrac{1}{9} = \dfrac{1}{x} \qquad \text{Multiply by the LCM, } 180x.$$

$$180x \cdot \dfrac{1}{10} + 180x \cdot \dfrac{1}{12} - 180x \cdot \dfrac{1}{9} = 180x \cdot \dfrac{1}{x}$$
$$18x + 15x - 20x = 180$$
$$13x = 180$$
$$x = \dfrac{180}{13} = 13\dfrac{11}{13}$$

It would take $13\dfrac{11}{13}$ hours ≈ 13.8 hours to fill the basin with all three pipes open.

10. Let $x =$ time for slower microprocessor to do the job. Then $\frac{3}{5}x =$ time for faster microprocessor to do the job.

$$\frac{1}{x} + \frac{1}{\frac{3}{5}x} = \frac{1}{2}$$

$$\frac{1}{x} + \frac{5}{3x} = \frac{1}{2} \qquad \text{Multiply by the LCM, } 6x.$$

$$6x \cdot \frac{1}{x} + 6x \cdot \frac{5}{3x} = 6x \cdot \frac{1}{2}$$

$$6 + 2 \cdot 5 = 3x$$

$$16 = 3x$$

$$\frac{16}{3} = x$$

$$\frac{16}{5} = \frac{3}{5}x$$

It would take the microprocessors $\frac{16}{3}$ or $5\frac{1}{3}$ milliseconds and $\frac{16}{5}$ or $3\frac{1}{5}$ milliseconds, respectively, to process the set of inputs individually.

14. Let $r =$ speed of the wind. Then $300 + r =$ speed of the plane with the wind. Then $300 - r =$ speed of the plane against the wind.

$$\text{Now } t_w \text{ (time with the wind)} = \frac{950}{300 + r}$$

$$t_a \text{ (time against the wind)} = \frac{650}{300 - r}$$

$$\text{Then } \frac{950}{300 + r} = \frac{650}{300 - r}.$$

Multiply each member by $(300 + r)(300 - r)$.

$$(300 + r)(300 - r) \cdot \frac{950}{300 + r} = (300 + r)(300 - r) \cdot \frac{650}{300 - r}$$

$$(300 - r) \cdot 950 = (300 + r) \cdot 650$$

$$285{,}000 - 950r = 195{,}000 + 650r$$

$$90{,}000 = 1{,}600r$$

$$r = \frac{90{,}000}{1{,}600}$$

$$r = \frac{900}{16} = 56\frac{1}{4}$$

The wind has a speed of $56\frac{1}{4}$ mph.

Chapter 4 review exercises

1. $\{x \mid x \in R, x \neq -7\}$ **2.** $\left\{x \mid x \in R, x \neq \frac{4}{3}\right\}$ **3.** $\{x \mid x \in R, x \neq 5\}$ **4.** $\left\{a \mid a \in R, a \neq -\frac{4}{3}, \frac{4}{3}\right\}$ **5.** $\left\{z \mid z \in R, z \neq -\frac{5}{2}, \frac{1}{3}\right\}$

6. $\left\{y \mid y \in R, y \neq -\frac{3}{2}\right\}$ **7.** $\frac{ab^3}{c^2}$ **8.** $\frac{-2n^2}{7mp^4}$ **9.** $\frac{5}{6}$ **10.** $\frac{3}{a-2}$ **11.** $\frac{-5}{2y+x}$ **12.** $\frac{y^2+4y+16}{y+4}$ **13.** $\frac{a-12}{a+1}$

14. $\frac{4x+3}{5x-1}$ **15.** $\frac{-(2y+3)}{2(3y+2)}$ **16.** $\frac{6y}{x}$ $(x \neq 0, y \neq 0)$ **17.** $12ay(a \neq 0, y \neq 0)$ **18.** $\frac{(4p+3)(p-4)}{3}$ $\left(p \neq -4, \frac{3}{4}\right)$

19. $\frac{z-3}{2(z+1)(z-1)}$ $(z \neq -1, 1, 3)$ **20.** $\frac{(m^2+2m+4)(m+6)}{m(m+5)}$ $(m \neq -5, 0, 2, 3)$ **21.** $1\left(a \neq -7, -3, -\frac{2}{5}, \frac{1}{2}\right)$

22. $\frac{(x+7)(2x-1)}{(4x^2-2x+1)(x+2)}$ $\left(x \neq -2, -\frac{1}{2}, \frac{1}{2}, 7\right)$ **23.** $\frac{x^2}{(4x+5)^2}$ $\left(x \neq -\frac{5}{4}, \frac{5}{4}\right)$ **24.** $-\frac{y+3}{y+2}\left(y \neq -3, -2, \frac{4}{7}\right)$

25. $\frac{m-n}{m+n}$ $\left(m \neq -n, \frac{n}{2} ; q \neq -p, p\right)$ **26.** $180x^3y^3$ **27.** $6x^2(x+2)(x+4)(x-4)$ **28.** $3a(a+5)(a-2)$

29. $p(p - 5)(p + 5)^2$ **30.** $\dfrac{41x}{12y}$ **31.** $\dfrac{-2n^2 - 28n - 25}{(n + 4)(n - 1)}$ **32.** $\dfrac{10p^2 + 29p + 81}{p(p + 9)(p + 2)(p - 2)}$ **33.** $\dfrac{6b^2 + 16b - 23}{3b - 2}$

34. $\dfrac{2y^2 + 12y + 11}{(y + 7)(y - 7)}$ **35.** $\dfrac{-(6x^2 + 27x + 4)}{(x - 6)(x + 6)(x + 4)}$ **36.** $\dfrac{37}{2(a - 2)}$ **37.** $\dfrac{-5x^2 + 63x - 102}{8(x - 7)(x + 4)}$ **38.** $\dfrac{4a + b}{(a - 2b)(a + 2b)}$

39. $\dfrac{1}{R_t} = \dfrac{I_1E_2E_3 + I_2E_1E_3 + I_3E_1E_2}{E_1E_2E_3}$ **40.** $\dfrac{2}{a}$ **41.** $\dfrac{5x}{4x - 12}$ **42.** $\dfrac{3x + 6}{x - 5}$ **43.** $\dfrac{7b - 38}{8b - 34}$ **44.** $\dfrac{x^2y - xy^2}{2y + 3x}$

45. $\dfrac{p^2 - 4p + 5}{p^2 - 4}$ **46.** $5a^6 + 3a^2 + 2$ **47.** $6a^3b^2c^2 - 3c^3 + 1$ **48.** $3x^2 - 3x + 4$ **49.** $2x^2 - x + 1$ **50.** $\left\{\dfrac{55}{216}\right\}$

51. $\{66\}$ **52.** $\left\{-\dfrac{37}{6}\right\}$ **53.** $\left\{\dfrac{7}{20}\right\}$ **54.** $\left\{-\dfrac{3}{29}\right\}$ **55.** $p = \dfrac{4m - 6n + 26}{3}$ **56.** $C = \dfrac{5}{9}(F - 32)$ **57.** $V_1 = \dfrac{P_2V_2T_1}{P_1T_2}$

58. $R_2 = \dfrac{R_tR_1}{R_1 - R_t}$ **59.** 2 days   **60.** 60 mph, auto; 90 mph, train   **61.** 3 mph

Chapter 4 cumulative test

1. 25   **2.** 23   **3.** $\dfrac{12}{45}$ **4.** 5   **5.** $\dfrac{13}{30}$ **6.** $7x + 11$ **7.** $-24a^6b^5$ **8.** $x^3 + 2x^2 - 3x + 20$ **9.** 2   **10.** $\dfrac{x^2 + 6x - 16}{x^2 + 3x}$

11. $\dfrac{-10y - 13}{24}$ **12.** $10x^2 + 39x - 27$ **13.** $16x^2 - 40x + 25$ **14.** $9y^2 - 25$ **15.** $2x^2 + 9x + 27 + \dfrac{80}{x - 3}$

16. $\dfrac{6a^2 + 11a - 10}{4a^2 + 23a - 35}$ **17.** $\dfrac{-2x + 19}{(2x - 3)(x + 1)(2x + 1)}$ **18.** $\left\{y \,\middle|\, y \in R, y \ne \dfrac{3}{2}\right\}$ **19.** $\left\{x \,\middle|\, x \in R, x \ne \dfrac{1}{2}, -\dfrac{1}{2}\right\}$ **20.** $\{x \,|\, x \in R\}$

21. $\{14\}$ **22.** $\left\{-\dfrac{11}{24}\right\}$ **23.** $\left\{-\dfrac{1}{8}\right\}$ **24.** $-\dfrac{13}{14}$ **25.** $\dfrac{4x^2 + x - 5}{5x^2 + 7x - 6}$ **26.** $\{x \,|\, x \le -1\}$ **27.** $\left\{y \,\middle|\, y < \dfrac{17}{2}\right\}$

28. $\{z \,|\, z > -5\}$ **29.** $\left\{x \,\middle|\, x \ge \dfrac{97}{23}\right\}$ **30.** $-\dfrac{3a}{2b^2}$ **31.** $\dfrac{p + 4}{p + 3}$ **32.** $-\dfrac{3}{2}$

Chapter 5

Exercise 5–1

Answers to odd-numbered problems

1. 4.243   **3.** -5.745 **5.** $\sqrt[3]{6}$ **7.** $\sqrt{x}$ **9.** $\sqrt[5]{b^4}$ **11.** 4   **13.** 16   **15.** 27   **17.** 64   **19.** $\dfrac{1}{2}$ **21.** $\dfrac{1}{2}$ **23.** $\dfrac{1}{9}$

25. $\dfrac{-1}{8}$ **27.** $\dfrac{1}{\sqrt[4]{x^3}}$ **29.** $a^{\frac{4}{7}}$ **31.** $x^{\frac{1}{5}}$ **33.** $|b|$ **35.** $|x^3|$ or $x^2|x|$ **37.** $|x + y|$ **39.** 2   **41.** $b^{\frac{17}{12}}$

43. $a^{\frac{8}{15}}$ **45.** x^3 **47.** $8y^3$ **49.** $a^2b^{\frac{2}{3}}$ **51.** $8a^{\frac{15}{2}}b^{\frac{3}{2}}$ **53.** $\dfrac{1}{y^2}$ **55.** $b^{\frac{1}{3}}$ **57.** $\dfrac{1}{a^{\frac{1}{2}}}$ **59.** $\dfrac{1}{x^{\frac{2}{3}}}$ **61.** $x^{\frac{1}{2}}y^{\frac{1}{2}}$

63. $x^{\frac{1}{6}}$ **65.** $c^{\frac{7}{4}}$ **67.** $x^{\frac{1}{6}}y^{\frac{1}{2}}$ **69.** -27 **71.** 14 feet   **73.** 6 meters   **75.** 30 mph   **77.** 18 miles
79. by taking the square root of the square root

Solutions to trial exercise problems

13. $(-64)^{\frac{2}{3}} = (\sqrt[3]{-64})^2 = (-4)^2 = 16$ **18.** $(-27)^{-\frac{1}{3}} = \dfrac{1}{(-27)^{\frac{1}{3}}} = \dfrac{1}{\sqrt[3]{-27}} = \dfrac{1}{-3} = -\dfrac{1}{3}$ **43.** $(a^{\frac{2}{3}})^{\frac{4}{5}} = a^{\frac{2}{3} \cdot \frac{4}{5}} = a^{\frac{8}{15}}$

47. $(16y^4)^{\frac{3}{4}} = (\sqrt[4]{16y^4})^3 = (2y)^3 = 8y^3$ **52.** $(x^{-\frac{1}{4}})^4 = x^{-\frac{1}{4} \cdot 4} = x^{-1} = \dfrac{1}{x}$ **56.** $\dfrac{y^{\frac{1}{4}}}{y^{\frac{1}{3}}} = y^{\frac{1}{4} - \frac{1}{3}} = y^{\frac{3}{12} - \frac{4}{12}} = y^{-\frac{1}{12}} = \dfrac{1}{y^{\frac{1}{12}}}$

66. $\dfrac{a^{-\frac{2}{3}}b^{\frac{1}{2}}}{a^{-\frac{1}{3}}b^{\frac{3}{4}}} = a^{-\frac{2}{3} - (-\frac{1}{3})}\, b^{\frac{1}{2} - \frac{3}{4}} = a^{-\frac{1}{3}}b^{\frac{2}{4} - \frac{3}{4}} = a^{-\frac{1}{3}}b^{-\frac{1}{4}} = \dfrac{1}{a^{\frac{1}{3}}b^{\frac{1}{4}}}$

Exercise 5–2

Answers to odd-numbered problems

1. $2\sqrt{5}$ **3.** $2\sqrt[3]{3}$ **5.** $a\sqrt[4]{a}$ **7.** a^2 **9.** c **11.** $5xy^4\sqrt{xy}$ **13.** $5a^3c^4\sqrt{2bc}$ **15.** $3ac^4\sqrt[3]{b^2}$ **17.** $3ab^3\sqrt[3]{3a^2b^2}$
19. $2a^2c^2\sqrt[5]{b^4c^2}$ **21.** $x + 3$ **23.** $3a + 1$ **25.** $9\sqrt{2}$ **27.** 7 **29.** 12 **31.** $2\sqrt[3]{9}$ **33.** $3a\sqrt{5}$ **35.** $a\sqrt[3]{a}$ **37.** x
39. $2x\sqrt[5]{x^2}$ **41.** $5ab\sqrt[3]{3a}$ **43.** $5x^2y^3\sqrt[3]{3y}$ **45.** $2xy\sqrt[4]{2}$ **47.** $\sqrt{y}$ **49.** $y\sqrt[4]{y^3}$ **51.** $\sqrt{2y}$ **53.** $\sqrt[3]{2ab^2}$ **55.** $4|x|$
57. $7|bc|$ **59.** $|2x - y|$ **61.** $3\sqrt[3]{3}$ in. **63.** 3 in. **65.** 6 units **67.** $5\sqrt{6}$ amperes **69.** 5 m **71.** 13 in. **73.** $\sqrt{41}$ in.
75. $3\sqrt{13}$ ft **77.** 20 mm **79.** $2\sqrt{21}$ cm **81.** 5 feet **83.** 44.27 meters per second

Solutions to trial exercise problems

10. $\sqrt{9x^2y^5} = \sqrt{9x^2y^2y^2y} = \sqrt{9}\,\sqrt{x^2}\,\sqrt{y^2}\,\sqrt{y^2}\,\sqrt{y} = 3xyy\sqrt{y} = 3xy^2\sqrt{y}$ **24.** $\sqrt{x^2 + y^2}$. Will not simplify because x^2 and y^2
are terms, not factors. **40.** $\sqrt[3]{4a^2b}\,\sqrt[3]{4a^2b^2} = \sqrt[3]{4a^2b \cdot 4a^2b^2} = \sqrt[3]{16a^4b^3} = \sqrt[3]{8 \cdot 2 \cdot a^3 \cdot a \cdot b^3} = \sqrt[3]{8}\,\sqrt[3]{2}\,\sqrt[3]{a^3}\,\sqrt[3]{a}\,\sqrt[3]{b^3}$

$= 2 \cdot \sqrt[3]{2} \cdot a \cdot \sqrt[3]{a} \cdot b = 2ab\sqrt[3]{2a}$ **48.** $\sqrt[6]{b^{10}} = b^{\frac{10}{6}} = b^{\frac{5}{3}} = \sqrt[3]{b^5} = b\sqrt[3]{b^2}$ **61.** $h = \sqrt[3]{\dfrac{12I}{b}} = \sqrt[3]{\dfrac{12(27)}{(4)}} = \sqrt[3]{3(27)}$

$= \sqrt[3]{27}\,\sqrt[3]{3} = 3\sqrt[3]{3}$ in. **73.** $c = \sqrt{a^2 + b^2} = \sqrt{(5)^2 + (4)^2} = \sqrt{25 + 16} = \sqrt{41}$ in.

Exercise 5–3

Answers to odd-numbered problems

1. $\dfrac{4}{5}$ **3.** $\dfrac{\sqrt{7}}{3}$ **5.** $\dfrac{2}{3}$ **7.** $\dfrac{a^3}{3}$ **9.** $\dfrac{\sqrt[3]{2x}}{y^5}$ **11.** $\dfrac{x^3}{yz^2}$ **13.** $\dfrac{2x}{y^2}$ **15.** $2x^2y\sqrt[5]{2}$ **17.** $\dfrac{\sqrt{2}}{2}$ **19.** $\dfrac{3\sqrt{10}}{10}$ **21.** $\dfrac{\sqrt{6}}{6}$ **23.** $\dfrac{3\sqrt{2}}{10}$

25. $\sqrt{2}$ **27.** $3\sqrt{2}$ **29.** $\dfrac{3\sqrt[3]{2}}{2}$ **31.** $\dfrac{2\sqrt[5]{3}}{3}$ **33.** $\dfrac{3\sqrt[3]{4}}{4} = \dfrac{3\sqrt{2}}{4}$ **35.** $\dfrac{x\sqrt{y}}{y}$ **37.** $\dfrac{\sqrt{c}}{c}$ **39.** $\dfrac{a\sqrt[3]{b}}{b}$ **41.** $\dfrac{\sqrt[5]{ab}}{b}$

43. $\dfrac{2x\sqrt[5]{y^3}}{y}$ **45.** $a\sqrt[5]{b}$ **47.** $\dfrac{\sqrt{2xyz}}{yz}$ **49.** $\dfrac{2\sqrt[3]{xyz^2}}{yz}$ **51.** $\dfrac{\sqrt[3]{8x^5y^4}}{2xy}$ **53.** $\dfrac{\sqrt[5]{x^3y}}{y}$ **55.** $\sqrt[3]{x^2y}$ **57.** $b\sqrt[3]{b^3c^4}$ **59.** $\dfrac{\sqrt{xy}}{2}$

61. $\dfrac{2a^2}{b}$ **63.** $\dfrac{y^2\sqrt[5]{z}}{xz}$ **65.** 6 units **67.** $\dfrac{2\sqrt{3gh}}{3}$ **69.** $\dfrac{c\sqrt{2}}{2}$ **71.** $\dfrac{2f\sqrt{3}}{3}$ **73.** $\dfrac{2\sqrt{2\pi kmT}}{\pi m}$

Solutions to trial exercise problems

23. $\sqrt{\dfrac{9}{50}} = \dfrac{\sqrt{9}}{\sqrt{50}} = \dfrac{3}{\sqrt{25 \cdot 2}} = \dfrac{3}{5\sqrt{2}} \cdot \dfrac{\sqrt{2}}{\sqrt{2}} = \dfrac{3\sqrt{2}}{5 \cdot 2} = \dfrac{3\sqrt{2}}{10}$ **44.** $\dfrac{x^2}{\sqrt[3]{x^2}} \cdot \dfrac{\sqrt[3]{x}}{\sqrt[3]{x}} = \dfrac{x^2\sqrt[3]{x}}{x} = x\sqrt[3]{x}$

51. $\sqrt[3]{\dfrac{1}{16x^2y^3}} = \dfrac{\sqrt[3]{1}}{\sqrt[3]{16x^2y^3}} = \dfrac{1}{\sqrt[3]{2^4x^2y^3}} \cdot \dfrac{\sqrt[3]{2^3x^5y^4}}{\sqrt[3]{2^3x^5y^4}} = \dfrac{\sqrt[3]{8x^5y^4}}{2xy}$ **59.** $\sqrt{\dfrac{2y}{x}}\sqrt{\dfrac{x^2}{8}} = \sqrt{\dfrac{2y}{x} \cdot \dfrac{x^2}{8}} = \sqrt{\dfrac{xy}{4}} = \dfrac{\sqrt{xy}}{\sqrt{4}} = \dfrac{\sqrt{xy}}{2}$

Exercise 5–4

Answers to odd-numbered problems

1. $11\sqrt{5}$ **3.** $7\sqrt{3}$ **5.** $8\sqrt{5}$ **7.** $-\sqrt{10}$ **9.** $7\sqrt[3]{4}$ **11.** $6\sqrt[4]{3}$ **13.** $5\sqrt[5]{12} - \sqrt[5]{16}$ **15.** $4\sqrt{3x} - 4\sqrt{2x}$
17. $-\sqrt{5}$ **19.** $-3\sqrt{3}$ **21.** $13\sqrt{3}$ **23.** $\sqrt{3}$ **25.** $5\sqrt[3]{2}$ **27.** $6\sqrt[3]{2} + 10\sqrt[3]{3}$ **29.** $3\sqrt[3]{3} + 10\sqrt[3]{2}$ **31.** $\sqrt{2x}$
33. $37a\sqrt{b}$ **35.** $70a\sqrt{b} - 11\sqrt{2b}$ **37.** $5\sqrt[4]{a}$ **39.** $-23\sqrt[3]{a^2}$ **41.** $4a^2\sqrt[3]{b}$ **43.** $2a^2\sqrt{ab}$ **45.** $2a^2b^2\sqrt{ab}$
47. $\dfrac{1 + 2\sqrt{5}}{5}$ **49.** $\dfrac{4 - 6\sqrt{3}}{9}$ **51.** $\dfrac{4\sqrt{5} + 5\sqrt{6}}{10}$ **53.** $\dfrac{6\sqrt{5} + \sqrt{10}}{5}$ **55.** $\dfrac{7\sqrt{3}}{12}$ **57.** $\dfrac{\sqrt{x}}{2x}$ **59.** $\dfrac{5\sqrt{xy} - 4y\sqrt{x}}{xy}$
61. 17 units **63.** $13\sqrt{13} \approx 46.87$ feet **65.** 18.23 feet

Solutions to trial exercise problems

12. $7\sqrt[3]{11} - 3\sqrt[3]{7} + 2\sqrt[3]{11} = (7\sqrt[3]{11} + 2\sqrt[3]{11}) - 3\sqrt[3]{7} = 9\sqrt[3]{11} - 3\sqrt[3]{7}$ **16.** $\sqrt{12} + 4\sqrt{3} = \sqrt{4 \cdot 3} + 4\sqrt{3}$
$= 2\sqrt{3} + 4\sqrt{3} = 6\sqrt{3}$ **41.** $\sqrt[3]{a^6b} + 3a^2\sqrt[3]{b} = \sqrt[3]{a^3a^3b} + 3a^2\sqrt[3]{b} = a \cdot a\sqrt[3]{b} + 3a^2\sqrt[3]{b} = a^2\sqrt[3]{b} + 3a^2\sqrt[3]{b}$

$= 4a^2\sqrt[3]{b}$ **52.** $\dfrac{4}{\sqrt{7}} - \dfrac{2}{\sqrt{14}} = \dfrac{4}{\sqrt{7}} \dfrac{\sqrt{7}}{\sqrt{7}} - \dfrac{2}{\sqrt{14}} \dfrac{\sqrt{14}}{\sqrt{14}} = \dfrac{4\sqrt{7}}{7} - \dfrac{2\sqrt{14}}{14} = \dfrac{4\sqrt{7}}{7} - \dfrac{\sqrt{14}}{7} = \dfrac{4\sqrt{7} - \sqrt{14}}{7}$

Exercise 5–5

Answers to odd-numbered problems

1. $3\sqrt{5} + 3\sqrt{3}$ **3.** $12\sqrt{7} + 4\sqrt{2}$ **5.** $\sqrt{6} + \sqrt{15}$ **7.** $3\sqrt{6} - \sqrt{22}$ **9.** $35\sqrt{10} - 20\sqrt{15}$ **11.** $5\sqrt{3} - 10$
13. $14\sqrt{5} - 56\sqrt{2}$ **15.** $x + \sqrt{xy}$ **17.** $6x\sqrt{y} - 15x$ **19.** $5x\sqrt{y} + 20y\sqrt{x}$ **21.** $9 + 5\sqrt{3}$ **23.** $20 + 9\sqrt{x} + x$
25. $3 + 2\sqrt{y} - 8y$ **27.** 7 **29.** -3 **31.** -2 **33.** $a - b^2$ **35.** $9x - 16y$ **37.** $11 - 4\sqrt{7}$ **39.** $91 - 40\sqrt{3}$
41. $4a + 4b\sqrt{a} + b^2$ **43.** $20x - \sqrt{xy} - y$ **45.** $2 - \sqrt{3}$ **47.** $\dfrac{12 - 3\sqrt{6}}{5}$ **49.** $\sqrt{10} + \sqrt{6}$ **51.** $\dfrac{6 + 3\sqrt{3}}{2}$

53. $\dfrac{-\sqrt{3} - 3\sqrt{2}}{5}$ **55.** $\dfrac{21\sqrt{2} + 4\sqrt{7}}{55}$ **57.** $\dfrac{x - \sqrt{xy}}{x - y}$ **59.** $\dfrac{a - \sqrt{a}}{a - 1}$ **61.** $\dfrac{x - 2y\sqrt{x} + y^2}{x - y^2}$ **63.** $\dfrac{\sqrt{ab} + \sqrt{a}}{b - 1}$

65. $\dfrac{2\sqrt{a} + 2\sqrt{ab}}{1 - b}$ **67.** $\dfrac{\sqrt{b} + 1}{b - 1}$ **69.** $\dfrac{T\sqrt{x^2 + r^2}}{(x^2 + r^2)^2}$

Solutions to trial exercise problems

10. $\sqrt{6}(\sqrt{2} + \sqrt{3}) = \sqrt{6 \cdot 2} + \sqrt{6 \cdot 3} = \sqrt{12} + \sqrt{18} = \sqrt{4 \cdot 3} + \sqrt{9 \cdot 2} = 2\sqrt{3} + 3\sqrt{2}$ **27.** $(3 - \sqrt{2})(3 + \sqrt{2})$
$= 9 + 3\sqrt{2} - 3\sqrt{2} - \sqrt{2}\sqrt{2} = 9 - 2 = 7.$ Since these are conjugate factors, we could have written $(3 - \sqrt{2})(3 + \sqrt{2})$

$= (3)^2 - (\sqrt{2})^2 = 9 - 2 = 7.$ **47.** $\dfrac{6}{4 + \sqrt{6}} = \dfrac{6}{4 + \sqrt{6}} \cdot \dfrac{4 - \sqrt{6}}{4 - \sqrt{6}} = \dfrac{6(4 - \sqrt{6})}{(4)^2 - (\sqrt{6})^2} = \dfrac{6(4 - \sqrt{6})}{16 - 6}$

$= \dfrac{6(4 - \sqrt{6})}{10} = \dfrac{3(4 - \sqrt{6})}{5} = \dfrac{12 - 3\sqrt{6}}{5}$ **53.** $\dfrac{\sqrt{6}}{\sqrt{2} - 2\sqrt{3}} = \dfrac{\sqrt{6}}{\sqrt{2} - 2\sqrt{3}} \cdot \dfrac{\sqrt{2} + 2\sqrt{3}}{\sqrt{2} + 2\sqrt{3}}$

$= \dfrac{\sqrt{6}(\sqrt{2} + 2\sqrt{3})}{(\sqrt{2})^2 - (2\sqrt{3})^2} = \dfrac{\sqrt{6 \cdot 2} + 2\sqrt{3 \cdot 6}}{2 - 4 \cdot 3} = \dfrac{\sqrt{12} + 2\sqrt{18}}{2 - 12} = \dfrac{\sqrt{4 \cdot 3} + 2\sqrt{9 \cdot 2}}{-10} = \dfrac{2\sqrt{3} + 2 \cdot 3\sqrt{2}}{-10} = \dfrac{2(\sqrt{3} + 3\sqrt{2})}{-10}$

$= \dfrac{-(\sqrt{3} + 3\sqrt{2})}{5} = \dfrac{-\sqrt{3} - 3\sqrt{2}}{5}$ **60.** $\dfrac{\sqrt{a} + b}{\sqrt{a} - b} = \dfrac{\sqrt{a} + b}{\sqrt{a} - b} \cdot \dfrac{\sqrt{a} + b}{\sqrt{a} + b} = \dfrac{(\sqrt{a} + b)(\sqrt{a} + b)}{(\sqrt{a})^2 - (b)^2}$

$= \dfrac{\sqrt{a}\sqrt{a} + b\sqrt{a} + b\sqrt{a} + b^2}{a - b^2} = \dfrac{a + 2b\sqrt{a} + b^2}{a - b^2}$ **63.** $\dfrac{a}{\sqrt{ab} - \sqrt{a}} = \dfrac{a}{\sqrt{ab} - \sqrt{a}} \cdot \dfrac{\sqrt{ab} + \sqrt{a}}{\sqrt{ab} + \sqrt{a}} = \dfrac{a(\sqrt{ab} + \sqrt{a})}{(\sqrt{ab})^2 - (\sqrt{a})^2}$

$= \dfrac{a(\sqrt{ab} + \sqrt{a})}{ab - a} = \dfrac{a(\sqrt{ab} + \sqrt{a})}{a(b - 1)} = \dfrac{\sqrt{ab} + \sqrt{a}}{b - 1}$

Exercise 5–6

Answers to odd-numbered problems

1. $3i$ **3.** $2i\sqrt{3}$ **5.** -9 **7.** -3 **9.** $-\sqrt{15}$ **11.** -2 **13.** -5 **15.** -6 **17.** $-3\sqrt{5}$ **19.** -1 **21.** 1 **23.** $-i$
25. $8 + 7i$ **27.** $-2 + 2i$ **29.** $1 + 8i$ **31.** i **33.** $5 - 2i$ **35.** $8 + 8i$ **37.** 13 **39.** $19 - 7i$ **41.** $19 + 17i$ **43.** 34
45. $7 - 24i$ **47.** $8i$ **49.** $5 - 4i$ **51.** $2 - i$ **53.** $\dfrac{7}{3} - \dfrac{5}{3}i$ **55.** $\dfrac{27}{26} - \dfrac{5}{26}i$ **57.** $\dfrac{4}{5} - \dfrac{3}{5}i$ **59.** $\dfrac{-4}{13} + \dfrac{19}{13}i$

61. $\dfrac{6}{5} - \dfrac{3}{5}i$ **63.** $\dfrac{4}{17} + \dfrac{18}{17}i$ **65.** $30 - 24i$ **67.** $0.571 - 0.143i$ **69.** $\dfrac{7}{5} + \dfrac{1}{5}i$ **71.** $\dfrac{60}{61} + \dfrac{50}{61}i$ **73.** $x \le 5$
75. $x < -11$

Solutions to trial exercise problems

5. $(3i)^2 = 3^2i^2 = 9(-1) = -9$ **28.** $(4 - 5i) - (3 - 7i) = 4 - 5i - 3 + 7i = 1 + 2i$ **29.** $(2 + \sqrt{-49}) - (1 - \sqrt{-1})$
$= (2 + i\sqrt{49}) - (1 - i) = (2 + 7i) - (1 - i) = 2 + 7i - 1 + i = 1 + 8i$ **32.** $[(2 + 5i) + (3 - 2i)] + (3 - i)$
$= [2 + 5i + 3 - 2i] + (3 - i) = [5 + 3i] + (3 - i) = 5 + 3i + 3 - i = 8 + 2i$ **44.** $(2 + i)^2 = (2 + i)(2 + i)$
$= 4 + 2i + 2i + i^2 = 4 + 4i + (-1) = 3 + 4i$ **48.** $\dfrac{3 - 2i}{i} = \dfrac{3 - 2i}{i} \dfrac{i}{i} = \dfrac{i(3 - 2i)}{i^2} = \dfrac{3i - 2i^2}{-1} = \dfrac{3i - 2(-1)}{-1} = \dfrac{3i + 2}{-1}$
$= -2 - 3i$ **54.** $\dfrac{4 - 3i}{1 + i} = \dfrac{4 - 3i}{1 + i} \cdot \dfrac{1 - i}{1 - i} = \dfrac{(4 - 3i)(1 - i)}{(1)^2 - (i)^2} = \dfrac{4 - 4i - 3i + 3i^2}{1 - (-1)} = \dfrac{4 - 7i + 3(-1)}{2}$
$= \dfrac{4 - 7i + (-3)}{2} = \dfrac{1 - 7i}{2} = \dfrac{1}{2} - \dfrac{7}{2}i$

Chapter 5 review exercises

1. 6   **2.** $\dfrac{1}{8}$ **3.** 9   **4.** $a^{\frac{11}{12}}$ **5.** $c^{\frac{1}{4}}$ **6.** $9x^2$ **7.** b^2 **8.** $a^{\frac{1}{6}}$ **9.** $4x^8y^4$ **10.** $x^{\frac{19}{6}}$ **11.** $a^{\frac{7}{6}}$ **12.** $2\sqrt{3}$ **13.** $5\sqrt{6}$
14. $x\sqrt[5]{x^2}$ **15.** $2ab\sqrt[3]{3b}$ **16.** $a\sqrt{a}$ **17.** $\sqrt[3]{2ab^2}$ **18.** 8 in.   **19.** $\dfrac{7}{8}$ **20.** $\dfrac{4\sqrt{3}}{9}$ **21.** $\dfrac{2x\sqrt[3]{2y^2}}{z^2}$ **22.** $\dfrac{\sqrt{2}}{4}$
23. $3\sqrt{2}$ **24.** $\dfrac{2\sqrt[3]{5}}{5}$ **25.** $\dfrac{x\sqrt{y}}{y}$ **26.** $\dfrac{\sqrt[3]{a^2b^2}}{b}$ **27.** $\sqrt[5]{x^3}$ **28.** $\dfrac{\sqrt[3]{ab^2c}}{bc}$ **29.** $\dfrac{\sqrt[4]{a^3b^2}}{b}$ **30.** $\dfrac{x\sqrt{y}}{y}$ **31.** $8\sqrt{3}$
32. $8\sqrt{2}$ **33.** $29\sqrt{2a}$ **34.** $3x^2\sqrt{xy}$ **35.** $\dfrac{5\sqrt{6} - 2\sqrt{3}}{6}$ **36.** $\dfrac{2\sqrt{ab} - b\sqrt{a}}{ab}$ **37.** $2\sqrt{3} - 2\sqrt{5}$ **38.** $2a\sqrt{b} + 4a$
39. $30 - 10\sqrt{5}$ **40.** 3   **41.** $4a - 9b$ **42.** $9x + 6y\sqrt{x} + y^2$ **43.** $\dfrac{\sqrt{6} - 2}{2}$ **44.** $4 - \sqrt{6}$ **45.** $\sqrt{2} + 1$
46. $\dfrac{a^2b\sqrt{a} + ab\sqrt{ab}}{a^2 - b}$ **47.** $7i$ **48.** $2i\sqrt{7}$ **49.** -4 **50.** -7 **51.** -6 **52.** -3 **53.** $-\sqrt{6}$ **54.** i **55.** $7 + 7i$
56. $-1 - 11i$ **57.** $18 - i$ **58.** $-21 + 20i$ **59.** $4 - 3i$ **60.** $-2 - \dfrac{7}{3}i$ **61.** $\dfrac{7}{5} - \dfrac{6}{5}i$ **62.** $\dfrac{69}{58} + \dfrac{13}{58}i$

Chapter 5 cumulative test

1. $(a - 8)(a + 1)$ **2.** $x(4x - 3)$ **3.** $9(x - 2)(x + 2)$ **4.** $(2x + 3)(x + 4)$ **5.** $(3a + 4)(a - 5)$ **6.** $(3x + 4)(2x + 3)$
7. (a) 28, (b) -8 **8.** $\left\{\dfrac{2}{5}\right\}$ **9.** $\left\{x \mid x > -\dfrac{11}{3}\right\}$ **10.** $-6y$ **11.** $\left\{-1, \dfrac{5}{3}\right\}$ **12.** $\left\{x \mid x < -\dfrac{11}{2} \text{ or } x > \dfrac{5}{2}\right\}$ **13.** $\left\{-\dfrac{13}{5}\right\}$
14. $\left\{x \mid -\dfrac{3}{2} \le x \le 2\right\}$ **15.** $2a^2b\sqrt[5]{2b^2}$ **16.** $10 - 5i$ **17.** $4\sqrt{3}$ **18.** $a^{\frac{7}{12}}$ **19.** $2ab^2\sqrt[3]{a}$ **20.** $8a^9b^{12}c^3$ **21.** $3i\sqrt{2}$
22. $\sqrt{10} - \sqrt{6}$ **23.** $-\dfrac{7}{13} - \dfrac{9}{13}i$ **24.** $2a^3$ **25.** $\dfrac{x^3y^2}{3}$ **26.** $\dfrac{\sqrt[3]{2a^2b^2c}}{2bc}$ **27.** 7 inches   **28.** 400 kg of 80% copper, 600 kg of
50% copper   **29.** 375 meters per second   **30.** 1,166.4 meters per second

Chapter 6

Exercise 6–1

Answers to odd-numbered problems

1. $\{3, -4\}$ **3.** $\left\{\dfrac{1}{3}, -\dfrac{5}{2}\right\}$ **5.** $\left\{0, -\dfrac{4}{3}, -10\right\}$ **7.** $\{2, 3\}$ **9.** $\{5\}$ **11.** $\{-3, 8\}$ **13.** $\{0, 1\}$ **15.** $\{-3, 3\}$ **17.** $\left\{-\dfrac{1}{2}, 2\right\}$
19. $\left\{-2, \dfrac{3}{4}\right\}$ **21.** $\{2\}$ **23.** $\{-8, 1\}$ **25.** $\{-5, 1\}$ **27.** $\left\{-\dfrac{1}{3}, 2\right\}$ **29.** $\left\{-\dfrac{5}{2}, 3\right\}$ **31.** $\{-6, 1\}$ **33.** $x = -2b, 12b$
35. $x = -\dfrac{7a}{4}, 2a$ **37.** $x = y$ **39.** $x = -\dfrac{4y}{3}, \dfrac{y}{2}$ **41.** $\{-11, 11\}$ **43.** $\{-7, 7\}$ **45.** $\{-4\sqrt{2}, 4\sqrt{2}\}$ **47.** $\{-6\sqrt{2}, 6\sqrt{2}\}$
49. $\{-2\sqrt{2}, 2\sqrt{2}\}$ **51.** $\{-5\sqrt{2}, 5\sqrt{2}\}$ **53.** $\{-1, -13\}$ **55.** $\{12 + 11i, 12 - 11i\}$ **57.** $\{-10 + 4\sqrt{3}, -10 - 4\sqrt{3}\}$

59. $\left\{2, -\dfrac{5}{2}\right\}$ **61.** $\left\{\dfrac{3 + 2i\sqrt{21}}{10}, \dfrac{3 - 2i\sqrt{21}}{10}\right\}$ **63.** $y = -8 \pm b$ **65.** (a) $t = 4$ sec, (b) $t = 2$ sec **67.** $t = 1$ sec

69. $n = 7$ **71.** $7m$ **73.** $8,15,17$ **75.** $6,8; -6,-8$ **77.** $7,9; -7,-9$

Solutions to trial exercise problems

5. $x(3x + 4)(x + 10) = 0$

$x = -\dfrac{4}{3}$ when $3x + 4 = 0$,

$x = -10$ when $x + 10 = 0$,

and $x = 0$, then

$S = \left\{0, -\dfrac{4}{3}, -10\right\}.$

23. $\dfrac{x}{2} + \dfrac{7}{2} = \dfrac{4}{x}$

Multiply each member by the LCM, $2x$.

$2x \cdot \dfrac{x}{2} + 2x \cdot \dfrac{7}{2} = 2x \cdot \dfrac{4}{x}$

$x^2 + 7x = 8$

$x^2 + 7x - 8 = 0$

$(x + 8)(x - 1) = 0$

$x = -8$ when $x + 8 = 0$ and

$x = 1$ when $x - 1 = 0$, so

$S = \{-8, 1\}.$

34. $3x^2 - 13xy + 4y^2 = 0$

$(3x - y)(x - 4y) = 0$

$x = \dfrac{y}{3}$ when $3x - y = 0$ and

$x = 4y$ when $x - 4y = 0$, so

$x = \dfrac{y}{3}$ or $x = 4y.$

62. $(x - 7)^2 = a^2, a > 0$

$x - 7 = \sqrt{a^2} = a$ or $x - 7 = -\sqrt{a^2} = -a$

Then $x = 7 + a$ or $x = 7 - a$

$x = 7 \pm a$

15. $-3y^2 + 27 = 0$

$-3(y^2 - 9) = 0$

$-3(y + 3)(y - 3) = 0$

$y = -3$ when $y + 3 = 0, y = 3$

when $y - 3 = 0$, so

$S = \{-3, 3\}.$

25. $(y + 6)(y - 2) = -7$

$y^2 + 4y - 12 = -7$

$y^2 + 4y - 5 = 0$

$(y + 5)(y - 1) = 0$

$y = -5$ when $y + 5 = 0$

and $y = 1$ when $y - 1 = 0$

$S = \{-5, 1\}$

54. $(x - 9)^2 = -144$

$x - 9 = \sqrt{-144} = 12i$ or $x - 9 = -\sqrt{-144} = -12i$

Then $x = 9 + 12i$ or $x = 9 - 12i$

$S = \{9 + 12i, 9 - 12i\}$

64. (a) $P = 100I - 5I^2$

$420 = 100I - 5I^2$

$5I^2 - 100I + 420 = 0$

$5(I^2 - 20I + 84) = 0$

$5(I - 6)(I - 14) = 0$

$I = 6$ when $I - 6 = 0$ and $I = 14$ when $I - 14 = 0$

So $P = 420$ when $I = 6$ amperes or $I = 14$ amperes.

Exercise 6–2

Answers to odd-numbered problems

1. $x^2 + 4x + 4 = (x + 2)^2$ **3.** $y^2 - 18y + 81 = (y - 9)^2$ **5.** $p^2 + 2p + 1 = (p + 1)^2$ **7.** $x^2 + 3x + \dfrac{9}{4} = \left(x + \dfrac{3}{2}\right)^2$

9. $w^2 - 11w + \dfrac{121}{4} = \left(w - \dfrac{11}{2}\right)^2$ **11.** $x^2 + 13x + \dfrac{169}{4} = \left(x + \dfrac{13}{2}\right)^2$ **13.** $\{-11, -1\}$ **15.** $\{1, 10\}$

17. $\{-4 + 3i, -4 - 3i\}$ **19.** $\{0, 8\}$ **21.** $\{-1, 0\}$ **23.** $\left\{\dfrac{-1 + \sqrt{13}}{2}, \dfrac{-1 - \sqrt{13}}{2}\right\}$ **25.** $\left\{-\dfrac{2}{3}, \dfrac{1}{2}\right\}$ **27.** $\left\{-\dfrac{1}{2}, 4\right\}$

29. $\left\{-2, \dfrac{1}{2}\right\}$ **31.** $\left\{\dfrac{1}{3}, 3\right\}$ **33.** $\left\{-\dfrac{1}{2}, \dfrac{3}{2}\right\}$ **35.** $\left\{\dfrac{3 - i\sqrt{11}}{10}, \dfrac{3 + i\sqrt{11}}{10}\right\}$ **37.** $\left\{\dfrac{1 + \sqrt{29}}{2}, \dfrac{1 - \sqrt{29}}{2}\right\}$ **39.** $\left\{\dfrac{1}{4}, -\dfrac{3}{2}\right\}$

41. $\left\{\dfrac{8 + \sqrt{74}}{10}, \dfrac{8 - \sqrt{74}}{10}\right\}$ **43.** $\left\{\dfrac{3 + \sqrt{41}}{4}, \dfrac{3 - \sqrt{41}}{4}\right\}$ **45.** $\left\{\dfrac{1 + 2i\sqrt{2}}{6}, \dfrac{1 - 2i\sqrt{2}}{6}\right\}$ **47.** $\left\{\dfrac{3 + \sqrt{137}}{8}, \dfrac{3 - \sqrt{137}}{8}\right\}$

49. $\left\{\dfrac{5}{2}, -1\right\}$ **51.** $\left\{-\dfrac{3}{5}, 1\right\}$ **53.** $t = \dfrac{-9 + \sqrt{6,481}}{32}$ sec ≈ 2.23 sec **55.** $p = -16 + 6\sqrt{21}$ ¢ ≈ 12¢ **57.** $h = 25$ or $h = 1$

59. $\dfrac{1 + \sqrt{65}}{2}, \dfrac{1 - \sqrt{65}}{2}$ **61.** $\sqrt{\dfrac{235}{10\pi}} = \dfrac{\sqrt{94\pi}}{2\pi}$

Solutions to trial exercise problems

7. $\left[\dfrac{1}{2}(3)\right]^2 = \left(\dfrac{3}{2}\right)^2 = \dfrac{9}{4}$

$x^2 + 3x + \dfrac{9}{4} = \left(x + \dfrac{3}{2}\right)^2$

20. $x^2 + 4x = 0$

$x^2 + 4x + 4 = 4$

$(x + 2)^2 = 4$

$x + 2 = \pm 2$

$x = -2 \pm 2$

$x = 0 \text{ or } x = -4$

$S = \{0, -4\}$

23. $y^2 = 3 - y$

$y^2 + y = 3$

$y^2 + y + \dfrac{1}{4} = 3 + \dfrac{1}{4}$

$\left(y + \dfrac{1}{2}\right)^2 = \dfrac{13}{4}$

$y + \dfrac{1}{2} = \pm \dfrac{\sqrt{13}}{2}$

$y = -\dfrac{1}{2} \pm \dfrac{\sqrt{13}}{2} = \dfrac{-1 \pm \sqrt{13}}{2}$

$S = \left\{\dfrac{-1 + \sqrt{13}}{2}, \dfrac{-1 - \sqrt{13}}{2}\right\}$

37. $(x + 2)(x - 3) = 1$

$x^2 - x - 6 = 1$

$x^2 - x = 7$

$x^2 - x + \dfrac{1}{4} = 7 + \dfrac{1}{4}$

$\left(x - \dfrac{1}{2}\right)^2 = \dfrac{29}{4}$

$x - \dfrac{1}{2} = \pm \dfrac{\sqrt{29}}{2}$

$x = \dfrac{1}{2} \pm \dfrac{\sqrt{29}}{2} = \dfrac{1 \pm \sqrt{29}}{2}$

$S = \left\{\dfrac{1 + \sqrt{29}}{2}, \dfrac{1 - \sqrt{29}}{2}\right\}$

43. $\dfrac{1}{2}x^2 - \dfrac{3}{4}x = 1$ Multiply by the LCM, 4.

$2x^2 - 3x = 4$ Multiply by $\dfrac{1}{2}$.

$x^2 - \dfrac{3}{2}x = 2$

$x^2 - \dfrac{3}{2}x + \dfrac{9}{16} = 2 + \dfrac{9}{16}$

$\left(x - \dfrac{3}{4}\right)^2 = \dfrac{41}{16}$

$x - \dfrac{3}{4} = \pm \dfrac{\sqrt{41}}{4}$

$x = \dfrac{3}{4} \pm \dfrac{\sqrt{41}}{4} = \dfrac{3 \pm \sqrt{41}}{4}$

$S = \left\{\dfrac{3 + \sqrt{41}}{4}, \dfrac{3 - \sqrt{41}}{4}\right\}$

49. $\dfrac{5}{x} - 2x + 3 = 0$ Multiply by the LCM, x.

$5 - 2x^2 + 3x = 0$ Multiply each member by -1.

$2x^2 - 3x - 5 = 0$

$2x^2 - 3x = 5$

$x^2 - \dfrac{3}{2}x = \dfrac{5}{2}$

$x^2 - \dfrac{3}{2}x + \dfrac{9}{16} = \dfrac{5}{2} + \dfrac{9}{16}$

$\left(x - \dfrac{3}{4}\right)^2 = \dfrac{49}{16}$

$x - \dfrac{3}{4} = \pm \dfrac{7}{4}$

$x = \dfrac{3}{4} \pm \dfrac{7}{4} = \dfrac{3 \pm 7}{4}$

$S = \left\{\dfrac{5}{2}, -1\right\}$

53. $s = 100$, so $100 = 9t + 16t^2$.

$16t^2 + 9t - 100 = 0$

$t^2 + \dfrac{9}{16}t - \dfrac{100}{16} = 0$

$t^2 + \dfrac{9}{16}t = \dfrac{25}{4}$

$t^2 + \dfrac{9}{16}t + \dfrac{81}{1,024} = \dfrac{25}{4} + \dfrac{81}{1,024}$

$\left(t + \dfrac{9}{32}\right)^2 = \dfrac{6,400 + 81}{1,024} = \dfrac{6,481}{1,024}$

$t + \dfrac{9}{32} = \pm \dfrac{\sqrt{6,481}}{32}$

$t = -\dfrac{9}{32} \pm \dfrac{\sqrt{6,481}}{32} = \dfrac{-9 \pm \sqrt{6,481}}{32}$

Therefore $t = \dfrac{-9 + \sqrt{6,481}}{32}$ sec ≈ 2.23 sec. (We discard the other value of t since $t > 0$.)

Exercise 6-3

Answers to odd-numbered problems

1. $\{-8,2\}$ **3.** $\{7\}$ **5.** $\left\{\dfrac{-2\sqrt{15}}{3}, \dfrac{2\sqrt{15}}{3}\right\}$ **7.** $\left\{0, \dfrac{2}{5}\right\}$ **9.** $\left\{\dfrac{-5 + i\sqrt{3}}{2}, \dfrac{-5 - i\sqrt{3}}{2}\right\}$ **11.** $\left\{\dfrac{3 - i\sqrt{11}}{2}, \dfrac{3 + i\sqrt{11}}{2}\right\}$

13. $\left\{2, \dfrac{3}{2}\right\}$ **15.** $\left\{\dfrac{2 + \sqrt{3}}{2}, \dfrac{2 - \sqrt{3}}{2}\right\}$ **17.** $\left\{\dfrac{2}{3}\right\}$ **19.** $\left\{\dfrac{1 - 2i\sqrt{5}}{3}, \dfrac{1 + 2i\sqrt{5}}{3}\right\}$ **21.** $\left\{\dfrac{2 - i}{3}, \dfrac{2 + i}{3}\right\}$

23. $\left\{\dfrac{-3 - 2\sqrt{5}}{3}, \dfrac{-3 + 2\sqrt{5}}{3}\right\}$ **25.** $\left\{\dfrac{1 - i\sqrt{55}}{4}, \dfrac{1 + i\sqrt{55}}{4}\right\}$ **27.** $\left\{\dfrac{1 + \sqrt{113}}{8}, \dfrac{1 - \sqrt{113}}{8}\right\}$ **29.** $\left\{\dfrac{3 + i\sqrt{3}}{4}, \dfrac{3 - i\sqrt{3}}{4}\right\}$

31. $\left\{-2, \dfrac{8}{3}\right\}$ **33.** $\left\{\dfrac{11 + \sqrt{145}}{6}, \dfrac{11 - \sqrt{145}}{6}\right\}$ **35.** $\left\{\dfrac{3 + \sqrt{21}}{2}, \dfrac{3 - \sqrt{21}}{2}\right\}$ **37.** $\left\{-4, \dfrac{2}{3}\right\}$ **39.** $\{6y, -3y\}$

41. $x = \dfrac{-1 \pm \sqrt{1 + 12y}}{4}$ **43.** $x = 2a \pm \sqrt{4a^2 - 3a}$ **45.** two, irrational   **47.** two, complex   **49.** one, rational

51. two, rational   **53.** two, complex   **55.** two, irrational   **57.** two, rational   **59.** two, irrational   **61.** two, irrational

63. $c = 15, a = 12, b = 9$ **65.** (a) $\dfrac{-3 + \sqrt{73}}{4}$ sec, (b) $\dfrac{-9 + \sqrt{401}}{8}$ sec   **67.** $r = 9.5\%$

Solutions to trial exercise problems

17. $9x^2 - 12x + 4 = 0$

$a = 9, b = -12, c = 4$

$x = \dfrac{-(-12) \pm \sqrt{(-12)^2 - 4(9)(4)}}{2(9)}$

$= \dfrac{12 \pm \sqrt{144 - 144}}{18} = \dfrac{12 \pm \sqrt{0}}{18} = \dfrac{12}{18} = \dfrac{2}{3}$

$S = \left\{\dfrac{2}{3}\right\}$

26. $3x - \dfrac{2}{x} + 5 = 0$

$3x^2 - 2 + 5x = 0$

$3x^2 + 5x - 2 = 0$

$a = 3, b = 5, c = -2$

$x = \dfrac{-5 \pm \sqrt{5^2 - 4(3)(-2)}}{2(3)}$

$x = \dfrac{-5 \pm \sqrt{25 + 24}}{6}$

$= \dfrac{-5 \pm \sqrt{49}}{6} = \dfrac{-5 \pm 7}{6}$

$x = \dfrac{-5 + 7}{6} = \dfrac{2}{6} = \dfrac{1}{3}$ or $x = \dfrac{-5 - 7}{6} = \dfrac{-12}{6} = -2$

$S = \left\{-2, \dfrac{1}{3}\right\}$

27. $2y^2 - \dfrac{7}{2} = \dfrac{y}{2}$

$4y^2 - 7 = y$

$4y^2 - y - 7 = 0$

$a = 4, b = -1, c = -7$

$y = \dfrac{-(-1) \pm \sqrt{(-1)^2 - 4(4)(-7)}}{2(4)}$

$= \dfrac{1 \pm \sqrt{1 + 112}}{8} = \dfrac{1 \pm \sqrt{113}}{8}$

$S = \left\{\dfrac{1 + \sqrt{113}}{8}, \dfrac{1 - \sqrt{113}}{8}\right\}$

32. $\dfrac{1}{x + 2} + \dfrac{1}{x - 3} - 2 = 0$ Multiply by the LCM, $(x + 2)(x - 3)$.

$(x - 3) + (x + 2) - 2(x + 2)(x - 3) = 0$

$x - 3 + x + 2 - 2(x^2 - x - 6) = 0$

$2x - 1 - 2x^2 + 2x + 12 = 0$

$-2x^2 + 4x + 11 = 0$

Multiply by -1 to get $2x^2 - 4x - 11 = 0$.

$a = 2, b = -4$, and $c = -11$

$x = \dfrac{-(-4) \pm \sqrt{(-4)^2 - 4(2)(-11)}}{2(2)}$

$= \dfrac{4 \pm \sqrt{16 + 88}}{4} = \dfrac{4 \pm \sqrt{104}}{4} = \dfrac{4 \pm 2\sqrt{26}}{4} = \dfrac{2(2 \pm \sqrt{26})}{4} = \dfrac{2 \pm \sqrt{26}}{2}$

$S = \left\{\dfrac{2 + \sqrt{26}}{2}, \dfrac{2 - \sqrt{26}}{2}\right\}$

35. $(z - 3)(z + 2) = 2z - 3$

$z^2 - z - 6 = 2z - 3$

$z^2 - 3z - 3 = 0$

$a = 1, b = -3, c = -3$

$z = \dfrac{-(-3) \pm \sqrt{(-3)^2 - 4(1)(-3)}}{2(1)}$

$= \dfrac{3 \pm \sqrt{9 + 12}}{2} = \dfrac{3 \pm \sqrt{21}}{2}$

$S = \left\{ \dfrac{3 + \sqrt{21}}{2}, \dfrac{3 - \sqrt{21}}{2} \right\}$

41. $4x^2 + 2x - 3y = 0$

$a = 4, b = 2, c = -3y$

$x = \dfrac{-2 \pm \sqrt{(2)^2 - 4(4)(-3y)}}{2(4)}$

$= \dfrac{-2 \pm \sqrt{4 + 48y}}{8} = \dfrac{-2 \pm \sqrt{4(1 + 12y)}}{8} = \dfrac{-2 \pm 2\sqrt{1 + 12y}}{8}$

$= \dfrac{-1 \pm \sqrt{1 + 12y}}{4}$

$x = \dfrac{-1 \pm \sqrt{1 + 12y}}{4}$

62. (b) $60 = \dfrac{1}{2}(32)t^2$

$60 = 16t^2$

$16t^2 - 60 = 0$

$a = 16, b = 0, c = -60$

$t = \dfrac{0 \pm \sqrt{0^2 - 4(16)(-60)}}{2(16)} = \dfrac{\pm \sqrt{3,840}}{32} = \dfrac{\pm 16\sqrt{15}}{32} = \dfrac{\pm \sqrt{15}}{2}$

Then $t = \dfrac{\sqrt{15}}{2} \approx 1.9$ sec. $\left(\text{Discard } t = \dfrac{-\sqrt{15}}{2} \text{ since } t > 0. \right)$

Exercise 6–4

Answers to odd-numbered problems

1. $t = 1$ sec   **3.** $t = 3$ sec   **5.** $t = 6$ sec   **7.** $t \approx 5.9$ sec   **9.** $t = 0$ sec and $t = 1.5$ sec   **11.** 30 amperes   **13.** 30 sides
15. $7 + 4\sqrt{7} \approx 17.6$ in.   **17.** 25 in.   **19.** $-15 + 15\sqrt{127}$ ft ≈ 154 ft by $15 + 15\sqrt{127}$ ft ≈ 184 ft   **21.** 5 ft   **23.** 13 in.
25. 50 yd by 100 yd   **27.** base $= 5$ in., altitude $= 6$ in.   **29.** $D = 4$ ft   **31.** \$3.00   **33.** (a) 15 pens, (b) 15 pens
35. $-1 + \sqrt{51} \approx 6$ units   **37.** 120 cakes   **39.** $20 \pm 8\sqrt{5} \approx 37.9$ or 2.1
41. Lisa, 1 hr 5 min (65 min); Debbie, 1 hr 44 min (104 min)   **43.** $1 + \sqrt{17} \approx 5.12$ hr

Solutions to trial exercise problems

4. $h = 80$ and $v_0 = 96$, so $80 = 96t - 16t^2$, then

$16t^2 - 96t + 80 = 0$

$16(t^2 - 6t + 5) = 0$

$16(t - 5)(t - 1) = 0$

The object reaches $h = 80$ feet at $t = 1$ sec.

Note: The object is at $h = 80$ feet again at $t = 5$ sec, on its way down to earth.

21. Let $s =$ the length of the side of the square.

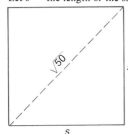

Using $s^2 + s^2 = (\sqrt{50})^2$

$2s^2 = 50$

so $2s^2 - 50 = 0$

$2(s^2 - 25) = 0$

$2(s - 5)(s + 5) = 0$, so $s = 5$ or $s = -5$.

Then $s = 5$ feet. (*Note:* Reject $s = -5$ since the length of the side cannot be negative.)

26. Let $x =$ the amount the dimensions are increased. Since the area of the original rectangle is 18 cm², and the new rectangle has area $3 \cdot 18 = 54$ cm², then

$(x + 6)(x + 3) = 54$

$x^2 + 9x + 18 = 54$

$x^2 + 9x - 36 = 0$

$(x + 12)(x - 3) = 0$, then $x = -12$ or $x = 3$.

Reject -12, so $x = 3$ cm and the dimensions are $3 + 6 = 9$ cm long and $3 + 3 = 6$ cm wide.

Appendix B   Answers and Solutions

36. Using $20,000 = \dfrac{1}{100}\, n^2 - 20\, n,$

$$2,000,000 = n^2 - 2,000\, n$$
$$n^2 - 2,000\, n - 2,000,000 = 0$$
$$n = \frac{-(-2,000) \pm \sqrt{(-2,000)^2 - 4(1)(-2,000,000)}}{2(1)}$$
$$= \frac{2,000 \pm \sqrt{4,000,000 + 8,000,000}}{2}$$
$$= \frac{2,000 \pm \sqrt{12,000,000}}{2} = \frac{2,000 \pm 2,000\sqrt{3}}{2}$$

Then $n = \dfrac{2,000 + 2,000\sqrt{3}}{2} = 1,000 + 1,000\sqrt{3} \approx 2,732\ (n > 0).$

Thus about 2,732 units must be produced to make a $20,000 profit.

41. Let $x =$ the time in minutes for Lisa to do the job. Then $x + 39 =$ the time in minutes for Debbie to do the job.

Then $\dfrac{1}{x} + \dfrac{1}{x + 39} = \dfrac{1}{40}$

$$40(x + 39) + 40x = x(x + 39)$$
$$40x + 1,560 + 40x = x^2 + 39x$$
$$x^2 - 41x - 1,560 = 0$$
$$x = \frac{-(-41) \pm \sqrt{(-41)^2 - 4(1)(-1,560)}}{2(1)}$$
$$= \frac{41 \pm \sqrt{1,681 + 6,240}}{2} = \frac{41 \pm \sqrt{7,921}}{2} = \frac{41 \pm 89}{2}$$

Thus $x = \dfrac{41 + 89}{2} = \dfrac{130}{2} = 65$ or $x = \dfrac{41 - 89}{2} = \dfrac{-48}{2} = -24$ (reject this answer).

Therefore Lisa can do the job in 65 minutes and Debbie can do the job in $65 + 39 = 104$ minutes.

Exercise 6-5

Answers to odd-numbered problems

1. $\{81\}$ **3.** $\emptyset$ **5.** $\{53\}$ **7.** $\{-5 + \sqrt{34}, -5 - \sqrt{34}\}$ **9.** $\{24\}$ **11.** $\{1\}$ **13.** $\{1\}$ **15.** $\{1\}$ **17.** $\{9\}$; -4 is extraneous

19. $\{6\}$, 3 is extraneous   **21.** $\{8\}$; 2 is extraneous   **23.** $\{1,4\}$ **25.** $\{5\}$; 2 is extraneous   **27.** $\left\{\dfrac{5}{2}\right\}$; 2 is extraneous   **29.** $\{0,3\}$

31. $\{3,11\}$ **33.** $\{0,8\}$ **35.** $\left\{\dfrac{3}{4}\right\}$ **37.** $\left\{-\dfrac{20}{9}, -4\right\}$ **39.** $\{34\}$ **41.** $\{-2,8\}$ **43.** $\left\{-\dfrac{5}{4}\right\}$ **45.** $y = \dfrac{9}{16x^3}$

47. $p = \pm\sqrt{q^2 + r^2}$ **49.** $A = 4\pi r^2 h$ **51.** $A = \pi(r^2 + R^2)$ **53.** $A = \dfrac{\pi D^3}{6}$ **55.** 6 ft   **57.** 75 ft   **59.** -27

Solutions to trial exercise problems

16.
$$\sqrt{p}\ \sqrt{p - 8} = 3$$
$$\text{Squaring, } p(p - 8) = 9$$
$$p^2 - 8p = 9$$
$$p^2 - 8p - 9 = 0$$
$$(p - 9)(p + 1) = 0$$
$$p = 9 \text{ or } p = -1$$

$S = \{9\}$
-1 is an extraneous solution.

26.
$$\sqrt{3x + 10} - 3x = 4$$
$$\sqrt{3x + 10} = 3x + 4$$
Square both sides.
$$3x + 10 = 9x^2 + 24x + 16$$
$$9x^2 + 21x + 6 = 0$$
$$3(3x^2 + 7x + 2) = 0$$
$$3(3x + 1)(x + 2) = 0$$
$$x = -\frac{1}{3} \text{ or } x = -2$$
$$S = \left\{-\frac{1}{3}\right\}$$
-2 is an extraneous solution.

29.
$$\sqrt{5x+1} - 1 = \sqrt{3x}$$
$$\sqrt{5x+1} = 1 + \sqrt{3x}$$
Squaring, $5x + 1 = 1 + 2\sqrt{3x} + 3x$
$$2x = 2\sqrt{3x}$$
$$x = \sqrt{3x}$$
$$x^2 = 3x$$
$$x^2 - 3x = 0$$
$$x(x - 3) = 0$$
$$x = 0 \text{ or } x = 3$$
$$S = \{0,3\}$$

34. $(2y + 3)^{\frac{1}{2}} - (4y - 1)^{\frac{1}{2}} = 1$
$$\sqrt{2y + 3} = 1 + \sqrt{4y - 1}$$
Squaring, $2y + 3 = 1 + 2\sqrt{4y - 1} + 4y - 1$
$$-2y + 3 = 2\sqrt{4y - 1}$$
$$4y^2 - 12y + 9 = 4(4y - 1)$$
$$4y^2 - 12y + 9 = 16y - 4$$
$$4y^2 - 28y + 13 = 0$$
$$(2y - 13)(2y - 1) = 0$$
$$y = \frac{13}{2} \text{ or } y = \frac{1}{2}$$
$$S = \left\{\frac{1}{2}\right\}$$
$\dfrac{13}{2}$ is an extraneous solution.

39.
$$\sqrt[3]{x - 7} = 3$$
Cubing, $x - 7 = 27$
$$x = 34$$
$$S = \{34\}$$

47.
$$\sqrt{p^2 - q^2} = r \text{ for } p$$
Squaring, $p^2 - q^2 = r^2$
$$p^2 = q^2 + r^2$$
$$p = \pm \sqrt{q^2 + r^2}$$

58. Let $x = $ the number.
Then $\sqrt{x} = 3i$ (Square each member.)
$$x = 9(-1)$$
$$x = -9.$$

Exercise 6–6

Answers to odd-numbered problems

1. $\{-\sqrt{5}, \sqrt{5}, -1, 1\}$ **3.** $\left\{-\dfrac{\sqrt{6}}{3}, -i, i, \dfrac{\sqrt{6}}{3}\right\}$ **5.** $\{2 \pm 2i\sqrt{2}, 2 \pm i\}$ **7.** $\{81\}$ **9.** $\emptyset$ **11.** $\{1,16\}$ **13.** $\{2\sqrt{13}, -2\sqrt{13}\}$

15. $\{-125, 8\}$ **17.** $\left\{-\dfrac{1}{125}, -8\right\}$ **19.** $\{1\}$ **21.** $\left\{\dfrac{1}{4}, -\dfrac{1}{3}\right\}$ **23.** $\left\{\dfrac{1}{4}, 2\right\}$ **25.** $\left\{-2, -\dfrac{1}{2}, \dfrac{1}{2}, 2\right\}$ **27.** $\{-5, -2, -1, 2\}$

Solutions to trial exercise problems

13. $(x^2 - 3) - 3\sqrt{x^2 - 3} - 28 = 0$
Let $u = \sqrt{x^2 - 3}$, then $u^2 = (\sqrt{x^2 - 3})^2 = x^2 - 3$ and so
$$u^2 - 3u - 28 = 0$$
$$(u - 7)(u + 4) = 0$$
So $u = 7$ or $u = -4$. Then substitute $\sqrt{x^2 - 3}$ for u.

$\sqrt{x^2 - 3} = 7$ or $\sqrt{x^2 - 3} = -4$
$x^2 - 3 = 49$ $x^2 - 3 = 16$
$x^2 = 52$ $x^2 = 19$
$x = \pm\sqrt{52} = \pm 2\sqrt{13}$ $x = \pm\sqrt{19}$
$S = \{2\sqrt{13}, -2\sqrt{13}\}$. *Note:* $\sqrt{19}$ and $-\sqrt{19}$ are extraneous solutions.

26. $(t^2 - t)^2 - 4(t^2 - t) - 12 = 0$
Let $u = t^2 - t$, then $u^2 = (t^2 - t)^2$ and so
$$u^2 - 4u - 12 = 0$$
$(u - 6)(u + 2) = 0$ so $u = 6$ or $u = -2$.

Then $t^2 - t = 6$ or $t^2 - t = -2$
$t^2 - t - 6 = 0$ $t^2 - t + 2 = 0$
$(t - 3)(t + 2) = 0$ $t = \dfrac{-(-1) \pm \sqrt{(-1)^2 - 4(1)(2)}}{2(1)}$
So $t = 3$ or $t = -2$.
                            $= \dfrac{1 \pm \sqrt{1 - 8}}{2} = \dfrac{1 \pm i\sqrt{7}}{2}$

$$S = \left\{3, -2, \frac{1 + i\sqrt{7}}{2}, \frac{1 - i\sqrt{7}}{2}\right\}$$

Answers to odd-numbered problems

1. $\{x \mid x < -3\} \cup \{x \mid x > 1\}$

3. $\{p \mid -7 \le p \le -2\}$

5. $\{r \mid r \le 0\} \cup \{r \ge 1\}$

7. $\{x \mid 1 < x < 4\}$

9. $\{q \mid -3 \le q \le 6\}$

11. $\{x \mid 1 < x < 2\}$

13. $\{u \mid u \le -6\} \cup \{u \mid u \ge 2\}$

15. $\{y \mid y < -2\} \cup \{y \mid y > 2\}$

17. $\{x \mid -\sqrt{3} < x < \sqrt{3}\}$

19. $\{x \mid 0 < x < 5\}$

21. $\{w \mid -3 \le w \le 0\}$

23. $\left\{y \mid -2 < y < \dfrac{3}{2}\right\}$

25. $\left\{q \mid q < -\dfrac{4}{3}\right\} \cup \left\{q \mid q > \dfrac{3}{2}\right\}$

27. $\left\{v \mid -\dfrac{2}{3} < v < \dfrac{4}{3}\right\}$

29. $\left\{x \mid x \le \dfrac{-1-\sqrt{13}}{2}\right\} \cup \left\{x \mid x \ge \dfrac{-1+\sqrt{13}}{2}\right\}$

31. $\{y \mid y < -6\} \cup \{y \mid -4 < y < 5\}$

33. $\{t \mid -5 < t < 0\} \cup \left\{t \mid t > \dfrac{8}{3}\right\}$

35. $\{p \mid p \le -3\} \cup \{p \mid p > -1\}$

37. $\left\{t \mid -\dfrac{4}{5} < t < \dfrac{3}{7}\right\}$

39. $\{y \mid y < 0\} \cup \left\{y \mid y \ge \dfrac{5}{3}\right\}$

41. $\{y \mid y \le 1\} \cup \{y \mid y > 6\}$

43. $\left\{t \mid -4 < t < -\dfrac{8}{3}\right\}$

45. $\left\{x \mid -\dfrac{15}{2} \le x < -5\right\}$

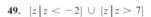

47. $\{x \mid x < 7\} \cup \left\{x \mid x \ge \dfrac{34}{3}\right\}$

49. $\{z \mid z < -2\} \cup \{z \mid z > 7\}$

51. $\{q \mid q < -4\} \cup \left\{q \mid -\dfrac{7}{2} \le q \le 0\right\}$

53. $\{t \mid t < 0\} \cup \left\{t \mid \dfrac{1}{9} < t < \dfrac{2}{3}\right\}$

Solutions to trial exercise problems

14.

$$x^2 - 1 < 0$$
$$(x + 1)(x - 1) < 0$$
The critical numbers are -1 and 1.
$$S = \{x \mid -1 < x < 1\}$$

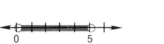

	$x + 1$	$x - 1$	product
$x < -1$	$-$	$-$	$+$
$-1 < x < 1$	$+$	$-$	$\ominus$
$x > 1$	$+$	$+$	$+$

16.

$$p^2 \ge 5$$
$$p^2 - 5 \ge 0$$
$$(p + \sqrt{5})(p - \sqrt{5}) \ge 0$$
The critical numbers are $-\sqrt{5}$ and $\sqrt{5}$.
$$S - \{p \mid p \le -\sqrt{5}\} \cup \{p \mid p \ge \sqrt{5}\}$$

	$p + \sqrt{5}$	$p - \sqrt{5}$	product
$p < -\sqrt{5}$	$-$	$-$	$\oplus$
$-\sqrt{5} < p < \sqrt{5}$	$+$	$-$	$\ominus$
$p > \sqrt{5}$	$+$	$+$	$\oplus$

19.

$$x^2 - 5x < 0$$
$$x(x - 5) < 0$$
The critical numbers are 0 and 5.
$$S = \{x \mid 0 < x < 5\}$$

	x	$x - 5$	product
$x < 0$	$-$	$-$	$+$
$0 < x < 5$	$+$	$-$	$\ominus$
$x > 5$	$+$	$+$	$+$

22.

$$2x^2 - 7x - 4 > 0$$
$$(2x + 1)(x - 4) > 0$$
The critical numbers are $-\dfrac{1}{2}$ and 4.

	$2x + 1$	$x - 4$	product
$x < -\dfrac{1}{2}$	$-$	$-$	$\oplus$
$-\dfrac{1}{2} < x < 4$	$+$	$-$	$-$
$x > 4$	$+$	$+$	$\oplus$

$$S = \left\{x \mid x < -\dfrac{1}{2}\right\} \cup \{x \mid x > 4\}$$

30. $(x - 3)(x + 1)(x - 2) > 0$ The critical numbers are -1, 2, and 3.

	$x - 3$	$x + 1$	$x - 2$	product
$x < -1$	$-$	$-$	$-$	$\ominus$
$-1 < x < 2$	$-$	$+$	$-$	$\oplus$
$2 < x < 3$	$-$	$+$	$+$	$\ominus$
$x > 3$	$+$	$+$	$+$	$\oplus$

$$S = \{x \mid -1 < x < 2\} \cup \{x \mid x > 3\}$$

38. $\dfrac{1}{x} \ge 2$

$\dfrac{1}{x} - 2 \ge 0$

$\dfrac{1 - 2x}{x} \ge 0$

	x	$1 - 2x$	quotient
$x < 0$	$-$	$+$	$-$
$0 < x < \dfrac{1}{2}$	$+$	$+$	$\oplus$
$x > \dfrac{1}{2}$	$+$	$-$	$-$

The critical numbers are 0 and $\dfrac{1}{2}$.

$S = \left\{ x \,\middle|\, 0 < x \le \dfrac{1}{2} \right\}$

45. $\dfrac{x}{x + 5} \ge 3$

$\dfrac{x}{x + 5} - 3 \ge 0$

$\dfrac{x - 3(x + 5)}{x + 5} \ge 0$

$\dfrac{x - 3x - 15}{x + 5} \ge 0$

$\dfrac{-2x - 15}{x + 5} \ge 0$

	$-2x - 15$	$x + 5$	quotient
$x < -\dfrac{15}{2}$	$+$	$-$	$-$
$-\dfrac{15}{2} < x < -5$	$-$	$-$	$\oplus$
$x > -5$	$-$	$+$	$-$

The critical numbers are -5 and $-\dfrac{15}{2}$.

$S = \left\{ x \,\middle|\, -\dfrac{15}{2} \le x < -5 \right\}$

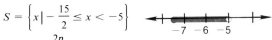

50. $\dfrac{2p}{p - 2} \le p$

$\dfrac{2p}{p - 2} - p \le 0$

$\dfrac{2p - p(p - 2)}{p - 2} \le 0$

$\dfrac{2p - p^2 + 2p}{p - 2} \le 0$

$\dfrac{-p^2 + 4p}{p - 2} \le 0$

$\dfrac{-p(p - 4)}{p - 2} \le 0$

	$-p$	$p - 2$	$p - 4$	quotient
$p < 0$	$+$	$-$	$-$	$+$
$0 < p < 2$	$-$	$-$	$-$	$\ominus$
$2 < p < 4$	$-$	$+$	$-$	$+$
$p > 4$	$-$	$+$	$+$	$\ominus$

The critical numbers are 0, 2, and 4
$S = \{ p \,|\, 0 \le p < 2 \} \cup \{ p \,|\, p \ge 4 \}$

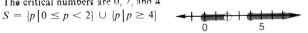

Chapter 6 review exercises

1. $\{10, -1\}$ **2.** $\{8, -4\}$ **3.** $\left\{ 0, \dfrac{7}{4} \right\}$ **4.** $\{-2, 2\}$ **5.** $\left\{ \dfrac{1}{2}, \dfrac{5}{2} \right\}$ **6.** $\left\{ \dfrac{5}{2}, -1 \right\}$ **7.** $\left\{ \dfrac{1}{3}, 2 \right\}$ **8.** $\{-3, 1\}$ **9.** $\{1, -15\}$

10. $\left\{ 0, \dfrac{8}{3} \right\}$ **11.** $20 \pm 10\sqrt{3} \approx 2.7$ sec or 37.3 sec; 40 sec **12.** $\left\{ \dfrac{-3 - \sqrt{41}}{2}, \dfrac{-3 + \sqrt{41}}{2} \right\}$ **13.** $\left\{ \dfrac{1}{2}, 1 \right\}$

14. $\left\{ \dfrac{-3 - i\sqrt{11}}{10}, \dfrac{-3 + i\sqrt{11}}{10} \right\}$ **15.** $\left\{ \dfrac{1 - \sqrt{85}}{14}, \dfrac{1 + \sqrt{85}}{14} \right\}$ **16.** $\left\{ \dfrac{3 - 5i\sqrt{15}}{16}, \dfrac{3 + 5i\sqrt{15}}{16} \right\}$

17. $x = \dfrac{-3a \pm a\sqrt{17}}{4}$ **18.** $\{10, 1\}$ **19.** $\left\{ \dfrac{-3 \pm \sqrt{33}}{6} \right\}$ **20.** $\left\{ -\dfrac{\sqrt{6}}{2}, \dfrac{\sqrt{6}}{2} \right\}$ **21.** $\{-7, 0\}$ **22.** $\left\{ \dfrac{-1 \pm i\sqrt{31}}{4} \right\}$

23. $\left\{-1,\dfrac{5}{3}\right\}$ **24.** $\left\{\dfrac{11}{4},1\right\}$ **25.** $x=\dfrac{-y\pm y\sqrt{33}}{8}$ **26.** $x=L$ or $x=\dfrac{1}{2}L$ **27.** $\dfrac{15\pm5\sqrt{6}}{4}\approx0.7$ sec or 6.8 sec **28.** 200 cm

by 50 cm or 2 m by $\dfrac{1}{2}$ m **29.** 15 ft, 8 ft, 17 ft **30.** Mary, $11+\sqrt{145}\approx23$ hr; Dick, $13+\sqrt{145}\approx25$ hr

31. $\{5\}$; -4 is extraneous **32.** $\left\{\dfrac{4}{3}\right\}$; $-\dfrac{1}{3}$ is extraneous **33.** $\left\{\dfrac{1}{4}\right\}$ **34.** $\{3,-2\}$ **35.** $\{5\}$; $\dfrac{5}{4}$ is extraneous **36.** $V=\dfrac{4\pi r^3}{3}$

37. $\{\sqrt{7},-\sqrt{7},i\sqrt{2},-i\sqrt{2}\}$ **38.** $\{0,4\}$ **39.** $\emptyset$; 1 and 64 are extraneous roots **40.** $\{4\}$; $\dfrac{1}{9}$ is extraneous root **41.** $\left\{-8,\dfrac{27}{8}\right\}$

42. $\left\{-1,\dfrac{1}{11}\right\}$ **43.** $\left\{-\dfrac{\sqrt{2}}{2},\dfrac{\sqrt{2}}{2},\dfrac{-i\sqrt{10}}{2},\dfrac{i\sqrt{10}}{2}\right\}$

44. $\{x\mid x<-7\}\cup\{x\mid x>3\}$

45. $\left\{x\mid -\dfrac{4}{3}\le x\le\dfrac{1}{2}\right\}$

46. $\{y\mid y\le0\}\cup\left\{y\mid y\ge\dfrac{3}{2}\right\}$

47. $\left\{z\mid -\dfrac{2}{3}<z<\dfrac{2}{3}\right\}$

48. $\{m\mid m<3-\sqrt{2}\}\cup\{m\mid m>3+\sqrt{2}\}$

49. $\left\{p\mid -\dfrac{3}{4}<p<2\right\}$

50. $\{x\mid x\le-2\sqrt{2}\}\cup\{x\mid x\ge2\sqrt{2}\}$

51. $\{y\mid 7-7\sqrt{2}\le y\le7+7\sqrt{2}\}$

52. $\{y\mid y<-4\}\cup\{y\mid 2<y<5\}$

53. $\left\{z\mid -1\le z\le-\dfrac{1}{2}\right\}\cup\left\{z\mid z\ge\dfrac{2}{3}\right\}$

54. $\{m\mid m\le-3\}\cup\{m\mid m>1\}$

55. $\{x\mid x\le-17\}\cup\{x\mid x>-7\}$

Chapter 6 cumulative test

1. -12 **2.** 2 **3.** 56 **4.** $7xy^2-6xy+6x^2y$ **5.** $9x^2+12xy+4y^2$ **6.** $16y^2-1$ **7.** $9x^3-16x^2-8$ **8.** $-3,375x^6y^9$

9. a^{20} **10.** $\dfrac{-2b^5}{a^5}$ **11.** (a) 6, (b) 1, (c) 41 **12.** -12 **13.** $\dfrac{2(a+1)}{a-4}$ **14.** $\dfrac{3}{4}$ **15.** $\dfrac{x+1}{x-3}$ **16.** $-\dfrac{2x^2-3x+1}{x^2+x-42}$

17. $\dfrac{12p-3}{(p-9)(p+2)(p-2)}$ **18.** $\dfrac{a+5}{a-6}$ **19.** $\left\{-\dfrac{1}{2},-3\right\}$ **20.** $\{x\mid 0\le x\le5\}$ **21.** $\left\{x\mid x<\dfrac{3}{2}\text{ or }x>\dfrac{11}{2}\right\}$ **22.** $\{-7\}$

23. $\{x \mid x \geq -33\}$ **24.** $\left\{-\dfrac{5}{3}\right\}$ **25.** $w = \dfrac{P - 2\ell}{2}$ **26.** $\dfrac{10y - 4}{4y + 5}$ **27.** $6\sqrt{3} - 3$ **28.** $16\sqrt{3}$ **29.** $-3\sqrt[3]{3}$ **30.** 13

31. $47 - 12\sqrt{15}$ **32.** $\dfrac{4\sqrt{5}}{5}$ **33.** $2\sqrt{6} - 5$ **34.** $\dfrac{-5 + 2i}{29}$ or $\dfrac{-5}{29} + \dfrac{2}{29}i$ **35.** $\{14,1\}$ **36.** $\left\{\dfrac{1 + i\sqrt{59}}{10}, \dfrac{1 - i\sqrt{59}}{10}\right\}$

37. $\left\{-\dfrac{3}{2}\right\}$ **38.** $\{z \mid -1 \leq z \leq 3\}$ **39.** $\{i\sqrt{5}, -i\sqrt{5}, \sqrt{10}, -\sqrt{10}\}$ **40.** $\{y \mid -7 < y < 10\}$.

Chapter 7

Exercise 7–1

Answers to odd-numbered problems

1. (2,4); quadrant I (see graph)   **3.** (−4,3); quadrant II (see graph)
5. (−1,−3); quadrant III (see graph)   **7.** (4,0); quadrantal (see graph)

9. (0,−1); quadrantal (see graph)   **11.** $\left(\dfrac{1}{2},3\right)$; quadrant I (see graph)

13. $\left(-\dfrac{7}{2},-\dfrac{5}{2}\right)$; quadrant III (see graph)

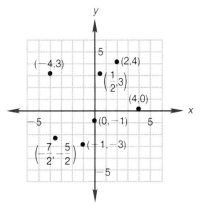

15. x-intercept, -6; y-intercept, 2   **17.** x-intercept, 4; y-intercept, 10   **19.** x-intercept, -2; y-intercept, 10

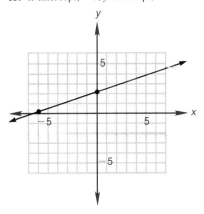

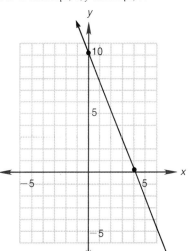

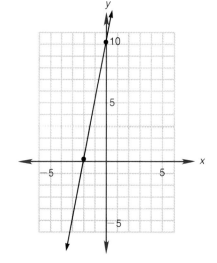

21. x-intercept, 0; y-intercept, 0

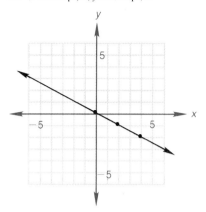

23. x-intercept, 0; y-intercept, 0

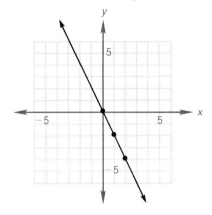

25. x-intercept, -2; y-intercept, $\dfrac{6}{5}$

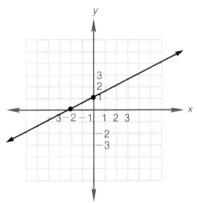

27. no x-intercept; y-intercept, -2

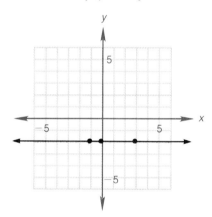

29. x-intercept, -1; no y-intercept

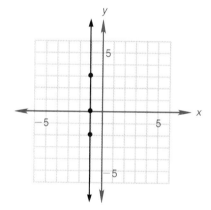

31. x-intercept, 0; all points on y-axis are y-intercepts

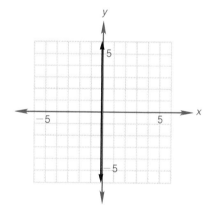

33. $(-2,7),(0,3),(2,-1),(4,-5)$;
$y = -2x + 3$

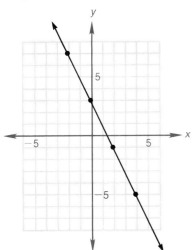

35. $(-5,9),(-3,7),(0,4),(4,0)$;
$y = -x + 4$

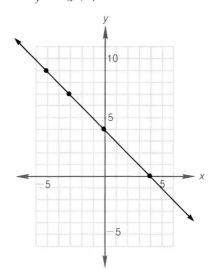

37. $(-3,-5),(0,-2),(2,0),(4,2)$;
$y = x - 2$

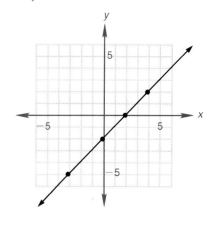

39. $\left(-1,-\dfrac{2}{3}\right),\left(0,-\dfrac{1}{3}\right),(1,0),\left(3,\dfrac{2}{3}\right)$;

$y = \dfrac{1}{3}x - \dfrac{1}{3}$

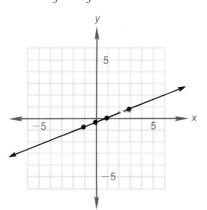

41. $(-2,7),(0,4),(2,1),(4,-2)$;

$y = -\dfrac{3}{2}x + 4$

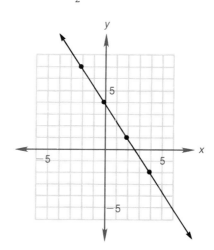

43. $y = x + 4$

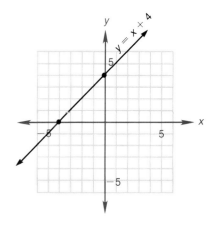

45. $3x - 2y = 12$

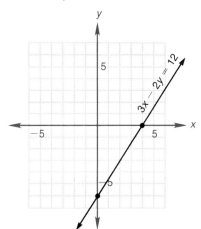

47. $C = \dfrac{5}{9}(F - 32)$

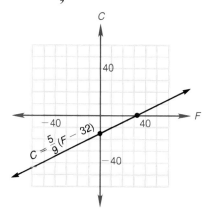

49.

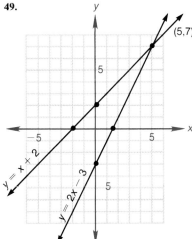

Solutions to trial exercise problems

20. $x = 3y$
When $x = 0$, $3y = 0$, $y = 0$,
the x- and y-intercepts are 0.
Choose second point $(3,1)$.

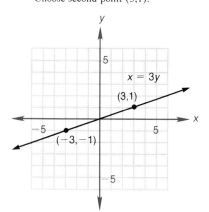

27. $y = -2$
There is no x-intercept
and the y-intercept is -2.

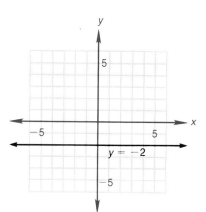

38. $x + 2y = 4$
Solving for y, $y = \dfrac{4 - x}{2}$.
When $x = -2$, $y = \dfrac{4 - (-2)}{2} = 3$

$x = 0$, $y = \dfrac{4 - 0}{2} = 2$

$x = 2$, $y = \dfrac{4 - 2}{2} = 1$

$x = 3$, $y = \dfrac{4 - 3}{2} = \dfrac{1}{2}$

$(-2,3),(0,2),(2,1),\left(3,\dfrac{1}{2}\right)$

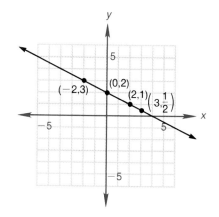

44. Two times x is $2x$, less 3 is $2x - 3$, is equal to y is $y = 2x - 3$.

x	-1	0	1	2	3
y	-5	-3	-1	1	3

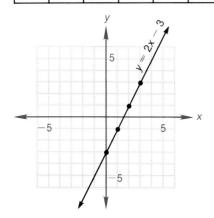

Exercise 7–2

Answers to odd-numbered problems

1. $d = \sqrt{41}$ units; $m = \dfrac{5}{4}$;

midpoint, $\left(4, \dfrac{9}{2}\right)$

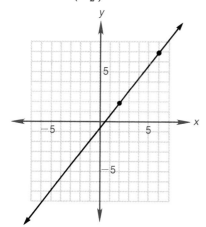

3. $d = \sqrt{29}$ units; $m = \dfrac{5}{2}$;

midpoint, $\left(-2, \dfrac{5}{2}\right)$

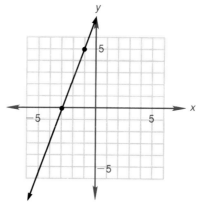

5. $d = 5$ units; undefined slope

midpoint, $\left(-1, \dfrac{13}{2}\right)$

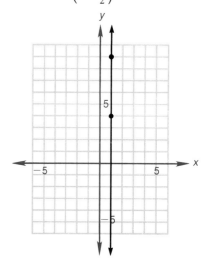

7. $d = 7$ units; $m = 0$;

midpoint, $\left(-\dfrac{1}{2}, 6\right)$

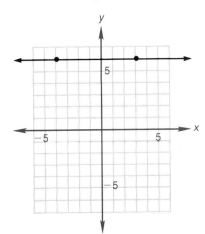

9. $d = \sqrt{2}$ units; $m = -1$;

midpoint, $\left(-\dfrac{7}{2}, -\dfrac{1}{2}\right)$

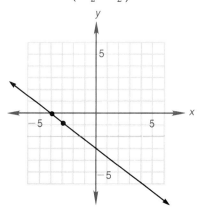

11. $d = \sqrt{170}$ units; $m = -\dfrac{7}{11}$

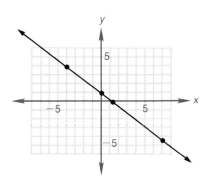

13. $d = \sqrt{85}$ units; $m = -\dfrac{9}{2}$

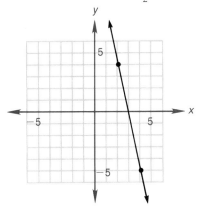

15. $d = 10$ units; undefined slope

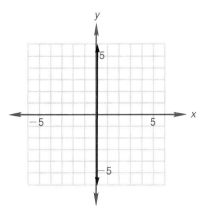

17. $m_1 = -3$, $m_2 = -3$, $m_1 = m_2$; parallel

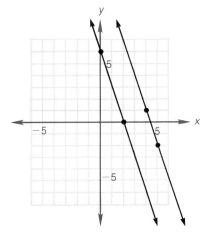

19. $m_1 = -2$, $m_2 = -\dfrac{7}{11}$, $m_1 \neq m_2$; not parallel

21. $m_1 = \dfrac{1}{8}$, $m_2 = -8$, $m_1 m_2 = -1$; perpendicular

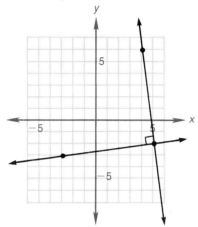

23. $m_1 = 1$, $m_2 = -1$, $m_1 m_2 = -1$; perpendicular

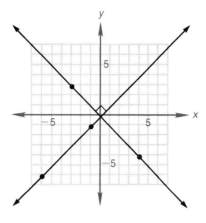

25. $m_1 = -2$, $m_2 = -2$, $m_1 = m_2$; parallel

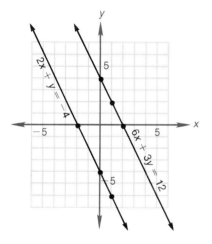

27. $m_1 = \dfrac{3}{4}$, $m_2 = -\dfrac{4}{3}$, $m_1 m_2 = -1$; perpendicular

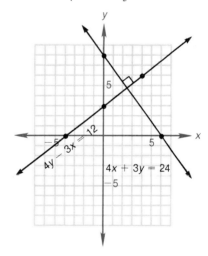

29. $m_1 = -\dfrac{1}{4}$, $m_2 = \dfrac{2}{5}$, $m_1 \neq m_2$, $m_1 m_2 \neq -1$; neither

31. $m_1 = \dfrac{1}{2}$, $m_2 = -2$, $m_1 m_2 = -1$; perpendicular

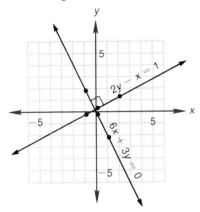

33. $m_1 = \dfrac{3}{7}$, $m_2 = -\dfrac{5}{3}$, $m_1 \neq m_2$, $m_1 m_2 \neq -1$; neither

35. $m_1 = \dfrac{4}{3}$, $m_2 = \dfrac{3}{4}$; neither

37. $m_1 = \dfrac{3}{4}$, $m_2 = -\dfrac{5}{3}$; neither

39. Pitch is $\dfrac{3}{5}$.

41. $m = \dfrac{2}{3}$

43. $m = \dfrac{1,000}{7}$

45. $m = \dfrac{15}{2}$

47. $m = -\dfrac{7}{3}$

49. Slopes of opposite sides are 2 and 0; opposite sides have lengths 4 and $\sqrt{5}$; $p = 8 + 2\sqrt{5}$. **51.** Slopes of two sides are $m_1 = 0$ and m_2 is undefined thus perpendicular; $6^2 + 5^2 = (\sqrt{61})^2$; $36 + 25 = 61$; $61 = 61$. **53.** Two sides have slope $m = 0$, so are parallel, while the other sides have unequal slopes. **55.** (a) The slope using any pair of points is $\dfrac{3}{2}$. (b) Distances between points are $\sqrt{13}$, $\sqrt{52}$, and $\sqrt{117}$; $\sqrt{13} + \sqrt{52} = \sqrt{117}$; $\sqrt{13} + 2\sqrt{13} = 3\sqrt{13}$; $3\sqrt{13} = 3\sqrt{13}$. **57.** $y = -3$ or -11 **59.** $(-2,12)$

Solutions to trial exercise problems

7. $(3,6)$ and $(-4,6)$

$\begin{aligned}
\text{distance} &= |3 - (-4)| \\
&= |3 + 4| \\
&= 7
\end{aligned}$

$m = \dfrac{6 - 6}{3 - (-4)} = \dfrac{0}{7} = 0$

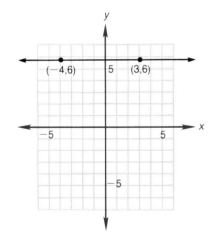

14. $(0,8)$ and $(0,-1)$

$\begin{aligned}
\text{distance} &= \sqrt{(0 - 0)^2 + [8 - (-1)]^2} \\
&= \sqrt{0 + 9^2} \\
&= \sqrt{81} \\
&= 9
\end{aligned}$

$m = \dfrac{8 - (-1)}{0 - 0} = \dfrac{9}{0}$ undefined

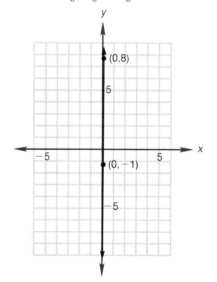

18. $m_1 = \dfrac{1-2}{5-(-4)} = \dfrac{-1}{9} = -\dfrac{1}{9}$

$m_2 = \dfrac{-3-1}{4-2} = \dfrac{-4}{2} = -2$

The lines are *not* parallel.

22. $m_1 = \dfrac{1-4}{1-4} = \dfrac{-3}{-3} = 1$

$m_2 = \dfrac{2-(-3)}{-2-3} = \dfrac{5}{-5} = -1$

Since $m_1 \cdot m_2 = 1 \cdot -1 = -1$, the lines are perpendicular.

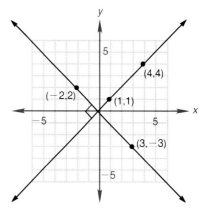

31. $2y - x = 1$

Using $\left(0, \dfrac{1}{2}\right)$ and $(-1,0)$,

$m_1 = \dfrac{\dfrac{1}{2} - 0}{0 - (-1)} = \dfrac{\dfrac{1}{2}}{1} = \dfrac{1}{2}$

Since $m_1 \cdot m_2 = \dfrac{1}{2} \cdot -2 = -1$,

the lines are perpendicular.

$6x + 3y = 0$

Using $(0,0)$ and $(1,-2)$

$m_2 = \dfrac{0-(-2)}{0-1} = \dfrac{2}{-1} = -2.$

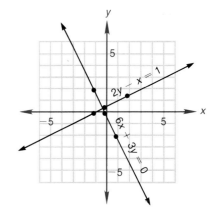

39. $m = \text{pitch} = \dfrac{9}{15} = \dfrac{3}{5}$

45. Using $(2,10)$ and $(4,25)$, $m = \dfrac{10-25}{2-4} = \dfrac{-15}{-2} = \dfrac{15}{2}$.

51. Using $(4,2)$ and $(4,-3)$,

$m = \dfrac{2-(-3)}{4-4} = \dfrac{5}{0}$ (undefined).

(vertical line)

Using $(-2,-3)$ and $(4,-3)$,

$m = \dfrac{-3-(-3)}{-2-4} = \dfrac{-3+3}{-6} = \dfrac{0}{-6} = 0.$

(horizontal line)

The two lines are perpendicular so the triangle has one right angle and is a right triangle.

Distance from $(4,2)$ to $(-2,-3) = \sqrt{[4-(-2)]^2 + [2-(-3)]^2} = \sqrt{6^2 + 5^2}$

$= \sqrt{36 + 25} = \sqrt{61}$

Distance from $(4,2)$ to $(4,-3) = \sqrt{(4-4)^2 + [2-(-3)]^2} = \sqrt{0^2 + 5^2}$

$= \sqrt{25} = 5$

Distance from $(-2,-3)$ to $(4,-3) = \sqrt{(-2-4)^2 + [-3-(-3)]^2}$

$= \sqrt{(-6)^2 + 0^2} = \sqrt{36} = 6$

Now $5^2 + 6^2 = (\sqrt{61})^2$

$25 + 36 = 61$

$61 = 61$

56. Let x be the abscissa. Then using $(x, -6)$ and $(4,5)$,

$$5\sqrt{5} = \sqrt{(x-4)^2 + (-6-5)^2}$$
$$5\sqrt{5} = \sqrt{x^2 - 8x + 16 + 121}$$
$$(5\sqrt{5})^2 = (\sqrt{x^2 - 8x + 137})^2$$
$$125 = x^2 - 8x + 137$$
$$0 = x^2 - 8x + 12$$
$$0 = (x-6)(x-2), \text{ so } x = 6 \text{ or } x = 2.$$

Thus the abscissa is 6 or 2.

58. Let x be the first component of the endpoint.
Let y be the second component of the endpoint.

Then $\dfrac{x + (-2)}{2} = 2$ $\qquad\qquad$ $\dfrac{y + 3}{2} = -3$

$\qquad\quad \dfrac{x - 2}{2} = 2$ $\qquad\qquad\qquad$ $y + 3 = -6$

$\qquad\qquad x - 2 = 4$ $\qquad\qquad\qquad\quad$ $y = -9$

$\qquad\qquad\qquad x = 6$

The other endpoint is $(6, -9)$.

Exercise 7–3

Answers to odd-numbered problems

1. $x - 2y = 5$ $\quad$ **3.** $5x - y = 7$ $\quad$ **5.** $5x + 6y - 0$ $\quad$ **7.** $x - y - 3$ $\quad$ **9.** $y = 4$ $\quad$ **11.** $x = -5$ $\quad$ **13.** $x = 5$ $\quad$ **15.** $x = 0$
17. $5x - 2y = 3$ $\quad$ **19.** $10x - 3y = 2$ $\quad$ **21.** $2x + y = 8$ $\quad$ **23.** $y = 4$ $\quad$ **25.** $x + y = 0$ $\quad$ **27.** $x = 4$
29. $y = -2x + 5; m = -2; b = 5$ $\qquad\qquad\qquad\qquad\qquad$ **31.** $y = 4x - 6; m = 4; b = -6$

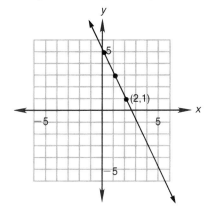

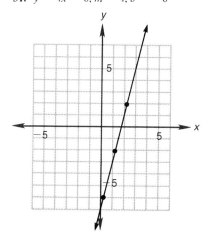

33. $y = \dfrac{2}{5}x; m = \dfrac{2}{5}; b = 0$ $\qquad$ **35.** $y = -\dfrac{8}{3}x; m = -\dfrac{8}{3}; b = 0$ $\qquad$ **37.** $y = -\dfrac{5}{3}; m = 0; b = -\dfrac{5}{3}$

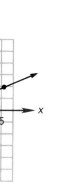

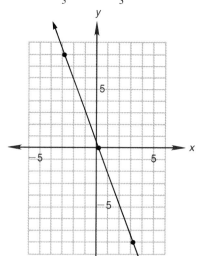

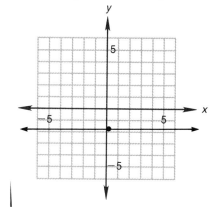

39. $3x + y = 5$ **41.** $3x - 2y = -7$ **43.** $2x + 9y = 0$ **45.** $x = 4$ **47.** neither **49.** perpendicular **51.** parallel

53. $y - b = \dfrac{b - 0}{0 - a}(x - 0); \ y - b = -\dfrac{b}{a}x; \ ay - ab = -bx; \ bx + ay = ab; \ \dfrac{x}{a} + \dfrac{y}{b} = 1$ **55.** $Ax + By = C; \ By = -Ax + C;$

$y = -\dfrac{A}{B}x + \dfrac{C}{B};$ the slope is $-\dfrac{A}{B};$ the y-intercept is $\dfrac{C}{B}$ **57.** $m = \dfrac{B}{A}$ **59.** $250x - 3y = 550$ **61.** $140,000

Solutions to trial exercise problems

3. $m = 5$ and $(x_1, y_1) = (0,7)$
$y - 7 = 5(x - 0)$
$y - 7 = 5x$
$5x - y = -7$

8. Horizontal line has slope 0,
then $y - (-3) = 0(x - 5)$
$y + 3 = 0; \ y = -3.$

20. $(5,0)$ and $(-2,-3)$
Now $m = \dfrac{0 - (-3)}{5 - (-2)} = \dfrac{3}{5 + 2} = \dfrac{3}{7}$

Using point $(5,0)$ and the point-slope form,

$y - 0 = \dfrac{3}{7}(x - 5)$

$y = \dfrac{3}{7}(x - 5)$

$7y = 3x - 15$
$-3x + 7y = -15.$

6. Using $y = mx + b, \ m = -6$
and $b = 2, \ y = -6x + 2$
$6x + y = 2.$

11. Having undefined slope, the line must be vertical and passing through $(-5,6)$, the *first component* of every point is -5. So $x = -5$ is the equation.

27. $(4,-3)$ and $(4,-7)$
$m = \dfrac{-3 - (-7)}{4 - 4} = \dfrac{4}{0} = $ undefined.

The slope is undefined, so the line is vertical and passes through $x = 4$. Thus the equation is $x = 4$.

29. $2x + y = 5$
$\quad\quad y = -2x + 5$
$m = -2; \ b = 5$

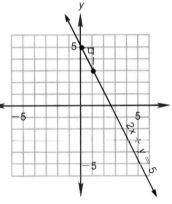

32. $3x - 7y = 0$
$\quad\quad\quad y = \dfrac{3}{7}x + 0$

$m = \dfrac{3}{7}; \ b = 0$

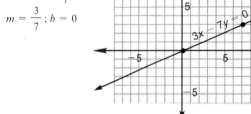

37. $3y + 5 = 0$

$y = -\dfrac{5}{3} = 0x - \dfrac{5}{3}$

$m = 0; \ b = -\dfrac{5}{3}$

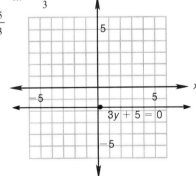

39. $3x + y = 6$
$\quad\quad y = -3x + 6$
So $m = -3.$ Using $(1,2),$
$y - 2 = -3(x - 1)$
$y - 2 = -3x + 3$
$3x + y = 5.$

48. $2y - 5x = -3$

$2y = 5x - 3$

$y = \dfrac{5}{2}x - \dfrac{3}{2}$

So $m = \dfrac{5}{2}$

$y + 3x = 4$

$y = -3x + 4$

So $m = -3$.

The lines are neither parallel nor perpendicular since

$\dfrac{5}{2} \neq -3$ and $\dfrac{5}{2} \cdot -3 \neq -1$.

52. (a) $(3,2)$ and $(4,1)$

$y - 2 = \dfrac{2 - 1}{3 - 4}(x - 3)$

$y - 2 = \dfrac{1}{-1}(x - 3)$

$y - 2 = -1(x - 3)$

$y - 2 = -x + 3$

$x + y = 5$

54. (a) $4x + 3y = 12$

Divide each member by 12.

$\dfrac{4x}{12} + \dfrac{3y}{12} = \dfrac{12}{12}$

$\dfrac{x}{3} + \dfrac{y}{4} = 1$

The x-intercept is 3 and the y-intercept is 4.

(d) $3x - 5y = 6$

Divide each member by 6.

$\dfrac{3x}{6} - \dfrac{5y}{6} = \dfrac{6}{6}$

$\dfrac{x}{2} + \dfrac{y}{-\dfrac{6}{5}} = 1$

The x-intercept is 2 and the y-intercept is $-\dfrac{6}{5}$.

58. $(300,150)$ and $(600,250)$

$y - 150 = \dfrac{250 - 150}{600 - 300}(x - 300)$

$y - 150 = \dfrac{100}{300}(x - 300)$

$y - 150 = \dfrac{1}{3}(x - 300)$

$3y - 450 = x - 300$

$x - 3y = -150$

Exercise 7–4

Answers to odd-numbered problems

1.

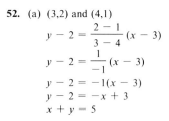

3.

5.

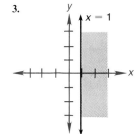

7.

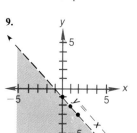

9.

11.

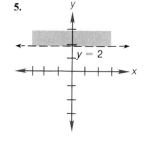

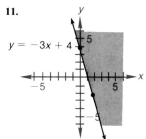

13.

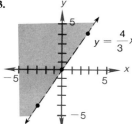

$y = \dfrac{4}{3}x$

15.

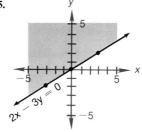

$2x - 3y = 0$

17.

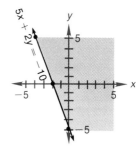

$5x + 2y = -10$

19.

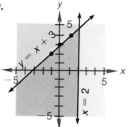

$y = x + 3$

$x = 2$

21.

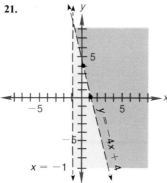

$x = -1$

$y = -4x + 4$

23.

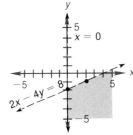

$x = 0$

$2x - 4y = 8$

25.

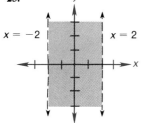

$x = -2$

$x = 2$

27.

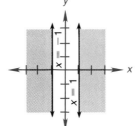

$x = -1$

$x = 1$

29.

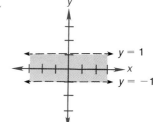

$y = 1$

$y = -1$

31.

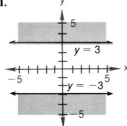

$y = 3$

$y = -3$

33.

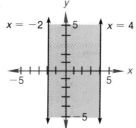

$x = -2$

$x = 4$

35.

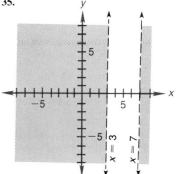

$x = 3$

$x = 7$

37.

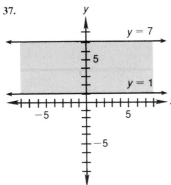

39.

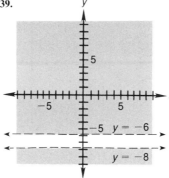

41.

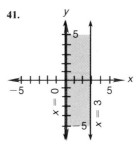

Solutions to trial exercise problems

1. $y < 3$

Since every point having $y < 3$ lies below the line $y = 3$, we dash the line $y = 3$ and shade the plane below this line.

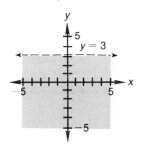

8. $x + y > 5$

Graph the line $x + y = 5$ (dashed) and since for (0,0) $0 + 0 \not> 5$, shade *above* this line.

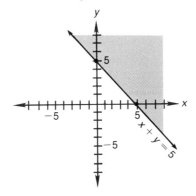

17. $5x + 2y \geq -10$

Graph the line $5x + 2y = -10$ (make this line *solid*). Since for (0,0), $5(0) + 2(0) \geq -10$ is true, shade the plane to the right of this line.

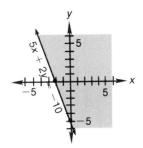

21. $\{(x,y) \mid 4x + y > 4\} \cup \{(x,y) \mid x > -1\}$

Since for $4x + y > 4$, the point (0,0) yields $4(0) + 0 > 4$ (which is not true), shade to the right of the dashed line $4x + y = 4$. For $x > -1$, shade to the right of the line $x = -1$. The double shading is the answer.

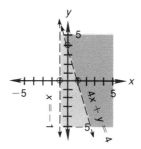

25. $|x| < 2, -2 < x < 2$

Shade the plane between the dashed vertical lines, $x = -2$ and $x = 2$.

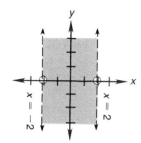

36. $|3 - x| \geq 4$

$3 - x \geq 4$ or $3 - x \leq -4$

$x \leq -1$ or $x \geq 7$

Shade to the left of the solid vertical line $x = -1$ and to the right of the solid vertical line $x = 7$.

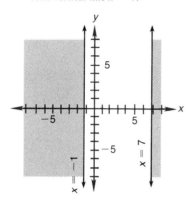

41. $|2x - 3| \leq 3$

$-3 \leq 2x - 3 \leq 3$

$0 \leq 2x \leq 6$

$0 \leq x \leq 3$

Shade between the solid lines $x = 0$ and $x = 3$.

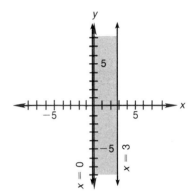

Chapter 7 review exercises

1. $(-3,5)$; quadrant II (see graph)

2. $(-2,-7)$; quadrant III (see graph)

3. $(0,-1)$; quadrantal (see graph)

4. $(5,0)$; quadrantal (see graph)

5. $\left(1, -\dfrac{3}{2}\right)$; quadrant IV (see graph)

6. $\left(\dfrac{1}{2}, \dfrac{11}{2}\right)$; quadrant I (see graph)

7. $(0,4)$; quadrantal (see graph)

8. $\left(-\dfrac{5}{2}, -4\right)$; quadrant III (see graph)

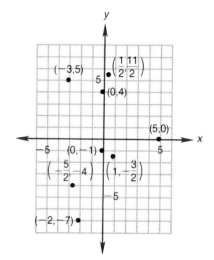

9. x-intercept, 6; y-intercept, -2

10. x-intercept, 4; y-intercept, -8

11. x-intercept, 2; y-intercept, -5

12. x-intercept, -6; no y-intercept

13. no x-intercept; y-intercept, 5

14. x-intercept, $\dfrac{7}{2}$; y-intercept, $\dfrac{7}{4}$

15. x-intercept, $\dfrac{3}{2}$; y-intercept, -3

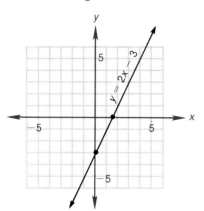

16. x-intercept, $\dfrac{1}{5}$; y-intercept, 1

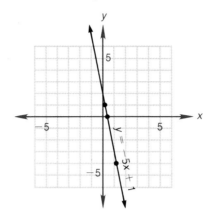

17. x-intercept, 0; y-intercept, 0

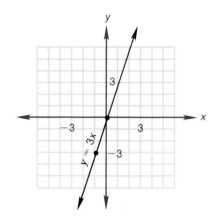

18. x-intercept, 4; y-intercept, -2

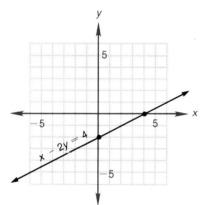

19. x-intercept, $-\dfrac{9}{2}$; y-intercept, 3

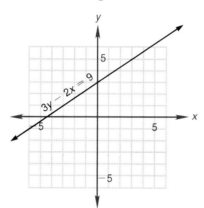

20. no x-intercept; y-intercept, -6

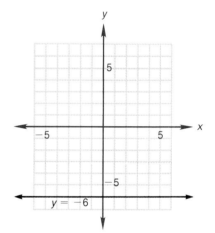

21. x-intercept, 1; no y-intercept

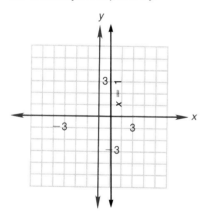

22. $\sqrt{13}$ units   **23.** 3 units   **24.** 4 units   **25.** neither; $m_1 = \dfrac{1}{4}$, $m_2 = \dfrac{-3}{2}$

26. perpendicular; $m_1 = 2$, $m_2 = \dfrac{-1}{2}$ **27.** parallel; $m = \dfrac{4}{7}$

28. neither; $m_1 = \dfrac{-1}{3}$, $m_2 = \dfrac{5}{9}$ **29.** parallel; $m = -2$

30. neither; $m_1 = \dfrac{1}{3}$, $m_2 = -\dfrac{2}{3}$ **31.** perpendicular; $m_1 = \dfrac{2}{5}$, $m_2 = \dfrac{-5}{2}$

32. perpendicular; $m_1 = \dfrac{3}{2}$, $m_2 = \dfrac{-2}{3}$ **33.** $m = \dfrac{18}{13}$

34. $7^2 + 3^2 = (\sqrt{58})^2$; $49 + 9 = 58$; $58 = 58$ **35.** $-2x + 3y = 17$

36. $y = 4$ **37.** $x = 1$ **38.** $8x - y = 21$

39. $y = 3x - 4$; $m = 3$; y-intercept $= -4$

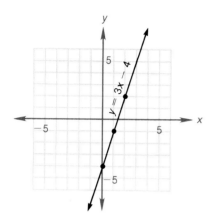

40. $y = \dfrac{-2}{3}x + 3$; $m = \dfrac{-2}{3}$; y-intercept $= 3$

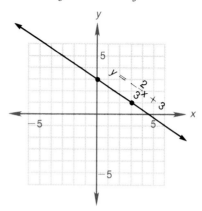

41. $y = \dfrac{3}{2}x$; $m = \dfrac{3}{2}$; y-intercept $= 0$

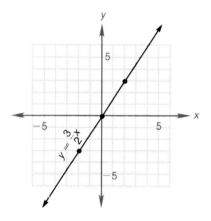

42. $y = \dfrac{-3}{2}$; $m = 0$; y-intercept $= \dfrac{-3}{2}$

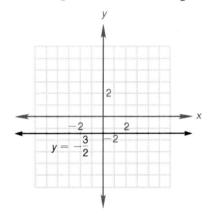

43. $x - 2y = -11$

44. $-5x + 4y = 12$

45.

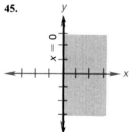

46.

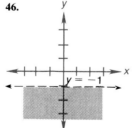

47.

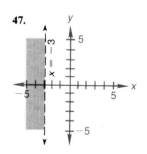

48.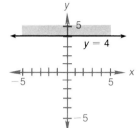

49.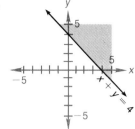

50.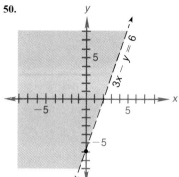

51.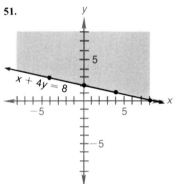

52.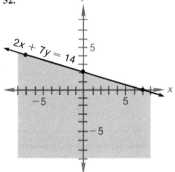

53.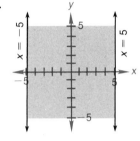

54.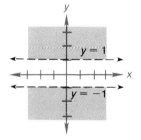

55.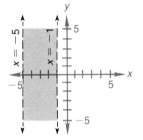

56.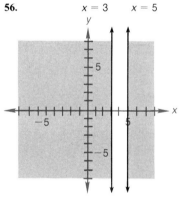

57.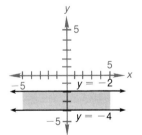

58.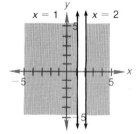

Chapter 7 cumulative test

1. $-18a^3b^5$ 2. $\dfrac{y^6}{x^5}$ 3. $\dfrac{a^6c^2}{16b^4}$ 4. $9y^2 - 3y - 3$ 5. $-3x^2 - 8xy + 3y^2$ 6. $12x^3y - 18xy^3 + 6x^3y^2 - 6x^4y^4$

7. $3(a + 2b)(a - 2b)$ 8. $(8x + 1)(x - 7)$ 9. $(2x + 3y)(4x^2 - 6xy + 9y^2)$ 10. $(4x - 5y)^2$ 11. $\left\{-\dfrac{45}{4}\right\}$ 12. $\{1, -4\}$

13. $\left\{y \mid -\dfrac{3}{2} < y < \dfrac{5}{2}\right\}$ 14. $\left\{y \mid y \le -4 \text{ or } y \ge -2\right\}$ 15. $\left\{0, \dfrac{1}{4}\right\}$ 16. $\{9, -2\}$ 17. $\left\{\dfrac{3}{2}, -1\right\}$ 18. $\left\{x \mid -\dfrac{1}{3} \le x < 3\right\}$

19. $\left\{\dfrac{3 + \sqrt{65}}{4}, \dfrac{3 - \sqrt{65}}{4}\right\}$ 20. $\dfrac{1}{y - 5}$ 21. $\dfrac{-3y^2 + 46y - 21}{2y(y + 7)(y - 7)}$ 22. 1 23. $\left\{\dfrac{11}{10}\right\}$ 24. $b = -\dfrac{a + 5}{24}$ 25. $6\sqrt{2}$

26. $21 + 8\sqrt{5}$ 27. $\dfrac{3\sqrt{5}}{5}$ 28. $6 + 3\sqrt{3}$ 29. 13 30. $2 - 7i$ 31. $\emptyset$; 7 is extraneous   32. $7x - y = 10$

33. $2x - 3y = -8$ 34. $x = 5$ 35. $m = \dfrac{3}{5}$, $b = -2$

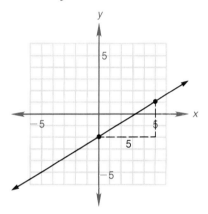

36. perpendicular   37. $d = \sqrt{65}$; midpoint, $\left(\dfrac{3}{2}, 1\right)$

Chapter 8

Exercise 8–1

Answers to odd-numbered problems

1. yes   3. yes   5. $\left\{\left(-\dfrac{3}{2}, -\dfrac{9}{2}\right)\right\}$ 7. $\{(-2,3)\}$ 9. $\{(2,3)\}$ 11. $\{(3,1)\}$ 13. $\left\{\left(\dfrac{27}{10}, \dfrac{1}{5}\right)\right\}$ 15. $\left\{\left(\dfrac{5}{68}, \dfrac{26}{17}\right)\right\}$

17. $S = \{(x,y) \mid 3x + y = 2\}$; dependent   19. $S = \emptyset$; inconsistent   21. $\left\{\left(\dfrac{22}{21}, \dfrac{13}{21}\right)\right\}$ 23. $\left\{\left(\dfrac{5}{3}, \dfrac{1}{2}\right)\right\}$

25. $\left\{\left(\dfrac{3}{5}, 0\right)\right\}$ 27. $\left\{\left(-\dfrac{31}{12}, \dfrac{49}{24}\right)\right\}$ 29. $\left\{\left(-\dfrac{711}{64}, \dfrac{431}{64}\right)\right\}$ or $\{(-11.1, 6.7)\}$ 31. $\{(7, -4)\}$ 33. $\left\{\left(\dfrac{7}{2}, -\dfrac{3}{2}\right)\right\}$ 35. $\left\{\left(\dfrac{5}{14}, \dfrac{6}{7}\right)\right\}$

37. $\left\{\left(\dfrac{17}{4}, \dfrac{5}{4}\right)\right\}$ 39. $\left\{\left(\dfrac{10}{3}, \dfrac{26}{3}\right)\right\}$ 41. $\{(-2, -5)\}$ 43. $\left\{\left(6, \dfrac{37}{6}\right)\right\}$ 45. $\{(-1,0)\}$ 47. $\left\{\left(-\dfrac{6}{29}, \dfrac{9}{29}\right)\right\}$ 49. $S = \emptyset$;

inconsistent   51. $S = \{(x,y) \mid 2x - y = 7\}$; dependent   53. $\left\{\left(\dfrac{5}{2}, 2\right)\right\}$ 55. $\left\{\left(\dfrac{3}{8}, \dfrac{33}{8}\right)\right\}$ 57. $\left\{\left(-\dfrac{12}{11}, -\dfrac{63}{11}\right)\right\}$ 59. $\left\{\left(-1, \dfrac{1}{4}\right)\right\}$

61. $\left\{\dfrac{7}{5}, -\dfrac{7}{4}\right\}$ 63. $-3x + 2y = -1$; $-5x + 2y = -18$; $\left\{\left(\dfrac{17}{2}, \dfrac{49}{4}\right)\right\}$ 65. $\left\{\left(\dfrac{11}{5}, -\dfrac{8}{5}\right)\right\}$ 67. $\left\{\left(\dfrac{3}{2}, -\dfrac{1}{2}\right)\right\}$

69. 8 ft by 12 ft   **71.** $6,000 at 6½%; $14,000 at 7%   **73.** $15,000 at 6%; $15,000 at 8%   **75.** $16,800 at 8%; $19,200 at 7%
77. $9,000 at 12%; $16,000 at 15%   **79.** 6 ft and 15 ft   **81.** 11 volts and 36 volts   **83.** 15 suits at $205; 17 suits at $152
85. 23 laborers (5 cat operators)   **87.** $56 for unskilled; $100 for skilled   **89.** 1,600 kg of 85% pure copper; 400 kg of 60% pure copper
91. 320 liters of 6% acid; 480 liters of 3.5% acid   **93.** speed of boat is 20 mph; speed of stream is 4 mph   **95.** The jogger runs
for $\dfrac{6}{7}$ of an hour or 51 min 26 sec.   **97.** The mother jogs at 2 mph; the daughter jogs at 4 mph.   **99.** cyclist's rate $= 12\dfrac{8}{9}$ mph;
pedestrian's rate $= 4\dfrac{8}{9}$ mph   **101.** 403 children tickets were sold (100 adult tickets were sold).   **103.** 19 quarters; 24 dimes

Solutions to trial exercise problems

11. $3x + 2y = 11$
$\underline{\quad x - \quad y = \quad 2}$ (times 2)

$3x + 2y = 11$
$\underline{2x - 2y = 4}$
$5x \qquad = 15$
$x = 3$

$3 - y = 2$
$-y = -1$ $S = \{(3,1)\}$
$y = 1$

16. $3x - \quad y = 10$ (times -2)
$\underline{6x - 2y = \quad 5}$

$-6x + 2y = -20$
$\underline{6x - 2y = \quad 5}$
$0 = -15$

The system is inconsistent. There are no common solutions. $S = \emptyset$.

23. $\dfrac{1}{2}x + \dfrac{1}{3}y = 1$ (times 6)
$\underline{\dfrac{1}{4}x - \dfrac{2}{3}y = \dfrac{1}{12}}$ (times 12)

$3x + 2y = 6$
$\underline{3x - 8y = 1}$ (times -1)

$3x + 2y = 6$
$\underline{-3x + 8y = -1}$
$10y = 5$
$y = \dfrac{1}{2}$

$\dfrac{1}{2}x + \dfrac{1}{3}\left(\dfrac{1}{2}\right) = 1$
$\dfrac{1}{2}x + \dfrac{1}{6} = 1$
$3x + 1 = 6$
$3x = 5$
$x = \dfrac{5}{3}$

$S = \left\{\left(\dfrac{5}{3}, \dfrac{1}{2}\right)\right\}$

28. $(0.3)x - (0.8)y = 0.9$ (times 10)
$\underline{(0.1)x + (0.4)y = 1.2}$ (times 10)

$3x - 8y = 9$
$\underline{x + 4y = 12}$ (times 2)

$3x - 8y = 9$
$\underline{2x + 8y = 24}$
$5x = 33$
$x = \dfrac{33}{5}$

$\dfrac{33}{5} + 4y = 12$
$4y = 12 - \dfrac{33}{5} = \dfrac{27}{5}$
$y = \dfrac{27}{20}$

$S = \left\{\left(\dfrac{33}{5}, \dfrac{27}{20}\right)\right\}$

31. $2x + y = 10$
$\underline{\qquad y = -x + 3}$

$2x + (-x + 3) = 10$
$x + 3 = 10$
$x = 7$

$y = -(7) + 3$
$y = -4$
$S = \{(7,-4)\}$

36. $3x - 5y = 4$
$\underline{x + 2y = -2}$

$x = -2y - 2$, substituting

$3(-2y - 2) - 5y = 4$
$-6y - 6 - 5y = 4$
$-11y = 10$
$y = -\dfrac{10}{11}$

Then $x = -2\left(-\dfrac{10}{11}\right) - 2 = \dfrac{20}{11} - 2 = -\dfrac{2}{11}$

$S = \left\{\left(-\dfrac{2}{11}, -\dfrac{10}{11}\right)\right\}$

41. $4x - 3y = 7$
$\underline{\qquad y = -5}$

Substituting,

$4x - 3(-5) = 7$
$4x + 15 = 7$ $S = \{(-2,-5)\}$
$4x = -8$
$x = -2$

55. $-\dfrac{1}{3}x + y = 4$ (3) $-x + 3y = 12$

$x = \dfrac{1}{3}y - 1$ (3) $3x = y - 3,$ then $y = 3x + 3.$

Substituting in $-x + 3y = 12,$ $-x + 3(3x + 3) = 12$

$$-x + 9x + 9 = 12$$
$$8x = 3$$
$$x = \frac{3}{8}$$

Then, $y = 3\left(\dfrac{3}{8}\right) + 3 = \dfrac{9}{8} + 3 = \dfrac{33}{8}.$

$S = \left\{\left(\dfrac{3}{8}, \dfrac{33}{8}\right)\right\}$

60. Let $p = \dfrac{1}{x}$ and $q = \dfrac{1}{y}.$

$\dfrac{2}{x} - \dfrac{3}{y} = 1$

$\dfrac{1}{x} + \dfrac{2}{y} = 2$

$\begin{aligned} 2p - 3q &= 1 \\ \underline{-2p - 4q} &= \underline{-2} \\ -7q &= -1 \\ q &= \dfrac{1}{7} \end{aligned}$

$2p - 3q = 1$
$p + 2q = 1$

$p + 2\left(\dfrac{1}{7}\right) = 1$

$p + \dfrac{2}{7} = 1$

$p = \dfrac{5}{7}$

$\dfrac{1}{x} = \dfrac{5}{7}$ implies $x = \dfrac{7}{5}$

$\dfrac{1}{y} = \dfrac{1}{7}$ implies $y = 7$

$S = \left\{\left(\dfrac{7}{5}, 7\right)\right\}$

63. (1) Using $(-1, -2)$ and $(3, 4),$ $m = \dfrac{4 - (-2)}{3 - (-1)} = \dfrac{6}{4} = \dfrac{3}{2}.$

Then $y - 4 = \dfrac{3}{2}(x - 3)$

$$2y - 8 = 3x - 9$$
$$-3x + 2y = -1$$

(2) Using $(4, 1)$ and $(2, -4),$ $m = \dfrac{1 - (-4)}{4 - 2} = \dfrac{5}{2}.$

Then $y - 1 = \dfrac{5}{2}(x - 4)$

$$2y - 2 = 5x - 20$$
$$-5x + 2y = -18$$

(3) Solving $-3x + 2y = -1$

$-5x + 2y = -18$ (-1)
$\begin{aligned} -3x + 2y &= -1 \\ \underline{5x - 2y} &= \underline{18} \\ 2x &= 17 \\ x &= \dfrac{17}{2} \end{aligned}$

Then $-3\left(\dfrac{17}{2}\right) + 2y = -1$

$$-\dfrac{51}{2} + 2y = -1$$
$$2y = \dfrac{49}{2}$$
$$y = \dfrac{49}{4}$$

$S = \left\{\left(\dfrac{17}{2}, \dfrac{49}{4}\right)\right\}$

69. Let w = width of the rectangle. Let ℓ = length of the rectangle.

(1) Then $2\ell + 3w = 48$

$2\ell + 2w = 40$ (times -1)

(2) $2\ell + 3w = \quad 48$

$-2\ell - 2w = -40$

$w = \quad 8$

(3) $2\ell + 2(8) = 40$

$2\ell + 16 = 40$

$2\ell = 24$

$\ell = 12$

The room is 8 feet wide and 12 feet long.

74. Let x = amount invested at 7%. Let y = amount invested at 9%.

Then $x + y = 16{,}000$

$0.07x = 0.09y$

(1) $x + \quad y = 16{,}000$ (times 9)

$7x - 9y = 0$

(2) $9x + 9y = 144{,}000$

$7x - 9y = 0$

$16x \quad = 144{,}000$

$x = 9{,}000$

Then $y \quad = 7{,}000$

Jamie invested \$9,000 at 7% and \$7,000 at 9%.

83. Let x = number of suits sold at \$152 each. Let y = number of suits sold at \$205 each.

Then $x + y = 32$ (times -152)

$152x + 205y = 5{,}659$

(1) $-152x - 152y = -4{,}864$

$152x + 205y = \quad 5{,}659$

$53y = 795$

$y = 15$

(2) $x + 15 = 32$

$x = 17$

15 suits were sold at \$205 each and 17 were sold at \$152 each.

94. Let x = number of hours at 4.5 mph. Let y = number of hours at 4 mph.

Then $x + y = 7$

$4.5x + 4y = 30$

(1) $-4x - 4y = -28$

$4.5x + 4y = 30$

$0.5x \quad = 2$

$x = 4$

(2) $4 + y = 7$

$y = 3$

They rowed 4 hours at 4.5 mph and 3 hours at 4 mph.

99. Let x = rate of the cyclist. Let y = rate of the pedestrian.

Then $2\frac{1}{4}x + 2\frac{1}{4}y = 40$

$5x = 5y + 40$ or $x = y + 8$

(1) $\frac{9}{4}x + \frac{9}{4}y = 40$ (times 4)

$x - \quad y = 8$ (times 9)

(2) $9x + 9y = 160$

$9x - 9y = \quad 72$

$18x \quad = 232$

$x = \dfrac{232}{18} = \dfrac{116}{9}$

(3) $\left(\dfrac{116}{9}\right) = y + 8$

$\dfrac{116}{9} = y + 8$

$y = \dfrac{116}{9} - 8 = \dfrac{44}{9}$

$y = \dfrac{44}{9}$

The cyclist travels at $\dfrac{116}{9} = 12\dfrac{8}{9}$ mph and the pedestrian walks at $\dfrac{44}{9} = 4\dfrac{8}{9}$ mph.

101. Let $x =$ the number of children's tickets sold. Let $y =$ the number of adult tickets sold.

$x + y = 503$

$1.25x + 3.50y = 853.75$

$$\begin{array}{l}(1) \qquad\qquad x + y = 503 \qquad \text{(times } -125) \qquad -125x - 125y = -62{,}875 \quad (2) \qquad \text{Then } x + 100 = 503 \\ \qquad\qquad 125x + 350y = 85{,}375 \qquad\qquad\qquad\qquad\qquad \underline{125x + 350y = 85{,}375} \qquad\qquad\qquad\qquad\qquad x = 403 \\ \qquad\qquad\qquad\qquad\qquad\qquad\qquad\qquad\qquad\qquad\qquad\qquad\qquad 225y = 22{,}500 \\ \qquad\qquad\qquad\qquad\qquad\qquad\qquad\qquad\qquad\qquad\qquad\qquad\qquad\quad y = 100 \end{array}$$

There were 403 children's tickets sold.

Exercise 8–2

Answers to odd-numbered problems

1. $\{(3,1,2)\}$ **3.** $\{(3,1,2)\}$ **5.** $\{(2,3,1)\}$ **7.** $\{(3,-1,2)\}$ **9.** $\{(5,-5,7)\}$ **11.** inconsistent; $S = \emptyset$

13. dependent; $S = \{(x,y,z) \mid x - 4y + z = -5\}$ **15.** $\{(1,-2,1)\}$ **17.** $\left\{\left(\dfrac{43}{3}, -3, \dfrac{122}{3}\right)\right\}$ **19.** $\{(8,1,-5)\}$ **21.** $\{(-1,-3,2)\}$

23. $\left\{\left(\dfrac{2}{5}, -\dfrac{23}{5}, -2\right)\right\}$ **25.** $\{(1,0,3)\}$ **27.** $\left\{\left(8, \dfrac{5}{2}, \dfrac{5}{2}\right)\right\}$ **29.** dependent; $S = \{(x,y,z) \mid 2x + 8y - 2z = 6\}$

31. $33°, 47°, 100°$ **33.** 25 m, 36 m, 61 m   **35.** p_1 (expensive) $= \$21.00$; p_2(middle-priced) $= \$15.00$; p_3(cheapest) $= \$11.00$

37. 16—\$750 stamps; 26—\$1,500 stamps; 4—\$25,000 stamps   **39.** $a = \dfrac{1}{2}$, $b = -\dfrac{9}{2}$, $c = 2$ **41.** 10—\$5 bills, 15—\$10 bills, 8—\$20 bills

Solutions to trial exercise problems

1.
$$\begin{array}{r} x + y + z = 6 \\ x - 2y - z = -1 \\ x + y - z = 2 \end{array}$$

(1)   Using
$$\begin{array}{r} x + y + z = 6 \\ \underline{x - 2y - z = -1} \\ 2x - y = 5 \end{array}$$

(3)   Solving $2x - y = 5$ (times 2)
$$2x + 2y = 8$$

(5)   Substituting
$$\begin{array}{r} 2(3) - y = 5 \\ 6 - y = 5 \\ -y = -1 \\ y = 1 \end{array}$$
$$S = \{(3,1,2)\}$$

(2)   Using
$$\begin{array}{r} x + y + z = 6 \\ \underline{x + y - z = 2} \\ 2x + 2y = 8 \end{array}$$

(4)   Then
$$\begin{array}{r} 4x - 2y = 10 \\ \underline{2x + 2y = 8} \\ 6x = 18 \\ x = 3 \end{array}$$

(6)   Substituting in $x + y + z = 6$
$$\begin{array}{r} 3 + 1 + z = 6 \\ 4 + z = 6 \\ z = 2 \end{array}$$

8.
$$\begin{array}{r} x + 2y + 3z = 5 \\ -x + y - z = -6 \\ 2x + y + 4z = 4 \end{array}$$

(1)   Using
$$\begin{array}{r} x + 2y + 3z = 5 \\ \underline{-x + y - z = -6} \\ 3y + 2z = -1 \end{array}$$

(2)   Using
$$\begin{array}{rl} -x + y - z = -6 & \text{(times 2)} \\ 2x + y + 4z = 4 & \end{array}$$
$$\begin{array}{r} -2x + 2y - 2z = -12 \\ \underline{2x + y + 4z = 4} \\ 3y + 2z = -8 \end{array}$$

(3)   Solving
$$\begin{array}{rl} 3y + 2z = -1 & \text{(times } -1) \\ 3y + 2z = -8 & \end{array}$$
$$\begin{array}{r} -3y - 2z = 1 \\ \underline{3y + 2z = -8} \\ 0 = -7 \end{array}$$

The system is *inconsistent*. There are no common solutions. $S = \emptyset$.

13.
$$\begin{array}{r} x - 4y + z = -5 \\ 3x - 12y + 3z = -15 \\ -2x + 8y - 2z = 10 \end{array}$$

Using
$$\begin{array}{rl} x - 4y + z = -5 & \text{(times } -3) \\ 3x - 12y + 3z = -15 & \end{array}$$
$$\begin{array}{r} -3x + 12y - 3z = 15 \\ \underline{3x - 12y + 3z = -15} \\ 0 = 0 \end{array}$$

The system is *dependent*. $S = \{(x,y,z) \mid x - 4y + z = -5\}$ (or either of the other equations).

18. $x - y = -1$
$x + z = -2$
$y - z = 2$

(1)   Using $x + z = 2$
$\underline{y - z = -2}$
$x + y = 0$

(2)   Using $x - y = -1$
$\underline{x + y = 0}$
$2x = -1$
$x = -\dfrac{1}{2}$

(3)   Using $x - y = -1$ and substituting $-\dfrac{1}{2}$ for x

$-\dfrac{1}{2} - y = -1$

$-y = -\dfrac{1}{2}$

$y = \dfrac{1}{2}$

(4)   Using $y - z = 2$, $\dfrac{1}{2} - z = 2$,

$-z = \dfrac{3}{2}$

$z = -\dfrac{3}{2}$

$S = \left\{ \left(-\dfrac{1}{2}, \dfrac{1}{2}, -\dfrac{3}{2} \right) \right\}$

33. Let $x =$ the length of the shortest side
$y =$ the length of the middle side
$z =$ the length of the longest side.
Then $\quad x + y + z = 122$
$z = x + y \qquad\qquad$ and
$2x + 11 = z$

$x + y + z = 122$
$-x - y + z = 0$
$2x - z = -11$

(1)   Add $\quad x + y + z = 122$
$\underline{-x - y + z = 0}$
$2z = 122$
$z = 61$

(2)   Substitute in $2x - z = -11$
$2x - 61 = -11$
$2x = 50$
$x = 25$

(3)   Substitute in $x + y + z = 122$
$25 + y + 61 = 122$
$y + 86 = 122$
$y = 36$

The sides have length 25 meters, 36 meters, and 61 meters.

38. (1)   Using $(0,5)$,
$5 = a(0)^2 + b(0) + c$
$c = 5$

(2)   Using $(-1,2)$,
$2 = a - b + c$

(3)   Using $(2, 17)$,
$17 = a(2)^2 + b(2) + c$
$17 = 4a + 2b + c$

(4)   Solve the system.
$a - b + c = 2$
$4a + 2b + c = 17$
$c = 5$

(5)   Substitute 5 for c.
$a - b + 5 = 2$
$\underline{4a + 2b + 5 = 17}$

then $a - b = -3$ (times 2) and
$4a + 2b = 12$

$2a - 2b = -6$
$\underline{4a + 2b = 12}$
$6a = 6$
$a = 1$

(6)   Substitute 1 for a in
$a - b = -3$
$1 - b = -3$
$-b = -4$
$b = 4$

Therefore $a = 1$, $b = 4$, $c = 5$.

Exercise 8–3

1. 11   **3.** -24 **5.** 5   **7.** -35 **9.** -12 **11.** -14 **13.** 0   **15.** -4 **17.** 29   **19.** -14 **21.** -24

23. a^3 **25.** $y^3 - x^2y$ **27.** -19 **29.** 37   **31.** $\left\{\left(\dfrac{5}{3}, -\dfrac{1}{3}\right)\right\}$

33. $\left\{\left(\dfrac{-2}{19}, \dfrac{24}{19}\right)\right\}$ **35.** $S = \emptyset$; inconsistent   **37.** $S = \{(x,y)\,|\,2x + 3y = -1\}$; dependent   **39.** $\{(0,-1)\}$ **41.** $\left\{\left(\dfrac{5}{2}, 4\right)\right\}$

43. $\left\{\left(-\dfrac{1}{19}, -\dfrac{6}{19}, -\dfrac{8}{19}\right)\right\}$ **45.** $\{(0,0,0)\}$ **47.** $S = \{(x,y,z)\,|\,3x - y - 6z = 5\}$; dependent

49. $S = \emptyset$; inconsistent   **51.** $S = \{(x,y,z)\,|\,3x + 2y + 4z = 3\}$; dependent   **53.** $\left\{\left(\dfrac{31}{9}, \dfrac{26}{9}, \dfrac{7}{9}\right)\right\}$ **55.** $\left\{\left(\dfrac{4}{3}, 3, \dfrac{8}{3}\right)\right\}$

Solutions to trial exercise problems

3. $\begin{vmatrix} 4 & -2 \\ -6 & -3 \end{vmatrix} = -12 - (12) = -24$

9. $\begin{vmatrix} 1 & 2 & 3 \\ 3 & 2 & 1 \\ 2 & 1 & 3 \end{vmatrix} = 1\begin{vmatrix} 2 & 1 \\ 1 & 3 \end{vmatrix} - 3\begin{vmatrix} 2 & 3 \\ 1 & 3 \end{vmatrix} + 2\begin{vmatrix} 2 & 3 \\ 2 & 1 \end{vmatrix}$ (Use first column.)

$= 1(6 - 1) - 3(6 - 3) + 2(2 - 6)$

$= 1(5) - 3(3) + 2(-4) = 5 - 9 - 8 = -12$

16. $\begin{vmatrix} -1 & -1 & -1 \\ 2 & 2 & 2 \\ 3 & -3 & 3 \end{vmatrix} = -2\begin{vmatrix} -1 & -1 \\ -3 & 3 \end{vmatrix} + 2\begin{vmatrix} -1 & -1 \\ 3 & 3 \end{vmatrix} - 2\begin{vmatrix} -1 & -1 \\ 3 & -3 \end{vmatrix}$ (Use second row.)

$= -2(-3 - 3) + 2(-3 + 3) - 2(3 + 3)$

$= -2(-6) + 2(0) - 2(6) = 12 + 0 - 12 = 0$

23. $\begin{vmatrix} a & 0 & a \\ 0 & a & 0 \\ 0 & 0 & a \end{vmatrix} = a\begin{vmatrix} a & 0 \\ 0 & a \end{vmatrix} - 0\begin{vmatrix} 0 & a \\ 0 & a \end{vmatrix} + 0\begin{vmatrix} 0 & a \\ a & 0 \end{vmatrix}$ (Use first column.)

$= a(a^2 - 0) - 0(0 - 0) + 0(0 - a^2)$

$= a^3 - 0 + 0 = a^3$

27. $\begin{vmatrix} 1 & 2 & 3 & -1 \\ 2 & 0 & 1 & 3 \\ -2 & 1 & 0 & -1 \\ 0 & 3 & 2 & 0 \end{vmatrix} = 1\begin{vmatrix} 0 & 1 & 3 \\ 1 & 0 & -1 \\ 3 & 2 & 0 \end{vmatrix} - 2\begin{vmatrix} 2 & 1 & 3 \\ -2 & 0 & -1 \\ 0 & 2 & 0 \end{vmatrix} + 3\begin{vmatrix} 2 & 0 & 3 \\ -2 & 1 & -1 \\ 0 & 3 & 0 \end{vmatrix} - (-1)\begin{vmatrix} 2 & 0 & 1 \\ -2 & 1 & 0 \\ 0 & 3 & 2 \end{vmatrix}$ (Use first row.)

$= 1\left[0\begin{vmatrix} 0 & -1 \\ 2 & 0 \end{vmatrix} - 1\begin{vmatrix} 1 & 3 \\ 2 & 0 \end{vmatrix} + 3\begin{vmatrix} 1 & 3 \\ 0 & -1 \end{vmatrix}\right] - 2\left[2\begin{vmatrix} 0 & -1 \\ 2 & 0 \end{vmatrix} + 2\begin{vmatrix} 1 & 3 \\ 2 & 0 \end{vmatrix} + 0\begin{vmatrix} 1 & 3 \\ 0 & -1 \end{vmatrix}\right]$

$+ 3\left[2\begin{vmatrix} 1 & -1 \\ 3 & 0 \end{vmatrix} + 2\begin{vmatrix} 0 & 3 \\ 3 & 0 \end{vmatrix} + 0\begin{vmatrix} 0 & 3 \\ 1 & -1 \end{vmatrix}\right] + 1\left[2\begin{vmatrix} 1 & 0 \\ 3 & 2 \end{vmatrix} + 2\begin{vmatrix} 0 & 1 \\ 3 & 2 \end{vmatrix} + 0\begin{vmatrix} 0 & 1 \\ 1 & 0 \end{vmatrix}\right]$

$= 1[0 - 1(-6) + 3(-1)] - 2[2(2) + 2(-6) + 0] + 3[2(3) + 2(-9) + 0] + 1[2(2) + 2(-3) + 0]$

$= 1[6 - 3] - 2[4 - 12] + 3[6 - 18] + 1[4 - 6]$

$= 3 + 16 - 36 - 2 = 19 - 38 = -19$

31. $x - y = 2$

$2x + y = 3$ $D = \begin{vmatrix} 1 & -1 \\ 2 & 1 \end{vmatrix} = 1 - (-2) = 3$

$D_x = \begin{vmatrix} 2 & -1 \\ 3 & 1 \end{vmatrix} = 2 - (-3) = 5$

$D_y = \begin{vmatrix} 1 & 2 \\ 2 & 3 \end{vmatrix} = 3 - 4 = -1$

Then $x = \dfrac{D_x}{D} = \dfrac{5}{3}$; $y = \dfrac{D_y}{D} = -\dfrac{1}{3}$

$S = \left\{\left(\dfrac{5}{3}, -\dfrac{1}{3}\right)\right\}$

34. $4x - y = 3$
$8x - 2y = 1$ $\qquad D = \begin{vmatrix} 4 & -1 \\ 8 & -2 \end{vmatrix} = -8 + 8 = 0$

$$D_x = \begin{vmatrix} 3 & -1 \\ 1 & -2 \end{vmatrix} = -6 + 1 = -5$$

$$D_y = \begin{vmatrix} 4 & 3 \\ 8 & 1 \end{vmatrix} = 4 - 24 = -20$$

Since $D = 0$ $D_x \neq 0$, and $D_y \neq 0$, the system is *inconsistent*. $S = \emptyset$.

41. $6x - 2y = 7$
$2y = 8$ $\qquad D = \begin{vmatrix} 6 & -2 \\ 0 & 2 \end{vmatrix} = 12 - 0 = 12$

$$D_x = \begin{vmatrix} 7 & -2 \\ 8 & 2 \end{vmatrix} = 14 - (-16) = 30$$

$$D_y = \begin{vmatrix} 6 & 7 \\ 0 & 8 \end{vmatrix} = 48 - 0 = 48$$

$$x = \frac{D_x}{D} = \frac{30}{12} = \frac{5}{2}; \quad y = \frac{D_y}{D} = \frac{48}{12} = 4$$

$$S = \left\{ \left(\frac{5}{2}, 4 \right) \right\}$$

43. $x - y + 3z = -1$
$2x + y - z = 0$
$-3x - 4y + z = 1$

$$D = \begin{vmatrix} 1 & -1 & 3 \\ 2 & 1 & -1 \\ -3 & -4 & 1 \end{vmatrix} = 1 \begin{vmatrix} 1 & -1 \\ -4 & 1 \end{vmatrix} - 2 \begin{vmatrix} -1 & 3 \\ -4 & 1 \end{vmatrix} + (-3) \begin{vmatrix} -1 & 3 \\ 1 & -1 \end{vmatrix}$$
$$= 1(1 - 4) - 2(-1 + 12) - 3(1 - 3)$$
$$= 1(-3) - 2(11) - 3(-2) = -3 - 22 + 6 = -19$$

$$D_x = \begin{vmatrix} -1 & -1 & 3 \\ 0 & 1 & -1 \\ 1 & -4 & 1 \end{vmatrix} = -1 \begin{vmatrix} 1 & -1 \\ -4 & 1 \end{vmatrix} - 0 \begin{vmatrix} -1 & 3 \\ -4 & 1 \end{vmatrix} + 1 \begin{vmatrix} -1 & 3 \\ 1 & -1 \end{vmatrix}$$
$$= -1(1 - 4) - 0(-1 + 12) + 1(1 - 3)$$
$$= -1(-3) - 0 + 1(-2)$$
$$= +3 + (-2) = 1$$

$$D_y = \begin{vmatrix} 1 & -1 & 3 \\ 2 & 0 & -1 \\ -3 & 1 & 1 \end{vmatrix} = 1 \begin{vmatrix} 0 & -1 \\ 1 & 1 \end{vmatrix} - 2 \begin{vmatrix} -1 & 3 \\ 1 & 1 \end{vmatrix} + (-3) \begin{vmatrix} -1 & 3 \\ 0 & -1 \end{vmatrix}$$
$$= 1(0 + 1) - 2(-1 - 3) - 3(1 - 0)$$
$$= 1(1) - 2(-4) - 3(1) = 1 + 8 - 3 = 6$$

$$D_z = \begin{vmatrix} 1 & -1 & -1 \\ 2 & 1 & 0 \\ -3 & -4 & 1 \end{vmatrix} = 1 \begin{vmatrix} 1 & 0 \\ -4 & 1 \end{vmatrix} - 2 \begin{vmatrix} -1 & -1 \\ -4 & 1 \end{vmatrix} + (-3) \begin{vmatrix} -1 & -1 \\ 1 & 0 \end{vmatrix}$$
$$= 1(1 - 0) - 2(-1 - 4) - 3(0 + 1)$$
$$= 1(1) - 2(-5) - 3(1) = 1 + 10 - 3 = 8$$

$$x = \frac{D_x}{D} = \frac{1}{-19} = -\frac{1}{19}; y = \frac{D_y}{D} = \frac{6}{-19} = -\frac{6}{19}; z = \frac{D_x}{D} = \frac{8}{-19} = -\frac{8}{19}$$

$$S = \left\{ \left(-\frac{1}{19}, -\frac{6}{19}, -\frac{8}{19} \right) \right\}$$

52. $x - 2y = 3$
$y + 3z = -2$
$2x - z = 4$

$$D = \begin{vmatrix} 1 & -2 & 0 \\ 0 & 1 & 3 \\ 2 & 0 & -1 \end{vmatrix} = 1 \begin{vmatrix} 1 & 3 \\ 0 & -1 \end{vmatrix} - 0 \begin{vmatrix} -2 & 0 \\ 0 & -1 \end{vmatrix} + 2 \begin{vmatrix} -2 & 0 \\ 1 & 3 \end{vmatrix}$$

$$= 1(-1 - 0) - 0(2 - 0) + 2(-6 - 0)$$
$$= 1(-1) - 0 + 2(-6) = -1 - 12 = -13$$

$$D_x = \begin{vmatrix} 3 & -2 & 0 \\ -2 & 1 & 3 \\ 4 & 0 & -1 \end{vmatrix} = 3 \begin{vmatrix} 1 & 3 \\ 0 & -1 \end{vmatrix} - (-2) \begin{vmatrix} -2 & 0 \\ 0 & -1 \end{vmatrix} + 4 \begin{vmatrix} -2 & 0 \\ 1 & 3 \end{vmatrix}$$

$$= 3(-1 - 0) + 2(2 - 0) + 4(-6 - 0)$$
$$= 3(-1) + 2(2) + 4(-6) = -3 + 4 - 24 = -23$$

$$D_y = \begin{vmatrix} 1 & 3 & 0 \\ 0 & -2 & 3 \\ 2 & 4 & -1 \end{vmatrix} = 1 \begin{vmatrix} -2 & 3 \\ 4 & -1 \end{vmatrix} - 0 \begin{vmatrix} 3 & 0 \\ 4 & -1 \end{vmatrix} + 2 \begin{vmatrix} 3 & 0 \\ -2 & 3 \end{vmatrix}$$

$$= 1(2 - 12) - 0 + 2(9 - 0)$$
$$= 1(-10) - 0 + 2(9) = -10 + 18 = 8$$

$$D_z = \begin{vmatrix} 1 & -2 & 3 \\ 0 & 1 & -2 \\ 2 & 0 & 4 \end{vmatrix} = 1 \begin{vmatrix} 1 & -2 \\ 0 & 4 \end{vmatrix} - 0 \begin{vmatrix} -2 & 3 \\ 0 & 4 \end{vmatrix} + 2 \begin{vmatrix} -2 & 3 \\ 1 & -2 \end{vmatrix}$$

$$= 1(4 - 0) - 0 + 2(4 - 3)$$
$$= 1(4) - 0 + 2(1) = 4 + 2 = 6$$

$$x = \frac{D_x}{D} = \frac{-23}{-13} = \frac{23}{13} ; y = \frac{D_y}{D} = \frac{8}{-13} = -\frac{8}{13} ; z = \frac{D_z}{D} = \frac{6}{-13} = -\frac{6}{13}$$

$$S = \left\{ \left(\frac{23}{13}, -\frac{8}{13}, -\frac{6}{13} \right) \right\}$$

Exercise 8–4

Answers to odd-numbered problems

1. $\{(2,3)\}$ **3.** $\{(-2,1)\}$ **5.** $\{(3,-1)\}$ **7.** $\left\{ \left(-\frac{1}{17}, -\frac{5}{17} \right) \right\}$ **9.** $\left\{ \left(-5, \frac{1}{2} \right) \right\}$ **11.** $\left\{ \left(\frac{1}{5}, \frac{4}{5}, -\frac{16}{5} \right) \right\}$ **13.** $\{(2,0,-1)\}$
15. $\{(4,1,0)\}$ **17.** $\emptyset$; inconsistent

Solutions to trial exercise problems

3. $x - 4y = -6$ augmented matrix $\begin{bmatrix} 1 & -4 & | & -6 \\ 3 & 1 & | & -5 \end{bmatrix}$
$3x + y = -5$

We want 0 in the second row, first column. Multiply row one by -3 and add to row two. We get

$\begin{bmatrix} 1 & -4 & | & -6 \\ 0 & 13 & | & 13 \end{bmatrix}$.

We now have the system $x - 4y = -6$
$ 13y = 13.$

Then $y = 1$ and replace y by 1 in the first equation $x - 4(1) = -6$
$ x - 4 = -6$
$ x = -2$

The solution set is $S = \{(-2,1)\}$.

10. $x + 3y - z = 5$ augmented matrix
$3x - y + 2z = 5$
$x + y + 2z = 7$

$$\begin{bmatrix} 1 & 3 & -1 & | & 5 \\ 3 & -1 & 2 & | & 5 \\ 1 & 1 & 2 & | & 7 \end{bmatrix}$$

Multiply row one by -3 and add to row two.

$$\begin{bmatrix} 1 & 3 & -1 & | & 5 \\ 0 & -10 & 5 & | & -10 \\ 1 & 1 & 2 & | & 7 \end{bmatrix}$$

Subtract row one from row three.

$$\begin{bmatrix} 1 & 3 & -1 & | & 5 \\ 0 & -10 & 5 & | & -10 \\ 0 & -2 & 3 & | & 2 \end{bmatrix}$$

Multiply row two by $-\dfrac{1}{10}$.

$$\begin{bmatrix} 1 & 3 & -1 & | & 5 \\ 0 & 1 & -\dfrac{1}{2} & | & 1 \\ 0 & -2 & 3 & | & 2 \end{bmatrix}$$

Multiply row two by 2 and add to row three.

$$\begin{bmatrix} 1 & 3 & -1 & | & 5 \\ 0 & 1 & -\dfrac{1}{2} & | & 1 \\ 0 & 0 & 2 & | & 4 \end{bmatrix}$$

$x + 3y - z = 5$
$y - \dfrac{1}{2}z = 1$
$2z = 4$
$z = 2$

Replace z by 2 in $y - \dfrac{1}{2}z = 1$
$y - 1 = 1$
$y = 2.$

Replace z by 2 and y by 2 in $x + 3y - z = 5$.
$x + 3(2) - 2 = 5$
$x + 6 - 2 = 5$
$x + 4 = 5$
$x = 1$

The solution set is $S = \{(1,2,2)\}$.

Chapter 8 review exercises

1. $\{(3,1)\}$ **2.** $\left\{\left(\dfrac{1}{2},\dfrac{1}{3}\right)\right\}$ **3.** $\{(x,y) \,|\, 2x + 3y = 4\}$; dependent   **4.** $\emptyset$; inconsistent   **5.** $\left\{\left(\dfrac{5}{4},-\dfrac{1}{4}\right)\right\}$ **6.** $\left\{\left(-\dfrac{12}{5},\dfrac{26}{5}\right)\right\}$

7. $\left(\dfrac{39}{5},\dfrac{3}{5}\right)$ **8.** $\{(3,-4)\}$ **9.** $\{(-1,-10)\}$ **10.** $\left\{\left(-\dfrac{2}{7},\dfrac{1}{7}\right)\right\}$ **11.** $\left\{\left(-\dfrac{11}{4},\dfrac{27}{4}\right)\right\}$ **12.** $\left\{\left(-\dfrac{7}{3},-\dfrac{5}{6}\right)\right\}$ **13.** $\left(\dfrac{11}{5},-\dfrac{3}{5}\right)$

14. $\ell = 60$ ft; $w = 30$ ft   **15.** 4 false alarms; 26 real alarms   **16.** \$1,800 at 6½%; \$1,200 at 7%

17. 16⅔ g of 20% tin; 33⅓ g of 5% tin   **18.** \$72 for topcoat; \$105 for suit   **19.** 5.2 hr   **20.** $\{(1,2,3)\}$

21. $\left\{\left(\dfrac{1}{10},\dfrac{1}{2},-\dfrac{3}{10}\right)\right\}$ **22.** $\{(3,-5,-1)\}$ **23.** $\{(3,2,4)\}$ **24.** $a = 3, b = -8, c = 2$ **25.** $F_1 = \dfrac{35}{2}, F_2 = -5, F_3 = 25$

26. $41°, 57°, 82°$ **27.** 135 nickels; 120 dimes; 30 quarters   **28.** 7   **29.** 24   **30.** 41   **31.** 108   **32.** 39   **33.** 88

34. $\{(1,-1)\}$ **35.** $\left\{\left(-\dfrac{5}{26},\dfrac{2}{13}\right)\right\}$ **36.** $\left\{\left(\dfrac{15}{4},9\right)\right\}$ **37.** $\left\{\left(\dfrac{8}{3},6\right)\right\}$ **38.** $\{(0,0,0)\}$ **39.** $\left\{\left(\dfrac{7}{6},\dfrac{-5}{6},\dfrac{-7}{6}\right)\right\}$ **40.** $\left\{\left(\dfrac{4}{7},-\dfrac{2}{7}\right)\right\}$

41. $\left\{\left(-3,\dfrac{16}{5},\dfrac{23}{5}\right)\right\}$

Chapter 8 cumulative test

1. -258 **2.** $V = 110$ **3.** $C = \dfrac{24}{5}$ **4.** $3a^2 - 2a + 6 + \dfrac{15}{a-3}$ **5.** $6a^5 - 5a^3 + 3a$ **6.** $4x^2 - 56x - 45$ **7.** $\dfrac{2x^4}{y^5}$

8. $\left\{\dfrac{14}{13}\right\}$ **9.** $\left\{\dfrac{1 + \sqrt{13}}{3}, \dfrac{1 - \sqrt{13}}{3}\right\}$ **10.** $\{-3 + \sqrt{14}, -3 - \sqrt{14}\}$ **11.** $\left\{\dfrac{-1 + \sqrt{41}}{4}, \dfrac{-1 - \sqrt{41}}{4}\right\}$ **12.** $\left\{z \,\middle|\, z \geq -\dfrac{4}{3}\right\}$

13. $\{x \mid -3 \leq x \leq 1\}$ **14.** $\left\{y \,\middle|\, y < -5 \text{ or } y > \dfrac{1}{2}\right\}$ **15.** $\left\{\dfrac{5}{2}, -\dfrac{3}{2}\right\}$ **16.** $\left\{x \,\middle|\, x < \dfrac{1}{2} \text{ or } x > \dfrac{7}{6}\right\}$ **17.** $\dfrac{4}{4y^2 - 6y + 9}$

18. $\dfrac{4x^2 - 13x - 12}{3}$ **19.** $\dfrac{4x + 2}{(x - 5)(x + 4)(x - 3)}$ **20.** $\dfrac{2y}{y - 7}$ **21.** $\dfrac{3p^2ab}{14q}$ **22.** 9 **23.** $\dfrac{1}{y^3}$ **24.** $a^{\frac{5}{4}}$ **25.** $4\sqrt{5}$

26. $\sqrt{12} + 3\sqrt{2} - 4\sqrt{6} - 12$ **27.** $11 + 40i$ **28.** $-2y\sqrt[3]{x}$ **29.** $\emptyset$; 36 and 1 are extraneous

30.

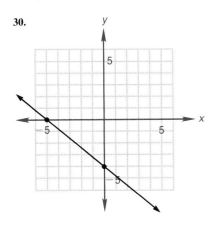

31.

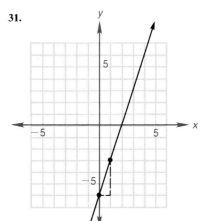

32. $5x + 7y = 28$

33. $4x - y = 1$

34.

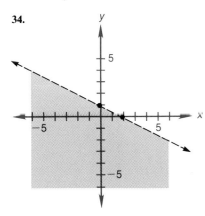

35. -31 **36.** $\{(2,0)\}$ **37.** $\left\{\left(\dfrac{9}{5}, -\dfrac{11}{5}\right)\right\}$

38. $\left\{\left(\dfrac{3}{17}, -\dfrac{4}{17}\right)\right\}$ **39.** $\left\{\left(-\dfrac{5}{11}, -\dfrac{7}{11}\right)\right\}$

40. $\left\{\left(-\dfrac{1}{6}, -\dfrac{2}{3}, \dfrac{1}{6}\right)\right\}$

Chapter 9

Exercise 9–1

Answers to odd-numbered problems

1. $(3,4)$ **3.** $(0,-16)$ **5.** $\left(\dfrac{3}{2}, -\dfrac{49}{4}\right)$ **7.** $(-11,-100)$ **9.** $\left(-\dfrac{1}{10}, \dfrac{21}{20}\right)$

11. x-intercept, $-1,1$; y-intercept, -1; $(0,-1)$, lowest point

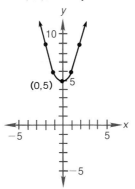

13. x-intercept, none; y-intercept, 5; $(0,5)$, lowest point

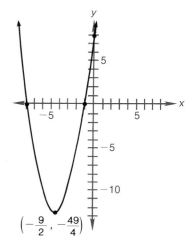

15. x-intercept, $-8,-1$; y-intercept, 8; $\left(-\dfrac{9}{2}, -\dfrac{49}{4}\right)$, lowest point

17. x-intercept, $-\dfrac{1}{2}, \dfrac{7}{2}$; y-intercept, -7; $\left(\dfrac{3}{2}, -16\right)$, lowest point

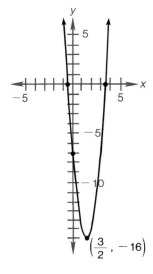

19. x-intercept, $-\dfrac{1}{3}, 5$; y-intercept, 5; $\left(\dfrac{7}{3}, \dfrac{64}{3}\right)$, highest point

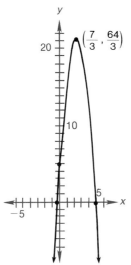

21. 1 second; 2 seconds

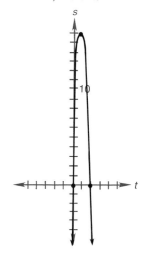

23. The maximum height is 144 feet; the arrow will reach the ground in 6 seconds.   **25.** 8 dresses must be sold daily.

27. 5 glasses of Kool Aid must be sold.   **29.** Maximum power $P = 245$ when current $I = 35$.   **31.** 28 and 28

33. $\ell = w = \dfrac{21}{2}$ or $10\dfrac{1}{2}$ m **35.** -16

Solutions to trial exercise problems

3. $y = x^2 - 16$. $y = (x - 0)^2 - 16$. The vertex is at $(0, -16)$ and since $a = 1$, $a > 0$, the graph opens up and the vertex is the lowest

point.   **6.** $y = -x^2 + 3x + 6$. $y = -1(x^2 - 3x) + 6 = -1\left(x^2 - 3x + \dfrac{9}{4}\right) + 6 + \dfrac{9}{4} = -1\left(x - \dfrac{3}{2}\right)^2 + \dfrac{33}{4}$. The vertex is

at $\left(\dfrac{3}{2}, \dfrac{33}{4}\right)$ and is the highest point since $a = -1$, $a < 0$, and the graph opens down.

16. $y = -x^2 + 4x + 10$

   (a) When $x = 0$, $y = 10$ (y-intercept) and when $y = 0$,

$$0 = -x^2 + 4x + 10$$
$$0 = x^2 - 4x - 10$$
$$x = \frac{-(-4) \pm \sqrt{(-4)^2 - 4(1)(-10)}}{2(1)} = \frac{4 \pm \sqrt{16 + 40}}{2} = \frac{4 \pm \sqrt{56}}{2}$$
$$= \frac{4 \pm 2\sqrt{14}}{2} = 2 \pm \sqrt{14} \approx 5.74 \text{ or } -1.74.$$

   (b) $y = -x^2 + 4x + 10$
$$= -1(x^2 - 4x) + 10$$
$$= -1(x^2 - 4x + 4) + 10 + 4$$
$$= -1(x - 2)^2 + 14$$
   The vertex is at $(2,14)$.

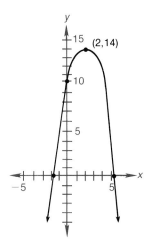

18. $y = -5x^2 + 6x - 1$

   (a) When $x = 0$, $y = -1$ (y-intercept) and when $y = 0$, then
$$0 = -5x^2 + 6x - 1$$
$$= 5x^2 - 6x + 1$$
$$= (5x - 1)(x - 1).$$
   So $x = \dfrac{1}{5}$ or $x = 1$ (x-intercepts).

   (b) $y = -5x^2 + 6x - 1$
$$= -5\left(x^2 - \frac{6}{5}x\right) - 1$$
$$= -5\left(x^2 - \frac{6}{5}x + \frac{9}{25}\right) - 1 + \frac{9}{5}$$
$$= -5\left(x - \frac{3}{5}\right)^2 + \frac{4}{5}$$
   The vertex is at $\left(\dfrac{3}{5}, \dfrac{4}{5}\right)$.

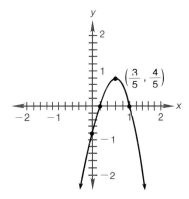

24. $P = -x^2 + 100x - 1,000$
$$= -1(x^2 - 100x) - 1,000$$
$$= -1(x^2 - 100x + 2,500) - 1,000 + 2,500$$
$$= -1(x - 50)^2 + 1,500$$
Since the vertex is at $(50;1,500)$, then 50 units must be produced to attain maximum profit.

31. Let x = one number and $56 - x$ = the other number.
   Then $y = x(56 - x) = 56x - x^2$
$$= -x^2 + 56x$$
$$= -1(x^2 - 56x + 784) + 784$$
$$= -1(x - 28)^2 + 784$$

The vertex is at $(28,784)$ and the numbers are 28 and $56 - 28 = 28$.

Exercise 9–2

Answers to odd-numbered problems

1. $(x - 1)^2 + (y - 2)^2 = 4$ **3.** $(x - 4)^2 + (y + 3)^2 = 6$ **5.** $x^2 + y^2 = 9$ **7.** $x^2 + y^2 + 10x - 4y + 28 = 0$
9. $C(3,2), r = 7$ **11.** $C(0,0), r = 6$ **13.** $C(-2,3), r = 6$ **15.** $C(-2,3), r = 5\sqrt{2}$

17.

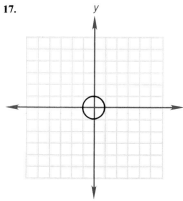

19.

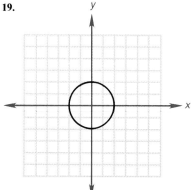

21.

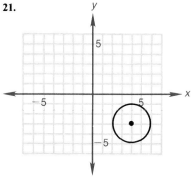

23.

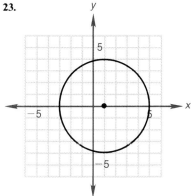

25.

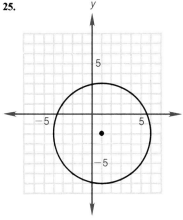

27.

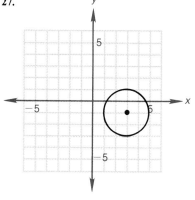

29. $r = \sqrt{(x - h)^2 + (y - k)^2}$
Square both sides.
$(x - h)^2 + (y - k)^2 = r^2$

Solutions to trial exercise problems

7. $(x + 5)^2 + (y - 2)^2 = 1^2; (x^2 + 10x + 25) + (y^2 - 4y + 4) = 1; x^2 + y^2 + 10x - 4y + 28 = 0$
15. $2x^2 + 2y^2 + 8x - 12y = 74; x^2 + y^2 + 4x - 6y = 37; (x^2 + 4x + 4) + (y^2 - 6y + 9) = 37 + 4 + 9; (x + 2)^2 + (y - 3)^2$
$= 50; C(-2,3), r = 5\sqrt{2}$

19. $3x^2 + 3y^2 = 12;\ x^2 + y^2 = 4;\ C(0,0),\ r = 2$

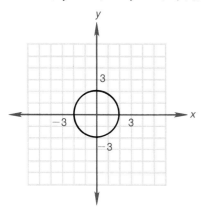

27. $2x^2 + 2y^2 - 12x + 4y = -12;\ x^2 + y^2 - 6x + 2y = -6;$
$(x^2 - 6x + 9) + (y^2 + 2y + 1) = -6 + 9 + 1;$
$(x - 3)^2 + (y + 1)^2 = 4;\ C(3,-1),\ r = 2$

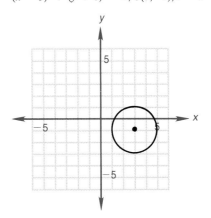

Exercise 9–3

Answers to odd-numbered problems

1. $x,\ (\pm 3,0);\ y,\ (0,\pm 5)$

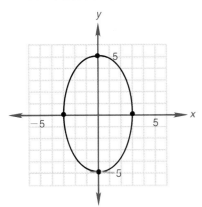

3. $x,\ (\pm 7,0);\ y,\ (0,\pm 5)$

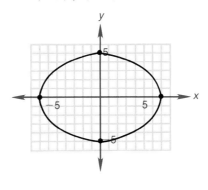

5. $x,\ (\pm 1,0);\ y,\ (0,\pm 3)$

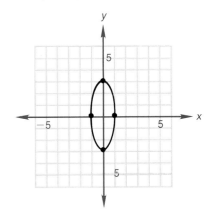

7. $x,\ (\pm 4,0);\ y,\ (0,\pm 1)$

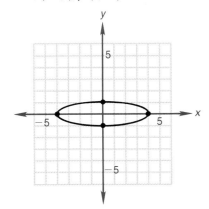

9. $x,\ (\pm 10,0);\ y,\ (0,\pm 2)$

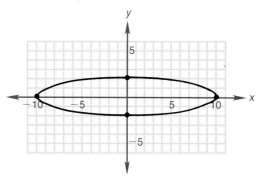

11. x, $(\pm 2,0)$; y, $(0, \pm \sqrt{3})$

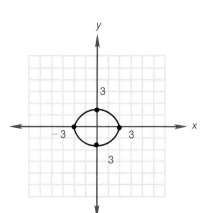

13. x, $(\pm \sqrt{2}, 0)$; y, $(0, \pm 4)$

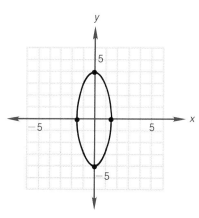

15. x, $(\pm 4, 0)$; asymptotes,
$$y = \pm \frac{3}{4}x$$

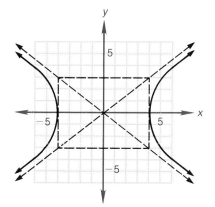

17. y, $(0, \pm 3)$; asymptotes,
$$y = \pm \frac{3}{2}x$$

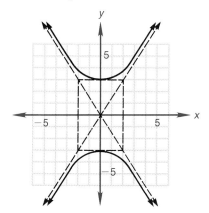

19. y, $(0, \pm 1)$; asymptotes,
$$y = \pm \frac{1}{3}x$$

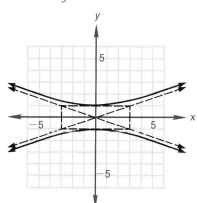

21. x, $(\pm 1, 0)$; asymptotes, $y = \pm 4x$

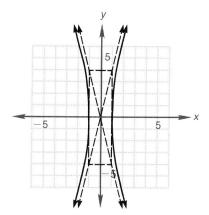

23. x, $(\pm 1, 0)$; asymptotes, $y = \pm 5x$

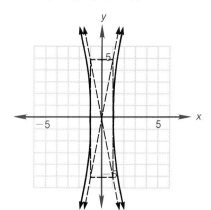

25. y, $(0, \pm 4)$; asymptotes, $y = \pm \frac{4}{5}x$

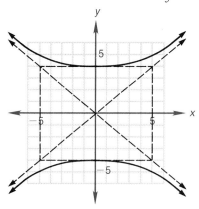

27. y, $(0, \pm 5)$; asymptotes, $y = \pm \frac{5}{2}x$

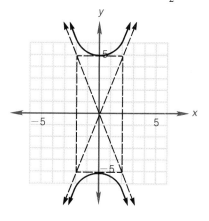

Appendix B Answers and Solutions

29. y, $(0, \pm 2)$; asymptotes, $y = \pm \dfrac{\sqrt{2}}{3}x$

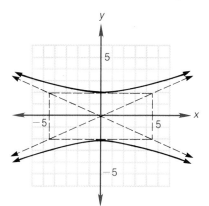

31. $x^2 + y^2 = 9$; circle

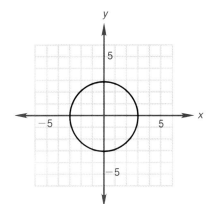

33. $x^2 + 3 = y$; parabola

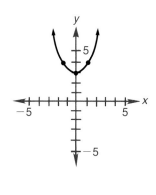

35. $x^2 + y^2 = \dfrac{25}{4}$; circle

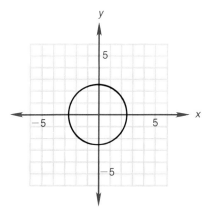

37. $x^2 + y^2 = 121$; circle

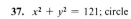
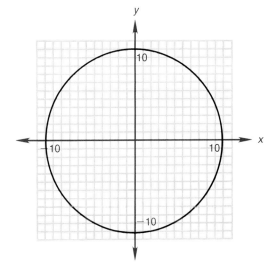

39. $y = -x^2 + 8$; parabola

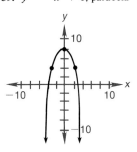

41.

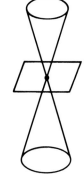

Solutions to trial exercise problems

5. $x^2 + \dfrac{y^2}{9} = 1$; $b^2 = 1$, so $b = 1$; $a^2 = 9$, so $a = 3$;

x-intercepts, $(1,0)$, $(-1,0)$; y-intercepts, $(0,3)$, $(0,-3)$

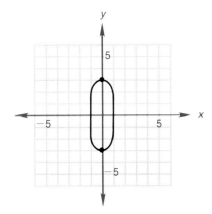

8. $36x^2 + 9y^2 = 324$; $\dfrac{x^2}{9} + \dfrac{y^2}{36} = 1$; $b^2 = 9$, $b = 3$;

$a^2 = 36$, $a = 6$; x-intercepts, $(3,0)$, $(-3,0)$; y-intercepts, $(0,6)$, $(0,-6)$

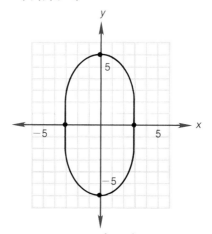

19. $y^2 - \dfrac{x^2}{9} = 1$; $a^2 = 9$, $a = 3$; $b^2 = 1$, $b = 1$;

y-intercepts, $(0,1)$, $(0,-1)$;

asymptotes, $y = \dfrac{1}{3}x$ and $y = -\dfrac{1}{3}x$

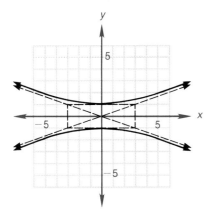

22. $9x^2 - y^2 = 36$; $\dfrac{x^2}{4} - \dfrac{y^2}{36} = 1$; $a^2 = 4$, $a = 2$; $b^2 = 36$,

$b = 6$; x-intercepts, $(2,0)$, $(-2,0)$;

asymptotes, $y = 3x$ and $y = -3x$

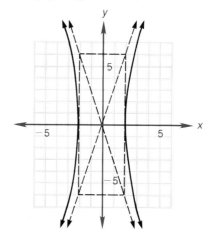

33. Since $x^2 + 3 = y$ has one variable linear and the other squared, it is a parabola.

Answers to odd-numbered problems

1. $\{(2,-1), (-1,2)\}$

3. $\{(7,-5), (-5,1)\}$

5. $\{(2,1), (-1,4)\}$

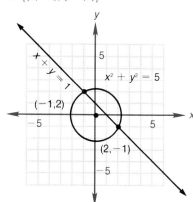

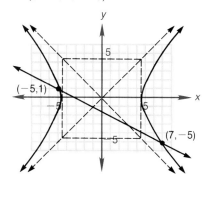

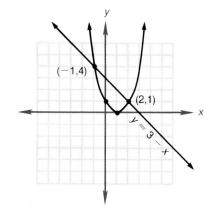

7. $\left\{\left(-\dfrac{2}{9},-\dfrac{22}{9}\right), (2,2)\right\}$ **9.** $\{(4,1), (-4,-1), (i,-4i), (-i,4i)\}$ **11.** $\{(2,2), (-2,-2)\}$ **13.** $\left\{\left(\dfrac{\sqrt{6}}{2},-1\right), \left(-\dfrac{\sqrt{6}}{2},-1\right)\right\}$

15. $\{(3,0), (-3,0)\}$ **17.** $\{(2\sqrt{3},\sqrt{13}), (2\sqrt{3},-\sqrt{13}), (-2\sqrt{3},\sqrt{13}), (-2\sqrt{3},-\sqrt{13})\}$ **19.** $\{(3,3), (3,-3), (-3,3), (-3,-3)\}$

21. $\{(3,0), (-3,0)\}$ **23.** $\{(\sqrt{7},i\sqrt{3}), (\sqrt{7},-i\sqrt{3}), (-\sqrt{7},i\sqrt{3}), (-\sqrt{7},-i\sqrt{3})\}$ **25.** $\{(\sqrt{3},\sqrt{3}), (-\sqrt{3},-\sqrt{3})\}$

27. 15 in. by 8 in.   **29.** 9 and 7   **31.** $x = 2$ when $p = 3$; $\left(x = \dfrac{15}{4}\text{ when }p = \dfrac{8}{5}; \text{ not logical}\right)$

Solutions to trial exercise problems

9. $x^2 - y^2 = 15$ (1) $y = \dfrac{4}{x}$ and substituting, $x^2 - \left(\dfrac{4}{x}\right)^2 = 15$

$\quad\; xy = 4$

$$x^2 - \dfrac{16}{x^2} = 15$$
$$x^4 - 15x^2 - 16 = 0$$
$$(x^2 - 16)(x^2 + 1) = 0$$
$$(x + 4)(x - 4)(x^2 + 1) = 0$$

Then $x = 4$ or $x = -4$; $x = \sqrt{-1} = i$ or $x = -\sqrt{-1} = -i$

(2) $y = \dfrac{4}{x}$, so $y = \dfrac{4}{4} = 1$ or $y = \dfrac{4}{-4} = -1$; or $y = \dfrac{4}{i} = -4i$ or $y = \dfrac{4}{-i} = 4i$

$S = \{(4,1), (-4, -1), (i, -4i), (-i, 4i)\}$

21. $2x^2 - 9y^2 = 18$ (1)   When $x = 3$ (2)   When $x = -3$

$\dfrac{4x^2 + 9y^2 = \;\;36}{6x^2 \qquad = \;\;54}$ $2(3)^2 - 9y^2 = 18$ $2(-3)^2 - 9y^2 = 18$

$\qquad\qquad x^2 = 9$ $18 - 9y^2 = 18$ $18 - 9y^2 = 18$

$\qquad\qquad x = \pm 3$ $-9y^2 = 0$ $-9y^2 = 0$

$S = \{(3,0), (-3,0)\}.$ $y = 0$ $y = 0$

24. $x^2 + y^2 = 10$ (times -1) $-x^2 \qquad - y^2 = -10$

$\dfrac{x^2 - 3xy + y^2 = 1}{}$ $\dfrac{x^2 - 3xy + y^2 = 1}{-3xy = -9}$

$xy = 3$

$y = \dfrac{3}{x}$

(2)   Substituting in $x^2 + y^2 = 10$

$$x^2 + \left(\frac{3}{x}\right)^2 = 10$$

$$x^2 + \frac{9}{x^2} = 10$$

$$x^4 + 9 = 10x^2$$

$$x^4 - 10x^2 + 9 = 0$$

$$(x^2 - 9)(x^2 - 1) = 0$$

So $x^2 - 9 = 0$; $x^2 = 9$; $x = \pm 3$

$x^2 - 1 = 0$; $x^2 = 1$; $x = \pm 1$

When $x = 3$, $y = \dfrac{3}{3} = 1$

$$x = -3, y = \frac{3}{-3} = -1$$

$$x = 1, y = \frac{3}{1} = 3$$

$$x = -1, y = \frac{3}{-1} = -3$$

Therefore

$S = \{(3,1), (-3,-1), (1,3), (-1,-3)\}.$

29. Let $x =$ the larger number.

Let $y =$ the smaller number.

Then $x + y = 16$

$x^2 - y^2 = 32$

(1)   Now $y = 16 - x$, and substituting, $x^2 - (16 - x)^2 = 32$

$$x^2 - 256 + 32x - x^2 = 32$$

$$32x = 288$$

$$x = 9$$

(2) Substituting, $9 + y = 16$, so $y = 7$.

The numbers are 9 and 7.

32.

Let $\ell =$ the length of the rectangle.

Let $w =$ the width of the rectangle.

Then $\qquad \ell w = 260$

$$2(\ell - 4)(w - 4) = 288$$

So,   (1) $\ell = \dfrac{260}{w}$, and

$$\ell w - 4\ell - 4w + 16 = 144$$

Then $\ell w - 4\ell - 4w \qquad = 128$

(2)   Substituting $\dfrac{260}{w}$ for ℓ,

$$w\left(\frac{260}{w}\right) - 4\left(\frac{260}{w}\right) - 4w = 128$$

$$260 - \frac{1,040}{w} - 4w = 128$$

$$-\frac{1,040}{w} - 4w = -132 \qquad (\text{times } -w)$$

$$1,040 + 4w^2 = 132w$$

So $4w^2 - 132w + 1,040 = 0$

$$4(w^2 - 33w + 260) = 0$$

$$4(w - 13)(w - 20) = 0$$

(3)   Then $\ell = \dfrac{260}{13} = 20$

or $\ell = \dfrac{260}{20} = 13$.

If $w = 13$, then $\ell = 20$ and if $w = 20$, then $\ell = 13$.

The rectangle has dimensions 13 inches by 20 inches.

Chapter 9 review exercises

1. vertex, $(-1,0)$; x-intercept, -1; y-intercept, -3

2. vertex, $(2,49)$; x-intercept, 9 and -5; y-intercepts, 45

3.

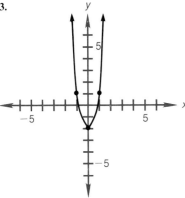

4.

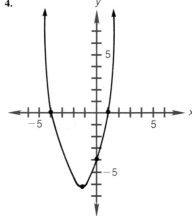

5.

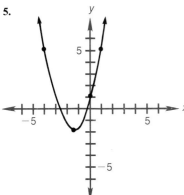

6. (a) $(x + 2)^2 + (y - 5)^2 = 25$, (b) $x^2 + y^2 + 4x - 10y + 4 = 0$

7. (a) $(x - 4)^2 + y^2 = 11$, (b) $x^2 + y^2 - 8x + 5 = 0$

8. (a) $x^2 + y^2 = 81$, (b) $x^2 + y^2 - 81 = 0$ **9.** $C(5,-1)$, $r = 6$

10. $C(0,0)$, $r = \sqrt{7}$ **11.** $C(-3,4)$, $r = 2\sqrt{6}$

12. $C\left(\dfrac{3}{2}, -\dfrac{5}{2}\right)$, $r = \dfrac{\sqrt{46}}{2}$ **13.** $C\left(-\dfrac{3}{2}, 2\right)$, $r = \dfrac{3\sqrt{5}}{2}$

14. $C(0,2)$, $r = \sqrt{19}$

15.

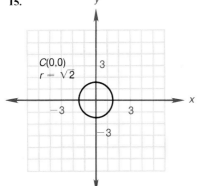

16.

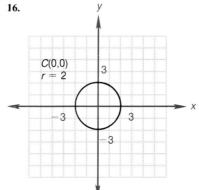

17.

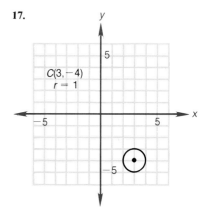

18.

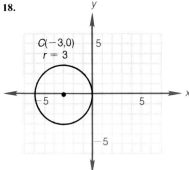

$C(-3,0)$
$r = 3$

19.

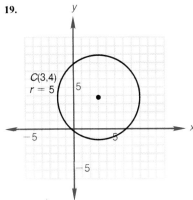

$C(3,4)$
$r = 5$

20.

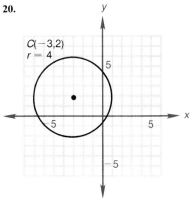

$C(-3,2)$
$r = 4$

21. $x, (\pm 4,0); y, (0, \pm 10)$

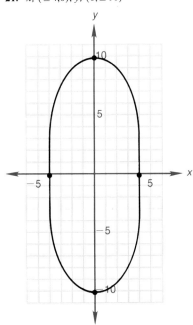

22. $x, (\pm 1,0); y, (0, \pm 5)$

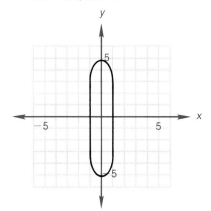

23. $x, (\pm 3,0); y, (0, \pm 4)$

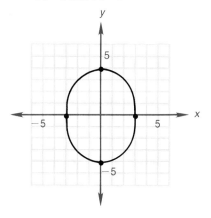

24. $x, (\pm 2,0); y, (0, \pm \sqrt{2})$

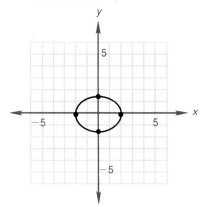

25. $x, (\pm 2,0); y, (0, \pm 2\sqrt{6})$

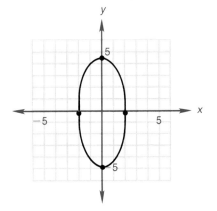

26. x, $(\pm 2\sqrt{2}, 0)$; y, $(0, \pm 1)$

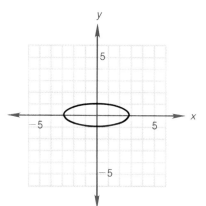

27. x, $(\pm 6, 0)$; asymptotes, $y = \pm\dfrac{7}{6}x$

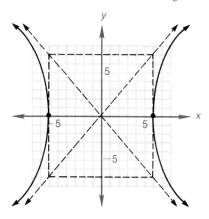

28. y, $(0, \pm 5)$; asymptotes, $y = \pm\dfrac{5}{3}x$

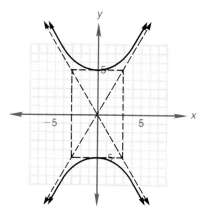

29. x, $(\pm 1, 0)$; asymptotes, $y = \pm 6x$

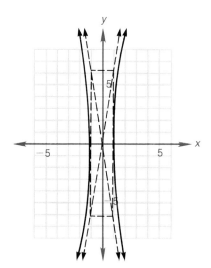

30. y, $(0, \pm 1)$; asymptotes, $y = \pm\dfrac{1}{3}x$

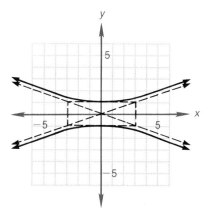

31. x, $(\pm 5, 0)$; asymptotes, $y = \pm\dfrac{4}{5}x$

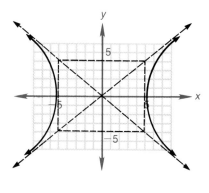

32. y, $(0, \pm 1)$; asymptotes, $y = \pm\dfrac{1}{3}x$

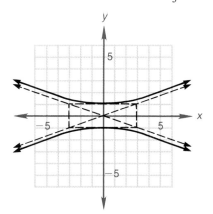

33. hyperbola

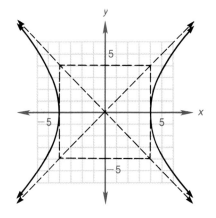

34. parabola

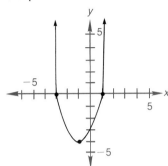

35. ellipse

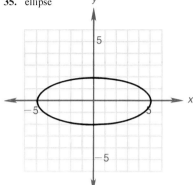

36. circle

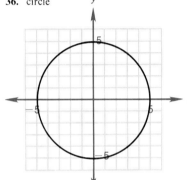

37. circle

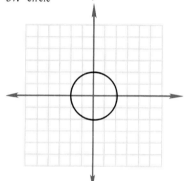

38. parabola

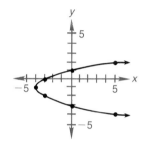

39. $\{(1,1)\}$ **40.** $\{(4,3),\ (3,4)\}$ **41.** $\{(3,2i),\ (3,-2i)\}$ **42.** $\{(2\sqrt{2},-2),\ (-2\sqrt{2},-2)\}$ **43.** $\left\{\left(\dfrac{4}{3},\dfrac{5}{3}\right)\right\}$

44. $\{(2,2i),\ (2,-2i),\ (-2,2i),\ (-2,-2i)\}$ **45.** $\{(\sqrt{19},\sqrt{3}),\ (\sqrt{19},-\sqrt{3}),\ (-\sqrt{19},\sqrt{3}),\ (-\sqrt{19},-\sqrt{3})\}$ **46.** 9 in. by 17 in.

Chapter 9 cumulative test

1. 13   **2.** 21   **3.** 57   **4.** $C = 100$ **5.** 4th   **6.** $6x^2 + 7x - 14$ **7.** $2x^2 - x + 3$ **8.** $7x - 5$ **9.** 17   **10.** -17

11. $\dfrac{49}{16}$ **12.** x^{4n+1} **13.** $6a^3 + 5a^2 - 6a + 1$ **14.** $4ab^2c^4 - 5b^2c^2$ **15.** $3x^2 - 10x + 15 + \dfrac{-34}{x+2}$ **16.** $\dfrac{2x^2 - 2x - 48}{x - 5}$

17. $\dfrac{4-b}{2b+1}$ **18.** $\dfrac{x+2}{x+3}$ **19.** $\dfrac{1}{2}$ **20.** $\varnothing$; 3 is extraneous   **21.** $\left\{\dfrac{1+i\sqrt{5}}{3},\dfrac{1-i\sqrt{5}}{3}\right\}$ **22.** $\left\{\dfrac{5}{2},-1\right\}$

23. $\{\sqrt{5},-\sqrt{5},i\sqrt{2},-i\sqrt{2}\}$ **24.** $\{13\}$ **25.**

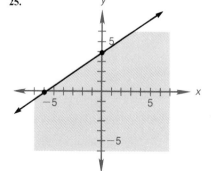

26. $m = -\dfrac{2}{3}$ **27.** $m = \dfrac{3}{5}$ **28.** neither   **29.** $y = 7$ **30.** $\left\{\left(-\dfrac{2}{11}, -\dfrac{7}{11}\right)\right\}$ **31.** inconsistent   **32.** -1 **33.** $C(3,-2)$,

$r = 4$ **34.** vertex, $(1,1)$; y-intercept, 3; no x-intercepts   **35.** x-intercepts, 2 and -2; y-intercepts, $\sqrt{6}$ and $-\sqrt{6}$ **36.** opens left

37. x-intercepts, 3 and -3; asymptotes, $y = \pm\dfrac{2}{3}x$

38. parabola   **39.** hyperbola, $\dfrac{y^2}{2} - \dfrac{x^2}{4} = 1$ **40.** circle, $x^2 + y^2 = 4$

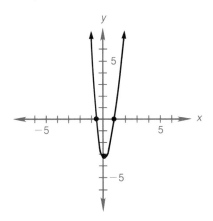   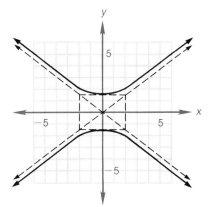   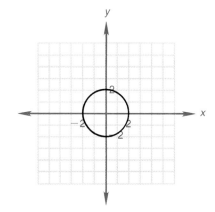

Chapter 10

Exercise 10–1

Answers to odd-numbered problems

1. function   **3.** not a function because two of the ordered pairs have the same first component $(2,3)$ and $(2,-7)$ **5.** not a function because two of the ordered pairs have the same first component $(-1,2)$ and $(-1,6)$ **7.** function   **9.** function   **11.** function
13. not a function because two of the ordered pairs have the same first component $(4,-2)$ and $(4,2)$ **15.** function   **17.** not a function since two ordered pairs have the same first component, $(-10,0)$ and $(-10,1)$ **19.** not a function since two ordered pairs have the same first component, $(3,1)$ and $(3,4)$
21. domain = $\{1,3,5,7\}$; range = $\{2,4,6,8\}$ **23.** domain = $\{-1\}$; range = $\{2,6,7,-8\}$ **25.** domain = $\{-3,2,3,-10\}$, range = $\{2\}$
27. domain = $\{-6,4,-1,9\}$, range = $\{7,3,8\}$ **29.** domain = $\{-3,-1,0,1,3\}$, range = $\{-9,-5,-3,-1,3\}$
31. domain = $\{-5,-3,0,3,5\}$, range = $\{-5,-3,0,3,5\}$ **33.** domain = $\{-5,-1,1,2,4\}$, range = $\left\{-\dfrac{1}{5},-1,1,\dfrac{1}{2},\dfrac{1}{4}\right\}$
35. domain = $\{x \,|\, x \in R\}$ **37.** domain = $\{x \,|\, x \in R\}$ **39.** domain = $\{x \,|\, x \geq 4\}$
41. domain = $\{x \,|\, x \in R, x \neq 0\}$ **43.** domain = $\left\{x \,|\, x \in R, x \neq -\dfrac{3}{2}\right\}$
45. domain = $\left\{x \,|\, x \geq -\dfrac{4}{3}\right\}$ **47.** function since any vertical line intersects the graph at only one point
49. function since any vertical line intersects the graph at only one point   **51.** not a function since at least one vertical line intersects more than one point on the graph   **53.** not a function since at least one vertical line intersects more than one point on the graph

Solutions to trial exercise problems

6. does define a function since no two ordered pairs have the same first component   **15.** defines a function—no two ordered pairs have the same first component   **25.** domain = $\{-3,2,3,-10\}$; range = $\{2\}$.   **40.** domain = $\{x \,|\, x \in R, x \neq 0\}$

Exercise 10–2

Answers to odd-numbered problems

1. -2 **3.** 0 **5.** 58 **7.** $-\dfrac{15}{4}$ **9.** $3a + 1$ **11.** $3a^2 - 2$ **13.** 27 **15.** 16 **17.** (a) $4x^2 + 8hx + 4h^2$, (b) $8hx + 4h^2$,

(c) $8x + 4h$ **19.** (a) $2x^2 + 4hx + 2h^2 + 3x + 3h + 2$, (b) $4hx + 2h^2 + 3h$, (c) $4x + 2h + 3$ **21.** $f(-5) = -17,\ (-5,-17)$;

$f(0) = -2,\ (0,-2);\ f\left(\dfrac{2}{3}\right) = 0,\ \left(\dfrac{2}{3},0\right)$ **23.** $h\left(-\dfrac{1}{2}\right) = \dfrac{11}{4},\ \left(-\dfrac{1}{2},\dfrac{11}{4}\right)$; $h(0) = 1,\ (0,1)$; $h(3) = 22,\ (3,22)$ **25.** $g(-15) = 10$,

$(-15,10);\ g(0) = 10,\ (0,10);\ g\left(\dfrac{6}{5}\right) = 10,\ \left(\dfrac{6}{5},10\right)$ **27.** 7 **29.** -6 **31.** $48x^2 + 48x + 7$ **33.** $27x^4 - 90x^2 + 70$

35. $(14,-10),\ (32,0),\ (212,100)$ **37.** $(1,1),\ (3,27),\ (5,125)$, domain $= \{s \mid s \geq 0\}$ **39.** $(2,37),\ (3,54),\ (5,88)$

Solutions to trial exercise problems

9. $f(a + 1) = 3(a + 1) - 2 = 3a + 3 - 2 = 3a + 1$ **12.** $f(5) - 3(5) - 2 - 15 - 2 - 13$ and $f(2) = 3(2)\quad 2 - 4$.
So $f(5) - f(2) = 13 - 4 = 9$.

19. (a) $f(x + h) = 2(x + h)^2 + 3(x + h) + 2$
$$= 2(x^2 + 2xh + h^2) + 3x + 3h + 2$$
$$= 2x^2 + 4xh + 2h^2 + 3x + 3h + 2$$

(b) $f(x + h) - f(x) = (2x^2 + 4xh + 2h^2 + 3x + 3h + 2) - (2x^2 + 3x + 2)$
$$= 2x^2 + 4xh + 2h^2 + 3x + 3h + 2 - 2x^2 - 3x - 2$$
$$= 4xh + 2h^2 + 3h$$

(c) $\dfrac{f(x + h) - f(x)}{h} = \dfrac{4xh + 2h^2 + 3h}{h} = \dfrac{h(4x + 2h + 3)}{h} = 4x + 2h + 3$ since $h \neq 0$

24. $g(x) = -7$
$$g\left(-\dfrac{7}{8}\right) = -7;\ \left(-\dfrac{7}{8},-7\right)$$
$g(0) = -7;\ (0,-7)$
$g(25) = -7;\ (25,-7)$

28. $f[g(4)] = f[4(4) + 2] = f(18) = 3(18)^2 - 5 = 967$

33. $f[f(x)] = f[3x^2 - 5] = 3(3x^2 - 5)^2 - 5 = 3(9x^4 - 30x^2 + 25) - 5 = 27x^4 - 90x^2 + 75 - 5 = 27x^4 - 90x^2 + 70$

35. $g(F) = \dfrac{5}{9}(F - 32)$

$$g(14) = \dfrac{5}{9}(14 - 32) = \dfrac{5}{9}(-18) = -10;\ (14,-10)$$

$$g(32) = \dfrac{5}{9}(32 - 32) = \dfrac{5}{9}(0) = 0;\ (32,0)$$

$$g(212) = \dfrac{5}{9}(212 - 32) = \dfrac{5}{9}(180) = 100;\ (212,100)$$

Exercise 10–3

Answers to odd-numbered problems

1. linear

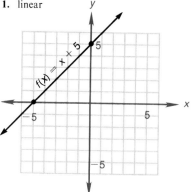

$f(x) = x + 5$

3. linear

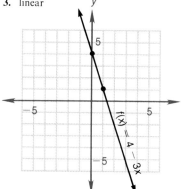

$f(x) = 4 - 3x$

5. linear

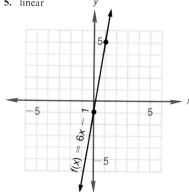

$f(x) = 6x - 1$

7. constant, linear

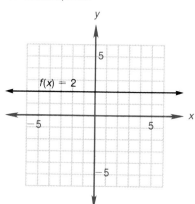

$f(x) = 2$

9. constant, linear

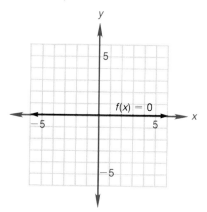

$f(x) = 0$

11. quadratic

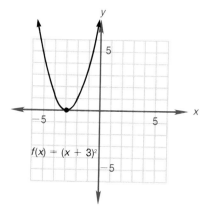

$f(x) = (x + 3)^2$

13. quadratic

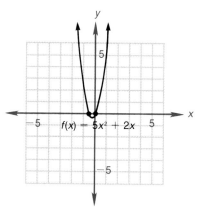

$f(x) = 5x^2 + 2x$

15. quadratic

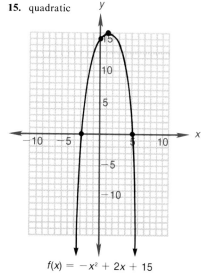

$f(x) = -x^2 + 2x + 15$

17. quadratic

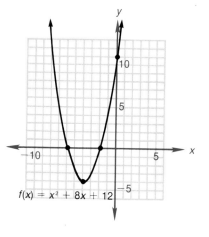

$f(x) = x^2 + 8x + 12$

19. square root

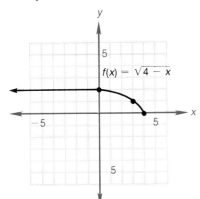

$f(x) = \sqrt{4 - x}$

21. square root

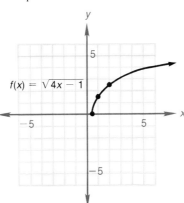

$f(x) = \sqrt{4x - 1}$

23. square root

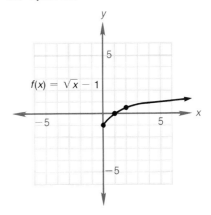

$f(x) = \sqrt{x} - 1$

25. square root

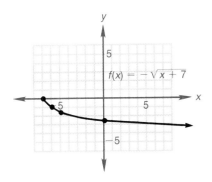

$f(x) = -\sqrt{x + 7}$

27. $f(-2) = 1, f(-1) = 0, f(0) = -1,$
$f(1) = 0, f(2) = 1$

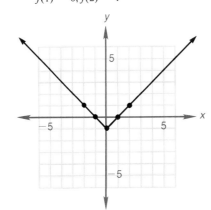

29. $f(-2) = 2, f(-1) = 1, f(0) = 0,$
$f(1) = 1, f(2) = 2$

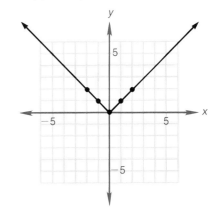

31. $f(-2) = 3, f(-1) = 4, f(0) = 5,$
$f(1) = 4, f(2) = 3$

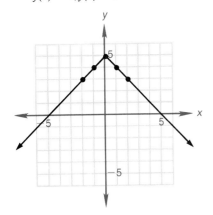

33. $f(-1) = 3, f(0) = 2, f(1) = 1,$
$f(2) = 0, f(3) = 1, f(4) = 2$

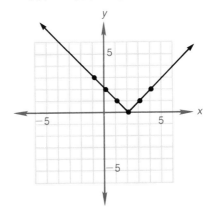

Solutions to trial exercise problems

4. $f(x) = 5 - x$ is a linear function since the largest power of the variable is one. Use y-intercept 5 and slope $m = -1$.

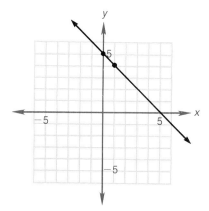

15. $f(x) = -x^2 + 2x + 15$ is a quadratic function whose graph is a parabola opening downward. The y-intercept is 15, x-intercepts are 5 and -3, and the vertex is $(1,16)$.

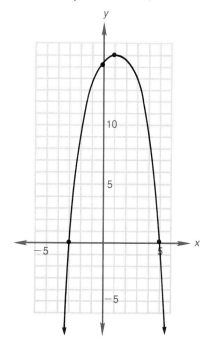

22. $f(x) = \sqrt{2x + 3}$ is a square root function with x-intercept $-\dfrac{3}{2}$, y-intercept $\sqrt{3} \approx 1.7$ and is part of a parabola (above the x-axis) opening to the right.

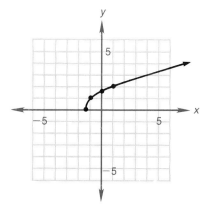

27. $f(x) = |x| - 1$ is an absolute value function with y-intercept -1, x-intercepts 1, -1 and is "V" shaped.

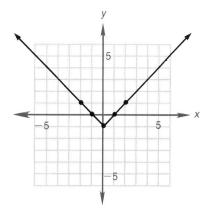

Exercise 10–4

Answers to odd-numbered problems

1. one-to-one function since no two ordered pairs have the same second component **3.** not a one-to-one function since $(1,-6)$ and $(4,-6)$ have the same second components **5.** one-to-one function since it is a linear function where $m \neq 0$ **7.** not a one-to-one function since $G(1) = -1$ and $G(2) = -1$ **9.** not a one-to-one function since at least two of the ordered pairs have the same second component $(1,2)$ and $(5,2)$ **11.** one-to-one function since $h(x)$ has a different value for every $x \geq 3$ **13.** one-to-one function since any horizontal line drawn in the plane intersects the graph at only one point **15.** not a one-to-one since a horizontal line intersects the graph in more than one point **17.** one-to-one function since any horizontal line does not intersect the graph in more than one point

19. $f^{-1}(x) = \{(4,-3), (-3, 2), (7,0)\}$ **21.** $h^{-1}(x) = \dfrac{x}{5}$ **23.** $G^{-1}(x) = \dfrac{x - 7}{3}$ **25.** $f^{-1}(x) = \sqrt[3]{x - 2}$

27. $h^{-1}(x) = x^2 - 5, x \geq 0$ **29.** $G^{-1}(x) = x^2 - 2, x \geq 0$ **31.** $f^{-1}(x) = (x + 2)^3$ **33.** $h^{-1}(x) = (x - 7)^3$

35.

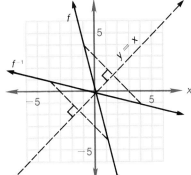

37.

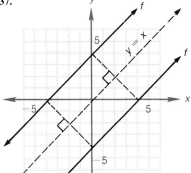

39.

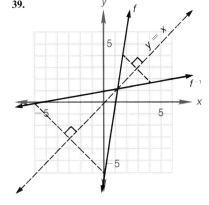

41.

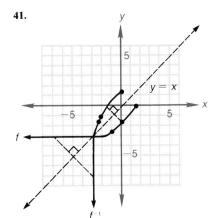

43.

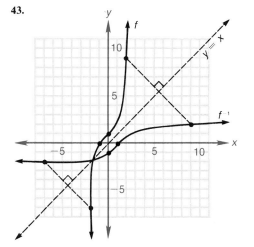

45.

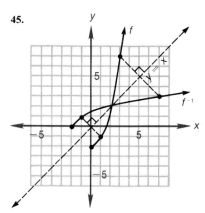

47.

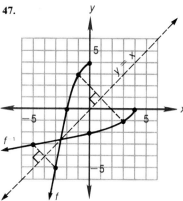

49. (a) $f[g(x)] = f\left(\dfrac{x-7}{4}\right) = 4\left(\dfrac{x-7}{4}\right) + 7 = x - 7 + 7 = x$, (b) $g[f(x)] = g(4x+7) = \dfrac{(4x+7)-7}{4} = \dfrac{4x}{4} = x$

51. (a) $f[g(x)] = f(x^2+2) = \sqrt{(x^2+2)-2} = \sqrt{x^2} = x$, (b) $g[f(x)] = g(\sqrt{x-2}) = (\sqrt{x-2})^2 + 2 = x - 2 + 2 = x$

Solutions to trial exercise problems

7. G is not a one-to-one function since $G(1) = 1 - 3(1) + 1 = -1$ and $G(2) = (2)^2 - 3(2) + 1 = 4 - 6 + 1 = -1$. Thus $(1,-1)$ and $(2,-1)$ are points on the graph. **11.** h is one-to-one since no two ordered pairs have the same second component.
25. Given $f(x) = x^3 + 2$, then $y = x^3 + 2$ and interchanging variables $x = y^3 + 2$, then $y^3 = x - 2$ and $y = \sqrt[3]{x-2}$. Thus $f^{-1}(x) = \sqrt[3]{x-2}$.
42. Graph the function $f(x) = \sqrt{x+2}$. Choose points on this curve and locate other points on the other side of the line $y = x$, the same distance away from this line. Connect these points.

45. Graph the function $f(x) = x^2 - 2$. Choose points on this curve that are on the other side of the line $y = x$, the same distance from this line. Connect these points to obtain the graph of f^{-1}.

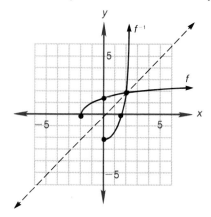

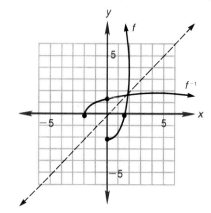

Exercise 10–5

Answers to odd-numbered problems

1. $S = kt$ **3.** $t = \dfrac{k}{r}$ **5.** $M = kmv$ **7.** $F = kAv^2$ **9.** $S_{max} = \dfrac{kT}{r^3}$ **11.** $F = \dfrac{kmv}{gt}$ **13.** $p = kh, k = 4$

15. $V = ks^3, k = 3$ **17.** $s = \dfrac{k}{T}, k = 120$ **19.** $p = \dfrac{kT}{V}, k = 20$ **21.** $A = kh(a+b), k = \dfrac{1}{2}$

23. $w = k\ell, k = \dfrac{3}{2}, w = \dfrac{45}{2}$ **25.** $y = \dfrac{k}{z^2}, k = 48, y = \dfrac{16}{3}$ **27.** $v = \dfrac{ks}{t}, k = 10, v = \dfrac{40}{3}$ **29.** $V = \dfrac{k}{P}; k = 6{,}000; P = \$2.00$

31. $E = \dfrac{k}{d^2}$, $k = 409.6$, $E = 11.38$ footcandles **33.** $R = \dfrac{kL}{d^2}$, $k = \dfrac{1}{144}$, $R = \dfrac{3}{2}$ ohms **35.** $F = \dfrac{kwh^2}{\ell}$, $k = 244{,}897.96$, $F = 2{,}204.08$ lb.

Solutions to trial exercise problems

5. Since M varies directly as the product of m and v, $M = kmv$. **8.** Since R varies directly as ℓ and inversely as A, $R = \dfrac{k\ell}{A}$.

18. $n = \dfrac{k}{p^2}$, $n = 14$ when $p = 9$, so $14 = \dfrac{k}{(9)^2} = \dfrac{k}{81}$, $k = 14 \cdot 81 = 1{,}134$. **27.** $v = \dfrac{ks}{t}$. $v = 20$ when $s = 4$ and $t = 2$,

so $20 = \dfrac{k \cdot 4}{2}$ and $k = 10$. Then $v = \dfrac{10 \cdot s}{t}$ and $s = 8$ when $t = 6$, so $v = \dfrac{10 \cdot 8}{6} = \dfrac{40}{3}$.

Chapter 10 review exercises

1. not a function since two of the ordered pairs have the same first component $(1,3)$ and $(1,0)$ **2.** function **3.** function **4.** function
5. function **6.** not a function since $x = k$; does *not* define a function; (vertical line) **7.** domain $= \{-3,4,0,-2\}$, range $= \{4,2,0,7\}$
8. domain $= \{-3,-2,-1,0,1,2,3\}$, range $= \{13,9,5,1,-3,-7,-11\}$

9. domain $= \{-3,0,1,4\}$, range $= \{-1,2,23\}$ **10.** domain $= \{-8,-2,0,3,7\}$, range $= \{-7\}$ **11.** domain $= \{x \mid x \in R\}$

12. domain $= \{x \mid x \in R\}$ **13.** domain $= \{x \mid x \in R, x \neq 0\}$ **14.** domain $= \left\{x \mid x \in R, x \neq \dfrac{4}{3}\right\}$

15. domain $= \{x \mid x \geq -2\}$ **16.** 0 **17.** -2 **18.** -3 **19.** 4 **20.** 8 **21.** 1 **22.** 1
23. linear   **24.** constant, linear   **25.** quadratic

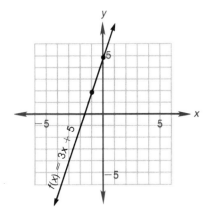

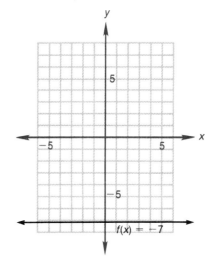

$f(x) = -7$

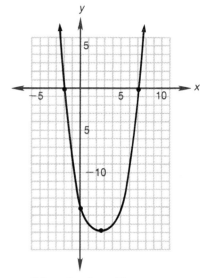

$f(x) = x^2 - 5x - 14$

26. square root

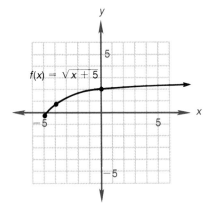

$f(x) = \sqrt{x} + 5$

27. absolute value

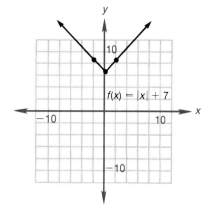

$f(x) = |x| + 7$

28. not one-to-one function   **29.** one-to-one function   **30.** one-to-one function   **31.** not one-to-one function   **32.** not one-to-one

function   **33.** one-to-one function   **34.** $f^{-1}(x) = \dfrac{x - 3}{4}$ **35.** $f^{-1}(x) = \sqrt{x + 2}$, $x \geq -2$ **36.** $f^{-1}(x) = \dfrac{1}{2}x^2 + \dfrac{3}{2}$, $x \geq 0$

37. (a) $f[g(x)] = f\left(\dfrac{5 - x}{2}\right) = 5 - 2\left(\dfrac{5 - x}{2}\right) = 5 - (5 - x) = x$, (b) $g[f(x)] = g(5 - 2x) = \dfrac{5 - (5 - 2x)}{2} = \dfrac{2x}{2} = x$

38. (a) $f[g(x)] = f\left(\sqrt[3]{\dfrac{x + 4}{3}}\right) = 3\left(\sqrt[3]{\dfrac{x + 4}{3}}\right)^3 - 4 = 3\left(\dfrac{x + 4}{3}\right) - 4 = x + 4 - 4 = x$,

(b) $g[f(x)] = g(3x^3 - 4) = \sqrt[3]{\dfrac{(3x^3 - 4) + 4}{3}} = \sqrt[3]{\dfrac{3x^3}{3}} = \sqrt[3]{x^3} = x$

39.

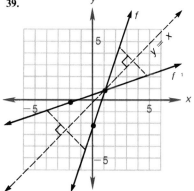

40.

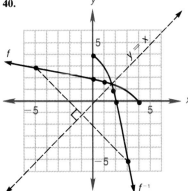

41.

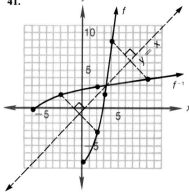

42. $x = ky^2$ **43.** $I = \dfrac{k}{R}$ **44.** $H = kvr$ **45.** $R = \dfrac{k\ell}{A}$ **46.** $s = kw^3, k = 2$ **47.** $y = \dfrac{k}{x^2}, k = 216$

48. $R = \dfrac{k\ell}{A}, k = .00495$ **49.** $P = kRI^2, k = 3, P = 960$ **50.** $d = \dfrac{kab}{c}, k = 10, d = 20$

Chapter 10 cumulative test

1. $\dfrac{3}{2}$ **2.** (a) -2, (b) -14, (c) 1 **3.** 204 **4.** $-x^{13}y^{18}$ **5.** x^{4n-3} **6.** 1 **7.** $\dfrac{1}{x^6y^7}$ **8.** $\dfrac{1}{16}$ **9.** 8 **10.** $\dfrac{1}{16}$

11. $3(a + 3b)(a - 3b)$ **12.** $(x^2 + y^2)(x + y)(y - x)$ **13.** $5(x - 2)(2x + 3)$ **14.** $(3a - 1)(2a + 3)$ **15.** $(ab - 1)(2ab + 1)$

16. $2(x - 3y)(x^2 + 3xy + 9y^2)$ **17.** $x = \dfrac{1 + by}{a - 2}$ **18.** $\left\{\dfrac{23}{5}\right\}$ **19.** $\{x \mid -2 \le x < 2\}$ **20.** $\{x \mid x \le -2 \text{ or } x \ge 8\}$ **21.** $\left\{\dfrac{15}{7}\right\}$

22. $\{3, -1\}$ **23.** $\dfrac{1}{a^{\frac{3}{8}}}$ **24.** $b^{\frac{1}{4}}$ **25.** x^2y^3 **26.** $\left\{\left(\dfrac{4}{7}, -\dfrac{19}{7}\right)\right\}$ **27.** $\left\{\left(\dfrac{9}{16}, \dfrac{3}{8}\right)\right\}$ **28.** 23 **29.** -25

30. 131 **31.** $f^{-1}(x) = \dfrac{x - 3}{4}$ **32.** (a) $f[g(x)] = f[\sqrt[3]{x - 1}] = (\sqrt[3]{x - 1})^3 + 1 = x - 1 + 1 = x$,

(b) $g[f(x)] = g[x^3 + 1] = \sqrt[3]{(x^3 + 1) - 1} = \sqrt[3]{x^3} = x$

33. y-intercept, -40
x-intercepts, -8 and 5

34. y-intercepts, 2 and -2
x-intercepts, $\sqrt{8}$ and $-\sqrt{8}$

35. x-intercepts, 4 and -4
no y-intercepts

36. constant

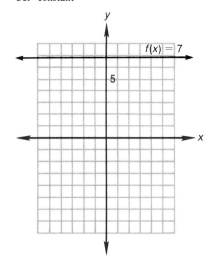

37. linear

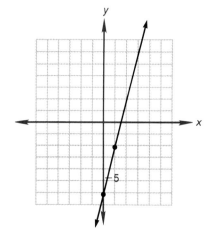

38. quadratic

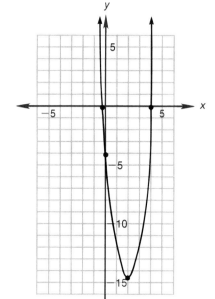

39. square root

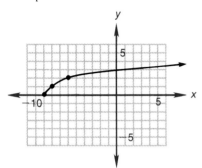

40. absolute value

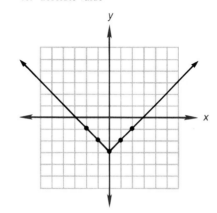

Chapter 11

Exercise 11–1

Answers to odd-numbered problems

1.

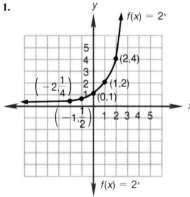

3.

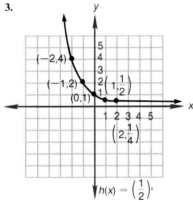

5.

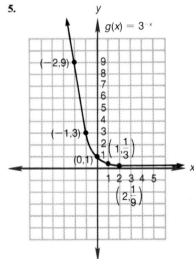

7.

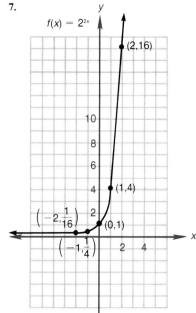

$f(x) = 2^{2x}$

(2,16)

10

8

(1,4)

$\left(-2,\frac{1}{16}\right)$

2

(0,1)

$\left(-1,\frac{1}{4}\right)$ 2 4

9.

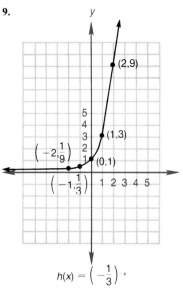

(2,9)

5
4
3 (1,3)
2
$\left(-2,\frac{1}{9}\right)$ 1
(0,1)
$\left(-1,\frac{1}{3}\right)$ 1 2 3 4 5

$h(x) = \left(-\frac{1}{3}\right)^{-x}$

11.

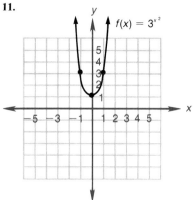

$f(x) = 3^{x^2}$

5
4
3
2
1

−5 −3 −1   1 2 3 4 5

13. $S = \{5\}$ **15.** $S = \left\{\dfrac{5}{2}\right\}$ **17.** $S = \left\{\dfrac{3}{2}\right\}$ **19.** $S = \{-2\}$ **21.** $S = \{-2\}$ **23.** $S = \left\{\dfrac{-5}{4}\right\}$ **25.** $S = \left\{\dfrac{2}{9}\right\}$

27. $S = \{-3\}$ **29.** $S = \{3\}$ **31.** $S = \left\{\dfrac{-1}{6}\right\}$ **33.** (a) 4,500; (b) 40,500; (c) 23,383; (d) $1,500(3)^{\frac{t}{12}}$ **35.** (a) \$20,000;

(b) \$31,821.46; (c) $5,000(2)^{\frac{t}{9}}$ **37.** (a) 1,000; (b) 250

Solutions to trial exercise problems

5. Using $g(x) = 3^{-x}$ when
  (1) $x = -2$,
    $g(-2) = 3^{-(-2)} = 3^2 = 9$
    $(-2,9)$
  (2) $x = -1$
    $g(-1) = 3^{-(-1)} = 3$
    $(-1,3)$
  (3) $x = 0,\ g(0) = 3^{-0} = 1$
    $(0,1)$
  (4) $x = 1,\ g(1) = 3^{-1} = \dfrac{1}{3}$
    $\left(1,\dfrac{1}{3}\right)$
  (5) $x = 2,\ g(2) = 3^{-2} = \dfrac{1}{3^2} = \dfrac{1}{9}$
    $\left(2,\dfrac{1}{9}\right)$

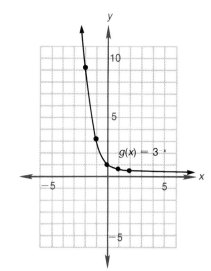

10

5

$g(x) = 3^{-x}$

−5   5

−5

10. (1) $f(-1) = 2^{-1-1} = 2^{-2} = \dfrac{1}{4} \left(-1, \dfrac{1}{4}\right)$

(2) $f(0) = 2^{-0-1} = 2^{-1} = \dfrac{1}{2} \left(0, \dfrac{1}{2}\right)$

(3) $f(1) = 2^{1-1} = 2^0 = 1$ $(1,1)$

(4) $f(2) = 2^{2-1} = 2^1 = 2$ $(2,2)$

(5) $f(3) = 2^{3-1} = 2^2 = 4$ $(3,4)$

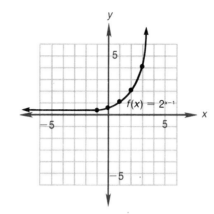

15. $4^x = 32$. Since $4 = 2^2$ and $32 = 2^5$, then $(2^2)^x = 2^5$ and $2^{2x} = 2^5$. Then $2x = 5$ and $x = \dfrac{5}{2}$. $S = \left\{\dfrac{5}{2}\right\}$. **24.** $9^{2x} = 27$.

Since $9 = 3^2$ and $27 = 3^3$, then $(3^2)^{2x} = 3^3$ and $3^{4x} = 3^3$. Thus $4x = 3$ and $x = \dfrac{3}{4}$. $S = \left\{\dfrac{3}{4}\right\}$. **31.** $27^{2x+1} = 9$. Since $27 = 3^3$

and $9 = 3^2$, then $(3^3)^{2x+1} = 3^2$ and $3^{6x+3} = 3^2$. Thus $6x + 3 = 2$, $6x = -1$, and $x = -\dfrac{1}{6}$. $S = \left\{-\dfrac{1}{6}\right\}$. **36.** (a) Let $t = 2$.

Then $A = 200(3)^{-0.5(2)} = 200(3)^{-1} = \dfrac{200}{3} = 66\dfrac{2}{3}$ grams. (b) Let $t = 12$. Then $A = 200(3)^{-0.5(12)} = 200(3)^{-6} = \dfrac{200}{3^6} = \dfrac{200}{729}$ grams.

Exercise 11–2

Answers to odd-numbered problems

1.

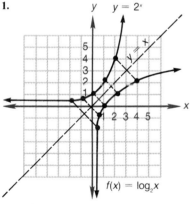

$f(x) = \log_2 x$

3.

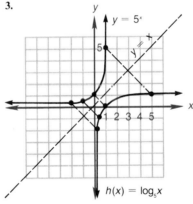

$h(x) = \log_5 x$

5.

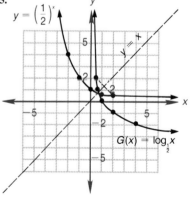

$G(x) = \log_2 x$

7. $\log_3 81 = 4$ **9.** $\log_2 64 = 6$ **11.** $\log_{\frac{1}{2}}\left(\dfrac{1}{8}\right) = 3$ **13.** $\log_{\frac{3}{2}}\left(\dfrac{8}{27}\right) = -3$ **15.** $2^4 = 16$ **17.** $10^3 = 1{,}000$

19. $4^{-2} = \dfrac{1}{16}$ **21.** $10^{-4} = 0.0001$ **23.** $\left(\dfrac{1}{2}\right)^{-3} = 8$ **25.** $\log_2 32 = 5$ **27.** $\log_5 25 = 2$ **29.** $\log_2\left(\dfrac{1}{4}\right) = -2$

31. $\log_6\left(\dfrac{1}{216}\right) = -3$ **33.** $\log_8 = 1$ **35.** $\log_{\frac{1}{3}}\left(\dfrac{1}{81}\right) = 4$ **37.** $\log_{\frac{5}{4}}\left(\dfrac{125}{64}\right) = 3$ **39.** $\log_7 \sqrt[3]{7} = \dfrac{1}{3}$ **41.** $S = \{2\}$

43. $S = \{3\}$ **45.** $S = \{7\}$ **47.** $S = \{b \mid b > 0, b \neq 1\}$ **49.** $S = \{81\}$ **51.** $S = \left\{\dfrac{1}{36}\right\}$ **53.** $S = \left\{\dfrac{1}{256}\right\}$ **55.** $\{p \mid p > -1\}$

57. $\{y \mid y < -3 \text{ or } y > 4\}$

Solutions to trial exercise problems

5. Graph the equation $y = \left(\frac{1}{2}\right)^x$ and reflect this curve about the line $y = x$ to obtain the graph of $G(x) = \log_{\frac{1}{2}} x$.

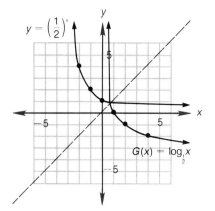

$y = \left(\frac{1}{2}\right)^x$

$G(x) = \log_{\frac{1}{2}} x$

11. $\left(\frac{1}{2}\right)^3 = \frac{1}{8}$ is equivalent to $\log_{\frac{1}{2}}\left(\frac{1}{8}\right) = 3.$ **21.** $\log_{10} 0.0001 = -4$ is equivalent to $10^{-4} = 0.0001.$ **29.** $\log_2\left(\frac{1}{4}\right) = x$

is equivalent to $2^x = \frac{1}{4}.$ Since $\frac{1}{4} = \left(\frac{1}{2}\right)^2 = 2^{-2}$, then $2^x = 2^{-2}$ and $x = -2.$ $\log_2\left(\frac{1}{4}\right) = -2.$ **34.** $\log_{\frac{1}{2}}\left(\frac{1}{8}\right) = x$ is equivalent

to $\left(\frac{1}{2}\right)^x = \frac{1}{8}.$ Since $\frac{1}{8} = \left(\frac{1}{2}\right)^3$, then $\left(\frac{1}{2}\right)^x = \left(\frac{1}{2}\right)^3$ and $x = 3.$ $\log_{\frac{1}{2}}\left(\frac{1}{8}\right) = 3.$ **43.** $\log_x\left(\frac{1}{27}\right) = -3$ is equivalent to $x^{-3} = \frac{1}{27}.$

Since $\frac{1}{27} = \frac{1}{3^3} = 3^{-3}$, then $x^{-3} = 3^{-3}$ and $x = 3.$ $S = \{3\}.$ **48.** $\log_{10} x = -2$ is equivalent to $10^{-2} = x.$ Then $x = \frac{1}{100}$ and $S = \left\{\frac{1}{100}\right\}.$

54. Since $x - 2 > 0$ for $x > 2$, then $\log_6(x - 2)$ is defined for $\{x \,|\, x > 2\} = (2, +\infty).$

Exercise 11-3

Answers to odd-numbered problems

1. $\log_b 3 + \log_b 5$ **3.** $\log_3 7 - \log_3 13$ **5.** $3 \log_b 5$ **7.** $4 \log_{10} 5$ **9.** $2 \log_{\frac{1}{2}} 2 + 2 \log_{\frac{1}{2}} 3$ **11.** $3 \log_5 2 + \log_5 3 - 2$

13. $3(\log_b x + \log_b y)$ **15.** $\frac{1}{2}\log_b x + \frac{1}{3}\log_b y$ **17.** $\log_b(2x + 3y)$ **19.** $\log_3 \frac{17}{3}$ **21.** $\log_e \frac{11}{16}$ **23.** $\log_6 64$ **25.** $\log_{10} 432$

27. $\log_4 6$ **29.** $\log_{10} \frac{200}{3}$ **31.** $\log_b x^{\frac{2}{3}}$ **33.** $\log_b \sqrt[4]{\frac{x^3}{y}}$ **35.** $\log_5\left(\frac{x^2 + 2x - 24}{9x^2}\right)$ **37.** $b^{1.1461} = 14$ **39.** $b^{1.6902} = 49$

41. $b^{1.6232} = 42$ **43.** $b^{-.1549} = \frac{7}{10}$ **45.** $b^{.4226} = \sqrt{7}$ **47.** $b^{.2386} = \sqrt[4]{9}$ **49.** $S = \left\{\frac{64}{5}\right\}$ **51.** $S = \{20\}$ **53.** $S = \{67\}$

55. $S = \{4\}$ **57.** $S = \{3\}$ **59.** *Proof:* Let $\log_b u = m$ and $\log_b v = n.$ Then $b^m = u$ and $b^n = v.$ $\frac{u}{v} = \frac{b^m}{b^n} = b^{m-n}.$ So $\log_b \frac{u}{v} = m - n$

$= \log_b u - \log_b v.$ **61.** Let $u = 2$, $v = 3$, and $b = 10.$ Now $\log_{10} 2 = 0.3010$, $\log_{10} 3 = 0.4771$, and $\log_{10} 5 = 0.6990.$
Then $\log_{10}(u + v) = \log_{10}(2 + 3) = \log_{10} 5 = 0.6990$, $\log_{10} u + \log_{10} v = \log_{10} 2 + \log_{10} 3 = 0.3010 + 0.4771 = 0.7781.$
Therefore $\log_{10} 5 \neq \log_{10} 2 + \log_{10} 3$ and $\log_b (u + v) \neq \log_b u + \log_b v.$

Solutions to trial exercise problems

8. Since $24 = 2^3 \cdot 3$, then $\log_b(24) = \log_b(2^3 \cdot 3) = \log_b 2^3 + \log_b 3 = 3\log_b 2 + \log_b 3.$ **10.** Since $\frac{15}{16} = \frac{3 \cdot 5}{2^4}$, then $\log_{10}\left(\frac{15}{16}\right)$

$= \log_{10}\left(\frac{3 \cdot 5}{2^4}\right) = \log_{10} 3 + \log_{10} 5 - \log_{10} 2^4 = \log_{10} 3 + \log_{10} 5 - 4\log_{10} 2.$ **12.** $\log_b\left(\frac{x^3}{y^2}\right) = \log_b x^3 - \log_b y^2 = 3\log_b x - 2\log_b y$

13. $\log_b(xy)^3 = 3\log_b(xy) = 3[\log_b x + \log_b y]$ **17.** $\log_b(2x + 3y) = \log_b(2x + 3y)$, since none of the properties apply to addition.
24. $2\log_2 7 + \log_2 5 = \log_2 7^2 + \log_2 5 = \log_2 49 + \log_2 5 = \log_2(49 \cdot 5) = \log_2(245)$ **28.** $\log_e 4 + 2\log_e 3 - 3\log_e 2$

$$= \log_e 4 + \log_e 3^2 - \log_e 2^3 = \log_e 4 + \log_e 9 - \log_e 8 = \log_e \frac{4 \cdot 9}{8} = \log_e\left(\frac{36}{8}\right) = \log_e\left(\frac{9}{2}\right)$$ **31.** $\frac{1}{3}\log_b(x^2) = \log_b(x^2)^{\frac{1}{3}} = \log_b x^{\frac{2}{3}}$

41. $\log_b(42) = \log_b(2 \cdot 3 \cdot 7) = \log_b 2 + \log_b 3 + \log_b 7 = (0.3010) + (0.4771) + (0.8451) = 1.6232$. Then $b^{1.6232} = 42$.

42. $\log_b\left(\frac{27}{4}\right) = \log_b 27 - \log_b 4 = \log_b 3^3 - \log_b 2^2 = 3\log_b 3 - 2\log_b 2 = 3(0.4771) - 2(0.3010) = 1.4313 - 0.6020 = 0.8293$.

Then $b^{0.8293} = \frac{27}{4}$. **54.** $\log_4(x - 5) - \log_4 3 = 2$ is equivalent to $\log_4\left(\frac{x - 5}{3}\right) = 2$, which is equivalent to $\frac{x - 5}{3} = 4^2$.

Then $\frac{x - 5}{3} = 16$, $x - 5 = 48$, and $x = 53$. Thus $S = \{53\}$. **55.** $\log_{10}(x + 21) + \log_{10} x = 2$ is equivalent to $\log_{10} x(x + 21) = 2$.
Then $x(x + 21) = 10^2$, $x^2 + 21x = 100$, and $x^2 + 21x - 100 = 0$. Factoring, $(x + 25)(x - 4) = 0$, so $x = -25$ or $x = 4$.
But if $x = -25$, then $\log_{10}(x + 21) = \log_{10}(-4)$, which does not exist. $S = \{4\}$.

Exercise 11–4

Answers to odd-numbered problems

1. $10^{0.9031} = 8$ **3.** $10^{1.7243} = 53$ **5.** $10^{4.9042} = 80{,}200$ **7.** $10^{8.8998} = 794{,}000{,}000$ **9.** $10^{7.9360-10} = .00863$ **11.** $10^{3.0294-10}$
$= 0.000000107$ **13.** 5.7024 **15.** (a) $5.8633 - 10$, (b) 37.9912 **17.** (a) 281.19, (b) 10.89 **19.** 2.7773 **21.** 14.85
23. 34.771 **25.** $6{,}990$ **27.** $82{,}000{,}000$ **29.** $.0000165$ **31.** 337.5 **33.** $.005887$ **35.** $.0000003368$ **37.** 48.37 **39.** 18.97
41. 3.3010 **43.** 1.585×10^{-6}

Solutions to trial exercise problems

1. $\log 8 = 0.9031$. Then $10^{0.9031} = 8$. **28.** Antilog $8.7528 - 10 = $ antilog -1.2472. Press $\boxed{1}\ \boxed{0}\ \boxed{.}\ \boxed{2}\ \boxed{4}\ \boxed{7}\ \boxed{2}\ \boxed{+/-}\ \boxed{=}$
and read "0.0565979 on the display. Antilog $8.7528 - 10 = 0.0566$. **30.** Antilog 8.8340 is found by pressing
$\boxed{1}\ \boxed{0}\ \boxed{y^x}\ \boxed{8}\ \boxed{.}\ \boxed{8}\ \boxed{3}\ \boxed{4}\ \boxed{0}\ \boxed{=}$. Read 6.823×10^8. Then antilog $8.8340 = 682{,}300{,}000$.
43. pH $= -\log(H^+)$. Then $5.8 = -\log(H^+)$; $-5.8 = \log(H^+)$. Find antilog -5.8 by pressing $\boxed{1}\ \boxed{0}\ \boxed{y^x}\ \boxed{5}\ \boxed{.}\ \boxed{8}\ \boxed{+/-}\ \boxed{=}$.
Read 0.0000016. Then $H^+ = 0.0000016$.

Exercise 11–5

Answers to odd-numbered problems

1. 1.2920 **3.** 2.3219 **5.** 6.2542 **7.** 1.2183 **9.** 1.0043 **11.** 1.9066 **13.** 4.0325 **15.** 5.8493 **17.** 1.6094
19. undefined **21.** 1.1333 **23.** 2.9459 **25.** 4.0986 **27.** 18.3 years **29.** 55.9 years **31.** $12{,}979.3$ years **33.** 269.9 ft
35. 13.5 days

Solutions to trial exercise problems

7. $\log_{23} 45.6 = \frac{\log 45.6}{\log 23} = \frac{1.6590}{1.3617} = 1.2183$ **10.** $\log_e 107 = 2.3026\log 107 = 2.3026(2.0294) = 4.6729$ **18.** $\log_e 3 = 2.3026\log 3$
$= 2.3026(0.4771) = 1.0986$ **23.** $\ln 7e = \ln 7 + \ln e = 2.3026\log 7 + 1 = 1.9459 + 1 = 2.9459$ **31.** $I = I_0e^{-kd}$, given
$k = 0.00853$, $I = 0.10I_0$, and we want d. Then $0.10I_0 = I_0e^{-0.00853d}$; $0.10 = e^{-0.00853d}$; $-0.00853d = \ln 0.10$; $d = \frac{\ln 0.10}{-0.00853} = 269.9$.

The light is reduced to 10% at a depth of 269.9 feet.

Exercise 11–6

Answers to odd-numbered problems

1. $S = \{3.17\}$ **3.** $S = \{0.29\}$ **5.** $S = \{-2.10\}$ **7.** $S = \{1.19\}$ **9.** $S = \{0.26\}$ **11.** $S = \{0.07\}$ **13.** $S = \{1.71\}$

15. $S = \{2.46\}$ **17.** $n = \frac{\log y}{3\log x}$ **19.** 4.2% **21.** 3.9 years **23.** 2.14 hours **25.** $.396$ hours or 24 minutes
27. 11.5 centuries **29.** 2.05 years **31.** 3.7%

Solutions to trial exercise problems

7. $6^{2x-1} = 12$; $\log 6^{2x-1} = \log 12$; $(2x - 1)\log 6 = \log 12$; $2x \log 6 - \log 6 = \log 12$; $2x \log 6 = \log 12 + \log 6$; $x = \dfrac{\log 12 + \log 6}{2 \log 6}$

$= \dfrac{1.0792 + 0.7782}{1.5564} = 1.19$. $S = \{1.19\}$. **14.** $2^{x^2} = 7$; $\log 2^{x^2} = \log 7$; $x^2 \log 2 = \log 7$; $x^2 = \dfrac{\log 7}{\log 2} = \dfrac{0.8451}{0.3010} = 2.81$. Then

$x = \pm 1.68$. $S = \{1.68, -1.68\}$. **17.** $y = x^{3n}$; $\log y = \log x^{3n}$; $\log y = 3n \log x$; $n = \dfrac{\log y}{3 \log x}$. **26.** $q = q_0(0.96)^t$.

Now $q = \dfrac{1}{2}q_0 = 0.5q_0$; so $0.5q_0 = q_0(0.96)^t$; $0.5 = (0.96)^t$; $\log 0.5 = \log(0.96)^t$; $\log (0.5) = t \log(0.96)$;

$t = \dfrac{\log(0.5)}{\log(0.96)} - \dfrac{-0.3010}{-0.0177} = 17.0$. The half-life of radium is 17.0 centuries.

Chapter 11 review exercises

1. (a)

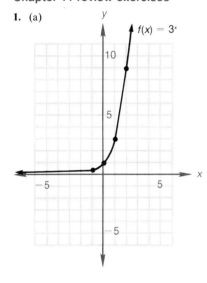

(b)

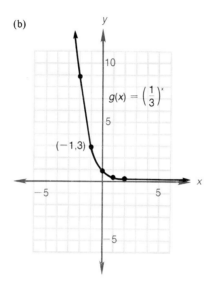

(c)

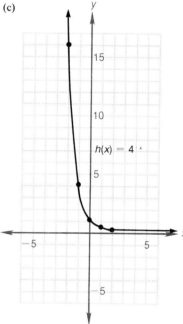

2. (a) $S = \{4\}$, (b) $S = \left\{\dfrac{7}{8}\right\}$

3.

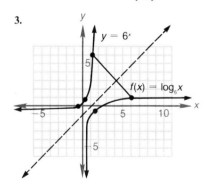

4. $\log_4 \dfrac{1}{64} = -3$ **5.** $\left(\dfrac{1}{3}\right)^{-3} = 27$ **6.** 3 **7.** -4 **8.** $\dfrac{1}{3}$ **9.** $\{5\}$ **10.** $\left\{\dfrac{1}{243}\right\}$

11. $\{4\sqrt{2}\}$ **12.** $x > 3$ or $x < -5$ **13.** $\log_b 7 + 3 \log_b 2$

14. $\log_4 5 - \log_4 2 - \log_4 3$ **15.** $2 \log_5 3 - 2 \log_5 2$ **16.** $\log_5 28$ **17.** $\log_6 5$

18. $\log_b \dfrac{x^4 y^2}{z}$ **19.** $\log_4 \left(\dfrac{x + 3}{x - 4}\right)$ **20.** $\log_b \left(\dfrac{x^3 (2x - 1)}{(x + 1)^3}\right)$

21. 1.3423; $b^{1.3423} = 22$ **22.** 2.0826; $b^{2.0826} = 121$ **23.** 1.9821; $b^{1.9821} = 96$

24. $0.3471; b^{.3471} = \sqrt[3]{11}$ **25.** $9.1348 - 10; b^{9.1348-10} = \dfrac{3}{22}$ **26.** $1.1710; b^{1.1710} = \dfrac{27}{\sqrt[4]{11}}$ **27.** $\{48\}$ **28.** $\left\{\dfrac{2}{9}\right\}$ **29.** $\left\{\dfrac{23}{4}\right\}$

30. $\left\{\dfrac{83}{26}\right\}$ **31.** 2.5340 **32.** 5.7050 **33.** $7.8669 - 10$ **34.** -0.7784 **35.** 15.2 yr   **36.** 1.09 **37.** 0.45 **38.** 3.86

39. 1.2 days   **40.** 2.0 days   **41.** $S = \{1.76\}$ **42.** $S = \{-0.79\}$ **43.** $S = \{2.81\}$ **44.** $S = \{0.89, -0.89\}$

45. $S = \{-1.04\}$ **46.** 2.20 hours

Chapter 11 cumulative test

1. $\{-3, -2, -1, 0, 1, 2, 3, 4, 5\}$ **2.** -8 **3.** -81 **4.** $4a^2 - 4ab + b^2$ **5.** $25x^2 - 9y^2$ **6.** $12y^2 - 14y - 10$ **7.** $-18a^5b^3$

8. $-\dfrac{27b^6}{a^3c^9}$ **9.** $\dfrac{a^{12}}{9b^{10}}$ **10.** $\{1\}$ **11.** $\{y \mid y \le 2\}$ **12.** $\left\{x \mid -2 \le x < \dfrac{5}{2}\right\}$ **13.** $\left\{\dfrac{5}{4}, \dfrac{1}{4}\right\}$ **14.** $\left\{x \mid -\dfrac{2}{3} \le x \le 2\right\}$

15. $\{x \mid x < -11 \text{ or } x > 1\}$ **16.** $\left\{\dfrac{3}{2}, -1\right\}$ **17.** $\left\{\dfrac{5}{4}\right\}$ **18.** $\dfrac{10a + 6b}{8a^2b^2}$ **19.** $\dfrac{15y}{y - 7}$ **20.** $\dfrac{x^2 - 9x + 20}{x^2 + 2x - 15}$ **21.** 1

22. $13 - 4\sqrt{3}$ **23.** 25 **24.** $-i$ **25.** $7\sqrt{2}$ **26.** $-(6 + 2\sqrt{2} + 3\sqrt{3} + \sqrt{6})$ **27.** $\left\{\dfrac{3 + \sqrt{37}}{2}, \dfrac{3 - \sqrt{37}}{2}\right\}$

28. $\{8\}$; -1 is extraneous   **29.** $\{z \mid -5 \le z \le 8\}$ **30.** (a) $2x - y = -6$, (b) $2x - 3y = -42$, (c) $4x - 3y = 29$

31. (a) 26, (b) -25, (c) -44, (d) 5 **32.**

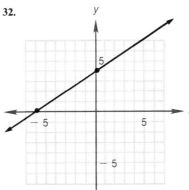

33.

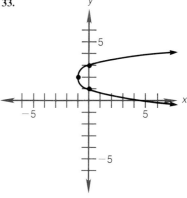

34. (a) circle, (b) parabola, (c) hyperbola, (d) ellipse   **35.** $\left\{\left(\dfrac{7}{26}, \dfrac{4}{13}\right)\right\}$ **36.** $\{2\}$ **37.** (a) 2, (b) $\dfrac{1}{36}$ **38.** 1.36 **39.** $\left\{\dfrac{17}{8}\right\}$

40. $\log_b\left(\dfrac{8}{125}\right)$

Chapter 12

Exercise 12-1

Answers to odd-numbered problems

1. $7, 11, 15, 19, 23$ **3.** $\dfrac{2}{3}, \dfrac{1}{3}, \dfrac{2}{9}, \dfrac{1}{6}, \dfrac{2}{15}$ **5.** $-6, \dfrac{7}{2}, \dfrac{8}{5}, \dfrac{9}{8}, \dfrac{10}{11}$ **7.** $\dfrac{2}{5}, \dfrac{2}{5}, \dfrac{8}{15}, \dfrac{4}{5}, \dfrac{32}{25}$ **9.** $-1, 7, -13, 19, -25$ **11.** $1, -\dfrac{9}{5}, 3, -\dfrac{81}{17}, \dfrac{81}{11}$

13. $1, 1, 1, 1, 1$ **15.** 32 **17.** $\dfrac{1}{111}$ **19.** $\dfrac{65}{41}$ **21.** 89 **23.** -247 **25.** $a_n = 2n + 4$ **27.** $a_n = 5n - 3$ **29.** $a_n = n^2 + 2$

31. $a_n = \dfrac{1}{2^{n-1}}$ **33.** $a_n = \dfrac{n + 2}{2n + 3}$ **35.** $a_n = (-1)^n(4n + 2)$ **37.** (a) $27{,}000$; (b) $243{,}000$; (c) $1{,}000(3^n)$ **39.** $\dfrac{5}{2}$ ft; $\dfrac{5}{32}$ ft; $10\dfrac{1}{2}$ ft

41. (a) $\$16{,}000$; $\$17{,}500$; $\$19{,}000$; $\$20{,}500$; $\$22{,}000$; $\$23{,}500$; (b) $\$16{,}000 + 1500(n - 1)$; (c) $\$44{,}500$

Solutions to trial exercise problems

11. $a_n = (-1)^{n-1} \cdot \dfrac{3^n}{2^n + 1}$; $a_1 = (-1)^{1-1} \cdot \dfrac{3}{2+1} = 1 \cdot \dfrac{3}{3} = 1$; $a_2 = (-1)^{2-1} \cdot \dfrac{3^2}{2^2+1} = (-1) \cdot \dfrac{9}{4+1} = -\dfrac{9}{5}$;

$a_3 = (-1)^{3-1} \cdot \dfrac{3^3}{2^3+1} = 1 \cdot \dfrac{27}{8+1} = 3$; $a_4 = (-1)^{4-1} \cdot \dfrac{3^4}{2^4+1} = (-1) \cdot \dfrac{81}{16+1} = -\dfrac{81}{17}$;

$a_5 = (-1)^{5-1} \cdot \dfrac{3^5}{2^5+1} = 1 \cdot \dfrac{243}{32+1} = \dfrac{243}{33} = \dfrac{81}{11}$. The sequence is $1, -\dfrac{9}{5}, 3, -\dfrac{81}{17}, \dfrac{81}{11}, \cdots$

21. $a_{14} = (-1)^{14}[6(14) + 5] = 1 \cdot (84 + 5) = 89$ **24.** $a_8 = 2(8)^2[3(8) - 1] = 2(64)(23) = 2{,}944$

30. $\dfrac{1}{3}, \dfrac{1}{9}, \dfrac{1}{27}, \dfrac{1}{81}, \dfrac{1}{243}, \cdots$ The numerators are all 1 and the denominators are powers of 3. So $a_n = \dfrac{1}{3^n}$.

35. $-6, 10, -14, 18, -22, \cdots$. The signs alternate so we have factor $(-1)^n$. Since $10 - 6 = 4$, $14 - 10 = 4$, $18 - 14 = 4$, there is a common difference of 4 so a term of the general term is $4n$. Since $4(1) + 2 = 6$ and $4(2) + 2 = 10$, we find the general term is $a_n = (-1)^n(4n + 2)$.   **37.** Since the culture triples every hour, we want (a) $a_3 = 1{,}000(3^3) = 27{,}000$; (b) $a_5 = 1{,}000(3^5) = 243{,}000$; (c) $a_n = 1{,}000(3^n)$.

Exercise 12-2

Answers to odd-numbered problems

1. $S_5 = 55$ **3.** $S_6 = 60$ **5.** $S_5 = 10$ **7.** $S_4 = 50$ **9.** $S_5 = 10$ **11.** $S_5 = \dfrac{743}{840}$ **13.** $S_5 = \dfrac{937}{168}$ **15.** $S_5 = \dfrac{5269}{900}$

17. $S_4 = \dfrac{7}{36}$ **19.** $S_3 = 24$ **21.** 45   **23.** $\dfrac{77}{60}$ **25.** $\dfrac{573}{60}$ **27.** -13 **29.** $\sum\limits_{i=1}^{5} i$ **31.** $\sum\limits_{i=1}^{6} i^3$ **33.** $\sum\limits_{i=1}^{4} \left(\dfrac{i+1}{i+2} \right)$

35. $\sum\limits_{i=1}^{4} \left(\dfrac{2i}{3^{i-1}} \right)$ **37.** $\sum\limits_{i=1}^{4} \left(\dfrac{2i+3}{3i+1} \right)$ **39.** $\sum\limits_{i=1}^{6} (-1)^i 2^i$

Solutions to trial exercise problems

5. $\sum\limits_{k=1}^{5} k(k-3) = 1(1-3) + 2(2-3) + 3(3-3) + 4(4-3) + 5(5-3) = -2 + (-2) + 0 + 4 + 10 = 10$

11. $\sum\limits_{k=1}^{5} \dfrac{1}{k+3} = \dfrac{1}{1+3} + \dfrac{1}{2+3} + \dfrac{1}{3+3} + \dfrac{1}{4+3} + \dfrac{1}{5+3}$

$= \dfrac{1}{4} + \dfrac{1}{5} + \dfrac{1}{6} + \dfrac{1}{7} + \dfrac{1}{8} = \dfrac{210 + 168 + 140 + 120 + 105}{840} = \dfrac{743}{840}$

16. $\sum\limits_{j=1}^{5} (-1)^j \cdot \dfrac{3}{2j} = (-1)^1 \cdot \dfrac{3}{2(1)} + (-1)^2 \cdot \dfrac{3}{2(2)} + (-1)^3 \cdot \dfrac{3}{2(3)} + (-1)^4 \cdot \dfrac{3}{2(4)} + (-1)^5 \cdot \dfrac{3}{2(5)}$

$= \left(-\dfrac{3}{2} \right) + \dfrac{3}{4} + \left(-\dfrac{1}{2} \right) + \dfrac{3}{8} + \left(-\dfrac{3}{10} \right)$

$= -\dfrac{23}{10} + \dfrac{9}{8} = \dfrac{-92 + 45}{40} = -\dfrac{47}{40}$

25. $\sum\limits_{j=0}^{5} \dfrac{2j+1}{j+1} = \dfrac{2(0)+1}{0+1} + \dfrac{2(1)+1}{1+1} + \dfrac{2(2)+1}{2+1} + \dfrac{2(3)+1}{3+1}$

$+ \dfrac{2(4)+1}{4+1} + \dfrac{2(5)+1}{5+1} = 1 + \left(\dfrac{3}{2} \right) + \left(\dfrac{5}{3} \right) + \left(\dfrac{7}{4} \right)$

$+ \left(\dfrac{9}{5} \right) + \left(\dfrac{11}{6} \right) = \dfrac{573}{60}$

33. $\dfrac{2}{3} + \dfrac{3}{4} + \dfrac{4}{5} + \dfrac{5}{6}$; The numerator is the number of the term plus 1, that is, $(n + 1)$, and the denominator is 2 plus the number of

the term. Then $a_n = \dfrac{n + 1}{n + 2}$ and we have 4 terms. $\displaystyle\sum_{i=1}^{4} \left(\dfrac{i + 1}{i + 2}\right)$ **38.** $2 - 5 + 8 - 11 + 14$; The signs alternate starting with the first

term positive, $(-1)^{n+1}$. Each term differs by 3 and the term is $3(1) - 1$, second term is $3(2) - 1$. So $a_n = (-1)^{n+1}(3n - 1)$ and we have

$\displaystyle\sum_{j=1}^{5} (-1)^{j+1}(3j - 1)$.

Exercise 12–3

Answers to odd-numbered problems

1. arithmetic; $d = 1$ **3.** arithmetic; $d = 2$ **5.** not arithmetic   **7.** arithmetic; $d = \dfrac{1}{2}$ **9.** arithmetic; $d = \dfrac{5}{3}$ **11.** $a_{16} = 79$

13. $a_{17} = -58$ **15.** $a_{12} = \dfrac{17}{3}$ **17.** $a_{14} = 9$ **19.** $a_{16} = 65$ **21.** $a_{25} = 105$ **23.** $n = 6$ terms   **25.** $n = 15$ terms

27. $n = 29$ terms   **29.** $S_{16} = 288$ **31.** $S_{14} = -175$ **33.** $S_{18} = 85\dfrac{1}{2}$ **35.** $S_{14} = 280$ **37.** $S_{19} = -95$ **39.** $S_{15} = 165$

41. $S_{22} = -440$ **43.** $S_{17} = 51$ **45.** $S_{10} = 13$ **47.** $S_{13} = 494$ **49.** $S_{15} = -165$ **51.** $S_{11} = -\dfrac{319}{6}$ **53.** 121 cans

55. 112 cans   **57.** 3,422   **59.** 240 ft; 1,600 ft   **61.** \$1,010 per month   **63.** \$82,080   **65.** \$540

Solutions to trial exercise problems

4. Since $6 - 4 = 2$, $8 - 6 = 2$, $10 - 8 = 2$, the sequence is arithmetic and $d = 2$.   **7.** Since $2 - \dfrac{3}{2} = \dfrac{1}{2}, \dfrac{5}{2} - 2 = \dfrac{1}{2}, 3 - \dfrac{5}{2}$

$= \dfrac{1}{2}$, the sequence is arithmetic and $d = \dfrac{1}{2}$.   **15.** Using $a_n = a_1 + (n - 1)d$, $a_{12} = 2 + (12 - 1)\dfrac{1}{3} = 2 + (11)\dfrac{1}{3} = 2 + \dfrac{11}{3} = \dfrac{17}{3}$.

28. Using $a_n = a_1 + (n - 1)d$, we want n when $a_n = -2$, $a_1 = \dfrac{5}{3}$, and $d = \dfrac{4}{3} - \dfrac{5}{3} = -\dfrac{1}{3}$. Then $-2 = \dfrac{5}{3} + (n - 1)\left(-\dfrac{1}{3}\right)$;

$-2 = \dfrac{5}{3} - \dfrac{1}{3}n + \dfrac{1}{3}$; $-2 = 2 - \dfrac{1}{3}n$; $-4 = -\dfrac{1}{3}n$; $n = 12$.   **33.** Using $S_n = \dfrac{n}{2}(a_1 + a_n)$, we want S_{18} when $n = 18$, $a_1 = \dfrac{1}{2}$,

and $a_{18} = 9$. So $S_{18} = \dfrac{18}{2}\left(\dfrac{1}{2} + 9\right) = 9\left(\dfrac{19}{2}\right) = \dfrac{171}{2}$ or $85\dfrac{1}{2}$.   **37.** Using $S_n = \dfrac{n}{2}(a_1 + a_n)$, we want S_{19} when $n = 19$,

$a_1 = 5 - 1 = 4$; $a_n = a_{19} = 5 - 19 = -14$. $S_{19} = \dfrac{19}{2}[4 + (-14)] = \dfrac{19}{2}(-10) = -\dfrac{190}{2} = -95$.   **44.** $\displaystyle\sum_{k=1}^{14} \dfrac{1}{2}k = \dfrac{14}{2}\left(\dfrac{1}{2} + 7\right)$

$= 7\left(\dfrac{15}{2}\right) = \dfrac{105}{2}$ or $52\dfrac{1}{2}$.   **51.** We want S_{11} using $S_n = \dfrac{n}{2}(a_1 + a_n)$. When $n = 11$, $a_1 = \dfrac{1}{6}$, and $a_{11} = \dfrac{1}{6} + (11 - 1)(-1)$

$= \dfrac{1}{6} + (10)(-1) = \dfrac{1}{6} + (-10) = -\dfrac{59}{6}$. So $S_{11} = \dfrac{11}{2}\left[\dfrac{1}{6} + \left(-\dfrac{59}{6}\right)\right] = \dfrac{11}{2}\left(-\dfrac{58}{6}\right) = \dfrac{11}{2}\left(-\dfrac{29}{3}\right) = -\dfrac{319}{6}$ or $-53\dfrac{1}{6}$.

54. We want S_n when $a_1 = 30$, $d = 27 - 30 = -3$, and $a_n = 3$. Now, using $a_n = a_1 + (n - 1)d$, we have $3 = 30 + (n - 1)(-3)$;

$-27 = -3n + 3$; $-30 = -3n$; $n = 10$. So $S_{10} = \dfrac{10}{2}(30 + 3) = 5(33) = 165$ boxes.   **62.** We want n when $S_n = 3,600$, $a_1 = 16$,

and $d = 32$. Using $S_n = \dfrac{n}{2}[2a_1 + (n - 1)d]$, $3,600 = \dfrac{n}{2}[2(16) + (n - 1) \cdot 32]$; $3,600 = \dfrac{n}{2}[32 + 32n - 32]$; $3,600 = \dfrac{n}{2}(32n)$;

$3,600 = 16n^2$; $n^2 = 225$; $n = 15$ sec.   **65.** We want a_{19} when $a_1 = 50$, $n = 19$, and $d = 30$. Using $a_n = a_1 + (n - 1)d$,

$a_{19} = 50 + (19 - 1)30 = 50 + 18(30) = 50 + 540 = 590$. Thus \$590 was deposited on her eighteenth birthday.

Exercise 12–4

Answers to odd-numbered problems

1. geometric; $27, 81, 243$; $r = 3$ **3.** geometric; $\dfrac{1}{54}, \dfrac{1}{162}, \dfrac{1}{486}$; $r = \dfrac{1}{3}$ **5.** geometric; $-\dfrac{1}{2}, \dfrac{1}{4}, -\dfrac{1}{8}$; $r = -\dfrac{1}{2}$

7. geometric; $\dfrac{-2}{9}, \dfrac{2}{27}, \dfrac{-2}{81}$; $r = -\dfrac{1}{3}$ **9.** not geometric   **11.** $a_n = 3(2)^{n-1}$ **13.** $a_n = 27\left(\dfrac{-2}{3}\right)^{n-1}$ **15.** $a_n = (\sqrt{3})^{n-1}$

17. $a_n = \dfrac{-1}{15}(-3)^{n-1}$ **19.** $a_5 = 162$ **21.** $a_4 = 3$ **23.** $a_6 = -160$ **25.** $a_5 = -\dfrac{1}{8}$ **27.** $a_7 = 576$ **29.** $a_9 = 1,792$

31. $S_5 = 1,694$ **33.** $S_4 = -\dfrac{312}{25}$ **35.** $S_6 = 27,993$ **37.** $S_9 = \dfrac{511}{512}$ **39.** $S_5 = -305$ **41.** $S_8 = 87,380$ **43.** $S_5 = -183$

45. $S_7 = \dfrac{4,118}{2,187}$ **47.** $S_6 = \dfrac{31,122}{15,625}$ **49.** $S_6 = -\dfrac{22,344}{3,125}$ **51.** $40\dfrac{7}{27}$ ft   **53.** \$512   **55.** \$10,737,418.24   **57.** $\dfrac{2,101}{3,125} \approx 0.67$ of the tank

Solutions to trial exercise problems

5. Since $\dfrac{-2}{4} = -\dfrac{1}{2}$ and $\dfrac{1}{-2} = -\dfrac{1}{2}$ the sequence is geometric with $r = -\dfrac{1}{2}$ and the next three terms are $-\dfrac{1}{2}, \dfrac{1}{4}, -\dfrac{1}{8}$.

15. Using $a_n = a_1 r^{n-1}$, since $\dfrac{\sqrt{3}}{1} = \sqrt{3}$ and $\dfrac{3}{\sqrt{3}} = \sqrt{3}$, then $r = \sqrt{3}$ and $a_1 = 1$. Thus $a_n = 1(\sqrt{3})^{n-1} = (\sqrt{3})^{n-1}$.

25. Since $a_1 = -32$ and $r = -\dfrac{1}{4}$, using $a_n = a_1 r^{n-1}$, $a_5 = (-32)\left(-\dfrac{1}{4}\right)^{5-1} = (-32)\left(-\dfrac{1}{4}\right)^4 = (-32)\left(\dfrac{1}{256}\right) = -\dfrac{1}{8}$.

26. Now $r = \dfrac{18}{3} = 6$ and $a_1 = 3$, then $a_6 = 3(6)^{6-1} = 3(6)^5 = 3(7,776) = 23,328$.   **33.** Using $S_n = \dfrac{a_1 - a_1 r^n}{1 - r}$, given $a_1 = -10$

and $r = \dfrac{1}{5}$, $S_4 = \dfrac{-10 - (-10)\left(\dfrac{1}{5}\right)^4}{1 - \dfrac{1}{5}} = \dfrac{-10 + 10\left(\dfrac{1}{625}\right)}{\dfrac{4}{5}} = \dfrac{-10 + \dfrac{2}{125}}{\dfrac{4}{5}} = \dfrac{-1,250 + 2}{100} = \dfrac{-1,248}{100} = -12\dfrac{12}{25}$.

34. Now $r = \dfrac{18}{9} = 2$ and $a_1 = 9$, so $S_7 = \dfrac{9 - 9(2)^7}{1 - 2} = \dfrac{9 - 9(128)}{-1} = \dfrac{9 - 1,152}{-1} = 1,143$.   **42.** Now $r = -2$ and $a_1 = -2$.

We want S_7. $S_7 = \dfrac{-2 - (-2)(-2)^7}{1 - (-2)} = \dfrac{-2 + 2(-2)^7}{3} = \dfrac{-2 + 2(-128)}{3} = \dfrac{-2 - 256}{3} = \dfrac{-258}{3} = -86$ **47.** Now $r = \dfrac{2}{5}$ and

$a_1 = 3\left(\dfrac{2}{5}\right) = \dfrac{6}{5}$. We want S_6. $S_6 = \dfrac{\dfrac{6}{5} - \dfrac{6}{5}\left(\dfrac{2}{5}\right)^6}{1 - \dfrac{2}{5}} = \dfrac{\dfrac{6}{5} - \dfrac{6}{5}\left(\dfrac{64}{15,625}\right)}{\dfrac{3}{5}} = \dfrac{6 - \dfrac{384}{15,625}}{3} = 2 - \dfrac{128}{15,625} = \dfrac{31,122}{15,625}$.

51. Since the ball will travel each height *twice*, after the initial drop of 9 ft, then we want $9 + 2\sum\limits_{i=1}^{5} 9\left(\dfrac{2}{3}\right)^i$. Now

$$2\sum_{i=1}^{5} 9\left(\dfrac{2}{3}\right)^i = 2\left[\dfrac{6 - 6\left(\dfrac{2}{3}\right)^5}{1 - \dfrac{2}{3}}\right] = 2\left[\dfrac{6 - 6\left(\dfrac{2}{3}\right)^5}{\dfrac{1}{3}}\right] = 2\left[\dfrac{6 - 6\left(\dfrac{32}{243}\right)}{\dfrac{1}{3}}\right]$$

$$= 2\left[\left(6 - \dfrac{192}{243}\right) \cdot 3\right] = 2\left[18 - \dfrac{192}{81}\right] = 2\left[\dfrac{1,458 - 192}{81}\right] = 2\left(\dfrac{1,266}{81}\right)$$

$$= 2\left(\dfrac{422}{27}\right) = \dfrac{844}{27} \text{ or } 31\dfrac{7}{27}. \text{ The ball has traveled } 9 + 31\dfrac{7}{27} = 40\dfrac{7}{27} \text{ ft at the sixth strike.}$$

Exercise 12–5

Answers to odd-numbered problems

1. 3　　**3.** −6　　**5.** $\dfrac{9}{10}$　　**7.** $-\dfrac{1}{2}$　　**9.** 18　　**11.** 5　　**13.** no sum　　**15.** 4　　**17.** $-\dfrac{2}{5}$　　**19.** $\dfrac{3}{4}$　　**21.** $\dfrac{1}{3}$　　**23.** $\dfrac{31}{110}$　　**25.** $\dfrac{2}{55}$

27. 570 in.　　**29.** 84 in.　　**31.** 25 mg

Solutions to trial exercise problems

3. Using $S_\infty = \dfrac{a_1}{1-r}$, $S_\infty = \dfrac{-3}{1-\dfrac{1}{2}} = \dfrac{-3}{\dfrac{1}{2}} = -6.$

6. Using $S_\infty = \dfrac{a_1}{1-r}$, $S_\infty = \dfrac{4}{1-\left(-\dfrac{1}{2}\right)} = \dfrac{4}{\dfrac{3}{2}} = 4 \cdot \dfrac{2}{3} = \dfrac{8}{3}.$

9. Now $a_1 = 12$ and $r = \dfrac{4}{12} = \dfrac{1}{3}$, so $S_\infty = \dfrac{12}{1-\dfrac{1}{3}} = \dfrac{12}{\dfrac{2}{3}} = 12 \cdot \dfrac{3}{2} = 18.$

13. $a_1 = 6$ and $r = \dfrac{-8}{6} = -\dfrac{4}{3}$. So S_∞ does not exist since $\left|-\dfrac{4}{3}\right| > 1.$

16. Now $a_1 = \left(\dfrac{7}{8}\right)^{1+1} = \left(\dfrac{7}{8}\right)^2 = \dfrac{49}{64}$ and $r = \dfrac{7}{8}$, so $\displaystyle\sum_{k=1}^{\infty}\left(\dfrac{7}{8}\right)^{k+1} = \dfrac{\dfrac{49}{64}}{1-\dfrac{7}{8}} = \dfrac{\dfrac{49}{64}}{\dfrac{1}{8}} = \dfrac{49}{64} \cdot \dfrac{8}{1} = \dfrac{49}{8}$ or $6\dfrac{1}{8}.$

23. $0.28181\overline{81} = 0.2 + 0.08181\overline{81}$. Now for $0.08181\overline{81}$ $a_1 = 0.081$ and $r = \dfrac{0.00081}{0.081} = 0.01$. Then $0.08181\overline{81} = \dfrac{0.081}{1-0.01} = \dfrac{0.081}{0.99}$

$= \dfrac{81}{990} = \dfrac{9}{110}$. Thus $0.28181\overline{81} = \dfrac{2}{10} + \dfrac{9}{110} = \dfrac{22+9}{110} = \dfrac{31}{110}.$

28. Now $a_1 = 16$ cm and $r = \dfrac{7}{8}$. We want $S_\infty = \dfrac{a_1}{1-r} = \dfrac{16}{1-\dfrac{7}{8}} = \dfrac{16}{\dfrac{1}{8}} = 16 \cdot 8 = 128.$

The bob travels 128 cm before coming to rest.

Exercise 12–6

1. 720　　**3.** 11,880　　**5.** 10　　**7.** 28　　**9.** 120　　**11.** $a^4 - 12a^3 + 54a^2 - 108a + 81$

13. $p^6 + 6p^5 q + 15 p^4q^2 + 20p^3q^3 + 15p^2q^4 + 6pq^5 + q^6$　　**15.** $16a^4 + 96a^3 + 216a^2 + 216a + 81$

17. $\dfrac{p^6}{64} - \dfrac{3}{16}p^5q + \dfrac{15}{16}p^4q^2 - \dfrac{5}{2}p^3q^3 + \dfrac{15}{4}p^2q^4 - 3pq^5 + q^6$　　**19.** $a^{10} + 5a^8 b^2 + 10a^6b^4 + 10a^4b^6 + 5a^2b^8 + b^{10}$

21. $-2,002a^9b^5$　　**23.** $7,920q^8$　　**25.** $18,144 k^6$　　**27.** 1.0263　　**29.** .8587　　**31.** $924a^9 (a > 0)$　　**33.** 70

Solutions to trial exercise problems

5. $\dfrac{10!}{9!} = \dfrac{10 \cdot 9!}{9!} = 10$　　**11.** $(a-3)^4 = a^4 + \dfrac{4}{1!} a^3(-3)^1 + \dfrac{4 \cdot 3}{2!}a^2(-3)^2 + \dfrac{4 \cdot 3 \cdot 2}{3!}a(-3)^3 + \dfrac{4 \cdot 3 \cdot 2 \cdot 1}{4!}(-3)^4$

$= a^4 - 12a^3 + 54a^2 - 108a + 81$　　**19.** $(a^2 + b^2)^5$

$= (a^2)^5 + \dfrac{5}{1!}(a^2)^4(b^2)^1 + \dfrac{5 \cdot 4}{2!}(a^2)^3(b^2)^2 + \dfrac{5 \cdot 4 \cdot 3}{3!}(a^2)^2(b^2)^3 + \dfrac{5 \cdot 4 \cdot 3 \cdot 2}{4!}(a^2)(b^2)^4 + \dfrac{5 \cdot 4 \cdot 3 \cdot 2 \cdot 1}{5!}(b^2)^5$

$= a^{10} + 5a^8b^2 + 10a^6b^4 + 10a^4b^6 + 5a^2b^8 + b^{10}$　　**25.** Given $(6 - k)^9$, we want the 7th term, where $n = 9, r = 7, x = 6,$

and $y = -k$. Using $\dfrac{n!}{[n-(r-1)]!(r-1)!}, x^{n-(r-1)}y^{r-1} = \dfrac{9!}{3!6!}(6)^3(-k)^6 = \dfrac{9 \cdot 8 \cdot 7}{3!}(216)(k)^6 = 84(216)k^6 = 18,144k^6.$

27. We want $(1.002)^{13} = (1 + 0.002)^{13} = 1^{13} + \dfrac{13}{1!}(1)^{12}(0.002) + \dfrac{13 \cdot 12}{2!}(1)^{11}(0.002)^2 + \dfrac{13 \cdot 12 \cdot 11}{3!}(1)^{10}(0.002)^3$

$= 1 + 0.026 + 0.000312 + \cdots = 1.026312 = 1.0263.$ **31.** We want the 7th term of $(a + \sqrt{a})^{12}$. Now $n = 12, r = 7, x = a$, and

$y = \sqrt{a}$. So, $\dfrac{n!}{[n - (r - 1)]!(r - 1)!} x^{n - (r-1)} y^{(r-1)} = \dfrac{12!}{6!6!} a^6(\sqrt{a})^6 = \dfrac{12 \cdot 11 \cdot 10 \cdot 9 \cdot 8 \cdot 7}{6 \cdot 5 \cdot 4 \cdot 3 \cdot 2 \cdot 1}(a^6)(a)^3 = 11 \cdot 3 \cdot 4 \cdot 7 \quad a^9 = 924a^9.$

The middle term of $(a + \sqrt{a})^{12}$ is $924a^9$. $(a > 0)$

Chapter 12 review exercises

1. $7,11,15,19,23$ **2.** $5, \dfrac{10}{3}, 3, \dfrac{20}{7}, \dfrac{25}{9}$ **3.** $\dfrac{-4}{7}, \dfrac{4}{9}, \dfrac{-4}{11}, \dfrac{4}{13}, \dfrac{-4}{15}$ **4.** $2, -4, 8, -16, 32$ **5.** $a_6 = -14$ **6.** $a_7 = -17$ **7.** $a_9 = \dfrac{6561}{2}$

8. $a_{11} = \dfrac{683}{59,049}$ **9.** $a_n = 2n + 3$ **10.** $a_n = 5n - 2$ **11.** $a_n = \dfrac{n+1}{4n-1}$ **12.** $a_n = (-1)^n (5n - 1)$ **13.** $a_n = 0.25n + 2.75;$

$a_8 = \$4.75$ **14.** $S_4 = 36$ **15.** $S_6 = 196$ **16.** $S_5 = \dfrac{229}{20}$ **17.** $S_6 = -\dfrac{37}{75}$ **18.** $\displaystyle\sum_{i=1}^{4} (3i + 2)$ **19.** $\displaystyle\sum_{k=1}^{5} (-1)^{k+1} \left(\dfrac{k+3}{k+4} \right)$

20. $a_{15} = 61$ **21.** $a_{17} = 77$ **22.** $a_{21} = 83$ **23.** $a_{19} = -42$ **24.** $n = 15$ **25.** $n = 21$ **26.** $S_{15} = 585$ **27.** $S_{21} = -420$

28. $S_{29} = 290$ **29.** $S_{25} = \dfrac{375}{2}$ **30.** $\$16,500$ **31.** $a_n = 3(2)^{n-1}$ **32.** $a_n = 8\left(\dfrac{-1}{2} \right)^{n-1}$ **33.** $a_5 = 405$ **34.** $a_4 = -\dfrac{8}{9}$

35. $a_3 = 16$ **36.** $a_7 = -576$ **37.** $S_5 = 363$ **38.** $S_6 = -\dfrac{189}{4}$ **39.** $S_5 = -\dfrac{543}{1,024}$ **40.** $S_7 = \dfrac{4,372}{2,187}$

41. $S_\infty = 12$ **42.** $S_\infty = -\dfrac{5}{2}$ **43.** $S_\infty = \dfrac{-3}{5}$ **44.** $S_\infty = -\dfrac{1}{45}$ **45.** $\dfrac{35}{99}$ **46.** $\dfrac{214}{495}$ **47.** 24 meters

48. $x^7 + 35x^6 + 525x^5 + 4375x^4 + 21,875x^3 + 65,625x^2 + 109,375x + 78,125$

49. $32a^5 - 240a^4b + 720a^3b^2 - 1080a^2b^3 + 810ab^4 - 243b^5$ **50.** $\dfrac{a^4}{16} - \dfrac{3a^3b}{2} + \dfrac{27a^2b^2}{2} - 54ab^3 + 81 b^4$ **51.** $84,480a^7$

52. $19,702,683a^8b^6$ **53.** $608,256x^{10}y^2$ **54.** $\dfrac{27}{32}$ ft. **55.** $1,287$

Final examination

1. $\{-7, -6, -5, -4, -3, -2, -1, 0, 1, 2, 3\}$ **2.** (a) $\{2,7\}$, (b) $\{-4, -1, 0, 2, 4, 7, 9\}$, (c) $\emptyset$ **3.** $\{-4, -3, -2, -1, 0, 1, 2, 3, 4, 5, 6, 7, 8\}$ **4.** 13

5. (a) $\dfrac{-2a^7}{b^5}$, (b) $\dfrac{a^6}{4b^4}$, (c) $\dfrac{b^4}{a^6}$ **6.** (a) $4y^2 + 28y + 49$, (b) $16x^2 - 4y^2$, (c) $x^3 - 8$ **7.** $4xy(3 - xy^2 + 2x^2y)$ **8.** $(7y + 1)(y - 5)$

9. $(2a - 5b)^2$ **10.** $2(2x + 5y)(2x - 5y)$ **11.** $2(2a - b)(4a^2 + 2ab + b^2)$ **12.** $(3a - b)(2x - y)$ **13.** $\left\{ \dfrac{21}{25} \right\}$

14. $\left\{ x \mid x < -\dfrac{29}{2} \right\}$ **15.** $\left\{ x \mid -\dfrac{5}{2} < x \le 1 \right\}$ **16.** $\{4,1\}$ **17.** $\left\{ x \mid \dfrac{1}{3} < x < 3 \right\}$ **18.** $\left\{ c \mid c \le -\dfrac{2}{5} \text{ or } c \ge 2 \right\}$ **19.** $\{8, -3\}$

20. $\left\{ -\dfrac{5}{4} \right\}$ **21.** $\dfrac{32a + 9}{(2a - 1)(4a + 3)}$ **22.** $\dfrac{y^2 - 5y + 1}{(y - 7)(y + 6)(y - 6)}$ **23.** $\dfrac{3a^2 - 4a + 1}{a^2 - 3a - 4}$ **24.** -2 **25.** $17\sqrt{3}$

26. $\dfrac{4 + 7i}{13}$ or $\dfrac{4}{13} + \dfrac{7}{13}i$ **27.** $12 - 8\sqrt{5} + 9\sqrt{2} - 6\sqrt{10}$ **28.** $\left\{ \dfrac{7 + \sqrt{73}}{6}, \dfrac{7 - \sqrt{73}}{6} \right\}$ **29.** $\{5\}$, 2 is extraneous

30. $\{y \mid 1 \le y \le 3\}$ **31.** (a) $3x - 2y = -1$, (b) $x + 4y = -8$, (c) $2x + 3y = 0$ **32.** $m = \dfrac{3}{5}$, $b = \dfrac{9}{5}$

33. (a) -18, (b) 13, (c) 5

34.

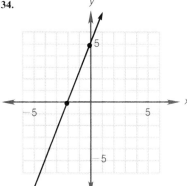

35.

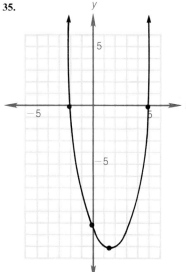

36.

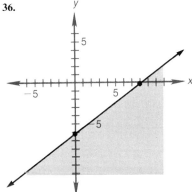

37.

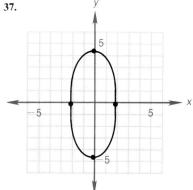

38. (a) circle, (b) hyperbola, (c) parabola, (d) ellipse   **39.** $\left\{\left(\dfrac{11}{19}, -\dfrac{12}{19}\right)\right\}$ **40.** 149   **41.** $\left\{\left(\dfrac{29}{11}, -\dfrac{5}{11}\right)\right\}$ **42.** $\log_b\left(\dfrac{30}{8}\right)$ or $\log_b\left(\dfrac{15}{4}\right)$

43. 1.77   **44.** 0.738   **45.** 3.58   **46.** $\displaystyle\sum_{i=1}^{5}(6i-3)$ **47.** 49   **48.** -44 **49.** $\dfrac{351}{2}$ **50.** $-\dfrac{1}{54}$ **51.** $-\dfrac{2}{3}$

52. $81x^4 - 216x^3y + 216x^2y^2 - 96xy^3 + 16y^4$ **53.** $924a^6$ **54.** $\dfrac{26}{111}$ **55.** 462

Index

Member of an equation, 49
Member of a set, 3
Midpoint of a line segment, 283
Mixture problems, 69
Monomial, 38
Multinomial, 38, 175
  multiplication of, 101–2
Multiplication, 17
  of fractions, 159
  of rational expressions, 157
Multiplication property of equality, 28, 51–52
Multiplication property of inequality, 72
Multiplicative inverse property, 25
Multiplicity, 234

N

Natural logarithms, 431
Natural numbers, 7
Negative numbers, 9
Negative reciprocal, 290
n factorial, 465
Nonlinear equations, systems of, 375–76
nth power property, 256
nth root, 194–95
Null set, 5
Number, 10
Number line, 9–10
Number problems, 62–64
Numerical coefficient, 38

O

One-to-one
  function, 398
Open sentence, 49
Opposite of, 12
Order, 10
Ordered pairs of numbers, 274
  components of, 275
Ordered triple of real numbers, 325
Order of operations, 32–35
Order relationship, 10, 73
Ordinate of a point, 275
Origin, 9, 274

P

Parabola, 354
  definition of, 354
  equation of a, 354
  of a parabola, 354–59
Parallel lines, 288–89
Partial sum of a series, 446
Pascal's triangle, 464–65
Perfect squares, 103
  trinomials, 136–37
Perpendicular lines, 289–90
Pi, 8, 37

Plane, 326
Point-slope form of a line, 294
Polynomial, 38
  division of, 175–78
  notation, 41
Positive numbers, 9
Primary
  denominator, 171
  numerator, 171
Prime, relatively, 154, 197
Prime factor form, 118
Prime numbers, 118
Principal nth root, 195
Product, 17, 118
Properties of a logarithm, 424–25, 431, 435
Pythagorean Theorem, 208, 282

Q

Quadrants, 274
Quadratic equation, 232
  applications of, 250–52
  in one variable, 233
  solution by completing the square,
    240–42
  solution by extracting roots, 235–36
  solution by factoring, 233
  solution by quadratic formula, 245–47
  standard form of, 233
Quadratic formula, 244–45
Quadratic function, 394
Quadratic inequalities, 264–67
  critical numbers of, 265
  test number of, 264
Quadratic-type equations, 261–63

R

Radical equations, 256
Radicals
  like, 215
  product property, 204
  quotient property, 209
  simplest form, 213
  standard form of, 213
Range
  of a function, 385
  of a relation, 383
Rational equations, 179
Rational exponents, 196–201
Rational expression
  definition, 151
  domain of a, 152
Rational inequality, 266
Rationalizing the denominator, 210–12,
  220–21
Rational number, 8
Real number, 7–9
Real number, properties of, 25
  additive inverse property of, 25

  associative property of addition, 25
  associative property of multiplication,
    25
  closure property of addition, 25
  closure property of multiplication, 25
  commutative property of addition, 25
  commutative property of
    multiplication, 25
  distributive property, 25
  identity property of addition, 25
  identity property of multiplication, 25
  multiplicative inverse property, 25
Real number line, 9
Reciprocal, 159
Rectangular coordinate system, 274
Reducing to lowest terms, 153
Reflexive property of equality, 24
Relation, 382
  domain of, 383
  range of, 383
Relatively prime, 154, 197
Replacement set, 7
Right member, 49
Root, 50
  nth, 194
  principal nth, 195
Roster method for sets, 3
rth term of a binomial expansion, 466

S

Scientific notation, 113–15
Secondary denominator, 171
Second component of a point, 275
Sense of an inequality, 73
Sequence, 442
  arithmetic, 449–53
  finite, 442
  general term of a, 443, 450
  geometric, 455
  infinite, 442
Series, 446
  infinite geometric, 460–62
Set, 3
  disjoint, 6
  element of, 3
  empty, 5
  intersection, 6
  member of, 3
  null, 5
  replacement, 7
  solution, 50
  union, 5
Set-builder notation, 7
Set of real numbers, 7–9
Set symbolism, 3–6
Sigma notation, 446
Sign, 14
Sign array, of a determinant, 336
Signed numbers, 14

Symbols

Symbol	Means or Is Read	Section Number
$\{a,b,c\}$	The set whose elements are a, b, and c	1–1
ϵ	Is an element of	1–1
$\subseteq$	Is a subset of	1–1
$\emptyset$	Empty set or null set	1–1
$\cup$	Union	1–1
$\cap$	Intersection	1–1
$N = \{1,2,3,\cdots\}$	The set of natural numbers	1–1
$W = \{0,1,2,3,\cdots\}$	The set of whole numbers	1–1
$J = \{\cdots,-3,-2,-1,0,1,2,3,\cdots\}$	The set of integers	1–1
$\{x\,|\,x$ satisfies some conditions$\}$	Set-builder notation	1–1
$Q = \left\{\dfrac{p}{q}\,\middle|\,p \text{ and } q \,\epsilon\, J, \text{ and } q \neq 0\right\}$	The set of rational numbers	1–1
H	The set of irrational numbers	1–1
R	The set of real numbers	1–1
$<$	Less than	1–1
$>$	Greater than	1–1
$\leq$	Less than or equal to	1–1
$\geq$	Greater than or equal to	1–1
$|x|$	Absolute value of x	1–1
$S = \{\ \ \}$	The solution set S	2–1
$[a,b]$	The closed interval from a to b	2–4
(a,b)	The open interval from a to b	2–4
∞	Infinity	2–4
a^n	The nth power of a	3–2
a^{-n}	a to the negative n power $\left(a^{-n} = \dfrac{1}{a^n}, a \neq 0\right)$	3–4
a^0	a to the zero power $(a^0 = 1, a \neq 0)$	3–4
$\sqrt[n]{a}$	The principal nth root of a	5–1
$a^{\frac{1}{n}}$	The principal nth root of a	5–1
i	Imaginary number ($i = \sqrt{-1}$)	5–6
$a + bi$	Complex number	5–6